Algorithms for
Computer Algebra

Algorithms for
Computer Algebra

K.O. Geddes
University of Waterloo

S.R. Czapor
Laurentian University

G. Labahn
University of Waterloo

Kluwer Academic Publishers
Boston/Dordrecht/London

Distributors for North America:
Kluwer Academic Publishers
101 Philip Drive
Assinippi Park
Norwell, Massachusetts 02061 USA

Distributors for all other countries:
Kluwer Academic Publishers Group
Distribution Centre
Post Office Box 322
3300 AH Dordrecht, THE NETHERLANDS

Library of Congress Cataloging-in-Publication Data

Geddes, K.O. (Keith O.), 1947-
Algorithms for computer algebra / K.O. Geddes, S.R. Czapor, G. Labahn.
p. cm.
Includes bibliographical references and index.
ISBN 0-7923-9259-0 (alk. paper)
1. Algebra--Data processing. 2. Algorithms. I. Czapor, S.R. (Stephen R.), 1957- . II. Labahn, G. (George), 1951- . III. Title.
QA155.7.E4G43 1992
512'.00285--dc20 92-25697
CIP

Printed on acid-free paper.

Printed in the United States of America

Dedication

For their support in numerous ways, this book is dedicated to

Debbie, Kim, Kathryn, and Kirsten

Alexandra, Merle, Serge, and Margaret

Laura, Claudia, Jeffrey, and Philip

CONTENTS

Chapter 8 Polynomial Factorization

Chapter 9 Solving Systems of Equations

Chapter 10 Gröbner Bases for Polynomial Ideals

Chapter 11 Integration of Rational Functions

LIST OF ALGORITHMS

LIST OF FIGURES

LIST OF TABLES

PREFACE

The field of computer algebra has gained widespread attention in recent years because of the increasing use of computer algebra systems in the scientific community. These systems, the best known of which are DERIVE, MACSYMA, MAPLE, MATHEMATICA, REDUCE, and SCRATCHPAD, differ markedly from programs that exist to perform only numerical scientific computations. Unlike the latter programs, computer algebra systems can manipulate symbolic mathematical objects in addition to numeric quantities. The operations that can be performed on these symbolic objects include most of the algebraic operations that one encounters in high school and university mathematics courses. The ability to perform calculations with symbolic objects opens up a vast array of applications for these systems. Polynomials, trigonometric functions, and other mathematical functions can be manipulated by these programs. Differentiation and integration can be performed for a wide class of functions. Polynomials can be factored, greatest common divisors can be determined, differential and recurrence equations can be solved. Indeed any mathematical operation that allows for an algebraic construction can, in principle, be performed in a computer algebra system.

The main advantage of a computer algebra system is its ability to handle large algebraic computations. As such, however, one cannot necessarily use the classical algorithms which appear in mathematics textbooks. Computing the greatest common divisor of two polynomials having rational number coefficients can be accomplished by the classical Euclidean algorithm. However, if one tries to use this algorithm on polynomials of even a moderate size one quickly realizes that the intermediate polynomials have coefficients which grow exponentially in size if one omits reducing the rationals to minimum denominators. An algorithm that exhibits exponential growth in the size of the intermediate expressions quickly becomes impractical as the problem size is increased. Unfortunately, such algorithms include a large number of familiar approaches such as the Euclidean algorithm. Programs written for purely numerical computation do not suffer from this difficulty since the numerics are all of a fixed size.

The use of exact arithmetic in computer algebra systems adds a second area of complexity to the algorithms. Algorithms designed for numerical computation have their complexity judged by a simple count of the number of arithmetic steps required. In these systems every arithmetic operation costs approximately the same. However, this is not true when exact arithmetic is used as can easily be seen by timing the addition of two rational numbers having 5 digit components and comparing it to the addition of two rational numbers having 5000 digit components.

The motivation for this book is twofold. On the one hand there is a definite need for a textbook to teach college and university students, as well as researchers new to the field, some basic information about computer algebra systems and the mathematical algorithms employed by them. At present, most information on algorithms for computer algebra can only be found by searching a wide range of research papers and Ph.D. theses. Some of this material is relatively easy to read, some is difficult; examples are scarce and descriptions of implementations are sometimes incomplete. This book hopes to fill this void.

The second reason for undertaking the writing of this book revolves around our interest in computer algebra system implementation. The authors are involved in the design and implementation of the MAPLE computer algebra system. The implementation of efficient algorithms requires a deep understanding of the mathematical basis of the algorithms. As mentioned previously, the efficiency of computer algebra algorithms is difficult to analyze mathematically. It is often necessary to implement a number of algorithms each with the same goal. Often two algorithms are each more efficient than the other depending on the type of input one encounters (e.g. sparse or dense polynomials). We found that gaining an in-depth understanding of the various algorithms was best accomplished by writing detailed descriptions of the underlying mathematics. Hence this book was a fairly natural step.

Major parts of this book have been used during the past decade in a computer algebra course at the University of Waterloo. The course is presented as an introduction to computer algebra for senior undergraduate and graduate students interested in the topic. The mathematical background assumed for this book is a level of mathematical maturity which would normally be achieved by at least two years of college or university mathematics courses. Specific course prerequisites would be first-year calculus, an introduction to linear algebra, and an introduction to computer science. Students normally have had some prior exposure to abstract algebra, but about one-quarter of the students we have taught had no prior knowledge of ring or field theory. This does not present a major obstacle, however, since the main algebraic objects encountered are polynomials and power series and students are usually quite comfortable with these. Ideals, which are first encountered in Chapter 5, provide the main challenge for these students. In other words, a course based on this book can be used to introduce students to ring and field theory rather than making these topics a prerequisite.

The breadth of topics covered in this book can be outlined as follows. Chapter 1 presents an introduction to computer algebra systems, including a brief historical sketch. Chapters 2 through 12 can be categorized into four parts:

I. Basic Algebra, Representation Issues, and Arithmetic Algorithms (Chapters 2, 3, 4): The fundamental concepts from the algebra of rings and fields are presented in Chapter 2, with particular reference to polynomials, rational functions, and power series. Chapter 3 presents the concepts of normal and canonical forms, and discusses data structure representations for algebraic objects in computer algebra systems. Chapter 4 presents some algorithms for performing arithmetic on polynomials and power series.

II. Homomorphisms and Lifting Algorithms (Chapters 5, 6): The next two chapters introduce the concept of a homomorphism as a mapping from a given domain to a simpler domain, and consider the inverse process of lifting one or more solutions in the image domain to the desired solution in the original domain. Two fundamental lifting processes are presented in detail: the Chinese remainder algorithm and the Hensel lifting algorithm.

III. Advanced Computations in Polynomial Domains (Chapters 7, 8, 9, 10): Building on the fundamental polynomial manipulation algorithms of the preceding chapters, Chapter 7 discusses various algorithms for polynomial GCD computation and Chapter 8 discusses polynomial factorization algorithms. Algorithms for solving linear systems of equations with coefficients from integer or polynomial domains are discussed in Chapter 9, followed by a consideration of the problem of solving systems of simultaneous polynomial equations. The latter problem leads to the topic of Gröbner bases for polynomial ideals, which is the topic of Chapter 10.

IV. Indefinite Integration (Chapters 11 and 12): The topic of the final two chapters is the indefinite integration problem of calculus. It would seem at first glance that this topic digresses from the primary emphasis on algorithms for polynomial computations in previous chapters. On the contrary, the Risch integration algorithm relies almost exclusively on algorithms for performing operations on multivariate polynomials. Indeed, these chapters serve as a focal point for most of the book in the sense that the development of the Risch integration algorithm relies on algorithms from each of the preceding chapters with the major exception of Chapter 10 (Gröbner Bases). Chapter 11 introduces some concepts from differential algebra and develops algorithms for integrating rational functions. Chapter 12 presents an in-depth treatment of the Liouville theory and the Risch algorithm for integrating transcendental elementary functions, followed by an overview of the case of integrating algebraic functions.

As is usually the case, the number of topics which could be covered in a book such as this is greater than what has been covered here. Some topics have been omitted because there is already a large body of literature on the topic. Prime factorization of integers and other integer arithmetic operations such as integer greatest common divisor fall in this category. Other topics which are of significant interest have not been included simply because one can only cover so much territory in a single book. Thus the major topics of differential equations, advanced linear algebra (e.g. Smith and Hermite normal forms), and simplification of radical expressions are examples of topics that are not covered in this book. The topic of indefinite integration over algebraic (or mixed algebraic and transcendental) extensions is only briefly summarized. An in-depth treatment of this subject would require a substantial introduction to the field of algebraic geometry. Indeed a complete account of the integration problem including algebraic functions and the requisite computational algebraic geometry background would constitute a substantial book by itself.

The diagram below shows how the various chapters in this book relate to each other. The dependency relationships do not imply that earlier chapters are absolute prerequisites for later chapters; for example, although Chapters 11 and 12 (Integration) have the strongest trail of precedent chapters in the diagram, it is entirely feasible to study these two chapters without the preceding chapters if one wishes simply to assume "it is known that" one can perform various computations such as polynomial GCD's, polynomial factorizations, and the solution of equations.

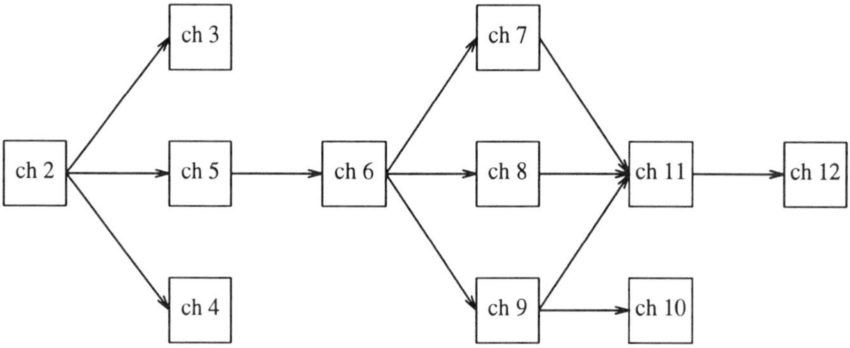

The paths through this diagram indicate some logical choices for sequences of chapters to be covered in a course where time constraints dictate that some chapters must be omitted. For example, a one-semester course might be based on material from one of the following two sets of chapters.

Course A: ch 2, 3, 4, 5, 6, 9, 10
..........(Algebraic algorithms including Gröbner bases)

Course B: ch 2, 5, 6, 7, 8, 11, 12
..........(Algebraic algorithms including the integration problem)

Acknowledgements

As with any significant project, there are a number of people who require special thanks. In particular, we would like to express our thanks to Manuel Bronstein, Stan Cabay, David Clark, Bruce Char, Greg Fee, Gaston Gonnet, Dominik Gruntz, Michael Monagan, and Bruno Salvy for reading earlier drafts of these chapters and making comments and suggestions. Needless to say, any errors of omission or commission which may remain are entirely the responsibility of the authors. Acknowledgement is due also to Richard Fateman, University of California at Berkeley, and to Jacques Calmet, Université de Grenoble (now at Universitaet Karlsruhe) for hosting Keith Geddes on sabbatical leaves in 1980 and 1986/87, respectively, during which several of the chapters were written.

Special mention is due to Robert (Bob) Moenck who, together with one of us (Keith Geddes), first conceived of this book more than a decade ago. While Bob's career went in a different direction before the project proceeded very far, his early influence on the project was significant.

Algorithms for
Computer Algebra

CHAPTER 1

INTRODUCTION TO COMPUTER ALGEBRA

1.1. INTRODUCTION

The desire to use a computer to perform a mathematical computation symbolically arises naturally whenever a long and tedious sequence of manipulations is required. We have all had the experience of working out a result which required page after page of algebraic manipulation and hours (perhaps days) of our time. This computation might have been to solve a linear system of equations exactly where an approximate numerical solution would not have been appropriate. Or it might have been to work out the indefinite integral of a fairly complicated function for which it was hoped that some transformation would put the integral into one of the forms appearing in a table of integrals. In the latter case, we might have stumbled upon an appropriate transformation or we might have eventually given up without knowing whether or not the integral could be expressed in terms of elementary functions. Or it might have been any one of numerous other problems requiring symbolic manipulation.

The idea of using computers for non-numerical computations is relatively old but the use of computers for the specific types of symbolic mathematical computations mentioned above is a fairly recent development. Some of the non-numerical computations which we do *not* deal with here include such processes as compilation of programming languages, word processing, logic programming, or artificial intelligence in its broadest sense. Rather, we are concerned here with the use of computers for specific mathematical computations which are to be performed symbolically. This subject area is referred to by various names including *algebraic manipulation, symbolic computation, algebraic algorithms,* and *computer algebra,* to name a few.

1.2. SYMBOLIC VERSUS NUMERIC COMPUTATION

It is perhaps useful to consider an example illustrating the contrast between numeric and symbolic computation. The Chebyshev polynomials which arise in numerical analysis are defined recursively as follows:

$$T_0(x) = 1; \quad T_1(x) = x;$$

$$T_k(x) = 2xT_{k-1}(x) - T_{k-2}(x) \quad \text{for } k \geq 2.$$

The first five Chebyshev polynomials are listed in Table 1.1.

Table 1.1. The first five Chebyshev polynomials.

k	$T_k(x)$
0	1
1	x
2	$2x^2 - 1$
3	$4x^3 - 3x$
4	$8x^4 - 8x^2 + 1$

As a typical numeric computation involving the Chebyshev polynomials, suppose that it is desired to compute the values of the first five Chebyshev polynomials at one or more values of the variable x. The FORTRAN program in Figure 1.1 might be used for this purpose. If the input for this program is the number 0.30 then the output will be the five numbers:

1.0000 0.3000 −0.8200 −0.7920 0.3448

Now suppose that we had left out the READ statement in the program of Figure 1.1. This would, of course, cause a run-time error in FORTRAN. However, a reasonable interpretation of this program without the READ statement might be that the first five Chebyshev polynomials are to be computed in symbolic form. The latter interpretation is precisely what can be accomplished in a language for symbolic computation. Figure 1.2 shows a program in one of the earliest symbolic manipulation languages, ALTRAN, which corresponds to this *symbolic* interpretation. The output from the ALTRAN program is presented in Figure 1.3. As this example indicates, FORTRAN was designed to manipulate numbers while ALTRAN was designed to manipulate polynomials.

```
C    PROGRAM FOR CHEBYSHEV POLYNOMIALS
     REAL T(5)
C
     READ(5,1) X
     T(1) = 1.0
     T(2) = X
     WRITE(6,2) T(1), T(2)
     DO 10 N = 3, 5
        T(N) = 2.0*X*T(N-1) - T(N-2)
        WRITE(6,2) T(N)
  10 CONTINUE
     STOP
C
   1 FORMAT(F5.2)
   2 FORMAT(F9.4)
     END
```

Figure 1.1. FORTRAN program involving Chebyshev polynomials.

```
PROCEDURE MAIN

 ALGEBRAIC (X:4) ARRAY (0:4) T
 INTEGER N

 T(0) = 1
 T(1) = X
 WRITE T(0), T(1)
 DO N = 2, 4
    T(N) = 2*X*T(N-1) - T(N-2)
    WRITE T(N)
 DOEND

END
```

Figure 1.2. ALTRAN program involving Chebyshev polynomials.

```
#T(0)
      1
#T(1)
      X
#T(2)
      2*X**2 - 1
#T(3)
      X * ( 4*X**2 - 3 )
#T(4)
      8*X**4 - 8*X**2 + 1
```

Figure 1.3. Output from ALTRAN program in Figure 1.2.

ALTRAN can be thought of as a variant of FORTRAN with the addition of an extra declaration, the "algebraic" type declaration. Even its name, ALTRAN, is derived from ALgebraic TRANslator, following the naming convention of FORTRAN (derived from FORmula TRANslator). Also, again following the spirit of FORTRAN, ALTRAN was designed for a batch processing mode. Later, with the advent of the hand-held numeric calculator in the mid-sixties, computer algebra systems began to be designed for interactive use, as a type of symbolic calculator. Since the early seventies, nearly all modern computer algebra systems have been designed for interactive use. In Figure 1.4 we present an example of this approach, showing an interactive session run in one of the modern systems, MAPLE, for performing the same computation with Chebyshev polynomials. We will return to the capabilities and usage of a modern computer algebra system in a later section.

1.3. A BRIEF HISTORICAL SKETCH

The development of systems for symbolic mathematical computation first became an active area of research and implementation during the decade 1961-1971. During this decade, the field progressed from birth through adolescence to at least some level of maturity. Of course, the process of maturing is a continuing process and the field of symbolic computation is now a recognized area of research and teaching in computer science and mathematics.

There are three recognizable, yet interdependent, forces in the development of this field. We may classify them under the headings *systems, algorithms,* and *applications*. By *systems* we mean the development of programming languages and the associated software for symbolic manipulation. *Algorithms* refers to the development of efficient mathematical algorithms for the manipulation of polynomials, rational functions, and more general classes of functions. The range of *applications* of symbolic computation is very broad and has provided the impetus for the development of systems and algorithms. In this section, the appearance of various programming languages (or systems) will be used as milestones in briefly tracing the history of symbolic mathematical computation. It should be remembered that crucial advances in mathematical algorithms (e.g. the computation of greatest common

```
> T[0] := 1;
                              T[0] := 1


> T[1] := x;
                              T[1] := x


> for n from 2 to 4 do
>    T[n] := expand( 2*x*T[n-1] - T[n-2] )
> od;


                                     2
                         T[2] := 2 x  - 1


                                    3
                        T[3] := 4 x  - 3 x


                                 4       2
                     T[4] := 8 x  - 8 x  + 1
```

Figure 1.4. MAPLE session involving Chebyshev polynomials.

divisors) were developing simultaneously, and that some of the systems were developed by and for people interested in specific applications.

To put the decade 1961-1971 into perspective, let us recall that FORTRAN appeared about 1958 and ALGOL in 1960. These two languages were designed primarily for numerical mathematical computation. Then in 1960/1961 came the development of LISP, a language for list processing. LISP was a major advancement on the road to languages for symbolic computation. An operation such as symbolic differentiation which is foreign to FORTRAN and ALGOL is relatively easy in LISP. (Indeed this is one of the standard programming assignments for students first learning LISP.) As will be noted later, several computer algebra systems were written in LISP.

1961-1966

In 1961, James Slagle at M.I.T. wrote a LISP program called SAINT for Symbolic Automatic INTegration. This was one of the earliest applications of LISP to symbolic computation and it was the first comprehensive attempt to program a computer to behave like a freshman calculus student. The program was based on a number of heuristics for indefinite integration and it performed about as well as a good calculus student.

One of the first systems for symbolic computation was FORMAC, developed by Jean Sammet, Robert Tobey, and others at IBM during the period 1962-1964. It was a FORTRAN preprocessor (a PL/I version appeared later) and it was designed for the manipulation of elementary functions including, of course, polynomials and rational functions. Another early system was ALPAK, a collection of FORTRAN-callable subroutines written in assembly language for the manipulation of polynomials and rational functions. It was designed by William S. Brown and others at Bell Laboratories and was generally available about 1964. A language now referred to as Early ALTRAN was designed at Bell Laboratories during the period 1964-1966. It used ALPAK as its package of computational procedures.

There were two other significant systems for symbolic computation developed during this period. George Collins at IBM and the University of Wisconsin (Madison) developed PM, a system for polynomial manipulation, an early version of which was operational in 1961 with improvements added to the system through 1966. The year 1965 marked the first appearance of MATHLAB, a LISP-based system for the manipulation of polynomials and rational functions, developed by Carl Engelman at M.I.T. It was the first interactive system designed to be used as a symbolic calculator. Included among its many firsts was the use of two-dimensional output to represent its mathematical output.

The work of this period culminated in the first ACM Symposium on Symbolic and Algebraic Manipulation held in March 1966 in Washington, D.C. That conference was summarized in the August 1966 issue of the Communications of the ACM [1].

1966-1971

In 1966/1967, Joel Moses at M.I.T. wrote a LISP program called SIN (for Symbolic INtegrator). Unlike the earlier SAINT program, SIN was algorithmic in approach and it was also much more efficient. In 1968, Tony Hearn at Stanford University developed REDUCE, an interactive LISP-based system for physics calculations. One of its principal design goals was portability over a wide range of platforms, and as such only a limited subset of LISP was actually used. The year 1968 also marked the appearance of Engelman's MATHLAB-68, an improved version of the earlier MATHLAB interactive system, and of the system known as Symbolic Mathematical Laboratory developed by William Martin at M.I.T. in 1967. The latter was a linking of several computers to do symbolic manipulation and to give good graphically formatted output on a CRT terminal.

The latter part of the decade saw the development of several important general purpose systems for symbolic computation. ALTRAN evolved from the earlier ALPAK and Early ALTRAN as a language and system for the efficient manipulation of polynomials and rational functions. George Collins developed SAC-1 (for Symbolic and Algebraic Calculations) as the successor of PM for the manipulation of polynomials and rational functions. CAMAL (CAMbridge ALgebra system) was developed by David Barton, Steve Bourne, and John Fitch at the University of Cambridge. It was implemented in the BCPL language, and was particularly geared to computations in celestial mechanics and general relativity. REDUCE was redesigned by 1970 into REDUCE 2, a general purpose system with special facilities for use in high-energy physics calculations. It was written in an ALGOL-like dialect called RLISP, avoiding the cumbersome parenthesized notation of LISP, while at the same time retaining its original design goal of being easily portable. SCRATCHPAD was developed by J. Griesmer and Richard Jenks at IBM Research as an interactive LISP-based system which incorporated significant portions of a number of previous systems and programs into its library, such as MATHLAB-68, REDUCE 2, Symbolic Mathematical Library, and SIN. Finally, the MACSYMA system first appeared about 1971. Designed by Joel Moses, William Martin, and others at M.I.T., MACSYMA was the most ambitious system of the decade. Besides the standard capabilities for algebraic manipulation, it included facilities to aid in such computations as limit calculations, symbolic integration, and the solution of equations.

The decade from 1961 to 1971 concluded with the Second Symposium on Symbolic and Algebraic Manipulation held in March 1971 in Los Angeles [4]. The proceedings of that conference constitute a remarkably comprehensive account of the state of the art of symbolic mathematical computation in 1971.

1971-1981

While all of the languages and systems of the sixties and seventies began as experiments, some of them were eventually put into ''production use'' by scientists, engineers, and applied mathematicians outside of the original group of developers. REDUCE, because of its early emphasis on portability, became one of the most widely available systems of this decade. As a result it was instrumental in bringing computer algebra to the attention of many new users. MACSYMA continued its strong development, especially with regard to algorithm development. Indeed, many of the standard techniques (e.g. integration of elementary functions, Hensel lifting, sparse modular algorithms) in use today either came from, or were strongly influenced by, the research group at M.I.T. It was by far the most powerful of the existing computer algebra systems.

SAC/ALDES by G. Collins and R. Loos was the follow-up to Collins' SAC-1. It was a non-interactive system consisting of modules written in the ALDES (ALgebraic DEScription) language, with a translator converting the results to ANSI FORTRAN. One of its most notable distinctions was in being the only major system to completely and carefully document its algorithms. A fourth general purpose system which made a significant mark in the late 1970's was muMATH. Developed by David Stoutemyer and Albert Rich at the University of Hawaii, it was written in a small subset of LISP and came with its own programming language, muSIMP. It was the first comprehensive computer algebra system which could actually run on the IBM family of PC computers. By being available on such small and widely accessible personal computers, muMATH opened up the possibility of widespread use of computer algebra systems for both research and teaching.

In addition to the systems mentioned above, a number of special purpose systems also generated some interest during the 1970's. Examples of these include: SHEEP, a system for tensor component manipulation designed by Inge Frick and others at the University of

Stockholm; TRIGMAN, specially designed for computation of Poisson series and written in FORTRAN by W. H. Jeffreys at University of Texas (Austin); and SCHOONSCHIP by M. Veltman of the Netherlands for computations in high-energy physics. Although the systems already mentioned have all been developed in North America and Europe, there were also a number of symbolic manipulation programs written in the U.S.S.R. One of these is ANALITIK, a system implemented in hardware by V. M. Glushkov and others at the Institute of Cybernetics, Kiev.

1981-1991

Due to the significant computer resource requirements of the major computer algebra systems, their widespread use remained (with the exception of muMATH) limited to researchers having access to considerable computing resources. With the introduction of microprocessor-based workstations, the possibility of relatively powerful desk-top computers became a reality. The introduction of a large number of different computing environments, coupled with the often nomadic life of researchers (at least in terms of workplace locations) caused a renewed emphasis on portability for the computer algebra systems of the 1980's. More efficiency (particularly memory space efficiency) was needed in order to run on the workstations that were becoming available at this time, or equivalently, to service significant numbers of users on the time-sharing environments of the day. This resulted in a movement towards the development of computer algebra systems based on newer "systems implementation" languages such as C, which allowed developers more flexibility to control the use of computer resources. The decade also marked a growth in the commercialization of computer algebra systems. This had both positive and negative effects on the field in general. On the negative side, users not only had to pay for these systems but also they were subjected to unrealistic claims as to what constituted the state of the art of these systems. However, on the positive side, commercialization brought about a marked increase in the usability of computer algebra systems, from major advances in user interfaces to improvements to their range of functionality in such areas as graphics and document preparation.

The beginning of the decade marked the origin of MAPLE. Initiated by Gaston Gonnet and Keith Geddes at the University of Waterloo, its primary motivation was to provide user accessibility to computer algebra. MAPLE was designed with a modular structure: a small compiled kernel of modest power, implemented completely in the systems implementation language C (originally B, another language in the "BCPL family") and a large mathematical library of routines written in the user-level MAPLE language to be interpreted by the kernel. Besides the command interpreter, the kernel also contained facilities such as integer and rational arithmetic, simple polynomial manipulation, and an efficient memory management system. The small size of the kernel allowed it to be implemented on a number of smaller platforms and allowed multiple users to access it on time-sharing systems. Its large mathematical library, on the other hand, allowed it to be powerful enough to meet the mathematical requirements of researchers.

Another system written in C was SMP (Symbolic Manipulation Program) by Stephen Wolfram at Caltech. It was portable over a wide range of machines and differed from existing systems by using a language interface that was rule-based. It took the point of view that the rule-based approach was the most natural language for humans to interface with a computer algebra program. This allowed it to present the user with a consistent, pattern-directed language for program development.

The newest of the computer algebra systems during this decade were MATHEMATICA and DERIVE. MATHEMATICA is a second system written by Stephen Wolfram (and others). It is best known as the first system to popularize an integrated environment supporting symbolics, numerics, and graphics. Indeed when MATHEMATICA first appeared in 1988, its graphical capabilities (2-D and 3-D plotting, including animation) far surpassed any of the graphics available on existing systems. MATHEMATICA was also one of the first systems to successfully illustrate the advantages of combining a computer algebra system with the easy-to-use editing features on machines designed to use graphical user-interfaces (i.e. window environments). Based on C, MATHEMATICA also comes with its own programming language which closely follows the rule-based approach of its predecessor, SMP.

DERIVE, written by David Stoutemyer and Albert Rich, is the follow-up to the successful muMATH system for personal computers. While lacking the wide range of symbolic capabilities of some other systems, DERIVE has an impressive range of applications considering the limitations of the 16-bit PC machines for which it was designed. It has a friendly user interface, with such added features as two-dimensional input editing of mathematical expressions and 3-D plotting facilities. It was designed to be used as an interactive system and not as a programming environment.

Along with the development of newer systems, there were also a number of changes to existing computer algebra systems. REDUCE 3 appeared in 1983, this time with a number of new packages added by outside developers. MACSYMA bifurcated into two versions, DOE-MACSYMA and one distributed by SYMBOLICS, a private company best known for its LISP machines. Both versions continued to develop, albeit in different directions, during this decade. AXIOM, (known originally as SCRATCHPAD II) was developed during this decade by Richard Jenks, Barry Trager, Stephen Watt and others at the IBM Thomas J. Watson Research Center. A successor to the first SCRATCHPAD language, it is the only "strongly typed" computer algebra system. Whereas other computer algebra systems develop algorithms for a specific collection of algebraic domains (such as, say, the field of rational numbers or the domain of polynomials over the integers), AXIOM allows users to write algorithms over general fields or domains.

As was the case in the previous decade, the eighties also found a number of specialized systems becoming available for general use. Probably the largest and most notable of these is the system CAYLEY, developed by John Cannon and others at the University of Sydney, Australia. CAYLEY can be thought of as a "MACSYMA for group theorists." It runs in large computing environments and provides a wide range of powerful commands for problems in computational group theory. An important feature of CAYLEY is a design geared to answering questions not only about individual elements of an algebraic structure, but more importantly, questions about the structure as a whole. Thus, while one could use a system such as MACSYMA or MAPLE to decide if an element in a given domain (such as a polynomial domain) has a given property (such as irreducibility), CAYLEY can be used to determine if a group structure is finite or infinite, or to list all the elements in the center of the structure (i.e. all elements which commute with all the elements of the structure).

Another system developed in this decade and designed to solve problems in computational group theory is GAP (Group Algorithms and Programming) developed by J. Neubüser and others at the University of Aachen, Germany. If CAYLEY can be considered to be the "MACSYMA of group theory," then GAP can be viewed as the "MAPLE of group theory." GAP follows the general design of MAPLE in implementing a small compiled kernel (in C) and a large group theory mathematical library written in its own programming language.

Examples of some other special purpose systems which appeared during this decade include FORM by J. Vermaseren, for high energy physics calculations, LiE, by A.M. Cohen for Lie Algebra calculations, MACAULAY, by Michael Stillman, a system specially built for computations in Algebraic Geometry and Commutative Algebra, and PARI by H. Cohen in France, a system oriented mainly for number theory calculations. As with most of the new systems of the eighties, these last two are also written in C for portability and efficiency.

Research Information about Computer Algebra

Research in computer algebra is a relatively young discipline, and the research literature is scattered throughout various journals devoted to mathematical computation. However, its state has advanced to the point where there are two research journals primarily devoted to this subject area: the *Journal of Symbolic Computation* published by Academic Press and *Applicable Algebra in Engineering, Communication and Computing* published by Springer-Verlag. Other than these two journals, the primary source of recent research advances and trends is a number of conference proceedings. Until recently, there was a sequence of North American conferences and a sequence of European conferences. The North American conferences, primarily organized by ACM SIGSAM (the ACM Special Interest Group on Symbolic and Algebraic Manipulation), include SYMSAM '66 (Washington, D.C.), SYMSAM '71 (Los Angeles), SYMSAC '76 (Yorktown Heights), SYMSAC '81 (Snowbird), and SYMSAC '86 (Waterloo). The European conferences, organized by SAME (Symbolic and Algebraic Manipulation in Europe) and ACM SIGSAM, include the following whose proceedings have appeared in the Springer-Verlag series *Lecture Notes in Computer Science:* EUROSAM '79 (Marseilles), EUROCAM '82 (Marseilles), EUROCAL '83 (London), EUROSAM '84 (Cambridge), EUROCAL '85 (Linz), and EUROCAL '87 (Leipzig). Starting in 1988, the two streams of conferences have been merged and they are now organized under the name ISSAC (International Symposium on Symbolic and Algebraic Computation), including ISSAC '88 (Rome), ISSAC '89 (Portland, Oregon), ISSAC '90 (Tokyo), ISSAC '91 (Bonn) and ISSAC '92 (Berkeley).

1.4. AN EXAMPLE OF A COMPUTER ALGEBRA SYSTEM: MAPLE

Traditional languages for scientific computation such as C, FORTRAN, or PASCAL are based on arithmetic of fixed-length integers and fixed-precision real (floating-point) numbers. Therefore, while various data and programming structures augment the usefulness of such systems, they still allow only a limited mode of computation. The inherent difficulty in obtaining meaningful insights from approximate results is often compounded by the difficulty of producing a reasonable approximation. Moreover, an indeterminate quantity (such as the variable x) may not be manipulated algebraically (as in the expression (x+1)*(x–1)).

In contrast, modern systems for symbolic computation support exact rational arithmetic, arbitrary-precision floating-point arithmetic, and algebraic manipulation of expressions containing indeterminates. The (ambitious) goal of such systems is to support mathematical manipulation in its full generality. In this section, we illustrate some computations performed in the computer algebra system MAPLE. In the examples, input to the system follows the left-justified prompt > and is terminated either by a semicolon or a colon (to display results, or not display them, respectively); system output is centered. Comments are preceded by a sharp sign #. A ditto " accesses the previous result. The output is displayed in a two-dimensional style which is typical of current computer algebra systems accessed through ordinary ASCII terminals. There are some mathematical user interfaces which exploit more sophisticated displays to support typeset-quality output (e.g. π instead of Pi, proper integral signs, et cetera).

To illustrate exact arithmetic and arbitrary-precision floating-point arithmetic, consider the following examples.

```
> 33!/2^31 + 41^41;
       1330877630632711998713399240963346255989932815022128910520902250516


> 43!/(2^43 -1);
              604152630633738356373551320685139975072645120000000000
              ------------------------------------------------------
                                  8796093022207
```

While the above operations take place in the field of rational numbers **Q**, arithmetic operations can also be performed in other numeric domains such as finite fields, complex numbers, and algebraic extension fields. For example,

```
> 483952545774574373476/ 122354323571234 mod 1000003;
                                        887782


> 10*(8+6*I)^(-1/2);
                                          10
                                     ------------
                                              1/2
                                     (8 + 6 I)
```

```
> evalc( " );
                                  3 - I
```

In the last calculation, the command "evalc" is used to place the result in standard complex number form, where "I" denotes the square root of -1. A similar approach is used by MAPLE when operating in domains such as algebraic extensions and Galois fields.

Computer algebra systems also allow for the use of other common mathematical constants and functions. Such expressions can be reduced to simplified forms.

```
> sqrt(15523/3 - 98/2);
                                          1/2
                                  124/3 3


> a:= sin(Pi/3) * exp(2 + ln(33));
                                  1/2
                       a := 1/2 3     exp(2 + ln(33))


> simplify(a);
                                      1/2
                               33/2 3     exp(2)


> evalf(a);
                                 211.1706396
```

In the above, the command "evalf" (evaluate in floating-point mode) provides a decimal expansion of the real value. The decimal expansion is computed to 10 digits by default, but the precision can be controlled by the user either by re-setting a system variable (the global variable "Digits") or by passing an additional parameter to the evaluation function as in

```
> evalf(a,60);
        211.170639624855418173457016949952935319763238458535271731859
```

Also among the numerical capabilities of computer algebra systems are a wide variety of arithmetic operations over the integers, such as factorization, primality testing, finding nearest prime numbers, and greatest common divisor (GCD) calculations.

```
> n:= 19380287199092196525608598055990942841820;
                 n := 19380287199092196525608598055990942841820


> isprime(n);
                                    false


> ifactor(n);
                   2    2          3      4                  2
                (2)  (3)  (5) (19)  (101)  (12282045523619)
```

```
> nextprime(n);
                         1938028719909219652560859805599094284204 3
```

```
> igcd(15990335972848346968323925788771404985, 15163659044370489780);
                                   12636382536975408 15
```

It should be apparent that in the previous examples each numerical expression is evaluated using the appropriate rules of algebra for the given number system. Notice that the arithmetic of rational numbers requires integer GCD computations because every result is automatically simplified by removing common factors from numerator and denominator. It should also be clear that such manipulations may require much more computer time (and memory space) than the corresponding numerical arithmetic in, say, FORTRAN. Hence, the efficiency of the algorithms which perform such tasks in computer algebra systems is a major concern.

Computer algebra systems can also perform standard arithmetic operations over domains requiring the use of symbolic variables. Examples of such algebraic structures include polynomial and power series domains along with their quotient fields. Typical examples of algebraic operations include expansion ("expand"), long division ("quo","rem"), normalization of rational functions ("normal"), and GCD calculation ("gcd").

```
> a := (x + y)^12 - (x - y)^12;
                                          12          12
                            a := (x + y)   - (x - y)
```

```
> expand(a);
           11          3  9          5  7           7  5          9  3         11
     24 y x   + 440 y  x  + 1584 y  x  + 1584 y  x  + 440 y  x  + 24 y   x
```

```
> quo( x^3*y-x^3*z+2*x^2*y^2-2*x^2*z^2+x*y^3+ x*y^2*z-x*z^3, x+y+z, x);
                                      2      2    2         2
                           (y - z) x  + (y  - z ) x + z  y
```

```
> gcd( x^3*y-x^3*z+2*x^2*y^2-2*x^2*z^2+x*y^3+ x*y^2*z-x*z^3, x+y+z );
                                         1
```

```
> b := (x^4 - y^4)/(x^3 + y^3) - (x^5 + y^5)/(x^4 - y^4);
                                      4    4    5    5
                                     x  - y    x  + y
                               b := ------- - -------
                                      3    3    4    4
                                     x  + y    x  - y
```

```
> normal(b);
                                    3  3
                                   x  y
          - --------------------------------------
              3    2       2    3     2         2
            (x  - x  y + x y  - y ) (x  - x y + y )

> f := (x + y)*(x - y)^6:    g := (x^2 - y^2)*(x - y)^3:    f / g;
                                         3
                              (x + y) (x - y)
                              ---------------
                                   2    2
                                  x  - y

> normal( f/g );
                                        2
                                 (x - y)
```

Notice that computer algebra systems perform automatic simplification in polynomial and rational function domains in a different manner than they do in numerical domains. There are indeed some *automatic* simplifications going on in the above MAPLE calculations. For example, the polynomial expansion automatically cancelled such terms as $12x - 12x$, while the quotient operation of f and g had some common factors cancel in the numerator and denominator. On the other hand, a user must explicitly ask that a polynomial expression be represented in expanded form, or that a rational expression be normalized by removing all common factors. Unlike the case of rational numbers, common factors of numerators and denominators of rational expressions are not computed automatically, although common factors which are ''obvious'' in the representation may cancel automatically. The amount of automatic cancellation depends on the type of representation used for the numerator and denominator. Chapter 3 investigates various issues of representation and simplification for polynomials and rational expressions.

In practice, the types of expressions shown above (namely, polynomials and rational functions) encompass much of everyday mathematics. However, we hasten to add that the classical algorithms known in mathematics may be neither optimal nor practical for many of the operations used by computer algebra systems. For example, in the previous collection of commands we performed a polynomial long division (with ''x'' as the main indeterminate); this, along with Euclid's algorithm, provides a basis *in principle* for the calculation of polynomial GCD's and hence for the arithmetic of rational functions. However, such an approach suffers serious drawbacks and is seldom used in practice. Indeed the problem of efficiently computing polynomial GCD's is a fundamental problem in computer algebra and is the topic of Chapter 7. For the case of the basic operations of addition, subtraction, multiplication, and division of polynomials, the standard ''high school'' algorithms *are* commonly used, with faster methods (e.g. methods based on the fast Fourier transform) applied only for special cases of very large, well-defined problems. Arithmetic algorithms are discussed in Chapter 4.

A fundamental operation in all computer algebra systems is the ability to factor polynomials (both univariate and multivariate) defined over various coefficient domains. Thus, for example, we have

```
> factor(x^6 - x^5 + x^2 + 1);
                                          6    5    2
                                         x  - x  + x  + 1

> factor(5*x^4 - 4*x^3 - 48*x^2 + 44*x + 3);
                                                 2
                          (x - 1) (x - 3) (5 x  + 16 x + 1)

> Factor(x^6 - x^5 + x^2 + 1) mod 13 ;
                       3       2               3      2
                     (x  + 10 x  + 8 x + 11) (x  + 2 x  + 11 x + 6)

> Factor(5*x^4 - 4*x^3 - 48*x^2 + 44*x + 3) mod 13 ;
                                                  2
                         5 (x + 12) (x + 10) (x  + 11 x + 8)

> factor(x^12 - y^12);
          2          2              2          2    2     2     4     2 2    4
  (x - y)(x  + x y + y )(x + y) (y  - x y + x ) (x  + y ) (x  - x  y  + y )

> alias( a = RootOf( x^4 - 2 ) ):

> factor( x^12 - 2*x^8 + 4*x^4 - 8 , a );
               4      2         4      2                              2    2
             (x  - 2 x  + 2) (x  + 2 x  + 2) (x - a) (x + a) (x  + a )

> Factor( x^6 - 2*x^4 + 4*x^2 - 8 , a ) mod 5;
                                                     2            2
                 (x + 4) (x + 2) (x + 1) (x + a ) (x + 4 a ) (x + 3)
```

In the previous examples, the first two factorizations are computed in the domain $\mathbf{Z}[x]$, the next two in the domain $\mathbf{Z}_{13}[x]$, and the fifth in $\mathbf{Z}[x,y]$. The sixth example asks for a factorization of a polynomial over $\mathbf{Q}(2^{1/4})$, an algebraic extension of $\mathbf{Q}$. The last example is a factorization of a polynomial over the coefficient domain $\mathbf{Z}_5(2^{1/4})$, a Galois field of order 625. The command ''alias'' in the above examples is used to keep the expressions readable (letting ''a'' denote a fourth root of 2).

While polynomial factorization is a computationally intensive operation for these systems, it is interesting to note that the basic tools require little more than polynomial GCD calculations, some matrix algebra over finite fields, and finding solutions of simple diophantine equations. Algorithms for polynomial factorization are discussed in Chapter 8 (with preliminaries in Chapter 6).

Computer algebra systems provide powerful tools for working with matrices. Operations such as Gaussian elimination, matrix inversion, calculation of determinants, eigenvalues, and eigenvectors can be performed. As with other computations, exact arithmetic is used and symbolic entries are allowed. For example, a general Vandermonde matrix (in three symbols x, y, z), its inverse and its determinant are given by

```
> V := vandermonde( [ x, y, z] );
                                   [           2 ]
                                   [ 1   x   x   ]
                                   [             ]
                                   [           2 ]
                              V := [ 1   y   y   ]
                                   [             ]
                                   [           2 ]
                                   [ 1   z   z   ]

> inverse(V);
[         y z                      z x                         y x               ]
[ -------------------   - ----------------------   ---------------------- ]
[            2                        2                          2            ]
[ y z - y x + x   - z x     - z x + y x - y   + y z   - z x + y x + z   - y z ]
[                                                                             ]
[         y + z                    x + z                       x + y          ]
[- -------------------   ----------------------   - ---------------------- ]
[             2                       2                           2           ]
[  y z - y x + x   - z x   - z x + y x - y   + y z    - z x + y x + z   - y z]
[                                                                             ]
[          1                         1                          1             ]
[ -------------------   - ----------------------   ---------------------- ]
[            2                        2                          2            ]
[ y z - y x + x   - z x     - z x + y x - y   + y z   - z x + y x + z   - y z ]

> det( V );
                        2    2        2    2       2    2
                     y z  - y  z - x z  + x  z + x y  - x  y

> factor(");
                      - (- z + y) (- z + x) (- y + x)
```

The latter command verifies a well-known mathematical property of the determinant of Vandermonde matrices (in the general 3×3 case).

Matrix algebra tools can be applied to solving systems of linear equations, one of the most common applications of mathematical software. In the case of computer algebra systems, the coefficients of the linear system are allowed to include symbolic variables. One can solve such a system once in terms of the unknown symbols (or ''parameters'') and then generate a number of numerical solutions by substituting numerical values for the parameters. For example,

```
> equation1 := ( 1   -  eps)*x + 2*y - 4*z - 1 = 0:

> equation2 := (3/2  -  eps)*x + 3*y - 5*z - 2 = 0:

> equation3 := (5/2 + eps)*x + 5*y - 7*z - 3 = 0:

> solutions := solve( {equation1, equation2, equation3}, {x, y, z} );
                                  1                      1 + 7 eps
            solutions := {x = - -----, z = 3/4, y = 1/4 ---------}
                                2 eps                       eps

> subs( eps=10^(-20), solutions );
      {z = 3/4, x = -50000000000000000000, y = 100000000000000000007/4}
```

where the last command substitutes the value eps $= 10^{-20}$ into the solution of the linear system. For numerical computation, the parameters must be given numerical values prior to solving the system, and the process must be repeated for any other parameter values of interest. The results generated from numerical computation are also subject to round-off error since exact arithmetic is not used. Indeed, the above example would cause difficulties for numerical techniques using standard-precision arithmetic. Of course, using exact arithmetic and allowing symbolic parameters requires specialized algorithms to achieve time and memory space efficiency. Chapter 9 discusses algorithms for solving equations.

Solving a linear system typically involves "eliminating" unknowns from equations to obtain a simplified system, usually a triangularized system which is then easy to solve. It is less well known that a similar approach is also available for systems of nonlinear polynomial equations. Chapters 9 and 10 provide an introduction to classical and modern techniques for solving systems of polynomial equations.

As an example of solving a nonlinear system of equations, consider the problem of finding the critical points of a bivariate function:

```
> f := x^2 *y*(1 - x - y)^3:

> equation1 := diff( f, x );  equation2 := diff( f, y );
                                          3       2             2
               equation1 := 2 x y (1 - x - y)  - 3 x  y (1 - x - y)
                                 2           3       2            2
                 equation2 := x  (1 - x - y)  - 3 x  y (1 - x - y)

> solve( {equation1,equation2}, {x,y} );
     {x = 0, y = y}, {y = 0, x = 1}, {y = 1/6, x = 1/3}, {y = y, x = 1 - y}
```

which gives two critical points along with two lines of singularity. One could then obtain more information about the types of singularities of this function by using MAPLE to calculate the Hessian of the function at all the critical points, or more simply by plotting a 3-D graph of the function.

One of the most useful features of computer algebra systems is their ability to solve problems from calculus. Operations such as computing limits, differentiation of functions, calculation of power series, definite and indefinite integration, and solving differential equations can be performed. We have already illustrated the use of MAPLE for calculus when we determined the critical points of a function. Some additional examples are

```
> limit( tan(x) / x , x=0 );
                                            1


> diff( ln(sec(x)), x );
                                          tan(x)


> series(  tan( sinh(x) ) - sinh( tan(x) ), x=0, 15 );
                     7      13  9      1451  11      6043    13        15
             1/90 x   +  --- x   +  ----- x    +  ------ x    + O(x   )
                         756        75600         332640


> series( BesselJ(0,x)/BesselJ(1,x), x, 12 );
       -1                    3               5                7         13     9          10
    2 x    - 1/4 x - 1/96 x    - 1/1536 x    - 1/23040 x    -  ------- x   + O(x   )
                                                               4423680
```

where the last example gives the Laurent series for the ratio of the two Bessel functions of the first kind, $J_0(x)$ and $J_1(x)$.

The ability to differentiate, take limits, or calculate Taylor (or Laurent) series as above does not surprise new users of computer algebra systems. These are mainly algebraic operations done easily by hand for simple cases. The role of the computer algebra system is to reduce the drudgery (and to eliminate the calculation errors!) of a straightforward, easy to understand, yet long calculation. The same cannot be said for solving the indefinite integration problem of calculus. Integration typically is not viewed as an algorithmic process, but rather as a collection of tricks which can only solve a limited number of integration problems. As such, the ability of computer algebra systems to calculate indefinite integrals is very impressive for most users. These systems do indeed begin their integration procedures by trying some heuristics of the type learned in traditional calculus courses. Indeed, until the late sixties this was the only approach available to computer algebra systems. However, in 1969 Robert Risch presented a decision procedure for the indefinite integration of a large class of functions known as the *elementary functions*. This class of functions includes the typical functions considered in calculus courses, such as the exponential, logarithm, trigonometric, inverse trigonometric, hyperbolic, and algebraic functions. (Additional research contributions in succeeding years have led to effective algorithms for integration in most cases, although the case of general algebraic functions can require a very large amount of computation.) The Risch algorithm either determines a closed formula expressing the integral as an elementary function, or else it proves that it is impossible to express the integral as an elementary function. Some examples of integration follow.

```
> int( ((3*x^2 - 7*x + 15)*exp(x) + 3*x^2 - 14)/(x - exp(x))^2, x );
                                           2
                                        3 x   - x + 14
                                        -------------
                                          x - exp(x)


> int( (3*x^3 - x + 14)/(x^2 + 4*x - 4), x );
         2                  2                     1/2                        1/2
  3/2 x  - 12 x + 59/2 ln(x + 4 x - 4) + 38 2   arctanh(1/8 (2 x + 4) 2    )



> int( x*exp( x^3 ), x );
                                          /
                                         |            3
                                         |   x exp(x ) dx
                                         |
                                        /
```

In the latter case, the output indicates that no closed form for the integral exists (as an elementary function). Chapters 11 and 12 develop algorithms for the integration problem.

Computer algebra systems are also useful for solving differential, or systems of differential, equations. There are a large number of techniques for solving differential equations which are entirely algebraic in nature, with the methods usually reducing to the computation of integrals (the simplest form of differential equations). Unlike the case of indefinite integration, there is no known complete decision procedure for solving differential equations. For differential equations where a closed form solution cannot be determined, often one can compute an approximate solution; for example, a series solution or a purely numerical solution. As an example, consider the following differential equation with initial conditions.

```
> diff_eqn := diff(y(x), x$2) + t*diff(y(x), x) - 2*t^2*y(x) = 0;
                          /  2        \
                          | d         |      / d       \       2
              diff_eqn := |----- y(x)| + t |---- y(x)| - 2 t  y(x) = 0
                          |  2        |      \ dx      /
                          \ dx        /


> init_conds := y(0) = t,  D(y)(0) = 2*t^2;
                                                                    2
                       init_conds := y(0) = t, D(y)(0) = 2 t


> dsolve( {diff_eqn, init_conds}, y(x));
                     y(x) = 4/3 t exp(t x) - 1/3 t exp(- 2 t x)
```

Designers of computer algebra systems cannot anticipate the needs of all users. Therefore it is important that systems include a facility for programming new functionality, so that individual users can expand the usefulness of the system in directions that they find necessary or interesting. The programming languages found in computer algebra systems typically provide a rich set of data structures and programming constructs which allow users to manipulate common mathematical objects (such as polynomials or trig functions) easily and efficiently. For example, returning to the problem of computing Chebyshev polynomials as considered earlier in this chapter, a program could be written in MAPLE as follows.

```
> Cheby   := proc(n,x)
>              local T,k;
>              T[0]:= 1; T[1] := x;
>              for k from 2 to n do
>                  T[k]:= expand( 2*x*T[k-1] - T[k-2]);
>              od;
>              RETURN(T[n]);
> end:
```

```
> Cheby(7,x);
                                      7         5        3
                                  64 x  - 112 x  + 56 x  - 7 x
```

In this section, we have highlighted the algebraic capabilities of computer algebra systems. We remark that many of the systems also provide graphical support (2-D and 3-D graphics), support for numerical routines (such as numerical root-finding, numerical integration, and numerical differential equation solvers), on-line ''help'' facilities, and many other features to support and extend their usability.

Exercises

1. Consider the following system of linear equations which depends on a parameter a:

$$\begin{aligned} ax_1 + a^2x_2 - x_3 &= 1, \\ a^2x_1 - x_2 + x_3 &= a, \\ -x_1 + x_2 + ax_3 &= a^2. \end{aligned}$$

   Solve this system by hand, in terms of the parameter a. (If the hand manipulation becomes too tedious for you, you may stop after obtaining an expression for *one* of the unknowns.) Check your result by using a procedure for solving linear equations in a computer algebra system. Note that the solution involves rational functions in the parameter a.

2. Calculate by hand the determinant of the coefficients in the linear system of Exercise 1. Check your result by using an appropriate computer routine. Note that the determinant is a polynomial in the parameter a.

3. For each of the following indefinite integrals, either state the answer or else state that you think the indefinite integral cannot be expressed in terms of elementary functions. You might wish to indicate the degree of confidence you have in each of your answers.

(i) $\int x \,/\, (1+e^x)\, dx$

(ii) $\int e^{x^2}\, dx$

(iii) $\int \sqrt{(x^2-1)(x^2-4)}\, dx$

(iv) $\int \sqrt{(x-1)(x-4)}\, dx$

(v) $\int \sqrt{(1+x)/(1-x)}\, dx$

(vi) $\int \log(x^2-5x+4)\, dx$

(vii) $\int \log(x)\,/(1+x)\, dx$

(viii) $\int 1 \,/\, \log(x)\, dx$

4. Give a brief overview of one of the following computer algebra systems: AXIOM, CAYLEY, DERIVE, MACSYMA, MAPLE, MATHEMATICA, REDUCE. Compare and contrast it with a computer algebra system with which you are familiar.

References

1. "Proc. of ACM Symposium on Symbolic and Algebraic Manipulation (SYMSAM '66), Washington D.C. (ed. R.W. Floyd)," *Comm. ACM*, **9** pp. 547-643 (1966).
2. "Symbol Manipulation Languages and Techniques," in *Proc. of the IFIP Working Conference on Symbol Manipulation Languages, Pisa, 1966*, ed. D.G. Bobrow, North-Holland (1968).
3. "Proc. of the 1968 Summer Institute on Symbolic Mathematical Computation (ed. R.G. Tobey)," I.B.M. Programming Lab. Rep. FSC69-0312 (1969).
4. *Proc. of the Second Symposium on Symbolic and Algebraic Manipulation (SYMSAM '71), Los Angeles*, ed. S.R. Petrick, ACM Press, New York (1971).
5. D. Barton and J.P. Fitch, "Applications of Algebraic Manipulation Programs in Physics," *Rep. Prog. Phys.*, **35** pp. 235-314 (1972).
6. W.S. Brown and A.C. Hearn, "Applications of Symbolic Algebraic Computation," Bell Laboratories Computing Science Technical Report #66 (1978).
7. B.W. Char, K.O. Geddes, G.H. Gonnet, B.L. Leong, M.B. Monagan, and S.M. Watt, *Maple V Language Reference Manual,*, Springer-Verlag (1991).
8. G.E. Collins, "Computer Algebra of Polynomials and Rational Functions," *Amer. Math. Monthly*, **80** pp. 725-755 (1973).
9. A.C. Hearn, "Scientific Applications of Symbolic Computation," pp. 83-108 in *Computer Science and Scientific Computing*, ed. J.M. Ortega, Academic Press, New York (1976).

10. A.D. Hall Jr., ''The Altran System for Rational Function Manipulation - A Survey,'' *Comm. ACM*, **14** pp. 517-521 (1971).

11. J. Moses, ''Algebraic Simplification: A Guide for the Perplexed,'' *Comm. ACM*, **14** pp. 527-537 (1971).

12. J. Moses, ''Algebraic Structures and their Algorithms,'' pp. 301-319 in *Algorithms and Complexity*, ed. J.F. Traub, Academic Press, New York (1976).

CHAPTER 2

ALGEBRA OF POLYNOMIALS, RATIONAL FUNCTIONS, AND POWER SERIES

2.1. INTRODUCTION

In this chapter we present some basic concepts from algebra which are of central importance in the development of algorithms and systems for symbolic mathematical computation. The main issues distinguishing various computer algebra systems arise out of the choice of algebraic structures to be manipulated and the choice of representations for the given algebraic structures.

2.2. RINGS AND FIELDS

A *group* (G; ∘) is a nonempty set G, closed under a binary operation ∘ satisfying the axioms:

A1: $a \circ (b \circ c) = (a \circ b) \circ c$ for all $a, b, c \in$ G (Associativity).

A2: There is an element e $\in$ G such that

$e \circ a = a \circ e = a$ for all $a \in$ G (Identity).

A3: For all $a \in$ G, there is an element $a^{-1} \in$ G such that

$a \circ a^{-1} = a^{-1} \circ a = e$ (Inverses).

An *abelian group* (or, *commutative group*) is a group in which the binary operation ∘ satisfies the additional axiom:

A4: $a \circ b = b \circ a$ for all $a, b \in$ G (Commutativity).

A *ring* (R; +, ·) is a nonempty set R closed under two binary operations + and · such that (R; +) is an abelian group (i.e. axioms A1-A4 hold with respect to +), · is associative and has an identity (i.e. axioms A1-A2 hold with respect to ·), and which satisfies the additional axiom:

A5: $a \cdot (b + c) = (a \cdot b) + (a \cdot c)$, and

$(a + b) \cdot c = (a \cdot c) + (b \cdot c)$

for all $a, b, c \in$ R (Distributivity).

A *commutative ring* is a ring in which $\cdot$ is commutative (i.e. axiom A4 holds with respect to $\cdot$). An *integral domain* is a commutative ring which satisfies the additional axiom:

A6: $a \cdot b = a \cdot c$ and $a \neq 0 \quad \Rightarrow \quad b = c$

for all $a, b, c \in$ R (Cancellation Law).

We note that for rings we normally denote the identity element with respect to + by 0, the identity element with respect to $\cdot$ by 1, and the inverse of a with respect to + by $-a$.

A *field* (F; +, $\cdot$) is a set F having two binary operations +, $\cdot$ such that (F; +) is an abelian group (i.e. axioms A1-A4 hold with respect to +), (F $-$ {0}; $\cdot$)[1] is an abelian group (i.e. axioms A1-A4 hold for all nonzero elements with respect to $\cdot$), and $\cdot$ is distributive over + (i.e. axiom A5 holds). In other words, a field is a commutative ring in which every nonzero element has a multiplicative inverse.

A concise summary of the definitions of these algebraic structures is given in Table 2.1. The algebraic structures of most interest in this book are integral domains and fields. Thus the basic underlying structure is the commutative ring. If multiplicative inverses exist then we have a field; otherwise we will at least have the cancellation law (axiom A6). Another axiom which is equivalent to the cancellation law and which is used by some authors in the definition of an integral domain is:

A6′: $a \cdot b = 0 \quad \Rightarrow \quad a = 0$ or $b = 0$

for all $a, b \in$ R (No Zero Divisors).

Of course, axioms A6 and A6′ hold in a field as a consequence of multiplicative inverses.

Some Number Algebras

The set of integers (positive, negative, and zero) forms an integral domain and is denoted by **Z.** The most familiar examples of fields are the rational numbers **Q,** the real numbers **R,** and the complex numbers **C.** The integers modulo n, $\mathbf{Z}_n$, is an example of a ring having only a finite set of elements. Here addition and multiplication are performed as in **Z** but all results are replaced by their remainders after division by n. This ring has exactly n elements and is an example of a residue ring (cf. Chapter 5).

When p is a prime the ring $\mathbf{Z}_p$ is actually an example of a finite field. As an example, $\mathbf{Z}_5$ consists of the set { 0,1,2,3,4 }; addition and multiplication tables for $\mathbf{Z}_5$ are presented in Table 2.2. Note that every nonzero element in $\mathbf{Z}_5$ has a multiplicative inverse, since $1 \cdot 1 = 1$,

1. The *set difference* of two sets A and B is defined by A $-$ B = { $a : a \in$ A and $a \notin$ B }.

Table 2.1. Definitions of algebraic structures.

Structure	Notation	Axioms
Group	(G; ∘)	A1; A2; A3
Abelian Group	(G; ∘)	A1; A2; A3; A4
Ring	(R; +, ·)	A1; A2; A3; A4 w.r.t. + A1; A2 w.r.t. · A5
Commutative Ring	(R; +, ·)	A1; A2; A3; A4 w.r.t. + A1; A2; A4 w.r.t. · A5
Integral Domain	(D; +, ·)	A1; A2; A3; A4 w.r.t. + A1; A2; A4 w.r.t. · A5; A6
Field	(F; +, ·)	A1; A2; A3; A4 w.r.t. + A1; A2; A3; A4 for F-{0} w.r.t. · A5 (Note: A6 holds as a consequence.)

$2 \cdot 3 = 1$, $3 \cdot 2 = 1$, and $4 \cdot 4 = 1$. If we consider the integers modulo n, $\mathbf{Z}_n$, for some non-prime integer n, then some nonzero elements will not have multiplicative inverses. $\mathbf{Z}_n$ is, in general, a commutative ring but not even an integral domain. For example, in $\mathbf{Z}_{12}$ we have $2 \cdot 6 = 0$. For finite rings, the concepts of the cancellation law (or, no zero divisors) and the existence of multiplicative inverses turn out to be equivalent; in other words, every finite integral domain is a field (cf. Exercise 2.7).

Table 2.2. Addition and multiplication tables for $\mathbf{Z}_5$.

+	0	1	2	3	4
0	0	1	2	3	4
1	1	2	3	4	0
2	2	3	4	0	1
3	3	4	0	1	2
4	4	0	1	2	3

·	0	1	2	3	4
0	0	0	0	0	0
1	0	1	2	3	4
2	0	2	4	1	3
3	0	3	1	4	2
4	0	4	3	2	1

2.3. DIVISIBILITY AND FACTORIZATION IN INTEGRAL DOMAINS

The concept of divisibility plays a central role in symbolic computation. Of course division is always possible in a field. In an integral domain division is not possible, in general, but the concept of factorization into primes which is familiar for the integers **Z** can be generalized to other integral domains. Throughout this section, D denotes an integral domain. Here and in the sequel, we adopt the standard mathematical convention of omitting the $\cdot$ symbol for multiplication.

Greatest Common Divisors

Definition 2.1. For $a, b \in$ D, a is called a *divisor* of b if $b = ax$ for some $x \in$ D, and we say that *a divides b* (notationally, $a \mid b$). Correspondingly, b is called a *multiple* of a.

●

Definition 2.2. For $a, b \in$ D, an element $c \in$ D is called a *greatest common divisor* (GCD) of a and b if $c \mid a$ and $c \mid b$ and c is a multiple of every other element which divides both a and b.

●

Definition 2.3. For $a, b \in$ D, an element $c \in$ D is called a *least common multiple* (LCM) of a and b if $a \mid c$ and $b \mid c$ and c is a divisor of every other element which is a multiple of both a and b.

●

The most familiar application of GCD's is in reducing rational numbers (i.e. quotients of integers) to "lowest terms". Another role of GCD's in symbolic computation is the corresponding problem of reducing rational functions (i.e. quotients of polynomials) to "lowest terms". The use of the phrase "a GCD" rather than "the GCD" is intentional. A GCD of two elements $a, b \in$ D, when it exists, is not unique (but almost).

Definition 2.4. Two elements $a, b \in$ D are called *associates* if $a \mid b$ and $b \mid a$.

●

Definition 2.5. An element $u \in$ D is called a *unit* (or *invertible)* if u has a multiplicative inverse in D.

●

Example 2.1.

In the integral domain **Z** of integers, note the following facts.

(i) The units in **Z** are 1 and −1.

(ii) 6 is a GCD of 18 and 30.

(iii) −6 is also a GCD of 18 and 30.

(iv) 6 and −6 are associates.

●

It can be easily proved that in any integral domain D, two elements c and d are associates if and only if $cu = d$ for some unit u. It is also easy to verify that if c is a GCD of a and b then so is any associate $d = cu$, and conversely if c and d are GCD's of a and b then c must be an associate of d. In the integral domains of interest in symbolic computation, it is conventional to impose an additional condition on the GCD in order to make it unique. This is accomplished by noting that the relation of associativity is an equivalence relation, which therefore decomposes an integral domain into *associate classes*. (For example, the associate classes in **Z** are $\{0\}$, $\{1, -1\}$, $\{2, -2\}$, $\ldots$.) For a particular integral domain, a criterion is chosen to single out one element of each associate class as its canonical representative and define it to be *unit normal*. In the integral domain **Z** we will define the nonnegative integers to be *unit normal*. In any field F, every nonzero element is an associate of every other nonzero element (in fact, every nonzero element is a unit). In this case we define the elements 0 and 1 to be *unit normal*.

Definition 2.6. In any integral domain D for which unit normal elements have been defined, an element c is called the *unit normal GCD* of $a, b \in$ D, denoted c=GCD(a, b), if c is a GCD of a and b and c is unit normal.

●

Clearly the unit normal GCD of two elements $a,b \in$ D is unique (once the unit normal elements have been defined). For each integral domain D of interest in this book, unit normal elements will be appropriately defined and the following properties will always hold:

(1) 0 is unit normal;

(2) 1 is the unit normal element for the associate class of units;

(3) if $a, b \in$ D are unit normal elements then their product ab is also a unit normal element in D.

In the sequel, whenever we refer to the GCD of $a, b \in$ D it is understood that we are referring to the unique unit normal GCD.

Example 2.2. In the integral domain **Z,** GCD(18,30) = 6.

●

Definition 2.7. Let D be an integral domain in which unit normal elements have been defined. The *normal part* of $a \in$ D, denoted n(a), is defined to be the unit normal representative of the associate class containing a. The *unit part* of $a \in$ D ($a \neq 0$), denoted u(a), is the unique unit in D such that

$$a = \mathrm{u}(a)\,\mathrm{n}(a)$$

Clearly n(0) = 0 and it is convenient to define u(0) = 1.

●

Example 2.3. In the integral domain **Z**, $\mathrm{n}(a) = |\,a\,|$ and $\mathrm{u}(a) = \mathrm{sign}(a)$ where the *sign* of an integer is defined by

$$\mathrm{sign}(a) = \begin{cases} -1 & \text{if } a < 0 \\ 1 & \text{if } a \geq 0 . \end{cases}$$

●

The LCM of two elements $a, b \in \mathrm{D}$, when it exists, can be made unique in a similar manner. It can be verified that a LCM of $a, b \in \mathrm{D}$ exists if and only if $\mathrm{GCD}(a,b)$ exists. Moreover, $\mathrm{GCD}(a,b)$ is clearly a divisor of the product ab and it easy to verify that the element

$$\frac{ab}{\mathrm{GCD}(a,b)}$$

is a LCM of a and b. We therefore define the unique *unit normal LCM* of $a, b \in \mathrm{D}$, denoted $\mathrm{LCM}(a,b)$ by

$$\mathrm{LCM}(a,b) = \frac{\mathrm{n}(ab)}{\mathrm{GCD}(a,b)} .$$

Unique Factorization Domains

Definition 2.8. An element $p \in \mathrm{D} - \{0\}$ is called a *prime* (or *irreducible)* if

(a) p is not a unit, and

(b) whenever $p = ab$ either a or b is a unit.

●

Definition 2.9. Two elements $a, b \in \mathrm{D}$ are called *relatively prime* if $\mathrm{GCD}(a,b)=1$.

●

Definition 2.10. An integral domain D is called a *unique factorization domain* (UFD) if for all $a \in \mathrm{D} - \{0\}$, either a is a unit or else a can be expressed as a finite product of primes (i.e. $a = p_1 p_2 \cdots p_n$ for some primes p_i, $1 \leq i \leq n$) such that this factorization into primes is unique up to associates and reordering.

●

The last statement in Definition 2.10 means that if $a = p_1 p_2 \cdots p_n$ and $a = q_1 q_2 \cdots q_m$ are two prime factorizations of the same element a then $n = m$ and there exists a reordering of the q_j's such that p_i is an associate of q_i for $1 \leq i \leq n$.

It follows from Definition 2.8 that if p is a prime in an integral domain D then so is any associate of p. If unit normal elements have been defined in D then we may restrict our attention to *unit normal primes* – i.e. primes which are unit normal. Clearly, every prime factorization can be put into the canonical form of the following definition.

Definition 2.11. Let D be a UFD in which unit normal elements have been defined. Then for $a \in$ D a prime factorization of the form

$$a = \mathrm{u}(a)\, p_1^{e_1} p_2^{e_2} \cdots p_n^{e_n}$$

is called a *unit normal factorization* if $p_i\ (1 \le i \le n)$ are unit normal primes, $e_i > 0\ (1 \le i \le n)$, and $p_i \ne p_j$ whenever $i \ne j$.

●

A basic property of primes in a UFD is that if $p \mid ab$ and p is a prime, then either $p \mid a$ or $p \mid b$ – i.e. p (or an associate of p) must appear as one of the factors in the prime factorization of a or of b. The integral domain **Z** of integers is the most familiar example of a UFD. It turns out that the integral domains of primary interest in symbolic computation, the polynomial domains to be introduced in the following sections, are also UFD's. (In the case of the polynomial domains, elements are usually referred to as irreducible rather than prime.) Exercise 2.11 shows that not every integral domain is a UFD and Exercise 2.12 shows that GCD's do not necessarily exist in an arbitrary integral domain. The following theorem assures us of the existence of GCD's in a UFD. Here and in the sequel, we assume without loss of generality that unit normal elements satisfying (1) - (3) have been defined for every integral domain D.

Theorem 2.1. If D is a UFD and if $a, b \in$ D are not both zero then GCD(a, b) exists and is unique.

Proof: The uniqueness has already been established. To show existence, first suppose that $a \ne 0$ and $b \ne 0$ and let their unique unit normal factorizations be

$$a = \mathrm{u}(a)\, p_1^{e_1} p_2^{e_2} \cdots p_n^{e_n} \quad \text{and} \quad b = \mathrm{u}(b)\, q_1^{f_1} q_2^{f_2} \cdots q_m^{f_m} \tag{2.1}$$

where p_i, q_j are unit normal primes. Let $r_1, \ldots, r_l$ denote the distinct elements in the set $\{p_1, \ldots, p_n, q_1, \ldots, q_m\}$. Then the factorizations (2.1) may be written in the form

$$a = \mathrm{u}(a) \prod_{i=1}^{l} r_i^{g_i} \text{ and } b = \mathrm{u}(b) \prod_{i=1}^{l} r_i^{h_i}$$

with some of the g_i's and h_i's zero. Clearly the element

$$d = \prod_{i=1}^{l} r_i^{\min(g_i,\, h_i)}$$

is the GCD of a and b. Finally, if one of a, b is zero assume without loss of generality that $a \ne 0$, $b = 0$. If a has the unique unit normal factorization as given in (2.1) then clearly the element

$$d = \prod_{i=1}^{n} p_i^{e_i}$$

is the GCD of a and b.

●

Euclidean Domains

There is a special class of integral domains in which the divisibility properties are particularly appealing. Unfortunately, most of the polynomial domains of interest to us will not belong to this class. The concepts are nonetheless of central importance and where a polynomial domain does not satisfy the "division property" discussed here we will be inventing a corresponding "pseudo-division property" in order to achieve our purposes.

Definition 2.12. A *Euclidean domain* is an integral domain D with a *valuation* $v: D - \{0\} \to \mathbf{N}$, where **N** denotes the set of nonnegative integers, having the following properties:

P1: For all $a, b \in D - \{0\}$, $v(ab) \ge v(a)$;

P2: For all $a, b \in D$ with $b \ne 0$, there exist elements $q, r \in D$ such that $a = bq + r$ where either $r = 0$ or $v(r) < v(b)$.

●

Example 2.4. The integers **Z** form a Euclidean domain with the valuation $v(a) = |a|$.

●

Property P2 of Definition 2.12 is known as the *division property* and is a familiar property of the integers. In the case of a polynomial domain, the valuation of a polynomial will be its degree. Note that the *quotient q* and the *remainder r* in P2 are not uniquely determined, in general, if $r \ne 0$. For example, in the Euclidean domain **Z** if $a = -8$, $b = 3$ then we have

$$-8 = (3)\,(-2) - 2 \quad \text{or} \quad -8 = (3)\,(-3) + 1$$

so that both pairs $q = -2, r = -2$ and $q = -3, r = 1$ satisfy P2. There are two different conventions which are adopted in various contexts to make the quotient and remainder unique in **Z**. One convention is to choose the pair q, r such that either $r = 0$ or $\text{sign}(r) = \text{sign}(a)$ (as in the first case above). The other convention is to choose the pair q, r such that either $r = 0$ or $\text{sign}(r) = \text{sign}(b)$ (as in the second case above). Fortunately, when we turn to polynomial domains the quotient and remainder will be uniquely determined.

Any Euclidean domain is a unique factorization domain and therefore GCD's exist (and are unique). Moreover, in a Euclidean domain the GCD can always be expressed in a special convenient form as stated in the following theorem.

Theorem 2.2. Let D be a Euclidean domain and let $a, b \in D$ (not both zero). If $g = \text{GCD}(a, b)$ then there exist elements $s, t \in D$ such that

$$g = sa + tb.$$

Proof: A constructive proof of Theorem 2.2 is presented in the following section.

●

Example 2.5. We stated in Example 2.2 that GCD(18,30) = 6. We have

$$6 = s \cdot 18 + t \cdot 30 \quad \text{where } s = 2 \text{ and } t = -1.$$

Note that in the Euclidean domain **Z** the elements s and t of Theorem 2.2 are not uniquely determined. Two other possible choices for s and t in this example are $s = -3$, $t = 2$ and $s = 7$, $t = -4$.

●

Hierarchy of Domains

In this section, we have introduced two new abstract structures intermediate to integral domains and fields. Table 2.3 shows the hierarchy of these domains. It is indicated there that a field F is a Euclidean domain, which can be seen by choosing the trivial valuation $v(a) = 1$ for all $a \in F - \{0\}$. (F is uninteresting as a Euclidean domain; for example, the remainder on division is always zero.) It also follows that a field F is a unique factorization domain. (F is a trivial UFD in which every nonzero element is a unit and therefore no element has a prime factorization – there are no primes in F.)

Table 2.3. Hierarchy of domains.

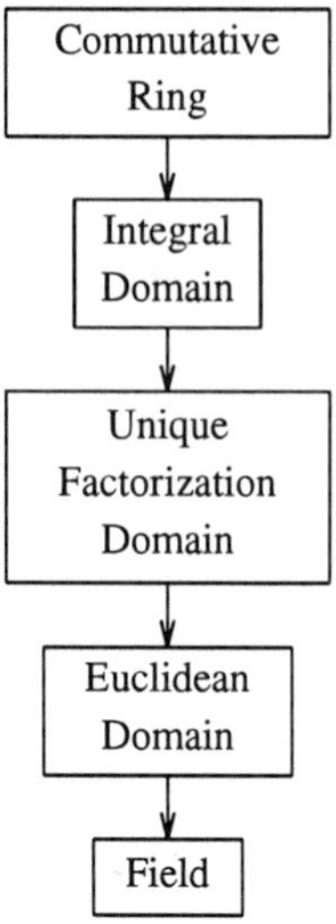

Notation: Downward pointing arrows indicate that a former domain becomes a latter domain if additional axioms are satisfied.

2.4. THE EUCLIDEAN ALGORITHM

From a computational point of view, we are interested not only in the existence of $g = \mathrm{GCD}(a,b)$ and the existence of elements s, t satisfying Theorem 2.2 in any Euclidean domain, but we are also interested in algorithms for computing these values. It might seem at first glance that the proof of Theorem 2.1 is a constructive proof yielding an algorithm for computing $\mathrm{GCD}(a,b)$ in any unique factorization domain. However the construction in that proof is based on prime factorizations of a and b and it is computationally much more difficult to determine a prime factorization than to compute $\mathrm{GCD}(a,b)$. A very effective algorithm for computing $\mathrm{GCD}(a,b)$ in any Euclidean domain will now be developed.

GCD's In Euclidean Domains

Theorem 2.3. Given $a, b \in \mathrm{D}$ $(b \neq 0)$ where D is a Euclidean domain, let q, r be a quotient and remainder satisfying

$$a = bq + r \text{ with } r = 0 \quad \text{or} \quad \mathrm{v}(r) < \mathrm{v}(b). \tag{2.2}$$

Then $\mathrm{GCD}(a,b) = \mathrm{GCD}(b,r)$.

Proof: Suppose that $g = \mathrm{GCD}(b,r)$ and $h = \mathrm{GCD}(a,b)$. From (2.2) we see that $g \mid a$ and therefore g is a common divisor of a and b. By definition of GCD, it follows that $g \mid h$. Rewriting equation (2.1) as

$$r = a - bq$$

we see that $h \mid r$ and so is a common divisor of b and r. Again by the definition of GCD, it follows that $h \mid g$. Thus g must be a *greatest* common divisor of a and b and hence an associate of h. Since g and h are both unit normal, we get that $g = h$.

●

In any integral domain D, it is useful to define

$$\mathrm{GCD}(0,0) = 0,$$

and obviously for any $a, b \in \mathrm{D}$:

$$\mathrm{GCD}(a,b) = \mathrm{GCD}(b,a).$$

It is also easy to show from the definitions that the following properties hold for any $a, b \in$ D:

$$\mathrm{GCD}(a,b) = \mathrm{GCD}(\mathrm{n}(a), \mathrm{n}(b));$$

$$\mathrm{GCD}(a,0) = \mathrm{n}(a),$$

where $\mathrm{n}(a)$ denotes the normal part of a as defined in Definition 2.7.

In any Euclidean domain D, if $a, b \in \mathrm{D}$ with $b \neq 0$ let q and r be a quotient and remainder such that

$$a = bq + r \text{ with } r = 0 \text{ or } \mathrm{v}(r) < \mathrm{v}(b)$$

and define the functions *quo* and *rem* by

$$\mathrm{quo}(a, b) = q;$$

$$\mathrm{rem}(a, b) = r.$$

(Note: The above functions are not well-defined, in general, because q and r are not uniquely determined. For the Euclidean domain **Z** we may adopt either of the two conventions mentioned in the preceding section in order to make the above functions well-defined. For the polynomial domains which will be of interest to us later we will see that q and r are uniquely determined by the division property.) For $a, b \in \mathrm{D}$ with $b \neq 0$ and $\mathrm{v}(a) \geq \mathrm{v}(b)$, by a *remainder sequence* for a and b we understand a sequence $\{r_i\}$ generated as follows:

$$\begin{aligned} &r_0 = a, \ r_1 = b, \\ &r_i = \mathrm{rem}(r_{i-2}, r_{i-1}), \quad i = 2, 3, 4, \ldots . \end{aligned} \tag{2.3}$$

(The sequence is undefined beyond a point where r_i=0 for some i.)

Theorem 2.4. Let D be a Euclidean domain with $a, b \in \mathrm{D}$ and $\mathrm{v}(a) \geq \mathrm{v}(b) > 0$. Let $\{r_i\}$ be a remainder sequence for a and b generated as in (2.3). Then there is a finite index $k \geq 1$ such that $r_{k+1} = 0$ and

$$\mathrm{GCD}(a, b) = \mathrm{n}(r_k). \tag{2.4}$$

Proof: Consider the sequence of valuations $\{\mathrm{v}(r_i)\}$ formed from the nonzero elements of the sequence $\{r_i\}$. By definition, $\{\mathrm{v}(r_i)\}$ is a strictly decreasing sequence of nonnegative integers. Since the first element of this sequence is $\mathrm{v}(a)$, there can be at most $\mathrm{v}(a)$+1 elements in the sequence. Therefore it must happen that r_{k+1}=0 for some $k \leq \mathrm{v}(b)$.

From Theorem 2.3 we have:

$$\mathrm{GCD}(a, b) = \mathrm{GCD}(r_0, r_1) = \mathrm{GCD}(r_1, r_2).$$

If r_2=0 then $k = 1$ and (2.4) holds by definition. Otherwise, we have from repeated use of Theorem 2.3:

$$\mathrm{GCD}(a, b) = \mathrm{GCD}(b, r_2) = \ \cdots \ = \mathrm{GCD}(r_k, r_{k+1}) = \mathrm{n}(r_k).$$

which is the desired result.

●

The Basic Algorithm

From Theorem 2.4, the GCD of $a, b \in \mathrm{D}$ $(b \neq 0)$ is simply the normal part of the last nonzero element of a remainder sequence $\{r_i\}$ generated as in (2.3). If $b = 0$ then $\mathrm{GCD}(a, b)$ is given by $\mathrm{n}(a)$. Thus we have a complete specification of the *Euclidean algorithm* to compute GCD's in any Euclidean domain, and it is given formally as Algorithm 2.1. We have chosen to take the normal parts of a and b initially in Algorithm 2.1 since this often simplifies the computation. For an actual implementation of Algorithm 2.1 we need only specify the functions rem(a,b) and n(a). Note that Algorithm 2.1 can also be applied to compute LCM(a,b) since if a and b are not both zero then

$$\mathrm{LCM}(a,b) = \frac{\mathrm{n}(ab)}{\mathrm{GCD}(a,b)}.$$

It is conventional to define

$$\mathrm{LCM}(0,0) = 0.$$

Example 2.6. In the Euclidean domain **Z**, the following function specifications are used for Algorithm 2.1. For any $a \in$ **Z**, n$(a) = |a|$ as noted in Example 2.3. The rem function for integers is defined as the remainder and is made unique by imposing one of the two conventions discussed in the preceding section. Note that since the **while**-loop in Algorithm 2.1 is entered with nonnegative integers and since either of the two conventions for defining rem will then produce a nonnegative remainder, the value of c on exit from the **while**-loop will be nonnegative. Therefore when applying Algorithm 2.1 in the particular Euclidean domain **Z** the final operation n(c) is unnecessary.

●

(Note: The essential ideas in Algorithm 2.1, as it applies to positive integers, date back to Euclid, circa 300 B.C.)

Algorithm 2.1. Euclidean Algorithm.

```
procedure Euclid(a, b)

   # Compute g = GCD(a, b), where a and b
   # are from a Euclidean domain D .

   c ← n(a); d ← n(b)

   while d ≠ 0 do {
       r ← rem(c, d)
       c ← d
       d ← r }

   g ← n(c)
   return(g)

end
```

Example 2.7. In the Euclidean domain **Z**, if $a = 18$ and $b = 30$ then the sequence of values computed for r, c, and d in Algorithm 2.1 is as follows:

iteration no.	r	c	d
–	–	18	30
1	18	30	18
2	12	18	12
3	6	12	6
4	0	6	0

Thus $g = 6$, and GCD(18,30) = 6 as noted in Example 2.2.

•

Extended Euclidean Algorithm (EEA)

The Euclidean algorithm can be readily extended so that while it computes $g = \text{GCD}(a,b)$ it will also compute the elements s, t of Theorem 2.2 which allow g to be expressed as a linear combination of a and b. We present the extended algorithm as Algorithm 2.2 and then justify it by giving a constructive proof of Theorem 2.2. Here and in the sequel, we employ the standard binary operation of *division* which is defined in any integral domain D as follows: if $a, b \in$ D and if a is a multiple of b then by definition, $a = b \cdot x$ for some $x \in$ D, and we define

$$a/b = x.$$

In particular, if b is a unit in D then any $a \in$ D is a multiple of b (i.e. $a = b(ab^{-1})$) and

$$a/b = ab^{-1}.$$

Note that the quo function is an extension of the division operation since if $a = b \cdot x$ then property P2 holds for a and b with $q = x$, $r = 0$ and hence $\text{quo}(a,b) = a/b$.

Note that the two divisions at the end of Algorithm 2.2 are valid in D because $\text{u}(a)$, $\text{u}(b)$, and $\text{u}(c)$ are units in D. Note also that the computation of $g = \text{GCD}(a,b)$ in Algorithm 2.2 is identical with the computation in Algorithm 2.1. The proof that the additional statements in Algorithm 2.2 correctly compute the elements s, t is contained in the constructive proof of Theorem 2.2 which we now present.

Proof of Theorem 2.2:

Let a, b be elements in a Euclidean domain D. Notice that the initial assignments before entering the **while**-loop in Algorithm 2.2 imply the relationships

$$c = c_1\,\text{n}(a) + c_2\,\text{n}(b) \tag{2.5}$$

and

$$d = d_1\,\text{n}(a) + d_2\,\text{n}(b). \tag{2.6}$$

Algorithm 2.2. Extended Euclidean Algorithm.

```
procedure EEA(a, b; s, t)

    # Given a and b in a Euclidean domain D, compute
    # g = GCD(a, b) and also compute elements s, t ∈ D
    # such that g = sa + tb .

    c ← n(a);   d ← n(b)
    c1 ← 1;     d1 ← 0
    c2 ← 0;     d2 ← 1

    while d ≠ 0 do {
        q ← quo(c, d);   r ← c − q·d
        r1 ← c1 − q·d1;  r2 ← c2 − q·d2
        c ← d;  c1 ← d1;  c2 ← d2
        d ← r;  d1 ← r1;  d2 ← r2 }
    # Normalize GCD
    g ← n(c)
    s ← c1 / (u(a) · u(c));   t ← c2 / (u(b) · u(c))
    return(g)
end
```

We claim that, as long as $d \neq 0$, equations (2.5) and (2.6) are invariant under the transformations of the **while**-loop in Algorithm 2.2 – i.e. if equations (2.5) and (2.6) hold at the beginning of the i-th iteration of the **while**-loop then they hold at the end of the i-th iteration. To see this, define $q = \text{quo}(c, d)$, multiply through in equation (2.6) by q, and subtract the result from equation (2.5). This gives

$$(c - qd) = (c_1 - qd_1)\,\text{n}(a) + (c_2 - qd_2)\,\text{n}(b) \tag{2.7}$$

which becomes

$$r = r_1\,\text{n}(a) + r_2\,\text{n}(b)$$

in the **while**-loop. The remaining transformations in the **while**-loop simply update c, c_1, c_2, d, d_1, and d_2 in such a way that equations (2.6) and (2.7) imply, at the end of the i-th iteration, equations (2.5) and (2.6), respectively. Thus (2.5) and (2.6) are loop invariant as claimed.

Now if we define

$$c = \mathrm{n}(a);\ d = \mathrm{n}(b);\ \ c_1 = 1;\ c_2 = 0;\ d_1 = 0;\ d_2 = 1,$$

then equations (2.5) and (2.6) clearly hold. If $d = 0$ then $b = 0$ and so

$$\mathrm{GCD}(a,b) = \mathrm{n}(a) = c$$

and

$$c = c_1\,\mathrm{n}(a) + c_2\,\mathrm{n}(b).$$

Otherwise, by Theorem 2.4, the transformations of the **while**-loop in Algorithm 2.2 may be applied some finite number, $k+1$, times yielding, at the end of the $(k+1)$-st iteration, elements c and d satisfying

$$d = 0 \ \text{ and } \ \mathrm{GCD}(a,b) = \mathrm{n}(c).$$

But since (2.5) is invariant, we also have elements $c_1, c_2 \in$ D such that

$$c = c_1\,\mathrm{n}(a) + c_2\,\mathrm{n}(b).$$

To complete the proof recall that for all $a \in$ D, $a = \mathrm{u}(a)\,\mathrm{n}(a)$ and $\mathrm{u}(a)$ is a unit (i.e. $\mathrm{u}(a)$ is invertible). Thus we can divide through by $\mathrm{u}(c)$ in (2.5), yielding

$$\mathrm{n}(c) = c_1 \frac{\mathrm{n}(a)}{\mathrm{u}(c)} + c_2 \frac{\mathrm{n}(b)}{\mathrm{u}(c)}.$$

Noting that $\mathrm{n}(a) = \dfrac{a}{\mathrm{u}(a)}$, $\mathrm{n}(b) = \dfrac{b}{\mathrm{u}(b)}$, we have from the previous five equations that, in all cases,

$$\mathrm{GCD}(a,b) = c_1 \frac{a}{\mathrm{u}(a)\,\mathrm{u}(c)} + c_2 \frac{b}{\mathrm{u}(b)\,\mathrm{u}(c)}.$$

Thus

$$\mathrm{GCD}(a,b) = sa + tb$$

as required, with $s = \dfrac{c_1}{\mathrm{u}(a)\,\mathrm{u}(c)}$ and $t = \dfrac{c_2}{\mathrm{u}(b)\,\mathrm{u}(c)}$.

•

Example 2.8. In the Euclidean domain **Z** if $a = 18$ and $b = 30$ then the sequence of values computed for q, c, c_1, c_2, d, d_1, and d_2 in Algorithm 2.2 is as follows.

iteration no.	q	c	c_1	c_2	d	d_1	d_2
–	–	18	1	0	30	0	1
1	0	30	0	1	18	1	0
2	1	18	1	0	12	–1	1
3	1	12	–1	1	6	2	–1
4	2	6	2	–1	0	–5	3

Thus $g = 6$, $s = 2$, and $t = -1$; i.e. $\mathrm{GCD}(18,30) = 6 = 2(18) - 1(30)$ as noted in Example 2.5.

•

2.5. UNIVARIATE POLYNOMIAL DOMAINS

For any commutative ring R, the notation R[x] denotes the set of all expressions of the form

$$a(x) = \sum_{k=0}^{m} a_k x^k$$

with $a_k \in$ R $(0 \le k \le m)$, where m is a nonnegative integer. Thus, R[x] denotes the set of all *polynomials* in the indeterminate x with coefficients lying in the ring R (or, more concisely, the set of all *univariate polynomials* over R). The *degree* deg($a(x)$) of a nonzero polynomial $a(x)$ is the largest integer n such that $a_n \ne 0$. The standard form of a polynomial $a(x)$ is

$$\sum_{k=0}^{n} a_k x^k \quad \text{with} \quad a_n \ne 0. \tag{2.8}$$

The exceptional case where a_k=0 for all k is called the *zero polynomial* and its standard form is 0. It is conventional to define deg(0) $= -\infty$. For a polynomial $a(x)$ in the standard form (2.8), $a_n x^n$ is called the *leading term*, a_n is called the *leading coefficient* (denoted functionally by lcoeff($a(x)$)), and a_0 is called the *constant term.* A polynomial with leading coefficient 1 is called a *monic polynomial.* A polynomial of degree 0 is called a *constant polynomial.* If l denotes the smallest integer such that $a_l \ne 0$ in (2.8) then the term $a_l x^l$ is called the *trailing term* and a_l is called the *trailing coefficient* (denoted functionally by tcoeff($a(x)$)). Note that if $a_0 \ne 0$ then the trailing term, trailing coefficient, and constant term are all identical.

The binary operations of addition and multiplication in the commutative ring R are extended to polynomials in the set R[x] as follows. If

$$a(x) = \sum_{k=0}^{m} a_k x^k \quad \text{and} \quad b(x) = \sum_{k=0}^{n} b_k x^k$$

then polynomial addition is defined by

$$c(x) = a(x) + b(x) = \sum_{k=0}^{\max(m,n)} c_k x^k$$

where

$$c_k = \begin{cases} a_k + b_k & \text{for } k \le \min(m,n) \\ a_k & \text{for } n < k \le m \text{ if } m > n \\ b_k & \text{for } m < k \le n \text{ if } m < n\,. \end{cases}$$

Similarly, if $a(x)$ and $b(x)$ are as above then polynomial multiplication is defined by

$$d(x) = a(x)\,b(x) = \sum_{k=0}^{m+n} d_k x^k$$

where $d_k = \sum_{i+j=k} a_i b_j$.

Algebraic Properties of R[x]

We now consider the properties of the algebraic structure R[x] under the operations of addition and multiplication defined above. Since addition and multiplication of polynomials in R[x] are defined in terms of addition and multiplication in the coefficient ring R, it is not surprising that the properties of R[x] are dependent on the properties of R. The following theorem summarizes a number of facts about univariate polynomial domains. The proofs are straightforward but tedious and will be omitted.

Theorem 2.5.

(i) If R is a commutative ring then R[x] is also a commutative ring. The zero (additive identity) in R[x] is the zero polynomial 0 and the (multiplicative) identity in R[x] is the constant polynomial 1.

(ii) If D is an integral domain then D[x] is also an integral domain. The units (invertibles) in D[x] are the constant polynomials a_0 such that a_0 is a unit in the coefficient domain D.

(iii) If D is a unique factorization domain (UFD) then D[x] is also a UFD. The primes (irreducibles) in D[x] are the polynomials which cannot be factored (apart from units and associates) with respect to the coefficient domain D.

(iv) If D is a Euclidean domain then D[x] is a UFD but not (necessarily) a Euclidean domain.

(v) If F is a field then F[x] is a Euclidean domain with the valuation

$$\mathrm{v}(a(x)) = \deg(a(x)).$$

●

Definition 2.13. In any polynomial domain D[x] over an integral domain D, the polynomials with unit normal leading coefficients are defined to be *unit normal.*

●

Example 2.9. In the polynomial domain **Z**[x] over the integers, the units are the constant polynomials 1 and –1. The unit normal polynomials in **Z**[x] are 0 and all polynomials with positive leading coefficients.

●

Example 2.10. In the polynomial domain **Q**[x] over the field of rational numbers, the units are all nonzero constant polynomials. The unit normal polynomials in **Q**[x] are all monic polynomials, and the 0 polynomial.

At this point let us note some properties which can be easily verified for the degree function in a polynomial domain D[x] over any integral domain D. For the degree of a sum we have

$$\deg(a(x) + b(x)) \le \max\{ \deg(a(x)), \deg(b(x)) \},$$

with equality holding if $\deg(a(x)) \ne \deg(b(x))$. For the degree of a product we have

$$\deg(a(x)\, b(x)) = \deg(a(x)) + \deg(b(x)). \tag{2.9}$$

For the degree of a quotient we have, assuming $b(x) \ne 0$,

$$\deg(\mathrm{quo}(a(x),b(x))) = \begin{cases} -\infty & \text{if } \deg(a(x)) < \deg(b(x)) \\ \deg(a(x)) - \deg(b(x)) & \text{otherwise.} \end{cases}$$

In particular note that if $b(x) \mid a(x)$ then we have

$$\deg(a(x)/b(x)) = \deg(a(x)) - \deg(b(x))$$

since when $b(x)$ divides $a(x)$ it follows that either $a(x) = 0$ or else $\deg(a(x)) \ge \deg(b(x))$.

We note from Theorem 2.5 that the algebraic structure of a coefficient domain D is inherited in full by the polynomial domain D[x] if D is an integral domain or a UFD, but if D is a Euclidean domain or a field then D[x] does not inherit the Euclidean axioms or the field axioms (see Example 2.12 and Example 2.13). However in the case of a field F, the polynomial domain F[x] becomes a Euclidean domain by choosing the valuation defined by the degree function. Since by definition $\deg(a(x)) \ge 0$ for any nonzero polynomial $a(x)$, this valuation is indeed a mapping from F[x] $-$ {0} into the nonnegative integers **N** as required by Definition 2.12. Property P1 of Definition 2.12 can be verified by using equation (2.9) since if $a(x), b(x) \in$ F[x] $-$ {0} then

$$\deg(a(x)\cdot b(x)) = \deg(a(x)) + \deg(b(x)) \ge \deg(a(x)).$$

Property P2 of Definition 2.12, the division property, is the familiar process of polynomial long division which can be carried out as long as the coefficient domain is a field F. Unlike the Euclidean domain **Z**, in the Euclidean domain F[x] the quotient q and remainder r of property P2 are *unique*.

Example 2.11. In the Euclidean domain **Q**[x] of polynomials over the field **Q** of rational numbers, let

$$a(x) = 3x^3 + x^2 + x + 5, \text{ and } \quad b(x) = 5x^2 - 3x + 1. \tag{2.10}$$

To find the quotient $q(x)$ and remainder $r(x)$ of the division property in Definition 2.12, we perform polynomial long division:

$$
\begin{array}{rrrrr}
 & \frac{3}{5}x + & \frac{14}{25} & & \\
\cline{2-5}
5x^2 - 3x + 1 & \sqrt{3x^3 +} & x^2 + & x + & 5 \\
 & 3x^3 - & \frac{9}{5}x^2 + & \frac{3}{5}x & \\
\cline{2-5}
 & & \frac{14}{5}x^2 + & \frac{2}{5}x + & 5 \\
 & & \frac{14}{5}x^2 - & \frac{42}{25}x + & \frac{14}{25} \\
\cline{3-5}
 & & & \frac{52}{25}x + & \frac{111}{25}
\end{array}
$$

Thus $a(x) = b(x)\, q(x) + r(x)$ where

$$q(x) = \frac{3}{5}x + \frac{14}{25}, \quad \text{and} \quad r(x) = \frac{52}{25}x + \frac{111}{25}.$$

●

Example 2.12. The polynomial domain $\mathbf{Z}[x]$ over the integers $\mathbf{Z}$ is an integral domain, in fact a UFD (because $\mathbf{Z}$ is a UFD), but $\mathbf{Z}[x]$ is not a Euclidean domain with the "natural" valuation $v(a(x)) = \deg(a(x))$. For consider the polynomials $a(x)$, $b(x)$ given in (2.10). Note that $a(x), b(x) \in \mathbf{Z}[x]$. Property P2 is not satisfied by using the polynomials $q(x)$, $r(x)$ of Example 2.11 because $q(x), r(x) \notin \mathbf{Z}[x]$. If we assume the existence of polynomials $q(x)$, $r(x) \in \mathbf{Z}[x]$ satisfying property P2 for the polynomials (2.10), then since $\deg(r(x)) < \deg(b(x)) = 2$ it is easy to argue that we must have

$$3x^3 + x^2 + x + 5 = (5x^2 - 3x + 1)\,(q_1 x + q_0) + (r_1 x + r_0)$$

for some coefficients $q_1, q_0, r_1, r_0 \in \mathbf{Z}$. But this implies

$$3 = 5q_1 \tag{2.11}$$

which is a contradiction since equation (2.11) has no solution in $\mathbf{Z}$. Thus property P2 does not hold in the domain $\mathbf{Z}[x]$ for the polynomials (2.10) and therefore $\mathbf{Z}[x]$ is not a Euclidean domain.

●

Example 2.12 shows that the coefficient domain must be a field in order to carry out polynomial long division because only in a field will equations of the form (2.11) always have a solution. A more concise argument for Example 2.12 could have been obtained by noting the uniqueness of $q(x)$, $r(x)$ in polynomial long division. The next example verifies that a polynomial domain F[x] over a field F is not itself a field.

Example 2.13. In a polynomial domain F[x] over any field F, the polynomial x has no inverse. For if it had an inverse, say $q(x)$, then

$$\begin{aligned} x\,q(x) = 1 &\Rightarrow \deg(x) + \deg(q(x)) = \deg(1) \\ &\Rightarrow 1 + \deg(q(x)) = 0 \\ &\Rightarrow \deg(q(x)) = -1 \end{aligned}$$

which is impossible. Therefore F[x] is not a field.

●

GCD Computation in F[x]

Since the univariate polynomial domain F[x] over a field F is a Euclidean domain, the Euclidean algorithm (Algorithm 2.1) and the extended Euclidean algorithm (Algorithm 2.2) can be used to compute GCD's in F[x]. For a nonzero polynomial $a(x) \in$ F[x] with leading coefficient a_n, the normal part and unit part of $a(x)$ satisfy:

$$\mathrm{n}(a(x)) = \frac{a(x)}{a_n}, \quad \mathrm{u}(a(x)) = a_n.$$

Note that $a_n \neq 0$ is a unit in F[x] because it is a unit in F. As usual, n(0) = 0 and u(0) = 1. For $a(x), b(x) \in$ F[x] with $b(x) \neq 0$, the quotient and remainder of property P2 are unique so the quo and rem functions are well-defined and the remainder sequence $\{r_i(x)\}$ defined by (2.3) is unique.

Example 2.14. In the Euclidean domain **Q**[x], let

$$a(x) = 48x^3 - 84x^2 + 42x - 36, \quad b(x) = -4x^3 - 10x^2 + 44x - 30. \tag{2.12}$$

The sequence of values computed for $r(x)$, $c(x)$, and $d(x)$ in Algorithm 2.1 is as follows. (Here $a(x)$, $b(x)$, $r(x)$, $c(x)$, and $d(x)$ are denoted by a, b, r, c, and d, respectively, in Algorithm 2.1. It is common practice to use the former notation, called ''functional notation'', for polynomials but clearly the latter notation is also acceptable when the underlying domain is understood.)

iteration no.	$r(x)$	$c(x)$	$d(x)$
–	–	$x^3-\frac{7}{4}x^2+\frac{7}{8}x-\frac{3}{4}$	$x^3+\frac{5}{2}x^2-11x+\frac{15}{2}$
1	$-\frac{17}{4}x^2+\frac{95}{8}x-\frac{33}{4}$	$x^3+\frac{5}{2}x^2-11x+\frac{15}{2}$	$-\frac{17}{4}x^2+\frac{95}{8}x-\frac{33}{4}$
2	$\frac{535}{289}x-\frac{1605}{578}$	$-\frac{17}{4}x^2+\frac{95}{8}x-\frac{33}{4}$	$\frac{535}{289}x-\frac{1605}{578}$
3	0	$\frac{535}{289}x-\frac{1605}{578}$	0

Thus $g(x) = \mathrm{n}(\frac{535}{289}x - \frac{1605}{578}) = x - \frac{3}{2}$.

●

Example 2.15. In the Euclidean domain $\mathbf{Q}[x]$, if Algorithm 2.2 is applied to the polynomials (2.12) of Example 2.14 then three iterations of the **while**-loop are required as in Example 2.14. At the end of the third iteration we have

$$r(x) = 0; \quad c(x) = \frac{535}{289}x - \frac{1605}{578}; \quad d(x) = 0$$

as before. We also have

$$c_1(x) = \frac{4}{17}x + \frac{360}{289};$$

$$c_2(x) = -\frac{4}{17}x - \frac{71}{289}.$$

Thus,

$$g(x) = \mathrm{n}(c(x)) = x - \frac{3}{2};$$

$$s(x) = \frac{c_1(x)}{48\left(\frac{535}{289}\right)} = \frac{17}{6420}x + \frac{3}{214};$$

$$t(x) = \frac{c_2(x)}{-4\left(\frac{535}{289}\right)} = \frac{17}{535}x + \frac{71}{2140}.$$

It is readily verified that

$$s(x)\,a(x) + t(x)\,b(x) = x - \frac{3}{2}.$$

●

In the Euclidean domain F[x] of univariate polynomials over a field F, an important application of the extended Euclidean algorithm in later chapters will be to solve the *polynomial diophantine equation*

$$\sigma(x)\,a(x) + \tau(x)\,b(x) = c(x)$$

where $a(x)$, $b(x)$, $c(x) \in$ F[x] are given polynomials and $\sigma(x)$, $\tau(x) \in$ F[x] are to be determined (if possible). The following theorem gives sufficient conditions for the existence and uniqueness of a solution to this polynomial diophantine equation and a constructive proof is given. Note that an important special case of the theorem occurs when $a(x)$ and $b(x)$ are relatively prime in which case the given polynomial diophantine equation can be solved for any given right hand side $c(x)$.

Theorem 2.6. Let F[x] be the Euclidean domain of univariate polynomials over a field F. Let $a(x)$, $b(x) \in$ F[x] be given nonzero polynomials and let $g(x) = \mathrm{GCD}(a(x), b(x)) \in$ F[x]. Then for any given polynomial $c(x) \in$ F[x] such that $g(x) \mid c(x)$ there exist unique polynomials $\sigma(x)$, $\tau(x) \in$ F[x] such that

$$\sigma(x)\, a(x) + \tau(x)\, b(x) = c(x) \quad \text{and} \tag{2.13}$$

$$\deg(\sigma(x)) < \deg(b(x)) - \deg(g(x)). \tag{2.14}$$

Moreover, if $\deg(c(x)) < \deg(a(x)) + \deg(b(x)) - \deg(g(x))$ then $\tau(x)$ satisfies

$$\deg(\tau(x)) < \deg(a(x)) - \deg(g(x)). \tag{2.15}$$

Proof: *Existence:* The extended Euclidean algorithm can be applied to compute polynomials $s(x)$, $t(x) \in$ F[x] satisfying the equation

$$s(x)\, a(x) + t(x)\, b(x) = g(x).$$

Then since $g(x) \mid c(x)$ it is easily seen that

$$(s(x)\, c(x) / g(x))\, a(x) + (t(x)\, c(x) / g(x))\, b(x) = c(x). \tag{2.16}$$

We therefore have a solution of equation (2.13), say $\tilde{\sigma}(x) = s(x)\, c(x) / g(x)$ and $\tilde{\tau}(x) = t(x)\, c(x) / g(x)$. However the degree constraint (2.14) will not in general be satisfied by this solution so we will proceed to show how to reduce the degree. Writing (2.16) in the form

$$\tilde{\sigma}(x)\, (a(x) / g(x)) + \tilde{\tau}(x)\, (b(x) / g(x)) = c(x) / g(x), \tag{2.17}$$

we apply Euclidean division of $\tilde{\sigma}(x)$ by $(b(x) / g(x))$ yielding $q(x)$, $r(x) \in$ F[x] such that

$$\tilde{\sigma}(x) = (b(x) / g(x))\, q(x) + r(x) \tag{2.18}$$

where $\deg(r(x)) < \deg(b(x)) - \deg(g(x))$. Now define $\sigma(x) = r(x)$ and note that (2.14) is satisfied. Also define $\tau(x) = \tilde{\tau}(x) + q(x)\, (a(x) / g(x))$. It is easily verified by using (2.17) and (2.18) that

$$\sigma(x)\, (a(x) / g(x)) + \tau(x)\, (b(x) / g(x)) = c(x) / g(x).$$

Equation (2.13) follows immediately.

Uniqueness: Let $\sigma_1(x)$, $\tau_1(x) \in$ F[x] and $\sigma_2(x)$, $\tau_2(x) \in$ F[x] be two pairs of polynomials satisfying (2.13) and (2.14). The two different solutions of (2.13) can be written in the form

$$\sigma_1(x)\, (a(x) / g(x)) + \tau_1(x)\, (b(x) / g(x)) = c(x) / g(x);$$

$$\sigma_2(x)\, (a(x) / g(x)) + \tau_2(x)\, (b(x) / g(x)) = c(x) / g(x)$$

which yields on subtraction

$$(\sigma_1(x) - \sigma_2(x))\, (a(x)/g(x)) = -(\tau_1(x) - \tau_2(x))\, (b(x)/g(x)). \tag{2.19}$$

Now since $a(x)/g(x)$ and $b(x)/g(x)$ are relatively prime it follows from equation (2.19) that

$$(b(x)/g(x)) \mid (\sigma_1(x) - \sigma_2(x)). \tag{2.20}$$

But from the degree constraint (2.14) satisfied by $\sigma_1(x)$ and $\sigma_2(x)$ it follows that

$$\deg(\sigma_1(x) - \sigma_2(x)) < \deg(b(x)/g(x)). \tag{2.21}$$

Now (2.20) and (2.21) together imply that $\sigma_1(x) - \sigma_2(x) = 0$. It then follows from (2.19) that $\tau_1(x) - \tau_2(x) = 0$ since $b(x)/g(x) \neq 0$. Therefore $\sigma_1(x) = \sigma_2(x)$ and $\tau_1(x) = \tau_2(x)$.

Final Degree Constraint: It remains to prove (2.15). From (2.13) we can write

$$\tau(x) = (c(x) - \sigma(x)\, a(x))\, /\, b(x)$$

so that

$$\deg(\tau(x)) = \deg(c(x) - \sigma(x)\, a(x)) - \deg(b(x)). \tag{2.22}$$

Now if $\deg(c(x)) \geq \deg(\sigma(x)\, a(x))$ then from (2.22)

$$\deg(\tau(x)) \leq \deg(c(x)) - \deg(b(x)) < \deg(a(x)) - \deg(g(x))$$

as long as $\deg(c(x)) < \deg(a(x)) + \deg(b(x)) - \deg(g(x))$ as stated. Otherwise if $\deg(c(x)) < \deg(\sigma(x)\, a(x))$ (in which case the stated degree bound for $c(x)$ also holds because of (2.14)) then from (2.22)

$$\deg(\tau(x)) = \deg(\sigma(x)\, a(x)) - \deg(b(x)) < \deg(a(x)) - \deg(g(x))$$

where the last inequality follows from (2.14). Thus (2.15) is proved.

●

The result of Theorem 2.6 is related to the concept of partial fraction decompositions of rational functions. For example, if $a(x)$ and $b(x)$ are relatively prime, then Theorem 2.6 states that we can always find polynomials $\sigma(x)$ and $\tau(x)$ satisfying

$$\frac{c(x)}{a(x)\cdot b(x)} = \frac{\tau(x)}{a(x)} + \frac{\sigma(x)}{b(x)} \tag{2.23}$$

with

$$\deg(\tau(x)) < \deg(a(x)), \quad \deg(\sigma(x)) < \deg(b(x)).$$

Here the degree inequalities follow from Theorem 2.6 since the degree of the GCD of $a(x)$ and $b(x)$ is 0 in this case. Equation (2.23) is what one looks for when constructing a partial fraction decomposition. Applications of Theorem 2.6 to the computation of partial fraction decompositions will arise later in Chapter 6 and in Chapters 11 and 12.

2.6. MULTIVARIATE POLYNOMIAL DOMAINS

The polynomial domains of most interest in symbolic computation are multivariate polynomials (i.e. polynomials in one or more indeterminates) over the integers **Z**, or over the rationals **Q**, or over a finite field F. We will see later in Chapters 11 and 12 that much of symbolic integration relies on computing efficiently in multivariate polynomial domains. In the previous section on univariate polynomial domains we have noted that $\mathbf{Q}[x]$ and $F[x]$ are Euclidean domains so that the Euclidean algorithm can be used to perform the important operation of computing GCD's. In the univariate polynomial domain $\mathbf{Z}[x]$ over the integers it would be possible to compute GCD's (and other important computations) by embedding $\mathbf{Z}[x]$ in the larger domain $\mathbf{Q}[x]$ so that the coefficient domain is a field. However, coefficient arithmetic in **Q** is considerably more expensive than arithmetic in **Z** so that in practice we prefer to develop GCD algorithms that are valid in the UFD $\mathbf{Z}[x]$. More significantly, when dealing with multivariate polynomials in two or more indeterminates it turns out that the multivariate polynomial domain is a UFD but not a Euclidean domain *even if* the coefficient domain is a field. Hence further discussion of GCD computation in $\mathbf{Z}[x]$ will be postponed to a later section after we have discussed multivariate polynomial domains, where the underlying algebraic structure will be the UFD rather than the Euclidean domain.

Bivariate Polynomials

For any commutative ring R, the notation $R[x_1,x_2]$ denotes the set of all expressions of the form

$$a(x_1,x_2) = \sum_{i=0}^{m_1} \sum_{j=0}^{m_2} a_{i,j} x_1^i x_2^j \tag{2.24}$$

with $a_{i,j} \in R$ for all relevant i,j, and where m_1 and m_2 are nonnegative integers. Thus, $R[x_1,x_2]$ denotes the set of *bivariate polynomials* over the ring R. For example, one polynomial in the set $\mathbf{Z}[x,y]$ is the bivariate polynomial

$$a(x,y) = 5x^3y^2 - x^2y^4 - 3x^2y^2 + 7xy^2 + 2xy - 2x + 4y^4 + 5. \tag{2.25}$$

In order to generalize our notation to polynomials in $v > 2$ indeterminates, it is convenient to introduce a vector notation for bivariate polynomials. Let $\mathbf{x} = (x_1,x_2)$ be the vector of indeterminates and for each term $a_{i,j} x_1^i x_2^j$ in (2.24) let $\mathbf{e} = (e_1,e_2)$ be the *exponent vector* with $e_1 = i$, $e_2 = j$. Note that the exponent vectors lie in the Cartesian product[2] set $\mathbf{N} \times \mathbf{N}$ (or $\mathbf{N}^2$) where **N** is the set of nonnegative integers. Then we may denote the set of all bivariate polynomials over the ring R by the notation $R[\mathbf{x}]$ and we may define it to be the set of all expressions of the form

2. The *Cartesian product* of n sets $A_1, A_2, \ldots, A_n$ is defined by $A_1 \times A_2 \times \cdots \times A_n = \{(a_1,a_2,\ldots,a_n) : a_i \in A_i\}$. If $A_1 = A_2 = \cdots = A_n = A$ then $A \times A \times \cdots \times A$ is also denoted by A^n.

$$a(\mathbf{x}) = \sum_{e \in \mathbf{N} \times \mathbf{N}} a_e \, \mathbf{x}^e \tag{2.26}$$

with $a_e \in$ R, where it is understood that only a finite number of coefficients a_e are nonzero. Note that a particular term $a_e \, \mathbf{x}^e$ in (2.26) is a shorthand notation for $a_{(e_1, e_2)} \, (x_1, x_2)^{(e_1, e_2)}$ which is our vector representation of the term $a_{e_1, e_2} x_1^{e_1} \, x_2^{e_2}$.

Multivariate Polynomials

Let us now consider the general case of $v \geq 1$ indeterminates. For any commutative ring R, the notation R$[x_1, \ldots, x_v]$, or R$[\mathbf{x}]$ where $\mathbf{x} = (x_1, \ldots, x_v)$, denotes the set of all expressions of the form

$$a(\mathbf{x}) = \sum_{e \in \mathbf{N}^v} a_e \, \mathbf{x}^e \tag{2.27}$$

with $a_e \in$ R, where it is understood that only a finite number of coefficients a_e are nonzero. In other words, R[$\mathbf{x}$] denotes the set of all *multivariate polynomials* over the ring R in the indeterminates $\mathbf{x}$. The exceptional case where there are no nonzero terms in (2.27) is called the *zero polynomial* and is denoted by 0.

Definition 2.14. The *lexicographical ordering of exponent vectors* $e \in \mathbf{N}^v$ is defined as follows. Let $\mathbf{d} = (d_1, \ldots, d_v)$ and $\mathbf{e} = (e_1, \ldots, e_v)$ be two exponent vectors in the set $\mathbf{N}^v$. If $d_i = e_i$ $(1 \leq i \leq v)$ then $\mathbf{d} = \mathbf{e}$. Otherwise, let j be the smallest integer such that $d_j \neq e_j$ and define:

$$\mathbf{d} < \mathbf{e} \quad \text{if} \quad d_j < e_j;$$

$$\mathbf{d} > \mathbf{e} \quad \text{if} \quad d_j > e_j.$$

●

Example 2.16. The terms in the bivariate polynomial $a(x,y) \in \mathbf{Z}[x,y]$ given in (2.25) are listed in lexicographically decreasing order of their exponent vectors.

●

Assuming that the terms in a nonzero multivariate polynomial $a(\mathbf{x})$ have been arranged in lexicographically decreasing order of their exponent vectors, the first term is called the *leading term,* its coefficient is called the *leading coefficient* (denoted functionally by lcoeff($a(\mathbf{x})$)), the last (nonzero) term is called the *trailing term* and its coefficient is called the *trailing coefficient* (denoted functionally by tcoeff($a(\mathbf{x})$)). A multivariate polynomial with leading coefficient 1 is called a monic polynomial. The *degree vector* $\partial(a(\mathbf{x}))$ of a multivariate polynomial $a(\mathbf{x}) \neq 0$ is the exponent vector of its leading term. The *total degree* deg($a_e \mathbf{x}^e$) *of a term* in a multivariate polynomial $a(\mathbf{x})$, where $\mathbf{e} = (e_1, \ldots, e_v)$, is the value $\sum_{i=1}^{v} e_i$. The *total degree* deg($a(\mathbf{x})$) *of a polynomial* $a(\mathbf{x}) \neq 0$ is the maximum of the total degrees of all of its nonzero terms. It is conventional to define $\deg(0) = -\infty$, while $\partial(0)$ is undefined. A polynomial with total degree 0 is called a *constant polynomial.*

A Recursive View of R[x]

It is convenient to define the operations of addition and multiplication on multivariate polynomials in $R[x_1, \ldots, x_v]$ in terms of the basic operations in a univariate polynomial ring. This can be done by using a different, but equivalent, definition of the set $R[x_1, \ldots, x_v]$. The new definition will be recursive. Let us first consider the case of bivariate polynomials in the indeterminates x_1 and x_2. Recalling that the set $R[x_2]$ of univariate polynomials over a commutative ring R forms a commutative ring, we may use it as a coefficient ring and define a new univariate polynomial ring $R[x_2][x_1]$ of polynomials in the indeterminate x_1, with coefficients lying in the commutative ring $R[x_2]$. By Theorem 2.5, $R[x_2][x_1]$ is a commutative ring with the operations of addition and multiplication defined in the previous section in terms of the operations in the coefficient ring $R[x_2]$. It is easy to see that the set of expressions in $R[x_2][x_1]$ is the set of all expressions of the form (2.24) which we have denoted by $R[x_1, x_2]$. Therefore we identify

$$R[x_1, x_2] = R[x_2][x_1] \tag{2.28}$$

and this identification serves to define the arithmetic operations on bivariate polynomials. (Clearly, we should be able to identify $R[x_1, x_2]$ as well with $R[x_1][x_2]$. The operations of addition and multiplication in $R[x_1][x_2]$ are defined differently than the operations in $R[x_2][x_1]$ but it is straightforward to prove that the commutative rings $R[x_1][x_2]$ and $R[x_2][x_1]$ are ring isomorphic.[3] Therefore we are justified in identifying all three of these rings.)

Turning now to multivariate polynomials in $v \geq 2$ indeterminates, a recursive definition of $R[x_1, \ldots, x_v]$ is given by

$$R[x_1, \ldots, x_v] = R[x_2, \ldots, x_v][x_1]. \tag{2.29}$$

Applying (2.29) recursively to $R[x_2, \ldots, x_v]$ leads to the identification

$$R[x_1, \ldots, x_v] = R[x_v][x_{v-1}] \cdots [x_1].$$

Thus from knowledge of the operations in $R[x_v]$ we define the operations in $R[x_v][x_{v-1}]$, and from $R[x_v][x_{v-1}]$ to $R[x_v][x_{v-1}][x_{v-2}]$, etc. (Again, the order of singling out indeterminates as in (2.29) is not important algebraically since the rings obtained by different orderings of the indeterminates can be shown to be isomorphic.) If the multivariate polynomial ring $R[x_1, \ldots, x_v]$ is viewed as in (2.29) then we refer to x_1 as the *main variable* and to $x_2, \ldots, x_v$ as the *auxiliary variables.* In this case we consider a polynomial

3. Two rings R_1 and R_2 are *isomorphic* if there is a mapping $\phi: R_1 \rightarrow R_2$ which is bijective (i.e. one-to-one and onto) and which preserves all of the ring operations. For a precise definition see Chapter 5.

$a(\mathbf{x}) \in R[x_1, \ldots, x_v]$ as a univariate polynomial in the main variable with coefficients lying in the ring of polynomials in the auxiliary variables.

Example 2.17. The polynomial $a(x,y) \in \mathbf{Z}[x,y]$ given in (2.25) may be viewed as a polynomial in the ring $\mathbf{Z}[y][x]$

$$a(x,y) = (5y^2)x^3 - (y^4+3y^2)x^2 + (7y^2+2y-2)x + (4y^4+5).$$

Considered as a polynomial in the ring $\mathbf{Z}[x][y]$ we have

$$a(x,y) = (-x^2+4)\, y^4 + (5x^3-3x^2+7x)\, y^2 + (2x)\, y + (-2x+5).$$

●

For a polynomial $a(\mathbf{x}) \in R[x_1, \ldots, x_v]$ we sometimes refer to the *degree of* $a(\mathbf{x})$ *in the i-th variable,* denoted $\deg_i(a(\mathbf{x}))$. This i-th degree is just the degree of $a(\mathbf{x})$ when considered as a univariate polynomial in the ring $R[x_1, \ldots, x_{i-1}, x_{i+1}, \ldots, x_v][x_i]$.

Example 2.18. Let $a(x,y) \in \mathbf{Z}[x,y]$ be the bivariate polynomial given in (2.25). The leading term of $a(x,y)$ is $5x^3y^2$ and the leading coefficient is 5. The values of the various degree functions are:

$$\partial(a(x,y)) = (3,2); \quad \deg(a(x,y)) = 6;$$

$$\deg_1(a(x,y)) = 3; \qquad \deg_2(a(x,y)) = 4.$$

●

Algebraic Properties of R[x]

The algebraic properties of a multivariate polynomial ring $R[\mathbf{x}]$, for various choices of algebraic structure R, can be deduced immediately from the recursive view of $R[\mathbf{x}]$ and Theorem 2.5. These properties are summarized in the following theorem whose proof is now trivial.

Theorem 2.7.

(i) If R is a commutative ring then $R[\mathbf{x}]$ is also a commutative ring. The zero in $R[\mathbf{x}]$ is the zero polynomial 0 and the identity in $R[\mathbf{x}]$ is the constant polynomial 1.

(ii) If D is an integral domain then $D[\mathbf{x}]$ is also an integral domain. The units in $D[\mathbf{x}]$ are the constant polynomials a_0 such that a_0 is a unit in the coefficient domain D.

(iii) If D is a UFD then $D[\mathbf{x}]$ is also a UFD.

(iv) If D is a Euclidean domain then $D[\mathbf{x}]$ is UFD but not a Euclidean domain.

(v) If F is field then $F[\mathbf{x}]$ is a UFD but not a Euclidean domain if the number of indeterminates is greater than one.

●

Definition 2.15. In any multivariate polynomial domain D[**x**] over an integral domain D, the polynomials with unit normal leading coefficients are defined to be *unit normal*.

●

At this point we note some of the properties of the various degree functions which have been introduced for multivariate polynomials. It can be readily verified that the following properties hold for nonzero polynomials in a domain D[**x**] over any integral domain D:

$$\partial(a(\mathbf{x}) + b(\mathbf{x})) \le \max\{\partial(a(\mathbf{x})), \partial(b(\mathbf{x}))\};$$

$$\partial(a(\mathbf{x})\, b(\mathbf{x})) = \partial(a(\mathbf{x})) + \partial(b(\mathbf{x}));$$

$$\deg_i(a(\mathbf{x}) + b(\mathbf{x})) \le \max\{\deg_i(a(\mathbf{x})), \deg_i(b(\mathbf{x}))\};$$

$$\deg_i(a(\mathbf{x})\, b(\mathbf{x})) = \deg_i(a(\mathbf{x})) + \deg_i(b(\mathbf{x}));$$

$$\deg(a(\mathbf{x}) + b(\mathbf{x})) \le \max\{\deg(a(\mathbf{x})), \deg(b(\mathbf{x}))\};$$

$$\deg(a(\mathbf{x})\, b(\mathbf{x})) = \deg(a(\mathbf{x})) + \deg(b(\mathbf{x})). \qquad (2.30)$$

The addition operation on degree vectors given above is the familiar operation of vector addition (component-by-component addition) and the "order" operations ≤ and max are well-defined by the lexicographical ordering of exponent vectors defined in Definition 2.14.

The concept of the derivative of a polynomial can be defined algebraically. For a univariate polynomial

$$a(x) = \sum_{k=0}^{n} a_k x^k \in D[x]$$

(where D is an arbitrary integral domain) the *derivative* of $a(x)$ is defined by

$$a'(x) = \sum_{k=0}^{n-1} (k+1)\, a_{k+1}\, x^k \in D[x].$$

It is straightforward to show (using completely algebraic arguments) that the familiar properties of derivatives hold:

(i) if $a(x) = b(x) + c(x)$ then $a'(x) = b'(x) + c'(x)$;

(ii) if $a(x) = b(x)\, c(x)$ then $a'(x) = b(x)\, c'(x) + b'(x)\, c(x)$;

(iii) if $a(x) = b(c(x))$ then $a'(x) = b'(c(x))\, c'(x)$.

For a multivariate polynomial $a(x_1, \ldots, x_v) \in D[x_1, \ldots, x_v]$ over an arbitrary integral domain D the *partial derivative* of $a(x_1, \ldots, x_v)$ *with respect to* x_i, denoted $a_{x_i}(x_1, \ldots, x_v)$, is simply the ordinary derivative of $a(x_1, \ldots, x_v)$ when considered as a univariate polynomial in the domain $D[x_1, \ldots, x_{i-1}, x_{i+1}, \ldots, x_v]\,[x_i]$. In later chapters it will be necessary to use the concept of a *Taylor series expansion* in the sense of the following Theorem 2.8 and also the bivariate version as presented in Theorem 2.9.

Theorem 2.8. Let $a(x) \in D[x]$ be a univariate polynomial over an arbitrary integral domain D. In the polynomial domain $D[x][y] = D[x,y]$,

$$a(x+y) = a(x) + a'(x)y + b(x,y)y^2 \tag{2.31}$$

for some polynomial $b(x,y) \in D[x,y]$.

Proof: First note that $x + y$ is a polynomial in the domain $D[x,y]$ and since $a(x) \in D[x]$ it follows that $a(x + y) \in D[x,y]$. Now any bivariate polynomial in $D[x,y]$, and in particular $a(x + y)$, may be expressed in the form:

$$a(x+y) = a_0(x) + a_1(x)y + b(x,y)y^2 \tag{2.32}$$

where $a_0(x), a_1(x) \in D[x]$ and $b(x,y) \in D[x,y]$. For we simply first write all terms independent of y, then all terms in which y appears linearly, and finally notice that what remains must have y^2 as a factor. It remains to show that $a_0(x) = a(x)$ and that $a_1(x) = a'(x)$.

Setting $y = 0$ in (2.32) immediately yields $a_0(x) = a(x)$. Taking the partial derivative with respect to y on both sides of equation (2.32) yields

$$a'(x+y) = a_1(x) + 2b(x,y)y + b_y(x,y)y^2.$$

Setting $y = 0$ then yields $a_1(x) = a'(x)$ and hence equation (2.31). ●

Theorem 2.9. Let $a(x,y) \in D[x,y]$ be a bivariate polynomial over an arbitrary integral domain D. In the polynomial domain $D[x,y][u,v] = D[x,y,u,v]$,

$$a(x+u,y+v) = a(x,y) + a_x(x,y)u + a_y(x,y)v + b_1(x,y,u,v)u^2$$
$$+ b_2(x,y,u,v)uv + b_3(x,y,u,v)v^2 \tag{2.33}$$

for some polynomials $b_1(x,y,u,v), b_2(x,y,u,v), b_3(x,y,u,v) \in D[x,y,u,v]$.

Proof: First consider the (univariate) polynomial

$$c(x) = a(x,y) \in D[y][x].$$

From Theorem 2.8 we have

$$c(x+u) = c(x) + c'(x)u + d(x,u)u^2$$

for some polynomial $d(x,u) \in D[y][x,u]$, or equivalently

$$a(x+u,y) = a(x,y) + a_x(x,y)u + e(x,y,u)u^2 \tag{2.34}$$

for some polynomial $e(x,y,u) \in D[x,y,u]$. Next consider the (univariate) polynomial

$$f(y) = a(x+u,y) \in D[x,u][y].$$

Applying Theorem 2.8 to express $f(y+v)$ we get

$$a(x+u,y+v) = a(x+u,y) + a_y(x+u,y)v + g(x,y,u,v)v^2 \tag{2.35}$$

for some polynomial $g(x,y,u,v) \in D[x,y,u,v]$. In (2.35), if we express the polynomial

$a(x+u,y)$ directly as given by (2.34) and if we express the polynomial $a_y(x+u,y)$ also in the form indicated by (2.34), we get

$$a(x+u,\, y+v) = a(x,y) + a_x(x,y)u + e(x,y,u)u^2 + a_y(x,y)v$$
$$+ a_{yx}(x,y)uv + \tilde{e}(x,y,u)u^2v + g(x,y,u,v)v^2 \qquad (2.36)$$

where $a_{yx}(x,y)$ denotes the partial derivative with respect to x of the polynomial $a_y(x,y) \in$ D$[x,y]$. Equation (2.36) can be put into the form of equation (2.33).

●

We see from Theorem 2.7 that a domain D[**x**] of multivariate polynomials forms a unique factorization domain (UFD) (as long as the coefficient domain D is a UFD) but that D[**x**] forms no higher algebraic structure in the hierarchy of Table 2.3 even if D is a higher algebraic structure (except in the case of univariate polynomials). Thus the UFD is the abstract structure which forms the setting for multivariate polynomial manipulation. In the next section we develop an algorithm for GCD computation in this new setting.

2.7. THE PRIMITIVE EUCLIDEAN ALGORITHM

The Euclidean algorithm of Section 2.3 cannot be used to compute GCD's in a multivariate polynomial domain D[**x**] because D[**x**] is not a Euclidean domain. However D[**x**] is a UFD (if D is a UFD) and we are assured by Theorem 2.1 that GCD's exist and are unique in any UFD.

Example 2.19. In the UFD **Z**$[x]$ let $a(x)$, $b(x)$ be the polynomials (2.12) defined in Example 2.14. Thus

$$a(x) = 48x^3 - 84x^2 + 42x - 36, \quad b(x) = -4x^3 - 10x^2 + 44x - 30.$$

The unique unit normal factorizations of $a(x)$ and $b(x)$ in **Z**$[x]$ are

$$a(x) = (2)\,(3)\,(2x-3)\,(4x^2 - x + 2);$$
$$b(x) = (-1)\,(2)\,(2x-3)\,(x-1)\,(x+5)$$

where we note that $\mathrm{u}(a(x)) = 1$ has not been explicitly written, and $\mathrm{u}(b(x)) = -1$. Thus

$$\mathrm{GCD}(a(x), b(x)) = 2\,(2x-3) = 4x - 6.$$

●

Example 2.20. Let $a(x)$, $b(x)$ be the polynomials from the previous example, but this time considered as polynomials in the Euclidean domain **Q**$[x]$. The unique unit normal factorizations of $a(x)$ and $b(x)$ in **Q**$[x]$ are

$$a(x) = (48)\,(x - \frac{3}{2})\,(x^2 - \frac{1}{4}x + \frac{1}{2});$$

$$b(x) = (-4)\,(x - \frac{3}{2})\,(x - 1)\,(x + 5)$$

where we note that $u(a(x)) = 48$ and $u(b(x)) = -4$. Thus

$$GCD(a(x), b(x)) = x - \frac{3}{2}.$$

as noted in Example 2.14.

●

As in the case of Euclidean domains, it is not practical to compute the GCD of $a(\mathbf{x})$, $b(\mathbf{x}) \in D[\mathbf{x}]$ by determining the prime factorizations of $a(\mathbf{x})$ and $b(\mathbf{x})$ but rather we will see that there is a GCD algorithm for the UFD $D[\mathbf{x}]$ which is very similar to the Euclidean algorithm. The new algorithm will be developed for the univariate polynomial domain $D[x]$ over a UFD D and then we will see that it applies immediately to the multivariate polynomial domain $D[\mathbf{x}]$ by the application of recursion.

Primitive Polynomials

We have noted in Section 2.4 that if elements in an integral domain are split into their unit parts and normal parts then the GCD of two elements is simply the GCD of their normal parts. It is convenient in a polynomial domain $D[x]$ to further split the normal part into a part lying in the coefficient domain D and a purely polynomial part. For example, the unit normal factorizations of $a(x)$, $b(x) \in \mathbf{Z}[x]$ in Example 2.19 consist of a unit followed by integer factors followed by polynomial factors and similarly $GCD(a(x), b(x))$ consists of integer factors followed by polynomial factors.

Definition 2.16. Let D be an integral domain D. The *GCD of n elements* $a_1, \ldots, a_n \in D$ is defined recursively for $n > 2$ by:

$$GCD(a_1, \ldots, a_n) = GCD(GCD(a_1, \ldots, a_{n-1}), a_n).$$

The $n \geq 2$ elements $a_1, \ldots, a_n \in D$ are called *relatively prime* if

$$GCD(a_1, \ldots, a_n) = 1.$$

●

Definition 2.17. A nonzero polynomial $a(x)$ in $D[x]$, D a UFD, is called *primitive* if it is a unit normal polynomial and its coefficients are relatively prime. In particular, if $a(x)$ has exactly one nonzero term then it is primitive if and only if it is monic.

●

Definition 2.18. In a polynomial domain $D[x]$ over a UFD D, the *content* of a nonzero polynomial $a(x)$, denoted $cont(a(x))$, is defined to be the (unique unit normal) GCD of the coefficients of $a(x)$. Any nonzero polynomial $a(x) \in D[x]$ has a unique representation in the form

$$a(x) = \mathrm{u}(a(x)) \cdot \mathrm{cont}(a(x)) \cdot \mathrm{pp}(a(x))$$

where $\mathrm{pp}(a(x))$ is a primitive polynomial called the *primitive part* of $a(x)$. It is convenient to define $\mathrm{cont}(0) = 0$ and $\mathrm{pp}(0) = 0$.

●

It is a classical result (known as Gauss' lemma) that the product of any two primitive polynomials is itself primitive. It follows from the above definitions that the GCD of two polynomials is the product of the GCD of their contents and the GCD of their primitive parts; notationally,

$$\mathrm{GCD}(a(x), b(x)) = \mathrm{GCD}(\mathrm{cont}(a(x)), \mathrm{cont}(b(x)))\ \mathrm{GCD}(\mathrm{pp}(a(x)), \mathrm{pp}(b(x))). \tag{2.37}$$

By definition, the computation of the GCD of the contents of $a(x), b(x) \in \mathrm{D}(x)$ is a computation in the coefficient domain D. Assuming that we know how to compute GCD's in D, we may restrict our attention to the computation of GCD's of *primitive* polynomials in $\mathrm{D}(x)$.

Example 2.21. For the polynomials $a(x), b(x) \in \mathbf{Z}[x]$ considered in Example 2.19 we have:

$$\mathrm{cont}(a(x)) = 6, \quad \mathrm{pp}(a(x)) = 8x^3 - 14x^2 + 7x - 6,$$

$$\mathrm{cont}(b(x)) = 2, \quad \mathrm{pp}(b(x)) = 2x^3 + 5x^2 - 22x + 15.$$

For the same polynomials considered as elements in the domain $\mathbf{Q}[x]$ as in Example 2.20 we have:

$$\mathrm{cont}(a(x)) = 1, \quad \mathrm{pp}(a(x)) = x^3 - \frac{7}{4}x^2 + \frac{7}{8}x - \frac{3}{4},$$

$$\mathrm{cont}(b(x)) = 1, \quad \mathrm{pp}(b(x)) = x^3 + \frac{5}{2}x^2 - 11x + \frac{15}{2}.$$

●

Pseudo-Division of Polynomials

The Euclidean algorithm is based on the computation of a sequence of remainders which is defined in terms of the division property in a Euclidean domain. For a non-Euclidean domain, $\mathrm{D}[x]$, the division property does not hold. However there is a very similar *pseudo-division* property which holds in any polynomial domain $\mathrm{D}[x]$, where D is a UFD. This new property can be understood by considering the UFD $\mathbf{Z}[x]$ of univariate polynomials over the integers.

Consider the polynomials $a(x), b(x)$ given by equation (2.12) in Example 2.11. As polynomials in the Euclidean domain $\mathbf{Q}[x]$, we found in Example 2.11 that the division property holds in the form:

$$(3x^3 + x^2 + x + 5) = (5x^2 - 3x + 1)\,(\frac{3}{5}x + \frac{14}{25}) + (\frac{52}{25}x + \frac{111}{25}). \tag{2.38}$$

Note that the leading coefficient of $b(x)$ is 5 and that the only denominators appearing in the coefficients of the quotient and remainder in (2.38) are 5 and 5^2. Therefore in this example, if we started with the polynomials $\tilde{a}(x)$ and $b(x)$ where

$$\tilde{a}(x) = 5^2\, a(x)$$

then we would have the following relationships among polynomials with *integer coefficients*:

$$5^2\,(3x^3 + x^2 + x + 5) = (5x^2 - 3x + 1)\,(15x + 14) + (52x + 111). \tag{2.39}$$

Equation (2.39) is an instance of the pseudo-division property which holds in any polynomial domain D[x] over a UFD D, just as equation (2.38) is an instance of the division property in a Euclidean domain. The generalization of (2.39) is obtained by close examination of the process of polynomial long division in a domain D[x]. If $\deg(a(x)) = m$, $\deg(b(x)) = n$, $m \ge n \ge 0$ and if the leading coefficient of $b(x)$ is β then viewing the division of $a(x)$ by $b(x)$ as operations in the coefficient domain D we find that the only divisions are divisions by β and such divisions occur $m - n + 1$ times. We thus have the following result.

Pseudo-Division Property (Property P3).

Let D[x] be a polynomial domain over a UFD D. For all $a(x),\ b(x) \in$ D[x] with $b(x) \ne 0$ and $\deg(a(x)) \ge \deg((b(x))$, there exist polynomials $q(x),\ r(x) \in$ D[x] such that

$$\text{P3:}\quad \beta^l\, a(x) = b(x)\, q(x) + r(x), \quad \deg(r(x)) < \deg(b(x))$$

where $\beta = \mathrm{lcoeff}(b(x))$ and $l = \deg(a(x)) - \deg(b(x)) + 1$.

●

For given polynomials $a(x),\ b(x) \in$ D[x] the polynomials $q(x)$ and $r(x)$ appearing in property P3 are called, respectively, the *pseudo-quotient* and *pseudo-remainder*. Functionally, we use the notation $\mathrm{pquo}(a(x),b(x))$ and $\mathrm{prem}(a(x),b(x))$ for the pseudo-quotient and pseudo-remainder, respectively, and we extend the definitions of these functions to the case $\deg(a(x)) < \deg(b(x))$ by defining in the latter case $\mathrm{pquo}(a(x),b(x)) = 0$ and $\mathrm{prem}(a(x),b(x)) = a(x)$. (Note that these special definitions satisfy the relationship P3 with $\beta = 1$ rather than with $\beta = \mathrm{lcoeff}(b(x))$.) Just as in the case of the division property (Property P2) for univariate polynomials over a field, the polynomials $q(x)$, $r(x)$ in property P3 are *unique*. We may note that for given $a(x),\ b(x) \in$ D[x], we obtain the pseudo-quotient $q(x)$ and pseudo-remainder $r(x)$ of property P3 by performing ordinary polynomial long division of $\beta^l\, a(x)$ by $b(x)$. In this process, all divisions will be exact in the coefficient domain D.

GCD Computation in D[x]

The pseudo-division property leads directly to an algorithm for computing GCD's in any polynomial domain D[x] over a UFD D. As previously noted, we may restrict our attention to primitive polynomials in D[x].

Theorem 2.10. Let D[x] be a polynomial domain over a UFD D. Given primitive polynomials $a(x)$, $b(x) \in$ D[x] with $b(x) \neq 0$ and $\deg(a(x)) \geq \deg(b(x))$, let $q(x)$, $r(x)$ be the pseudo-quotient and pseudo-remainder satisfying property P3.
Then

$$\mathrm{GCD}(a(x), b(x)) = \mathrm{GCD}(b(x), \mathrm{pp}(r(x))). \qquad (2.40)$$

Proof: From property P3 we have

$$\beta^l\, a(x) = b(x)\, q(x) + r(x)$$

and applying to this equation the same argument as in the proof of Theorem 2.3 yields

$$\mathrm{GCD}(\beta^l\, a(x),\, b(x)) = \mathrm{GCD}(b(x),\, r(x)). \qquad (2.41)$$

Applying (2.37) to the left side of (2.41) yields

$$\mathrm{GCD}(\beta^l\, a(x), b(x)) = \mathrm{GCD}(\beta^l, 1)\, \mathrm{GCD}(a(x), b(x))$$
$$= \mathrm{GCD}(a(x),\, b(x))$$

where we have used the fact that $a(x)$, $b(x)$ are primitive polynomials. Similarly, applying (2.37) to the right side of (2.41) yields

$$\mathrm{GCD}(b(x), r(x)) = \mathrm{GCD}(1, \mathrm{cont}(r(x)))\, \mathrm{GCD}(b(x), \mathrm{pp}(r(x)))$$
$$= \mathrm{GCD}(b(x), \mathrm{pp}(r(x))).$$

The result follows.

●

It is obvious that for primitive polynomials $a(x)$, $b(x)$ we can define an iteration for GCD computation in D[x] based on equation (2.40) and this iteration must terminate since $\deg(r(x)) < \deg(b(x))$ at each step. This result is the basis of Algorithm 2.3. In Algorithm 2.3 the sequence of remainders which is generated is such that the remainder computed in each iteration is normalized to be primitive, so the algorithm is commonly referred to as the primitive Euclidean algorithm. Algorithm 2.3 uses the *prem* function (in the sense of the extended definition given above) and it also assumes the existence of an algorithm for GCD computation in the coefficient domain D which would be used to compute contents, and hence primitive parts, and also to compute the quantity γ in that algorithm.

Example 2.22. In the UFD **Z**[x], let $a(x)$, $b(x)$ be the polynomials considered variously in Examples 2.14 - 2.15 and Examples 2.19 - 2.21. Thus

$$a(x) = 48x^3 - 84x^2 + 42x - 36, \quad b(x) = -4x^3 - 10x^2 + 44x - 30.$$

The sequence of values computed for $r(x)$, $c(x)$, and $d(x)$ in Algorithm 2.3 is as follows:

Algorithm 2.3. Primitive Euclidean Algorithm.

```
procedure PrimitiveEuclidean(a(x),b(x))
    # Given polynomials a(x), b(x) ∈ D[x]
    # where D is a UFD, we compute
    # g(x) = GCD(a(x), b(x)) .

    c(x) ← pp(a(x)); d(x) ← pp(b(x))

    while d(x) ≠ 0 do {
        r(x) ← prem(c(x),d(x))
        c(x) ← d(x)
        d(x) ← pp(r(x)) }

    γ ← GCD(cont(a(x)),cont(b(x)))
    g(x) ← γ c(x)
    return(g(x))
end
```

iteration	$r(x)$	$c(x)$	$d(x)$
0	–	$8x^3-14x^2+7x-6$	$2x^3+5x^2-22x+15$
1	$-68x^2+190x-132$	$2x^3+5x^2-22x+15$	$34x^2-95x+66$
2	$4280x-6420$	$34x^2-95x+66$	$2x-3$
3	0	$2x-3$	0

Then $\gamma = \text{GCD}(6,2) = 2$ and $g(x) = 2(2x-3) = 4x-6$ as noted in Example 2.19.

•

Multivariate GCD Computation

The primary significance of Algorithm 2.3 is that it may be applied to compute GCD's in a *multivariate* polynomial domain D[**x**] over a UFD. Choosing x_1 as the main variable, we identify $D[x_1, \ldots, x_v]$ with the univariate polynomial domain $D[x_2, \ldots, x_v][x_1]$ over the UFD $D[x_2, \ldots, x_v]$. In order to apply Algorithm 2.3, we must be able to compute GCD's in the "coefficient domain" $D[x_2, \ldots, x_v]$ – but this may be accomplished by recursively applying Algorithm 2.3, identifying $D[x_2, \ldots, x_v]$ with $D[x_3, \ldots, x_v][x_2]$, etc. Thus the recursive view of a multivariate polynomial domain leads naturally to a recursive algorithm for GCD computation.

Example 2.23. In the UFD $\mathbf{Z}[x,y]$ let $a(x,y)$ and $b(x,y)$ be given by

$$a(x,y) = -30x^3y + 90x^2y^2 + 15x^2 - 60xy + 45y^2,$$

$$b(x,y) = 100x^2y - 140x^2 - 250xy^2 + 350xy - 150y^3 + 210y^2.$$

Choosing x as the main variable, we view $a(x,y)$ and $b(x,y)$ as elements in the domain $\mathbf{Z}[y][x]$:

$$a(x,y) = (-30y)\,x^3 + (90y^2 + 15)\,x^2 - (60y)\,x + (45y^2),$$

$$b(x,y) = (100y - 140)\,x^2 - (250y^2 - 350y)\,x - (150y^3 - 210y^2).$$

The first step in Algorithm 2.3 requires that we remove the unit part and the content from each polynomial; this requires a recursive application of Algorithm 2.3 to compute GCD's in the domain $\mathbf{Z}[y]$. We find:

$$\mathrm{u}(a(x,y)) = -1,$$

$$\mathrm{cont}(a(x,y)) = \mathrm{GCD}(30y, -(90y^2 + 15), 60y, -45y^2) = 15;$$

$$\mathrm{pp}(a(x,y)) = (2y)\,x^3 - (6y^2 + 1)\,x^2 + (4y)\,x - (3y^2);$$

and

$$\mathrm{u}(b(x,y)) = 1,$$

$$\mathrm{cont}(b(x,y)) = \mathrm{GCD}(100y - 140, -(250y^2 - 350y), -(150y^3 - 210y^2))$$

$$= 50y - 70.$$

$$\mathrm{pp}(b(x,y)) = (2)\,x^2 - (5y)\,x - (3y^2).$$

The sequence of values computed for $r(x)$, $c(x)$, and $d(x)$ in Algorithm 2.3 is then as follows:

iteration	$r(x)$	$c(x)$	$d(x)$
0	–	$(2y)x^3-(6y^2+1)x^2+(4y)x-(3y^2)$	$2x^2-(5y)x-(3y^2)$
1	$(2y^3+6y)x-(6y^4+18y^2)$	$2x^2-(5y)x-(3y^2)$	$x-(3y)$
2	0	$x-(3y)$	0

Thus,

$$\gamma = \mathrm{GCD}(15, 50y - 70) = 5$$

and

$$g(x) = 5(x - (3y)) = 5x - (15y);$$

that is,

$$\text{GCD}(a(x,y), b(x,y)) = 5x - 15y.$$

●

The Euclidean Algorithm Revisited

Algorithm 2.3 is a generalization of Algorithm 2.1 and we may apply Algorithm 2.3 to compute GCD's in a Euclidean domain F[x] over a field F. In this regard, note that the GCD of any two elements (not both zero) in a field F is 1 since every nonzero element in a field is a unit. In particular, cont($a(x)$) = 1 for all nonzero $a(x) \in$ F[x] and hence

$$\text{pp}(a(x)) = \text{n}(a(x)) \text{ for all } a(x) \in \text{F}[x].$$

Functionally, the operations pp(·) and n(·) when applied in a Euclidean domain F[x] both specify that their argument is to be made monic. The prem function can be seen to be identical with the standard rem function when applied to *primitive* polynomials in F[x] since $\beta = 1$ in property P3 when $b(x)$ is monic.

A comparison of Algorithm 2.3 with Algorithm 2.1 thus shows that when applied in a polynomial domain F[x] over a field F, both algorithms perform the same computation except that in Algorithm 2.3 the remainder is normalized, that is, made monic in each iteration. This additional normalization within each iteration serves to simplify the computation somewhat and may be considered a useful improvement to Algorithm 2.1.

Example 2.24. In the Euclidean domain **Q**[x], let $a(x)$, $b(x)$ be the polynomials of Example 2.14. The sequence of values computed for $r(x)$, $c(x)$, and $d(x)$ in Algorithm 2.3 is as follows:

iteration	$r(x)$	$c(x)$	$d(x)$
0	–	$x^3 - \frac{7}{4}x^2 + \frac{7}{8}x - \frac{3}{4}$	$x^3 + \frac{5}{2}x^2 - 11x + \frac{15}{2}$
1	$-\frac{17}{4}x^2 + \frac{95}{8}x - \frac{33}{4}$	$x^3 + \frac{5}{2}x^2 - 11x + \frac{15}{2}$	$x^2 - \frac{95}{34}x + \frac{33}{17}$
2	$\frac{535}{289}x - \frac{1605}{578}$	$x^2 - \frac{95}{34}x + \frac{33}{17}$	$x - \frac{3}{2}$
3	0	$x - \frac{3}{2}$	0

Then $\gamma = 1$ and $g(x) = x - \frac{3}{2}$ as computed by Algorithm 2.1 in Example 2.14.

●

2.8. QUOTIENT FIELDS AND RATIONAL FUNCTIONS

An important property of an integral domain is that it can be extended to a field in a very simple manner. One reason for wanting a field is, for example, to be able to solve linear equations – a process which requires division. The most familiar example of extending an integral domain to a field is the process of constructing the field **Q** of rationals from the integral domain **Z** of integers. This particular construction extends immediately to any integral domain.

Quotient Fields

Let D be an integral domain and consider the set of quotients

$$S = \{ a/b : a \in D, b \in D - \{0\} \}.$$

Keeping in mind the usual properties of field arithmetic, we define the following relation on S:

$$a/b \sim c/d \text{ if and only if } ad = bc.$$

It is readily verified that the relation ~ is an equivalence relation on S and it therefore divides S into equivalence classes $[a/b]$. The set of equivalence classes is called a *quotient set,* denoted by

$$S/\sim = \{ [a/b] : a \in D, b \in D - \{0\} \}$$

(read "S modulo the equivalence relation ~"). In dealing with the quotient set S/~, any member of an equivalence class may serve as its *representative*. Thus when we write a/b we really mean the equivalence class $[a/b]$ containing the particular quotient a/b. The operations of addition and multiplication in the integral domain D are extended to the quotient set S/~ as follows: if a/b and c/d are in S/~ (in the above sense) then

$$(a/b) + (c/d) = (ad + bc) / bd; \tag{2.42}$$

$$(a/b)\,(c/d) = ac/bd. \tag{2.43}$$

(It is straightforward to show that the operations of addition and multiplication on equivalence classes in S/~ are well-defined by (2.42) and (2.43) in the sense that the sum or product of equivalence classes is independent of the particular representatives used for the equivalence classes.) The quotient set S/~ with the operations of addition and multiplication defined by (2.42) and (2.43) is a field, called the *quotient field* (or *field of quotients*) of the integral domain D and denoted variously by Q(D) or F_D.

The quotient field Q(D) contains (an isomorphic copy of) the integral domain D. Specifically, the integral domain D is identified with the subset of Q(D) defined by

$$\{ a/1 : a \in D \}$$

using the natural relationship $a \leftrightarrow a/1$. Indeed the quotient field Q(D) is the *smallest* field which contains the integral domain D. The zero in the field Q(D) is the quotient 0/1 and the multiplicative identity is 1/1. By convention, a quotient $a/1 \in$ Q(D) with denominator 1 is denoted by a; in particular, the zero and identity are denoted by 0 and 1.

When dealing with an algebraic system whose constituent elements are equivalence classes, it is fine in principle to note that any member of an equivalence class may serve as its representative but in practice we need a *canonical form* for the equivalence classes so that the representation is unique. Otherwise, a problem such as determining when two expressions are equal becomes very nontrivial. If GCD's exist in the integral domain D and if a canonical form (i.e. unique representation) for elements of D has been determined, then a common means of defining a canonical form for elements in the quotient field Q(D) is as follows: the representative a/b of $[a/b] \in$ Q(D) is canonical if

$$\mathrm{GCD}(a,b) = 1, \tag{2.44}$$

$$b \text{ is unit normal in D}, \tag{2.45}$$

$$a \text{ and } b \text{ are canonical in D}. \tag{2.46}$$

Any representative c/d may be put in this canonical form by a straightforward computational procedure: compute GCD(c,d) and divide it out of numerator and denominator, multiply numerator and denominator by the inverse of the unit u(d), and put the resulting numerator and denominator into their canonical forms as elements of D. It can be verified (see Exercise 2.20) that for each equivalence class in Q(D) there is one and only one representative satisfying (2.44), (2.45) and (2.46).

Example 2.25. If D is the domain **Z** of integers then the quotient field Q(**Z**) is the field of rational numbers, denoted by **Q**. A rational number (representative) a/b is canonical if a and b have no common factors and b is positive. The following rational numbers all belong to the same equivalence class:

$$-2/4,\ 2/{-4},\ 100/{-200},\ -600/1200;$$

their canonical representative is $-1/2$.

•

Rational Functions

For a polynomial domain D[**x**] over a UFD D, the quotient field Q(D[**x**]) is called the field of *rational functions* (or *rational expressions*) over D in the indeterminates **x**, and is denoted by D(**x**). Elements of D(**x**) are (equivalence classes of) quotients of the form.

$$a(\mathbf{x}) / b(\mathbf{x}) \ \text{ where } a(\mathbf{x}), b(\mathbf{x}) \in \mathrm{D}[\mathbf{x}] \text{ with } b(\mathbf{x}) \neq 0.$$

The canonical form of a rational function (representative) $a(\mathbf{x})/b(\mathbf{x}) \in$ D(**x**) depends on the canonical form chosen for multivariate polynomials in D[**x**] (canonical forms for multivariate polynomials are discussed in Chapter 3) but the definition of canonical forms for rational functions will always include conditions (2.44) and (2.45) – namely, $a(\mathbf{x})$ and $b(\mathbf{x})$ have no common factors and the leading coefficient of $b(\mathbf{x})$ is unit normal in the coefficient domain D.

The operation of addition in a quotient field is a relatively complex operation. From (2.42) we see that to add two quotients requires three multiplications and one addition in the underlying integral domain. Additionally, a GCD computation will be required to obtain the canonical form of the sum. It is the latter operation which is the most expensive and its cost is a dominating factor in many computations. For the field D(**x**) of rational functions, we try to minimize the cost of GCD computation by intelligently choosing the representation for rational functions (see Chapter 3) and by using an efficient GCD algorithm (see Chapter 7). On the other hand, the operation of multiplication in a quotient field is less expensive than addition. From (2.43) we see that to multiply two quotients requires only two multiplications in the underlying integral domain, but more significantly, with an appropriate choice of representation (namely, factored normal form as defined in Chapter 3) it is possible to greatly reduce the amount of GCD computation required in performing the operation (2.43) compared with the operation (2.42).

Two polynomial domains of interest in symbolic computation are the domains $\mathbf{Z}[x]$ and $\mathbf{Q}[x]$. Let us consider for a moment the corresponding fields of rational functions $\mathbf{Z}(x)$ and $\mathbf{Q}(x)$. In the univariate case, a typical example of a rational function (representative) in $\mathbf{Q}(x)$ is

$$a(x) / b(x) = (\frac{17}{100}x^2 - \frac{3}{112}x + \frac{1}{2}) / (\frac{5}{9}x^2 + \frac{4}{5}). \tag{2.47}$$

But note that the equivalence class $[a(x)/b(x)]$ also contains representatives with integer coefficients. The simplest such representative is obtained by multiplying numerator and denominator in (2.47) by the least common multiple (LCM) of all coefficient denominators; in this case:[4]

$$\text{LCM}\,(100, 112, 2, 9, 5) = 25200.$$

Thus another representative for the rational function (2.47) in $\mathbf{Q}(x)$ is

$$a(x) / b(x) = (4284x^2 - 675x + 12600) / (14000x^2 + 20160) \tag{2.48}$$

which is also a rational function (representative) in the domain $\mathbf{Z}(x)$. The argument just posed leads to a very general result which we will not prove more formally here; namely, if D is any integral domain and if F_D denotes the quotient field of D, then the fields of rational functions D(**x**) and F_D(**x**) are isomorphic. More specifically, there is a natural one-to-one correspondence between the equivalence classes in D(**x**) and the equivalence classes in F_D(**x**). The only difference between the two fields is that each equivalence class has many more representatives in F_D(**x**) than in D(**x**).

Example 2.26. In the field $\mathbf{Q}(x)$, a canonical form for the rational function (2.47) satisfying conditions (2.44) and (2.45) is obtained by making the denominator unit normal (i.e. monic):

4. The LCM of n elements $a_1, \ldots, a_n$ in an integral domain D is defined recursively for $n > 2$ by:

$$\text{LCM}(a_1, \ldots, a_{n-1}, a_n) = \text{LCM}(\text{LCM}(a_1, \ldots, a_{n-1}), a_n).$$

$$a(x)/b(x) = (\frac{153}{500}x^2 - \frac{27}{560}x + \frac{9}{10})/(x^2 + \frac{36}{25})$$

(since there are already no common factors). In the field $\mathbf{Z}(x)$, the same rational function has (2.48) as a canonical form since the denominator in (2.48) is unit normal in $\mathbf{Z}[x]$ and there are no common factors (including integer common factors).

●

2.9. POWER SERIES AND EXTENDED POWER SERIES

Ordinary Power Series

Whereas algebraists are interested in such objects as univariate polynomials

$$a_0 + a_1x + \cdots + a_nx^n$$

a more useful object for analysts occurs when we do not stop at a_nx^n, that is, when we have a power series. The definition of univariate polynomials can be readily extended to a definition of univariate power series. For any commutative ring R, the notation R[[x]] denotes the set of all expressions of the form

$$a(x) = \sum_{k=0}^{\infty} a_kx^k \tag{2.49}$$

with $a_k \in$ R. In other words, R[[x]] denotes the set of all *power series* in the indeterminate x over the ring R. The *order* ord($a(x)$) of a nonzero power series $a(x)$ as in (2.49) is the least integer k such that $a_k \neq 0$. The exceptional case where $a_k = 0$ for all k is called the *zero power series* and is denoted by 0. It is conventional to define ord(0) $= \infty$. For a nonzero power series $a(x)$ as in (2.49) with ord($a(x)$) $= l$, the term a_lx^l is called the *low order term* of $a(x)$, a_l is called the *low order coefficient,* and a_0 is called the *constant term.* A power series in which $a_k = 0$ for all $k \geq 1$ is called a *constant power series.*

The binary operations of addition and multiplication in the commutative ring R are extended to power series in the set R[[x]] as follows. If

$$a(x) = \sum_{k=0}^{\infty} a_kx^k \text{ and } b(x) = \sum_{k=0}^{\infty} b_kx^k$$

then power series addition is defined by

$$c(x) = a(x) + b(x) = \sum_{k=0}^{\infty} c_kx^k \tag{2.50}$$

where

$$c_k = a_k + b_k \text{ for all } k \geq 0;$$

power series multiplication is defined by

$$d(x) = a(x)\,b(x) = \sum_{k=0}^{\infty} d_k x^k \tag{2.51}$$

where

$$d_k = a_0 b_k + \cdots + a_k b_0, \qquad \text{for all } k \geq 0.$$

Note that the set R$[x]$ of univariate polynomials over R is the subset of R$[[x]]$ consisting of all power series with only a finite number of nonzero terms. Equations (2.50) and (2.51) reduce to the definitions of polynomial addition and multiplication when $a(x)$ and $b(x)$ have only a finite number of nonzero terms. Just as in the case of polynomials, the set R$[[x]]$ of power series inherits a ring structure from its coefficient ring R under the operations (2.50) and (2.51). The following theorem states the basic results.

Theorem 2.11.

(i) If R is a commutative ring then R$[[x]]$ is also a commutative ring. The zero in R$[[x]]$ is the zero power series 0 ($= 0 + 0x + 0x^2 + \cdots$) and the identity in R$[[x]]$ is the constant power series 1 ($= 1 + 0x + 0x^2 + \cdots$).

(ii) If D is an integral domain then D$[[x]]$ is also an integral domain. The units (invertibles) in D$[[x]]$ are all power series whose constant term a_0 is a unit in the coefficient domain D.

(iii) If F is a field then F$[[x]]$ is a Euclidean domain with the valuation

$$v(a(x)) = \operatorname{ord}(a(x)). \tag{2.52}$$

●

It is instructive to note the following constructive proof of the second statement in part (ii) of Theorem 2.11. If $a(x) = \sum_{k=0}^{\infty} a_k x^k$ is a unit in D$[[x]]$ then there must exist a power series $b(x) = \sum_{k=0}^{\infty} b_k x^k$ such that $a(x)\,b(x) = 1$. By the definitions of power series multiplication, we must have

$$1 = a_0 b_0,$$
$$0 = a_0 b_1 + a_1 b_0,$$
$$\vdots$$
$$0 = a_0 b_n + a_1 b_{n-1} + \cdots + a_n b_0,$$
$$\vdots$$

Thus, a_0 is a unit in D with $a_0^{-1} = b_0$. Conversely, if a_0 is a unit in D then the above equations can be solved for the b_k's as follows:

$$b_0 = a_0^{-1},$$

$$b_1 = -a_0^{-1}\,(a_1 b_0),$$

$$\vdots$$

$$b_n = -a_0^{-1}\,(a_1 b_{n-1} + \cdots + a_n b_0),$$

$$\vdots$$

Thus we can construct $b(x)$ such that $a(x)\,b(x) = 1$, which implies that $a(x)$ is a unit in D[[x]].

Example 2.27. In the polynomial domain **Z**[x] the only units are 1 and –1. In the power series domain **Z**[[x]], any power series with constant term 1 or –1 is a unit in **Z**[[x]]. For example, the power series $1 - x$ is a unit in **Z**[[x]] with

$$(1-x)^{-1} = 1 + x + x^2 + x^3 + \cdots .$$

●

Example 2.28. In any power series domain F[[x]] over a field F, every power series of order 0 is a unit in F[[x]]. For if $a(x) \in$ F[[x]] is of order 0 then its constant term $a_0 \neq 0$ is a unit in the coefficient field F.

●

The order function defined on power series has properties similar to the degree functions defined on polynomials. It can be readily verified that the following properties hold for power series in a domain D[[x]] over any integral domain D:

$$\mathrm{ord}(a(x) + b(x)) \geq \min\{\mathrm{ord}(a(x)), \mathrm{ord}(b(x))\}; \tag{2.53}$$

$$\mathrm{ord}(a(x)\,b(x)) = \mathrm{ord}(a(x)) + \mathrm{ord}(b(x)). \tag{2.54}$$

Using (2.54) we can verify that in a power series domain F[[x]] over a field F, (2.52) is a valid valuation according to Definition 2.12. Since by definition $\mathrm{ord}(a(x)) \geq 0$ for any nonzero power series $a(x)$, the valuation (2.52) is indeed a mapping from F[[x]]–{0} into the nonnegative integers **N** as required by Definition 2.12. Property P1 can be verified by using (2.54) since if $a(x), b(x) \in$ F[[x]]–{0} then

$$\mathrm{ord}(a(x)\,b(x)) = \mathrm{ord}(a(x)) + \mathrm{ord}(b(x)) \geq \mathrm{ord}(a(x)).$$

In order to verify property P2 first note that for nonzero $a(x), b(x) \in$ F[[x]] either $a(x) \mid b(x)$ or $b(x) \mid a(x)$. To see this let $\mathrm{ord}(a(x)) = l$ and $\mathrm{ord}(b(x)) = m$ so that

$$a(x) = x^l\,\tilde{a}(x) \;\text{ and }\; b(x) = x^m\,\tilde{b}(x)$$

where $\tilde{a}(x)$ and $\tilde{b}(x)$ are units in F[[x]]. Then if $l \geq m$ we have

$$a(x) / b(x) = x^{l-m}\,\tilde{a}(x)\,[\tilde{b}(x)]^{-1} \in \mathrm{F}[[x]]$$

and similarly if $l < m$ then

$$b(x)/a(x) = x^{m-l}\,\tilde{b}(x)\,[\tilde{a}(x)]^{-1} \in \mathrm{F}[[x]].$$

Therefore given $a(x), b(x) \in \mathrm{F}[[x]]$ with $b(x) \neq 0$ we have

$$a(x) = b(x)\,q(x) + r(x)$$

where if $\mathrm{ord}(a(x)) \geq \mathrm{ord}(b(x))$ then $q(x) = a(x)/b(x)$, $r(x) = 0$ while if $\mathrm{ord}(a(x)) < \mathrm{ord}(b(x))$ then $q(x) = 0$, $r(x) = a(x)$. This verifies property P2 proving that F[[x]] is a Euclidean domain if F is a field.

The Quotient Field D((x))

For a power series domain D[[x]] over an integral domain D, the quotient field Q(D[[x]]) is called the field of *power series rational functions* over D and is denoted by D((x)). Elements of D((x)) are (equivalence classes of) quotients of the form

$$a(x)/b(x) \text{ where } a(x), b(x) \in \mathrm{D}[[x]] \text{ with } b(x) \neq 0.$$

Unlike ordinary (polynomial) rational functions, power series rational functions cannot in general be put into a canonical form by removing "common factors" since the power series domain D[[x]] is not a unique factorization domain. Indeed it is not even clear how to define "unit normal" elements in the integral domain D[[x]]. Recall that in any integral domain the relation of associativity is an equivalence relation and the idea of "unit normal" elements is to single out one element from each associate class as its canonical representative. In D[[x]], two power series are in the same associate class if one can be obtained from the other by multiplying it by a power series whose constant term is a unit in D.

Example 2.29. In the power series domain **Z**[[x]], the following power series all belong to the same associate class:

$$a(x) = 2 + 2x + 2x^2 + 3x^3 + 4x^4 + \cdots;$$

$$b(x) = 2 + 4x + 6x^2 + 9x^3 + 13x^4 + \cdots;$$

$$c(x) = 2 + x^3 + x^4 + x^5 + x^6 + \cdots.$$

This can be seen by noting that

$$b(x) = a(x)\,(1 + x + x^2 + x^3 + x^4 + \cdots)$$

and

$$c(x) = a(x)\,(1 - x).$$

It is not clear how to single out one of $a(x)$, $b(x)$, $c(x)$, or some other associate of these, as the unit normal element.

●

The Quotient Field F((x))

The case of a power series domain F[[x]] over a field F and its corresponding quotient field F((x)) can be dealt with in a manner just like polynomials and ordinary (polynomial) rational functions. For if $a(x) \in \mathrm{F}((x))$ is a nonzero power series then $a(x)$ can be expressed in the form

$$a(x) = x^l\, b(x)$$

where l = ord($a(x)$) and

$$b(x) = a_l + a_{l+1}x + a_{l+2}x^2 + \cdots .$$

Then $a_l \neq 0$ and hence $b(x)$ is a unit power series in F[[x]]. This leads us to the following definition.

Definition 2.19. In any power series domain F[[x]] over a field F, the monomials x^l ($l \geq 0$) and the zero power series 0 are defined to be *unit normal*.

●

From the above definition we have the following "functional specifications" for the normal part n($a(x)$) and the unit part u($a(x)$) of a nonzero power series $a(x) \in$ F[[x]]:

$$\mathrm{n}(a(x)) = x^{\mathrm{ord}(a(x))}; \tag{2.55}$$

$$\mathrm{u}(a(x)) = a(x) / x^{\mathrm{ord}(a(x))}. \tag{2.56}$$

(Note that the monomial x^0 is identified with the constant power series 1 and therefore the unit normal element for the associate class of units is 1 as usual.) With this definition of unit normal elements it becomes straightforward to define the GCD of any two power series $a(x)$, $b(x) \in$ F[[x]] (not both zero); namely,

$$\mathrm{GCD}(a(x), b(x)) = x^{\min\{\mathrm{ord}(a(x)),\, \mathrm{ord}(b(x))\}}. \tag{2.57}$$

To see that (2.57) is valid, recall that we may restrict our attention to the *unit normal GCD* which must be a monomial x^l and clearly the "greatest" monomial which divides both $a(x)$ and $b(x)$ is that given by formula (2.57).

Canonical forms for elements of the quotient field F((x)) can now be defined to satisfy conditions (2.44) - (2.46) just as in the case of ordinary (polynomial) rational functions. Namely, if a representation for power series in the domain F[[x]] has been chosen then the canonical form of a power series rational function (representative) $a(x) / b(x) \in$ F((x)) is obtained by dividing out GCD($a(x), b(x)$) and then making the denominator unit normal. It follows that the canonical form of a power series rational function over a field F is always of the form

$$\tilde{a}(x) / x^n \tag{2.58}$$

where $\tilde{a}(x) \in$ F[[x]] and $n \geq 0$; moreover if $n > 0$ then ord($\tilde{a}(x)$) = 0. Clearly the representation of canonical quotient (2.58) is only trivially more complicated than the representation of a power series in the domain F[[x]], and similarly the arithmetic operations on canonical quotients of the form (2.58) are only slightly more complicated then the operations in the domain F[[x]].

Since the power series rational functions in a field F((x)) over any field F have the simple canonical representation (2.58) while the elements in a field D((x)) over a general integral domain D have a much more complicated representation, we will always embed the field D((x)) into the larger field $F_D((x))$ for computational purposes (where F_D denotes the quotient field of the coefficient domain D). Thus we will never need to represent quotients $a(x)/b(x)$ where $a(x)$ and $b(x)$ are both power series. We have noted earlier that for ordinary (polynomial) rational functions the fields D(x) and $F_D(x)$ are isomorphic. The following example indicates that for power series rational functions, the field D((x)) is a proper subset of (i.e. not isomorphic to) the field $F_D((x))$ when D is not a field.

Example 2.30. In the domain **Q**((x)) of power series rational functions over the field **Q**, let

$$a(x)/b(x) = (1 + x + \frac{1}{2}x^2 + \frac{1}{3}x^3 + \frac{1}{4}x^4 + \cdots)/(1 - x).$$

The power series rational function $a(x)/b(x)$ has no representation with integer coefficients because the denominators of the coefficients in the numerator power series grow without bound. Thus the equivalence class $[a(x)/b(x)] \in \mathbf{Q}((x))$ has no corresponding equivalence class in the field **Z**((x)). Note that the reduced form of $a(x)/b(x)$ in the field **Q**((x)) is a power series since $(1-x)$ is a unit in **Q**((x)); specifically, the reduced form is

$$a(x)/b(x) = 1 + 2x + \frac{5}{2}x^2 + \frac{17}{6}x^3 + \frac{37}{12}x^4 + \cdots .$$

●

Extended Power Series

We have seen that to represent the elements of a field F((x)) of power series rational functions over a field F, we need only to represent expressions of the form

$$(\sum_{k=0}^{\infty} a_k x^k)/x^n \qquad (2.59)$$

where n is a nonnegative integer. One way to represent such expressions is in the form of "extended power series" which we now define.

For any field F, the set F<x> of *extended power series* over F is defined to be the set of all expressions of the form

$$a(x) = \sum_{k=m}^{\infty} a_k x^k \qquad (2.60)$$

with $a_k \in$ F $(k \ge m)$, where $m \in Z$ (i.e. m is any finite integer, positive, negative or zero). As in the case of ordinary power series, we define the *order* ord($a(x)$) of a nonzero extended power series $a(x)$ as in (2.60) to be the least integer k such that $a_k \ne 0$. Thus ord($a(x)$) < 0 for many extended power series $a(x) \in$ F<x> but clearly the set F<x> also contains the set F[[x]] of ordinary power series satisfying ord($a(x)$) ≥ 0. As with ordinary power series, the *zero extended power series* is denoted by 0, ord(0) = ∞ by definition, and if $a(x)$ is a nonzero extended power series as in (2.60) with ord($a(x)$) = m then $a_m x^m$ is the *low order term*, a_m is the *low order coefficient*, and a_0 is the *constant term*. An extended power series in which a_k = 0 for all $k \ge 1$ and for all $k < 0$ is called a *constant extended power series*.

Addition and multiplication of extended power series are defined exactly as for ordinary power series as follows. If

$$a(x) = \sum_{k=m}^{\infty} a_k x^k \quad \text{and} \quad b(x) = \sum_{k=n}^{\infty} b_k x^k$$

then addition is defined by

$$c(x) = a(x) + b(x) = \sum_{k=\min(m,n)}^{\infty} c_k x^k \tag{2.61}$$

where

$$c_k = \begin{cases} a_k + b_k & \text{for } k \ge \max(m,n) \\ a_k & \text{for } m \le k < n \ \text{ if } m < n \\ b_k & \text{for } n \le k < m \ \text{ if } m > n\,. \end{cases}$$

Similarly, multiplication is defined by

$$d(x) = a(x)\,b(x) = \sum_{k=m+n}^{\infty} d_k x^k \tag{2.62}$$

where

$$d_k = \sum_{i+j=k} a_i b_j .$$

It is easy to verify that the order function defined on extended power series satisfies properties (2.53) and (2.54) for the order of a sum and product, just as for ordinary power series. Under the operations (2.61) and (2.62), $\mathrm{F}<x>$ is a field with zero element the zero extended power series 0 and with identity the constant extended power series 1.

Let us consider a constructive proof that every nonzero extended power series $a(x) \in \mathrm{F}<x>$ has an inverse in $\mathrm{F}<x>$. Firstly, if $\mathrm{ord}(a(x)) = 0$ then $a(x)$ is a unit in the power series domain $\mathrm{F}[[x]]$ and the inverse power series $[a(x)]^{-1} \in \mathrm{F}[[x]]$ may be considered an element of $\mathrm{F}<x>$. Then $[a(x)]^{-1}$ is the desired inverse in $\mathrm{F}<x>$ because power series multiplication is defined the same in $\mathrm{F}<x>$ as in $\mathrm{F}[[x]]$. More generally, if $\mathrm{ord}(a(x)) = m$ (which may be positive, negative, or zero) then $a(x) = x^m\, b(x)$ where $\mathrm{ord}(b(x)) = 0$. Then it is easily verified that the inverse of $a(x)$ in $\mathrm{F}<x>$ is given by

$$[a(x)]^{-1} = x^{-m}\,[b(x)]^{-1}.$$

Note in particular that

$$\mathrm{ord}([a(x)]^{-1}) = -\mathrm{ord}(a(x)).$$

Example 2.31. In the field Q<x> let

$$a(x) = x^2 + \frac{1}{2}x^3 + \frac{1}{4}x^4 + \frac{1}{8}x^5 + \frac{1}{16}x^6 + \cdots .$$

The inverse of $a(x)$ can be determined by noting that

$$a(x) = x^2 \left(1 + \frac{1}{2}x + \frac{1}{4}x^2 + \frac{1}{8}x^3 + \frac{1}{16}x^4 + \cdots\right)$$

and

$$\left(1 + \frac{1}{2}x + \frac{1}{4}x^2 + \frac{1}{8}x^3 + \frac{1}{16}x^4 + \cdots\right)^{-1} = 1 - \frac{1}{2}x.$$

Thus,

$$[a(x)]^{-1} = x^{-2}\left(1 - \frac{1}{2}x\right) = x^{-2} - \frac{1}{2}x^{-1}.$$

●

As we have already implied, a power series rational function in the canonical form (2.59) may be represented as an extended power series. Specifically, we may identify the quotient (2.59) in the field F((x)) with the extended power series $a(x) \in$ F<x> defined by

$$a(x) = \sum_{k=-n}^{\infty} a_{k+n} x^k. \tag{2.63}$$

Formally, it can be proved that the mapping between the fields F((x)) and F<x> defined by identifying (2.59) with (2.63) is an isomorphism. Thus F<x> is not a new algebraic system but rather it is simply a convenient representation for the quotient field F((x)).

2.10. RELATIONSHIPS AMONG DOMAINS

As we come to the close of this chapter it is appropriate to consider the relationships which exist among the various extensions of polynomial domains which have been introduced.

Given an arbitrary integral domain D, we have introduced univariate domains of polynomials, rational functions, power series, and power series rational functions, denoted respectively by D[x], D(x), D[[x]], and D((x)). Several relationships among these four domains are obvious; for example,

$$D[x] \subset D(x) \subset D((x)), \text{ and}$$

$$D[x] \subset D[[x]] \subset D((x)).$$

The notation $S \subset R$ used here denotes not only that S is a *subset* of R but moreover that S is a *subring*[5] of the ring R. The diagram in Figure 2.1 summarizes these simple relationships. The only pair for which the relationship is unclear is the "diagonal" pair D(x) and D[[x]]. The relationship between rational functions and power series will be considered shortly.

5. If [R; +, ×] is a ring then a subset S of R is a *subring* (more formally, [S; +, ×] is a subring) if S is closed under the ring operations defined on R. (See Chapter 5.)

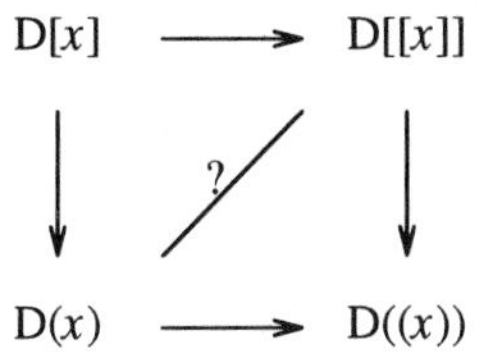

Figure 2.1. Relationships among four domains over an integral domain.

If F_D denotes the quotient field of the integral domain D we may consider, along with the four domains of Figure 2.1, the corresponding domains $F_D[x]$, $F_D(x)$, $F_D[[x]]$, and $F_D((x))$; we also have the field $F_D\langle x\rangle$ of extended power series. These domains satisfy a diagram like that in Figure 2.1. The diagram in Figure 2.2 shows the relationships among the latter domains and also shows their relationships with the domains of Figure 2.1.

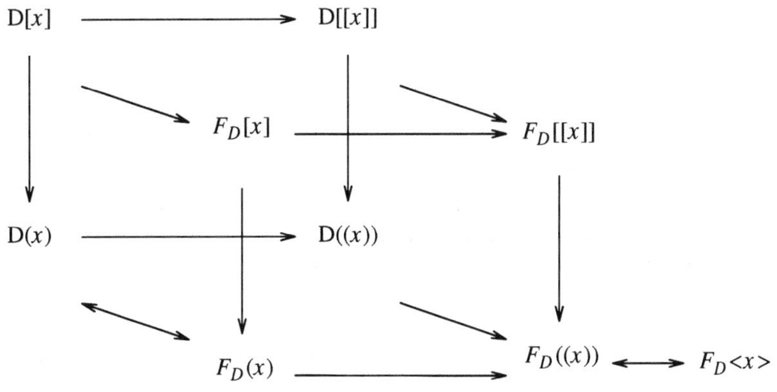

Notation: $\rightarrow$ *denotes subring of;* $\leftrightarrow$ *denotes isomorphism*

Figure 2.2. Relationships among nine domains.

Along with the unspecified relationship noted in Figure 2.1, there are three additional unspecified relationships in Figure 2.2:

(i) $F_D[x]$ _?_ $D[[x]]$;

(ii) $D((x))$ _?_ $F_D[[x]]$;

(iii) $D(x)$ _?_ $F_D[[x]]$.

In order to determine the relationship between a pair of domains A and B, we may consider a larger domain C which contains both of them and pose the question: In the domain C, what is the intersection of the subset A with the subset B? Thus for (i) - (iii) above we may pose the question in the domain $F_D((x))$. Relationship (i) is trivial and uninteresting; namely,

$$\{F_D[x] \cap D[[x]]\} = D[x].$$

Relationship (ii) is a little more complicated; for example,

$$D[[x]] \subset \{D((x)) \cap F_D[[x]]\}$$

and

$$F_D[x] \subset \{D((x)) \cap F_D[[x]]\}$$

but the intersection contains more than just $D[[x]] \cup F_D[x]$. Since the domain $D((x))$ is avoided for computational purposes (by embedding it in $F_D((x))$), relationship (ii) is not of practical interest and will not be pursued further. (See Exercise 2.24.)

Relationship (iii) leads to an interesting pair of questions. In one direction, we are asking under what conditions a rational function $a(x)/b(x) \in D(x)$ can be expressed as a power series $c(x) \in F_D[[x]]$. By putting $a(x)/b(x)$ into the canonical form (2.59) as an element in $F_D((x))$, we see that $a(x)/b(x)$ is a power series in $F_D[[x]]$ if and only if $\mathrm{ord}(b(x)) \le \mathrm{ord}(a(x))$ – i.e. if and only if the rational function $a(x)/b(x) \in D(x)$ has a canonical representative with denominator of order 0. In the other direction, we are asking under what conditions a power series $c(x) \in F_D[[x]]$ can be expressed as a rational function $a(x)/b(x) \in D(x)$. This question is of considerable practical interest because it is asking when an infinite expression (a power series) can be represented by a finite expression (a rational function). By examining the formula for the coefficients in the power series expansion of a rational function, we obtain the following answer. A power series $c(x) = \sum_{k=0}^{\infty} c_k x^k \in F_D[[x]]$ is equal in $F_D((x))$ to a rational function $a(x)/b(x) \in D(x)$ if and only if the c_k's ultimately satisfy a finite linear recurrence; specifically, there must exist nonnegative integers l, n and constants $d_1, d_2, \ldots, d_n \in F_D$ such that

$$c_k = d_1 c_{k-1} + d_2 c_{k-2} + \cdots + d_n c_{k-n} \text{ for all } k > l. \tag{2.64}$$

More specifically, if the power series $c(x)$ satisfies (2.64) then in $F_D((x))$,

$$c(x) = a(x)/(1 - d_1 x - d_2 x^2 - \cdots - d_n x^n)$$

where $\deg(a(x)) \le l$. (Of course, the rational function can be normalized so that its coefficients lie in D since $D(x) = F_D(x)$.)

Let us finally return to the relationship marked by a question mark in Figure 2.1, namely, the relationship between $D(x)$ and $D[[x]]$. In view of the relationship between $D(x)$ and $F_D[[x]]$ stated above, the following statements are easily verified. A rational function $a(x)/b(x) \in D(x)$ can be expressed as a power series $c(x) \in D[[x]]$ if and only if the rational function has a canonical representative in which the constant term of the denominator is a unit in D. A power series $c(x) = \sum_{k=0}^{\infty} c_k x^k \in \mathrm{D}[[x]]$ can be expressed as a rational function $a(x)/b(x) \in \mathrm{D}(x)$ if and only if the c_k's ultimately satisfy a finite linear recurrence of the form

$$d_0 c_k + d_1 c_{k-1} + \cdots + d_n c_{k-n} = 0 \text{ for all } k > l, \quad (2.65)$$

for some nonnegative integers l, n and some constants $d_0, d_1, \ldots, d_n \in \mathrm{D}$. Note that the recurrence (2.65) expressed over D is equivalent to the recurrence (2.64) expressed over $\mathrm{F_D}$.

Exercises

1. Let M denote the set of all 2×2 matrices

$$\begin{bmatrix} a & b \\ c & d \end{bmatrix}$$

with entries $a, b, c, d \in \mathbf{R}$. Verify that the algebraic system [M; +, .], where + and . denote the standard operations of matrix addition and matrix multiplication, is a ring. Give a counter-example to show that [M; +, .] is not a commutative ring.

2. Prove that in any commutative ring, axiom A6 (Cancellation Law) implies and is implied by axiom A6′ (No Zero Divisors).

3. Form addition and multiplication tables for the commutative ring $\mathbf{Z}_6$. Show that $\mathbf{Z}_6$ is not an integral domain by displaying counter-examples for axioms A6 and A6′. Show that $\mathbf{Z}_6$ is not a field by explicitly displaying a counter-example for one of the field axioms.

4. Make a table of inverses for the field $\mathbf{Z}_{37}$. *Hint:* Determine the inverses of 2 and 3, and then use the following law which holds in any field:

$$(xy)^{-1} = x^{-1}y^{-1}.$$

5. Prove that in any integral domain D, elements $c, d \in \mathrm{D}$ are associates if and only if $cu = d$ for some unit u.

6. Prove that in any integral domain D, if $p \in \mathrm{D}$ is a prime then so is any associate of p.

7. Prove that every finite integral domain is a field.

8. The *characteristic* of a field F is the smallest integer k such that $k \cdot f = 0$ for all f $\in$ F. If no such integer exists, then the field is said to have characteristic 0. Let F be a finite field.

   (a) Prove that F must have a prime characteristic.

   (b) If p is the characteristic of F, prove that F is a *vector space* over the field $\mathbf{Z}_p$.

   (c) If n is the dimension of F as a vector space over $\mathbf{Z}_p$, show that F must have p^n elements.

9. The set G of *Gaussian integers* is the subset of the complex numbers **C** defined by

$$G = \{a + b\sqrt{-1} : \ a, b \in \mathbf{Z}\}$$

   (where we usually use the notation $\sqrt{-1} = i$). Verify that G, with the standard operations of addition and multiplication of complex numbers, is an integral domain. Further, verify that G is a Euclidean domain with the valuation

$$v(a + b\sqrt{-1}) = a^2 + b^2.$$

10. Let S be the subset of the complex numbers **C** defined by

$$S = \{a + b\sqrt{-5} : a, b \in \mathbf{Z}\}$$

   (where we may take $\sqrt{-5} = \sqrt{5}i$). Verify that S, with the usual operations, is an integral domain. Prove that the only units in S are 1 and -1.

11. In the integral domain S defined in Exercise 10, show that the element 21 has two different factorizations into primes. *Hint:* For one of the factorizations, let one of the primes be $1 - 2\sqrt{-5}$. Why is this a prime?

12. In the integral domain S defined in Exercise 10, show that the elements 147 and $21 - 42\sqrt{-5}$ have no greatest common divisor. *Hint:* First show that 21 is a common divisor and that $7 - 14\sqrt{-5}$ is a common divisor.

13. (a) Apply Algorithm 2.1 (by hand) to compute, in the Euclidean domain **Z**,

$$g = \text{GCD}\,(3801, 525).$$

   (b) Apply Algorithm 2.2 (by hand) to compute g as in part (a) and thus determine integers s and t such that

$$g = s(3801) + t(525).$$

14. Let F be a field and $a(x)$ a polynomial in F[x]. Define an equivalence relation on F[x] by

$$r(x) \approx s(x) \quad \Leftrightarrow \quad a(x) \mid r(x) - s(x).$$

We write $r(x) \equiv s(x) \bmod a(x)$ when $r(x) \approx s(x)$ and we denote the equivalence class of $r(x)$ by $[r(x)]$.

(a) Show that this indeed defines an equivalence relation on $F[x]$.

(b) Denote the set of equivalence classes by $F[x]/<a(x)>$. Define addition and multiplication operations by

$$[r(x)] + [s(x)] = [r(x) + s(x)], \quad [r(x)]\cdot[s(x)] = [r(x)\cdot s(x)].$$

Show that these arithmetic operations are well defined and that the set of equivalence classes becomes a ring under these operations.

(c) If $a(x)$ is *irreducible* then show that $F[x]/<a(x)>$ forms a field under the operations defined in part (b) (cf. the field $\mathbf{Z}_p$ for a prime p).

If α is a root of $a(x)$ in some field (not in F of course) then we usually write

$$F(\alpha) = F[x]/<a(x)>$$

and call this an *extension* field of F. More specifically, we say that $F(\alpha)$ is the field F extended by α. Thus, for example, the field $\mathbf{Q}(\sqrt{2})$ is the field of rationals extended by the square root of 2, and

$$\mathbf{Q}(\sqrt{2}) = \mathbf{Q}[x]/<x^2 - 2>.$$

15. Let F be a field and $a(x)$ an irreducible polynomial in $F[x]$ of degree n. Show that the field $F[x]/<a(x)>$ is also a *vector space* of dimension n over the field F.

16. Show that, if p is a prime integer and $p(x)$ is an irreducible polynomial in $\mathbf{Z}_p[x]$ of degree m, then $\mathbf{Z}_p[x]/<p(x)>$ is a finite field containing p^m elements. This field is called the *Galois* field of order p^m and is denoted by $GF(p^m)$.

17. (Primitive Element Theorem.) Prove that the multiplicative group of any finite field must be cyclic. The generator of such a group is usually referred to as a *primitive* element of the finite field.

18. (a) Apply Algorithm 2.1 (by hand) to compute, in the Euclidean domain $\mathbf{Q}[x]$,

$$\mathrm{GCD}(4x^4 + 13x^3 + 15x^2 + 7x + 1,\ 2x^3 + x^2 - 4x - 3).$$

(b) Apply Algorithm 2.3 (by hand) to compute, in the Euclidean domain $\mathbf{Q}[x]$, the GCD of the polynomials in part (a).

(c) Apply Algorithm 2.3 (by hand) to compute, in the UFD $\mathbf{Z}[x]$, the GCD of the polynomials in part (a).

19. Apply Algorithm 2.3 (by hand) to compute, in the UFD $\mathbf{Z}[x, y]$,

$$\mathrm{GCD}\,(15xy - 21x - 15y^2 + 21y,\ 6x^2 - 3xy - 3y^2).$$

20. Let F[x] be the univariate Euclidean domain of univariate polynomials over a field F. Let $a_1(x), \ldots, a_r(x)$ be r irreducible, pairwise relatively prime polynomials in F[x]. Define r polynomials $v_1(x), \ldots, v_r(x)$ by

$$v_i(x) = \prod_{j=1,\, j \neq i}^{r} a_j(x).$$

Show that there must exist r unique polynomials $\sigma_i(x)$ with $\deg(\sigma_i(x)) < \deg(a_i(x))$ satisfying

$$\sigma_1(x) \cdot v_1(x) + \cdots + \sigma_r(x) \cdot v_r(x) = 1.$$

21. In the quotient field Q(D) of any integral domain D in which GCD's exist, prove that each equivalence class $[a \,/\, b] \in$ Q(D) has one and only one representative $a \,/\, b$ satisfying properties (2.44) - (2.45) of Section 2.7.

22. (a) In the field $\mathbf{Z}(x)$ of rational functions over $\mathbf{Z}$, let

$$a(x)/b(x) = (1080x^3 - 3204x^2 + 1620x - 900)/(-264x^2 + 348x + 780);$$

$$c(x)/d(x) = (10x^2 - 10)/(165x^2 + 360x + 195).$$

Put $a(x)\,/\,b(x)$ and $c(x)\,/\,d(x)$ into their canonical forms satisfying properties (2.44)-(2.45) of Section 2.7.

(b) Let $a(x)\,/\,b(x)$ and $c(x)\,/\,d(x)$ be the rational functions defined in part (a). Calculate:

$$[a(x)\,/\,b(x)] + [c(x)\,/\,d(x)] \quad \text{and}$$

$$[a(x)\,/\,b(x)][c(x)\,/\,d(x)]$$

and put the results into their canonical forms as elements of the field $\mathbf{Z}(x)$.

(c) What are the canonical forms of the two rational functions in part (a) as elements of the field $\mathbf{Q}(x)$? What are the canonical forms of the sum and product of these two rational functions as elements of $\mathbf{Q}(x)$?

23. Determine the inverse in the power series domain $\mathbf{Z}[[x]]$ of the unit power series

$$a(x) = 1 + x + 2x^2 + 3x^3 + 5x^4 + \cdots$$

where $a_k = a_{k-1} + a_{k-2}$ $(k \geq 2)$. (Note: The sequence $\{a_k\}$ is the famous *Fibonacci sequence.*)

24. (a) In the field $\mathbf{Q}((x))$ of power series rational functions over $\mathbf{Q}$, let

$$\frac{a(x)}{b(x)} = \frac{1 + x + x^2 + x^3 + x^4 + \cdots}{2x^4 + 2x^5 + 4x^6 + 6x^7 + 10x^8 + \cdots}$$

where $b_k = b_{k-1} + b_{k-2}$ $(k \geq 6)$. Put $a(x)/b(x)$ into its canonical form satisfying properties (2.44)-(2.45) of Section 2.7.

(b) Express $a(x)/b(x)$ of part (a) as an extended power series in the field $\mathbf{Q}<x>$.

25. Give a complete specification of the elements in the intersection of the domains $\mathrm{D}((x))$ and $\mathrm{F_D}[[x]]$, as subsets of $\mathrm{F_D}((x))$.

26. Determine a rational function representation in $\mathbf{Z}(x)$ for the following power series in $\mathbf{Z}[[x]]$:

$$c(x) = 1 + x + 2x^2 + 3x^3 + 4x^4 + 5x^5 + \cdots .$$

Hint: Noting that $c_k = k$ does not lead directly to a finite linear recurrence of the form (2.65), use the fact that $k = 2(k-1) - (k-2)$.

References

1. G. Birkoff and S. MacLane, *A Survey of Modern Algebra (3rd ed.),* Macmillian (1965).
2. G. Birkoff and T.C. Bartee, *Modern Applied Algebra,* McGraw-Hill (1970).
3. W.S. Brown, "On Euclid's Algorithm and the Computation of Polynomial Greatest Divisors," *J. ACM,* **18** pp. 476-504 (1971).
4. D.E. Knuth, *The Art of Computer Programming, Volume 2: Seminumerical Algorithms (second edition),* Addison-Wesley (1981).
5. J.D. Lipson, *Elements of Algebra and Algebraic Computing,* Addision-Wesley (1981).
6. B.L. van der Waerden, *Modern Algebra (Vols. I and II),* Ungar (1970).

CHAPTER 3

NORMAL FORMS

AND

ALGEBRAIC REPRESENTATIONS

3.1. INTRODUCTION

This chapter is concerned with the computer representation of the algebraic objects discussed in Chapter 2. The zero equivalence problem is introduced and the important concepts of normal form and canonical form are defined. Various normal forms are presented for polynomials, rational functions, and power series. Finally data structures are considered for the representation of multiprecision integers, rational numbers, polynomials, rational functions, and power series.

3.2. LEVELS OF ABSTRACTION

In Chapter 2 we discussed domains of polynomials, rational functions, and power series in an abstract setting. That is to say, a polynomial (for example) was considered to be a basic object in a domain D[**x**] in the same sense that an integer is considered to be a basic object when discussing the properties of the domain **Z**. The ring operations of + and · were considered to be primitive operations on the objects in the domain under consideration. However when we consider a computer implementation for representing and manipulating the objects in these various domains, we find that there is a great difference in the complexity of the data structures required to represent a polynomial, rational function, or power series compared with the representation of an integer. Also, at the machine level the complexity of the algorithms defining the ring operations is very dependent on the actual objects being manipulated.

While the point of view used in Chapter 2 is too abstract for purposes of understanding issues such as the complexity of polynomial multiplication, the data structure level (where a distinction is made, for example, between a linked list representation and an array representation) is too low-level for convenience. It is useful to consider an intermediate level of abstraction between these two extremes. Three levels of abstraction will be identified as follows:

(i) The *object* level is the abstract level where the elements of a domain are considered to be primitive objects.

(ii) The *form* level is the level of abstraction in which we are concerned with how an object is represented in terms of some chosen "basic symbols", recognizing that a particular object may have many different valid representations in terms of the chosen symbols. For example, at this level we would distinguish between the following different representations of the same bivariate polynomial in the domain $\mathbf{Z}[x,y]$:

$$a(x,y) = 12x^2y - 4xy + 9x - 3; \tag{3.1}$$

$$a(x,y) = (3x - 1)(4xy + 3); \tag{3.2}$$

$$a(x,y) = (12y)x^2 + (-4y + 9)x - 3. \tag{3.3}$$

(iii) The *data structure* level is where we are concerned with the organization of computer memory used in representing an object in a particular form. For example, the polynomial $a(x,y)$ in the form (3.1) could be represented by a linked list consisting of four links (one for each term), or $a(x,y)$ could be represented by an array of length six containing the integers 12, 0, –4, 9, 0, –3 as the coefficients of a bivariate polynomial with implied exponent vectors (2, 1), (2, 0), (1, 1), (1, 0), (0, 1), (0, 0) in that order, or $a(x,y)$ could be represented by some other data structure.

In a high-level computer language for symbolic computation the operations such as + and $\cdot$ will be used as primitive operations in the spirit of the object level of abstraction. However in succeeding chapters we will be discussing algorithms for various operations on polynomials, rational functions, and power series and these algorithms will be described at the form level of abstraction. The next three sections discuss in more detail the various issues of form which arise. Then choices of data structures for each of the above classes of objects will be considered.

3.3. NORMAL FORM AND CANONICAL FORM

The Problem of Simplification

When symbolic expressions are formed and manipulated, there soon arises the general problem of *simplification*. For example, the manipulation of bivariate polynomials might lead to the expression

$$(12x^2y - 4xy + 9x - 3) - (3x - 1)(4xy + 3). \tag{3.4}$$

Comparing with (3.1) and (3.2) it can be seen that the expression (3.4) is the zero polynomial as an object in the domain $\mathbf{Z}[x,y]$. Clearly it would be a desirable property of a system for polynomial manipulation to replace the expression (3.4) by the expression 0 as soon as it is encountered. There are two important aspects to this problem:

(i) a large amount of computer resources (memory space and execution time) may be wasted storing and manipulating unsimplified expressions (indeed a computation may exhaust the allocated computer resources before completion because of the space and time consumed by unsimplified expressions); and

(ii) from a human engineering point of view, we would like results to be expressed in their simplest possible form.

The problem of algorithmically specifying the "simplest" form for a given expression is a very difficult problem. For example, when manipulating polynomials from the domain $\mathbf{Z}[x,y]$ we could demand that all polynomials be fully expanded (with like terms combined appropriately), in which case the expression (3.4) would be represented as the zero polynomial. However consider the expression

$$(x+y)^{1000} - y^{1000}; \tag{3.5}$$

the expanded form of this polynomial will contain a thousand terms and from either the human engineering point of view or computer resource considerations, expression (3.5) would be considered "simpler" as it stands than in expanded form. Similarly, the expression

$$x^{1000} - y^{1000} \tag{3.6}$$

which is in expanded form is "simpler" than a corresponding factored form in which $(x-y)$ is factored out.

Zero Equivalence

The *zero equivalence* problem is the special case of the general simplification problem in which we are concerned with recognizing when an expression is equivalent to zero. This special case is singled out because it is a well-defined problem (whereas "simplification" is not well-defined until an ordering is imposed on the expressions to indicate when one expression is to be considered simpler than another) and also because an algorithm for determining zero equivalence is considered to be a sufficient "simplification" algorithm in some practical situations. However even this well-defined subproblem is a very difficult problem. For example, when manipulating the more general functions to be considered later in this book one might encounter the expression

$$\log(\tan(\frac{x}{2}) + \sec(\frac{x}{2})) - \sinh^{-1}(\frac{\sin(x)}{1+\cos(x)}), \quad -1 \le x \le 1 \tag{3.7}$$

which can be recognized as zero only after very nontrivial transformations. The zero equivalence problem at this level of generality is known to be unsolvable by any algorithm in a "sufficiently rich" class of expressions (cf. Richardson [9]). Fortunately though, the zero equivalence problem can be solved in many classes of expressions of practical interest. In particular the cases of polynomials, rational functions, and power series do not pose any serious difficulties.

Transformation Functions

The simplification problem can be treated in a general way as follows. Consider a set E of expressions and let ~ be an equivalence relation defined on E. Then ~ partitions E into equivalence classes and the quotient set E/~ denotes the set of all equivalence classes. (This terminology has already been introduced in Chapter 2 for the special case where E is the set of quotients of elements from an integral domain.) The simplification problem can then be treated by specifying a transformation f: E $\rightarrow$ E such that for any expression $a \in$ E, the transformed expression f(a) belongs to the same equivalence class as a in the quotient set E/~. Ideally it would be desired that f(a) be "simpler" than a.

In stating the definitions and theorems in this section we will use the symbol $\equiv$ to denote the relation "is identical to" at the form level of abstraction (i.e. identical as strings of symbols). For example, the standard mathematical use of the symbol = denotes the relation "is equal to" at the object level of abstraction so that

$$12x^2y - 4xy + 9x - 3 = (3x - 1)(4xy + 3),$$

whereas the above relation is not true if = is replaced by $\equiv$. In fact the relation = of mathematical equality is precisely the equivalence relation ~ which we have in mind here. (In future sections there will be no confusion in reverting to the more general use of = for both $\equiv$ and ~ since the appropriate meaning will be clear from the context.)

Definition 3.1. Let E be a set of expressions and let ~ be an equivalence relation on E. A *normal function* for [E; ~] is a computable function f: E $\rightarrow$ E which satisfies the following properties:

(i) $\quad$ f(a) ~ a $\quad$ for all $a \in$ E;

(ii) $\quad$ $a \sim 0 \;\Rightarrow\;$ f(a) $\equiv$ f(0) for all $a \in$ E.

●

Definition 3.2. Let [E; ~] be as above. A *canonical function* for [E; ~] is a normal function f: E $\rightarrow$ E which satisfies the additional property:

(iii) $\quad$ $a \sim b \;\Rightarrow\;$ f(a) $\equiv$ f(b) $\;$ for all $a,b \in$ E.

●

Definition 3.3. If f is a normal function for [E; ~] then an expression $\tilde{a} \in$ E is said to be a *normal form* if f($\tilde{a}$) $\equiv \tilde{a}$. If f is a canonical function for [E; ~] then an expression $\tilde{a} \in$ E is said to be a *canonical form* if f($\tilde{a}$) $\equiv \tilde{a}$.

●

Example 3.1. Let E be the domain **Z**[x] of univariate polynomials over the integers. Consider the normal functions f_1 and f_2 specified as follows:

f_1: (i) $\quad$ multiply out all products of polynomials;

(ii) collect terms of the same degree.

f_2: (i) multiply out all products of polynomials;

(ii) collect terms of the same degree;

(iii) rearrange the terms into descending order of their degrees.

Then f_1 is a normal function which is not a canonical function and f_2 is a canonical function. A normal form for polynomials in $\mathbf{Z}[x]$ corresponding to f_1 is

$$a_1x^{e_1} + a_2x^{e_2} + \cdots + a_mx^{e_m} \quad \text{with } e_i \neq e_j \text{ when } i \neq j.$$

A canonical form for polynomials in $\mathbf{Z}[x]$ corresponding to f_2 is

$$a_1x^{e_1} + a_2x^{e_2} + \cdots + a_mx^{e_m} \quad \text{with } e_i < e_j \text{ when } i > j.$$

●

It is obvious that in a class E of expressions for which a normal function has been defined, the zero equivalence problem is solved. However there is not a unique normal form for all expressions in a particular equivalence class in E/~ unless the equivalence class contains 0. A canonical form, in contrast, provides a unique representative for each equivalence class in E/~, as the following theorem proves.

Theorem 3.1. If f is a canonical function for [E; ~] then the following properties hold:

(i) f is idempotent (i.e. $f \circ f = f$ where $\circ$ denotes composition of functions);

(ii) $f(a) \equiv f(b)$ if and only if $a \sim b$;

(iii) in each equivalence class in E/~ there exists a unique canonical form.

Proof:

(i) $f(a) \sim a$ for all $a \in E$, by Definition 3.1 (i)

$\Rightarrow$ $f(f(a)) \equiv f(a)$ for all $a \in E$, by Definition 3.2 (iii).

(ii) "if": This holds by Definition 3.2 (iii).
"only if": Let $f(a) \equiv f(b)$. Then

$$a \sim f(a) \equiv f(b) \sim b \quad \text{by Definition 3.1 (i)}$$

$$\Rightarrow \quad a \sim b.$$

(iii) "Existence": Let a be any element of a particular equivalence class. Define $\tilde{a} \equiv f(a)$. Then

$$\begin{aligned} f(\tilde{a}) &\equiv f(f(a)) \\ &\equiv f(a), \text{ by idempotency} \\ &\equiv \tilde{a}. \end{aligned}$$

"Uniqueness": Suppose $\tilde{a}_1$ and $\tilde{a}_2$ are two canonical forms in the same equivalence class in E/~. Then

$$\begin{aligned} &\tilde{a}_1 \sim \tilde{a}_2 \\ \Rightarrow\ & \mathrm{f}(\tilde{a}_1) \equiv \mathrm{f}(\tilde{a}_2) \text{ by Definition 3.2 (iii)} \\ \Rightarrow\ & \tilde{a}_1 \equiv \tilde{a}_2 \text{ by definition of canonical form.} \end{aligned}$$

●

3.4. NORMAL FORMS FOR POLYNOMIALS

Multivariate Polynomial Representations

The problem of representing multivariate polynomials gives rise to several important issues at the form level of abstraction. One such issue was briefly encountered in Chapter 2, namely the choice between recursive representation and distributive representation. In the *recursive representation* a polynomial $a(x_1, \ldots, x_\nu) \in D[x_1, \ldots, x_\nu]$ is represented as

$$a(x_1, \ldots, x_\nu) = \sum_{i=0}^{\deg_1(a(\mathbf{x}))} a_i(x_2, \ldots, x_\nu)\, x_1^i$$

(i.e. as an element of the domain $D[x_2, \ldots, x_\nu][x_1]$) where, recursively, the polynomial coefficients $a_i(x_2, \ldots, x_\nu)$ are represented as elements of the domain $D[x_3, \ldots, x_\nu][x_2]$, and so on so that ultimately the polynomial $a(x_1, \ldots, x_\nu)$ is viewed as an element of the domain $D[x_\nu][x_{\nu-1}] \cdots [x_1]$. An example of a polynomial from the domain $\mathbf{Z}[x,y,z]$ expressed in the recursive representation is:

$$a(x,y,z) = (3y^2+(-2z^3)y+5z^2)x^2 + 4x + ((-6z+1)y^3+3y^2+(z^4+1)). \tag{3.8}$$

In the *distributive representation* a polynomial $a(\mathbf{x}) \in D[\mathbf{x}]$ is represented as

$$a(\mathbf{x}) = \sum_{\mathbf{e} \in \mathbf{N}^\nu} a_{\mathbf{e}} \mathbf{x}^{\mathbf{e}}$$

where $a_{\mathbf{e}} \in D$. For example, the polynomial $a(x,y,z) \in \mathbf{Z}[x,y,z]$ given in (3.8) could be expressed in the distributive representation as

$$a(x,y,z) = 3x^2y^2-2x^2yz^3+5x^2z^2+4x-6y^3z+y^3+3y^2+z^4+1. \tag{3.9}$$

Another representation issue which arises is the question of sparse versus dense representation. In the *sparse representation* only the terms with nonzero coefficients are represented while in the *dense representation* all terms which could possibly appear in a polynomial of the specified degree are represented (whether or not some of these terms have zero coefficients in a specific case). For example, a natural representation of univariate polynomials $\sum_{i=0}^{n} a_i x^i \in \mathbf{Z}[x]$ of specified maximum degree n using arrays is to store the $(n+1)$-array $(a_0, \ldots, a_n)$; this is a dense representation since zero coefficients will be explicitly stored. While it is quite possible to generalize this example to obtain a corresponding dense representation for multivariate polynomials (e.g. by imposing the lexicographical ordering of exponent vectors), such a representation is highly impractical. For if a particular computation involves the manipulation of polynomials in v indeterminates with maximum degree d in each indeterminate then the number of coefficients which must be stored for each specific polynomial is $(d+1)^v$. It is not an uncommonly large problem to have, for example, $v = 5$ and $d = 15$ in which case each polynomial requires the storage of over a million coefficients. In a practical problem with $v = 5$ and $d = 15$ most of the million coefficients will be zero, since otherwise the computation being attempted is beyond the capacity of present-day computers. (And who would be interested in looking at an expression containing one million terms!) All of the major systems for symbolic computation therefore use the sparse representation for multivariate polynomials.

A similar choice must be made about storing zero exponents (i.e. whether or not to store them). When we express polynomials such as (3.8) and (3.9) on the written page we do not explicitly write those monomials which have zero exponents (just as we naturally use the sparse representation of polynomials on the written page). However the penalty in memory space for choosing to store zero exponents is not excessive while such a choice may result in more efficient algorithms for manipulating polynomials. Some computer algebra systems store zero exponents and others do not.

The polynomial representation issues discussed so far do not address the specification of normal or canonical forms. While all of these issues can be considered at the form level of abstraction, the three issues discussed above involve concepts that are closer to the data structure level than the issue of normal/canonical forms. A hierarchy of the levels of abstraction of these various representation issues is illustrated in Figure 3.1.

Definitions of Normal Forms

Definition 3.4. An *expanded normal form* for polynomial expressions in a domain $D[x_1, \ldots, x_v]$ can be specified by the normal function

f_1: (i) multiply out all products of polynomials;

(ii) collect terms of the same degree.

An *expanded canonical form* for polynomial expressions in a domain $D[x_1, \ldots, x_v]$ can be specified by the canonical function

f_2: apply f_1, then

(iii) rearrange the terms into descending order of their degrees.

●

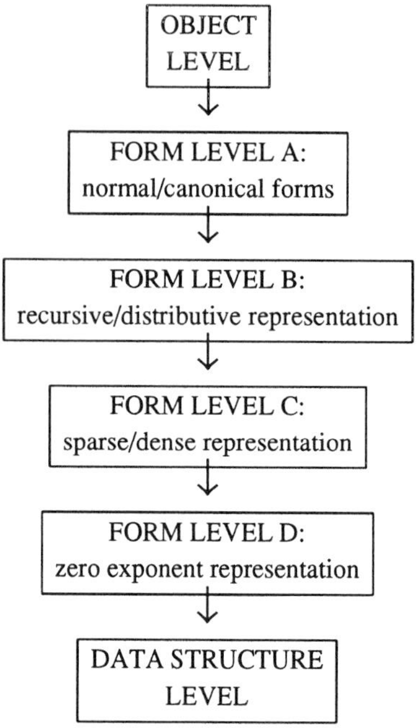

Figure 3.1.
Levels of abstraction for multivariate polynomial representations.

Definition 3.5. A *factored normal form* for polynomial expressions in a domain $D[x_1, \ldots, x_v]$ can be specified by the normal function

f_3: if the expression is in the product form $\prod_{i=1}^{k} p_i$, $p_i \in D[x_1, \ldots, x_v]$ for $i = 1,2, \ldots, k$, where no p_i is itself in product form, then replace the expression by $\prod_{i=1}^{k} f_2(p_i)$ where f_2 is the canonical function defined in Definition 3.4 and where the latter product is understood to be zero if any of its factors is zero.

A *factored canonical form* for polynomial expressions in a domain $D[x_1, \ldots, x_v]$ (assuming that D is a UFD) can be specified by the canonical function

f_4: apply f_3 and if the result is nonzero then factorize each $f_2(p_i)$ into its unit normal factorization (according to Definition 2.13) and collect factors to obtain the unit normal factorization of the complete expression (made unique by imposing a pre-specified ordering on the factors).

●

In an implementation of the transformation functions f_1 and f_2 of Definition 3.4, the concept of "degree" means "exponent vector" if the distributive representation is used and it means "univariate degree" applied to each successive level of recursion if the recursive representation is used. Note also that in the specification of the transformation function f_3 of Definition 3.5, the canonical function f_2 could be replaced by the normal function f_1 and f_3 would still be a valid normal function. However the normal function f_3 as specified is more often used in practical systems. Finally note that the factors p_i $(1 \le i \le k)$ appearing in Definition 3.5 are not necessarily distinct (i.e. there may be some repeated factors).

Example 3.2. The polynomial $a(x,y,z)$ in (3.8) is expressed in expanded canonical form using the recursive representation for the domain $\mathbf{Z}[x,y,z]$. The same polynomial is expressed in (3.9) in expanded canonical form using the distributive representation for the domain $\mathbf{Z}[x,y,z]$.

●

Example 3.3. Let $a(x,y) \in \mathbf{Z}[x,y]$ be the expression

$$a(x,y) = ((x^2 - xy + x) + (x^2 + 3)\,(x - y + 1)) \cdot$$
$$((y^3 - 3y^2 - 9y - 5) + x^4\,(y^2 + 2y + 1)).$$

Using the distributive representation for writing polynomials, an expanded normal form obtained by applying f_1 of Definition 3.4 to $a(x,y)$ might be (depending on the order in which the multiplication algorithm produces the terms):

$$f_1(a(x,y)) = 5x^2y^3 + 3x^2y^2 - 13x^2y - 10x^2 + 3x^6y + 2x^6 - xy^4 + 7xy^3$$
$$- 3xy^2 - 31xy - x^5y^3 + 2x^5y^2 + 7x^5y - 20x + 4x^5 + x^3y^3$$
$$- 3x^3y^2 - 9x^3y - 5x^3 + x^7y^2 + 2x^7y + x^7 - x^2y^4 - x^6y^3$$
$$- 3y^4 + 12y^3 + 18y^2 - 12y - 3x^4y^3 - 3x^4y^2 + 3x^4y - 15 + 3x^4.$$

The expanded canonical form obtained by applying f_2 of Definition 3.4 to $a(x,y)$ is

$$f_2\,(a(x,y)) = x^7y^2 + 2x^7y + x^7 - x^6y^3 + 3x^6y + 2x^6 - x^5y^3 + 2x^5y^2$$
$$+ 7x^5y + 4x^5 - 3x^4y^3 - 3x^4y^2 + 3x^4y + 3x^4 + x^3y^3 - 3x^3y^2$$
$$- 9x^3y - 5x^3 - x^2y^4 + 5x^2y^3 + 3x^2y^2 - 13x^2y - 10x^2 - xy^4$$
$$+ 7xy^3 - 3xy^2 - 31xy - 20x - 3y^4 + 12y^3 + 18y^2 - 12y - 15.$$

Applying, respectively, f_3 and f_4 of Definition 3.5 to $a(x,y)$ yields the factored normal form

$$f_3(a(x,y)) = (x^3 - x^2y + 2x^2 - xy + 4x - 3y + 3) \cdot$$
$$(x^4y^2 + 2x^4y + x^4 + y^3 - 3y^2 - 9y - 5)$$

and the factored canonical form

$$f_4(a(x,y)) = (x - y + 1)\,(x^2 + x + 3)\,(x^4 + y - 5)\,(y + 1)^2 .$$

●

Some Practical Observations

The normal and canonical functions f_1, f_2, f_3, and f_4 of Definitions 3.4-3.5 are not all practical functions to implement in a system for symbolic computation. Specifically the canonical function f_4 (factored canonical form) is rarely used in a practical system because polynomial factorization is a relatively expensive operation (see Chapter 8). On the other hand, one probably would not choose to implement the normal function f_1 since the canonical function f_2 requires little additional cost and yields a canonical (unique) form. One therefore finds in several systems for symbolic computation a variation of the canonical function f_2 (expanded canonical form) and also a variation of the normal function f_3 (factored normal form), usually with some form of user control over which "simplification" function is to be used for a given computation. It can be seen from Example 3.3 that the normal function f_3 might sometimes be preferable to the canonical function f_2 because it may leave the expression in a more compact form, thus saving space and also possibly saving execution time in subsequent operations. On the other hand, the canonical function f_2 would "simplify" the expression

$$\begin{aligned} a(x,y) = (x - y)\,(&x^{19} + x^{18}y + x^{17}y^2 + x^{16}y^3 + x^{15}y^4 + x^{14}y^5 \\ &+ x^{13}y^6 + x^{12}y^7 + x^{11}y^8 + x^{10}y^9 + x^9y^{10} + x^8y^{11} + x^7y^{12} \\ &+ x^6y^{13} + x^5y^{14} + x^4y^{15} + x^3y^{16} + x^2y^{17} + xy^{18} + y^{19}) \end{aligned}$$

into the expression

$$f_2(a(x,y)) = x^{20} - y^{20}$$

while the normal function f_3 would leave $a(x,y)$ unchanged. Finally it should be noted that both f_2 and f_3 would transform the expression (3.5) into an expression containing a thousand terms and therefore it is desirable to also have in a system a weaker "simplifying" function which would not apply any transformation to an expression like (3.5). The latter type of transformation function would be neither a canonical function nor a normal function.

3.5. NORMAL FORMS FOR RATIONAL FUNCTIONS AND POWER SERIES

Rational Functions

Recall that a field $D(x_1, \ldots, x_v)$ of rational functions is simply the quotient field of a polynomial domain $D[x_1, \ldots, x_v]$. The choice of normal forms for rational functions therefore follows quite naturally from the polynomial forms that are chosen. The general concept of a canonical form for elements in a quotient field was defined in Section 2.7 by conditions

(2.44)-(2.46), which becomes the following definition for the case of rational functions if we choose the expanded canonical form of Definition 3.4 for the underlying polynomial domain. (We will assume that D is a UFD so that GCD's exist.)

Definition 3.6. An *expanded canonical form* for rational expressions in a field $D(x_1, \ldots, x_v)$ can be specified by the canonical function

f_5: (i) [form common denominator] put the expression into the form a/b where $a, b \in D[x_1, \ldots, x_v]$ by performing the arithmetic operations according to equations (2.42)-(2.43);

(ii) [satisfy condition (2.44): remove GCD] compute $g = \mathrm{GCD}(a,b) \in D[x_1, \ldots, x_v]$ (e.g. by using Algorithm 2.3) and replace the expression a/b by a'/b' where $a = a'g$ and $b = b'g$;

(iii) [satisfy condition (2.45): unit normalize] replace the expression a'/b' by a''/b'' where $a'' = a' \cdot (\mathrm{u}(b'))^{-1}$ and $b'' = b' \cdot (\mathrm{u}(b'))^{-1}$;

(iv) [satisfy condition (2.46): make polynomials canonical] replace the expression a''/b'' by $f_2(a'')/f_2(b'')$ where f_2 is the canonical function of Definition 3.4.

●

It is not made explicit in the above definition of canonical function f_5 whether or not some normal or canonical function would be applied to the numerator and denominator polynomials (a and b) computed in step (i). It might seem that in order to apply Algorithm 2.3 in step (ii) the polynomials a and b need to be in expanded canonical form, but as a practical observation it should be noted that if instead a and b are put into factored normal form (for example) then step (ii) can be carried out by applying Algorithm 2.3 separately to the various factors.

As in the case of polynomials, it can be useful to consider non-canonical normal forms for rational functions (and indeed more general forms which are neither canonical nor normal). We will not set out formal definitions of normal forms for rational expressions but several possible normal forms can be outlined as follows:

factored/factored: numerator and denominator both in factored normal form;

factored/expanded: numerator in factored normal form and denominator in expanded canonical form;

expanded/factored: numerator in expanded canonical form and denominator in factored normal form.

In this notation the expanded canonical form of Definition 3.6 would be denoted as *expanded/expanded.* In the above we are assuming that conditions (2.44) and (2.45) are satisfied but that condition (2.46) is not necessarily satisfied. Noting that to satisfy condition (2.44) (i.e. to remove the GCD of numerator and denominator) requires a (relatively expensive) GCD computation, it can be useful to consider four more normal forms for rational functions obtained from the above four numerator/denominator combinations with the additional stipulation that condition (2.44) is not necessarily satisfied.

Among these various normal forms one that has been found to be particularly useful for the efficient manipulation of rational expressions is the *expanded/factored* normal form, with condition (2.44) satisfied by an efficient scheme for GCD computation which exploits the presence of explicit factors whenever an arithmetic operation is performed. One of the original computer algebra systems, ALTRAN, used just such a choice as its the default mode, with the other normal forms available by user specification. Finally noting that step (i) of function f_5 (Definition 3.6) is itself nontrivial, a weaker "simplifying" function normally chooses to leave expressions in a less transformed state yielding neither a canonical nor a normal form (until the user requests a "rational canonicalization").

Power Series: The TPS Representation

The representation of power series poses the problem of finding a finite representation for an infinite expression. Obviously we cannot represent a power series in a form directly analogous to the expanded canonical form for polynomials because there are an infinite number of terms. One common solution to this problem is to use the *truncated power series* (TPS) representation in which a power series

$$a(x) = \sum_{i=0}^{\infty} a_k x^k \quad \in \mathrm{D}[[x]] \tag{3.10}$$

is represented as

$$\sum_{k=0}^{t} a_k x^k \tag{3.11}$$

where t is a specified *truncation degree*. Thus only a finite number of terms are actually represented and a TPS such as (3.11) looks exactly like a polynomial. However a distinction must be made between a polynomial of degree t and TPS with truncation degree t because the results of arithmetic operations on the two types of objects are not identical. In order to make this distinction it is convenient to use the following notation for the TPS representation (with truncation degree t) of the power series (3.10):

$$a(x) = \sum_{k=0}^{t} a_k x^k + \mathrm{O}(x^{t+1}), \tag{3.12}$$

where in general the expression $\mathrm{O}(x^p)$ denotes an unspecified power series $\alpha(x)$ with $\mathrm{ord}(\alpha(x)) \geq p$.

The non-exact nature of the TPS representation of power series poses a problem when we consider normal forms for power series. For if in Definition 3.1 we consider the set E of expressions to be the set of all (infinite) power series then the transformation performed by representing a power series in its TPS representation (with specified truncation degree t) violates the first property of a normal function. Specifically, two power series which are not equivalent will be transformed into the same TPS representation if they happen to have identical coefficients through degree t. On the other hand, we can take a more practical point of view and consider the set E_t of all TPS expressions of the form (3.12) with specified truncation degree t. Since we are only considering univariate power series domains with the most general coefficient domain being a field of rational functions, it follows immediately that (3.12) is a normal form for the set E_t if we choose a normal form for the coefficients and it is a canonical form for the set E_t if we choose a canonical form for the coefficients.

Power Series: Non-Truncated Representations

The TPS representation is not the only approach by which an infinite power series can be finitely represented. There are representations which are both finite and exact. For example, the Taylor series expansion about the point $x = 0$ of the function e^x might be written as

$$\sum_{k=0}^{\infty} \frac{1}{k!} x^k \tag{3.13}$$

which is an exact representation of the complete infinite series using only a finite number of symbols. The form (3.13) is a special instance of what we shall name the *non-truncated power series* (NTPS) representation of a power series $a(x) \in D[[x]]$ which takes the general form

$$a(x) = \sum_{k=0}^{\infty} \mathrm{f}_a(k) x^k, \tag{3.14}$$

where $\mathrm{f}_a(k)$ is a specified *coefficient function* (defined for all nonnegative integers k) which computes the k-th coefficient. By representing the coefficient function $\mathrm{f}_a(k)$, the infinite power series $a(x)$ is fully represented. The fact that only a finite number of coefficients could ever be explicitly computed is in this way separated from the representation issue and, unlike the TPS representation, there is no need to pre-specify the maximum number of coefficients which may eventually be explicitly computed.

If $a(x)$ is the power series (3.13) then the coefficient function can be specified by

$$\mathrm{f}_a(k) := \frac{1}{k!} . \tag{3.15}$$

In a practical implementation of the NTPS representation it would be wise to store coefficients that are explicitly computed so that they need not be re-computed when and if they are required again later in a computation. Thus at a particular point in a computation if the first l coefficients for $a(x)$ have previously been explicitly computed and stored in a linear list $\mathbf{a} = (a_0, a_1, \ldots, a_{l-1})$ then the specification of the coefficient function should be changed from (3.15) to

$$\mathrm{f}_a(k) := \textbf{if } k < l \textbf{ then } \mathbf{a}[k] \textbf{ else } \frac{1}{k!} \tag{3.16}$$

where $\mathbf{a}[k]$ denotes an element access in the linear list $\mathbf{a}$. Initially $l = 0$ in specification (3.16) and in general l is the current length of $\mathbf{a}$. It can also be seen that from the point of view of computational efficiency it might be better to change specification (3.16) to

$$\mathrm{f}_a(k) := \textbf{if } k = 0 \textbf{ then } 1$$

$$\textbf{else if } k < l \textbf{ then } \mathrm{a}[k] \textbf{ else } \frac{\mathrm{f}_a(k-1)}{k} \tag{3.17}$$

where the recurrence $a_k = \dfrac{a_{k-1}}{k}$ (for $k > 0$) will be used to compute successive coefficients.

The specification of the coefficient function can become even more complex than indicated above. For example if $a(x)$ and $b(x)$ are two power series with coefficient functions $\mathrm{f}_a(k)$ and $\mathrm{f}_b(k)$ then the sum

$$c(x) = a(x) + b(x)$$

can be specified by the coefficient function

$$\mathrm{f}_c(k) := \mathrm{f}_a(k) + \mathrm{f}_b(k) \tag{3.18}$$

and the product

$$d(x) = a(x)\,b(x)$$

can be specified by the coefficient function

$$\mathrm{f}_d(k) := \sum_{i=0}^{k} \mathrm{f}_a(i)\,\mathrm{f}_b(k-i). \tag{3.19}$$

If $\mathrm{f}_a(k)$ and $\mathrm{f}_b(k)$ are explicit expressions in k then the coefficient function $\mathrm{f}_c(k)$ in (3.18) can be expressed as an explicit expression in k but the coefficient function $\mathrm{f}_d(k)$ in (3.19) cannot in general be simplified to an explicit expression in k.

The problem of specifying normal forms or canonical forms for the NTPS representation has not received any attention in the literature. A practical implementation of the NTPS representation has been described by Norman [8], and is implemented in the SCRATCHPAD system. The NTPS representation is seen to offer some advantages over the TPS representation. However the question of normal forms is left at the TPS level in the sense that the objects ultimately seen by the user are TPS representations, and we have already seen that TPS normal or canonical forms are readily obtained. While a true normal form for the NTPS representation in its most general form is impossible (because such a normal form would imply a solution to the zero equivalence problem for a very general class of expressions), it would be of considerable practical interest to have a canonical form for some reasonable subset of all possible coefficient function specifications. (For example, see Exercises 3.7 and 3.8 for some special forms of coefficient function specifications which can arise.)

Extended Power Series

Recall that a field F<x> of extended power series over a coefficient field F can be identified with the quotient field F((x)) of a power series domain F[[x]]. It was shown in Section 2.9 that a canonical form for the quotient field F((x)) satisfying conditions (2.44)-(2.46) takes the form

$$\frac{a(x)}{x^n} \tag{3.20}$$

where $a(x) \in$ F[[x]] and $n \geq 0$. Thus normal and canonical forms for extended power series are obtained directly from the forms chosen for representation of ordinary power series. The representation of an extended power series can be viewed as the representation of an ordinary power series plus an additional piece of information specifying the value of n in (3.20).

3.6. DATA STRUCTURES FOR MULTIPRECISION INTEGERS AND RATIONAL NUMBERS

We turn now to the data structure level of abstraction. Before discussing data structures for polynomials in a domain D[**x**] or rational functions in a field D(**x**), it is necessary to determine what data structures will be used for the representation of objects in the coefficient domain D. We consider two possible choices for D: the integral domain **Z** of integers and the field **Q** of rational numbers.

Multiprecision Integers

A typical digital computer has hardware facilities for storing and performing arithmetic operations upon a basic data type which is usually called "integer" and which we shall call *single-precision integer*. The range of values for a single-precision integer is limited by the number of distinct encodings that can be made in the computer word, which is typically 8, 16, 32, 36, 48, or 64 bits in length. Thus the value of a signed single-precision integer cannot exceed about 9 or 10 decimal digits in length for the middle-range word sizes listed above or about 19 decimal digits for the largest word size listed above. These restricted representations of objects in the integral domain **Z** are not sufficient for the purposes of symbolic computation.

A more useful representation of integers can be obtained by imposing a data structure on top of the basic data type of "single-precision integer". A *multiprecision integer* is a linear list $(d_0, d_1, \ldots, d_{l-1})$ of single-precision integers and a sign, s, which can take on the value plus or minus one. This represents the value

$$s \sum_{i=0}^{l-1} d_i \beta^i$$

where the base β has been pre-specified. The sign is usually stored within the list $(d_0, \ldots, d_{l-1})$, possibly as the sign of d_0 or one or more of the other entries.

The base β could be, in principle, any positive integer greater than 1 such that $\beta-1$ is a single-precision integer, but for efficiency β would be chosen to be a *large* such integer. Two common choices for β are (i) β such that $\beta-1$ is the largest positive single-precision integer (e.g. $\beta = 2^{31}$ if the (signed) word size is 32 bits), and (ii) $\beta = 10^p$ where p is chosen as large as possible such that $\beta-1$ is a single-precision integer (e.g. $\beta = 10^9$ if the word size is 32 bits). The *length* l of the linear list used to represent a multiprecision integer may be dynamic (i.e. chosen appropriately for the particular integer being represented) or static (i.e. a pre-specified fixed length), depending on whether the linear list is implemented using linked allocation or using array (sequential) allocation.

Linked List Representation and Array Representation

One common method of implementing the linear list data structure for multiprecision integers uses a linked list where each *node* in the linked list is of the form

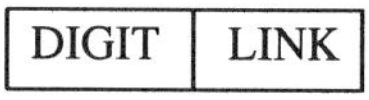

The DIGIT field contains one base-β digit (a single-precision integer) and the LINK field contains a pointer to the next node in the linked list (or an "end of list" pointer). Thus the multiprecision integer $d = (d_0, d_1, \ldots, d_{l-1})$ with value

$$d = s \sum_{i=0}^{l-1} d_i \beta^i \tag{3.21}$$

is represented by the linked list

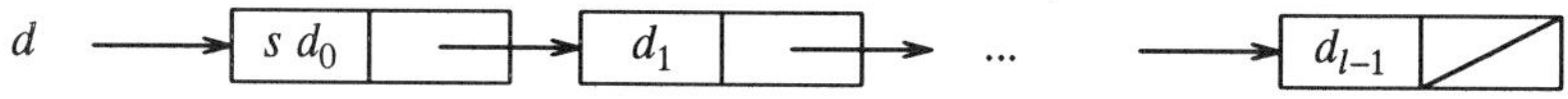

where the sign has been stored with d_0. Note that the order in which the β-digits d_i are linked is in reverse order compared with the conventional way of writing numbers. For example if $\beta = 10^3$ then the decimal number

$$N = 1234567890 \tag{3.22}$$

is represented by the linked list

This ordering corresponds to the natural way of writing the base-β expansion of a number as in (3.21) and, more significantly, it corresponds to the order in which the digits are accessed when performing the operations of addition and multiplication on integers.

A second possible implementation for multiprecision integers is *dynamic array allocation*. In this scheme a multi-precision integer is stored as a variable length array containing the length of the integer in computer words, and the digits of the integer. In such a scheme, the sign s is typically encoded in the header word along with the length l

$s\ l$	d_0	d_1		...		d_{l-1}

For example if $\beta = 10^3$ then the number N in (3.22) would be represented as

$N \longrightarrow$

4	890	567	234	1

A third possibility for implementing linear lists is to use fixed-length arrays. In this scheme the length l of the allowable multiprecision integers is a pre-specified constant and every multiprecision integer is allocated an array of length l (i.e. l sequential words in the computer memory). Thus the multiprecision integer d in (3.21) is represented by the array

$s\ d_0$	d_1		...		d_{l-1}

where it should be noted that every multiprecision integer must be expressed using l β-digits (by artificially introducing zeros for the high-order terms in (3.21) if necessary). For example if $\beta = 10^3$ and $l = 10$ then integers not exceeding 30 decimal digits in length can be represented and the particular decimal number N in (3.22) is represented by the array

$N \longrightarrow$

890	567	234	1	0	0	0	0	0	0

Advantages and Disadvantages

There are several factors affecting the choice of internal representation for integers. The main disadvantage of a fixed-length array representation is the requirement that the length l be pre-specified. This leads to two significant disadvantages: (i) a decision must be made as to the maximum length of integers that will be required (with a system failure occurring whenever this maximum is exceeded); and (ii) a considerable amount of memory space is wasted storing high-order zero digits (with a corresponding waste in processor time accessing irrelevant zero digits). Linked list representation and dynamic array representation both avoid these problems since irrelevant high-order zero digits are not represented and the length of the list is limited only by the total storage pool available to the system.

On the other hand the use of linked lists also involves at least two disadvantages: (i) a considerable amount of memory space is required for the pointers; and (ii) the processing time required to access successive digits is significantly higher than for array accesses. These two disadvantages would seem to be especially serious for this particular application of linked lists because the need for pointers could potentially double the amount of memory used by multiprecision integers and also because the digits of an integer will *always* be stored and accessed in sequence. Dynamic array representation uses less storage and has faster processing time for access to digits. However this method requires a sophisticated storage management algorithm.

The advantage of *indefinite-precision integers* (i.e. multiprecision integers with dynamically determined length l) makes the first two allocation schemes the only practical methods for modern computer algebra systems. The LISP-based systems such as MACSYMA, REDUCE, and SCRATCHPAD all use the multiprecision integers supported by the LISP systems upon which they are built, which in most cases is linked list representation. MAPLE, which is C-based, is an example of a system which uses dynamic array representation. The older system, ALTRAN, is an example of a system which used fixed-length arrays.

Rational Numbers

The field **Q** of rational numbers is the quotient field of the integral domain **Z** of integers. A natural representation for rational numbers is therefore the pair (*numerator*, *denominator*) where each of *numerator* and *denominator* is a multiprecision integer. The basic data structure is a list of length two each element of which is itself a linear list. The representation is made canonical by imposing conditions (2.44) - (2.45) of Section 2.8 (condition (2.46) is automatic since the representation for multiprecision integers will be unique).

If multiprecision integers are represented by either linked allocation or array allocation, a rational number can be represented by a node

LINK1	LINK2

where LINK1 is a pointer to the numerator multiprecision integer (either a linked list or an array) and similarly LINK2 is a pointer to the denominator. In the case of array allocation for multiprecision integers, it is also possible to represent a rational number by a two-dimensional array (e.g. with l rows and 2 columns) since the length l of the numerator and denominator is a fixed constant.

3.7. DATA STRUCTURES FOR POLYNOMIALS, RATIONAL FUNCTIONS, AND POWER SERIES

Relationships between Form and Data Structure

The data structures used to represent multivariate polynomials in a particular system influence (or conversely, are influenced by) some of the choices made at the form level of abstraction. Referring to the hierarchy illustrated in Figure 3.1 of Section 3.4, the choice made at form level A (normal/canonical forms) is independent of the basic data structure to be used. At form level B the choice between the recursive representation and the distributive representation is in practice closely related to the choice of basic data structure. The recursive representation is the common choice in systems using a linked list data structure while the distributive representation is found in systems using an array data structure. (Note however that these particular combinations of choice at form level B and the data structure level are not the only possible combinations.) At form level C the sparse representation is the choice in all of the major systems for reasons previously noted and this fact is reflected in the details of the data structure. The choice at form level D regarding the representation of zero exponents is more variable among systems.

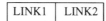

In this section we describe three possible data structures for multivariate polynomials. The first is a linked list data structure using the recursive, sparse representation. The second is a dynamic array data structure using the distributive, sparse representation. The third structure is used in ALTRAN and is referred to as a *descriptor block* implementation. We describe these data structures as they apply to multivariate polynomials in expanded canonical form. Then we describe the additional structure which can be imposed on these basic data structures to allow for the implementation of the factored normal form.

A Linked List Data Structure

Using the recursive representation of multivariate polynomials in expanded canonical form, a polynomial domain $D[x_1, \ldots, x_v]$ is viewed as the domain $D[x_2, \ldots, x_v][x_1]$ and this view is applied recursively to the "coefficient domain" $D[x_2, \ldots, x_v]$. With this point of view, a polynomial $a(x_1, \ldots, x_v) \in D[x_1, \ldots, x_v]$ is considered at the "highest level" to be a univariate polynomial in x_1 and it can be represented using a linked list where each node in the linked list is of the form

COEF_LINK	EXPONENT	NEXT_LINK

Each such node represents one polynomial term $a_i x_1^i$ with $a_i \in D[x_2, \ldots, x_v]$, where the EXPONENT field contains the value i (as a single-precision integer), the COEF_LINK field contains a pointer to the coefficient a_i of x_1^i, and the NEXT_LINK field contains a pointer to the next term in the polynomial (or an "end of list" pointer). This representation is applied recursively. In order to know the name of the indeterminate being distinguished at each level of this recursive representation, we can use a "header node"

INDET_LINK	FIRST_LINK

where the INDET_LINK field contains a pointer to the name of the indeterminate and the FIRST_LINK field contains a pointer to the first term in the polynomial (at this specific level of recursion).

Example 3.4. Let $a(x,y,z) \in \mathbf{Z}[x,y,z]$ be the polynomial

$$a(x,y,z) = 3x^2y^2 - 2x^2yz^3 + 5x^2z^2 + 4x - z^4 + 1$$

or, in recursive representation,

$$a(x,y,z) = (3y^2 + (-2z^3)y + 5z^2)x^2 + 4x + (-z^4 + 1).$$

Using the linked list data structure just described, the recursive form of the polynomial $a(x,y,z)$ is represented as shown in Figure 3.2.

•

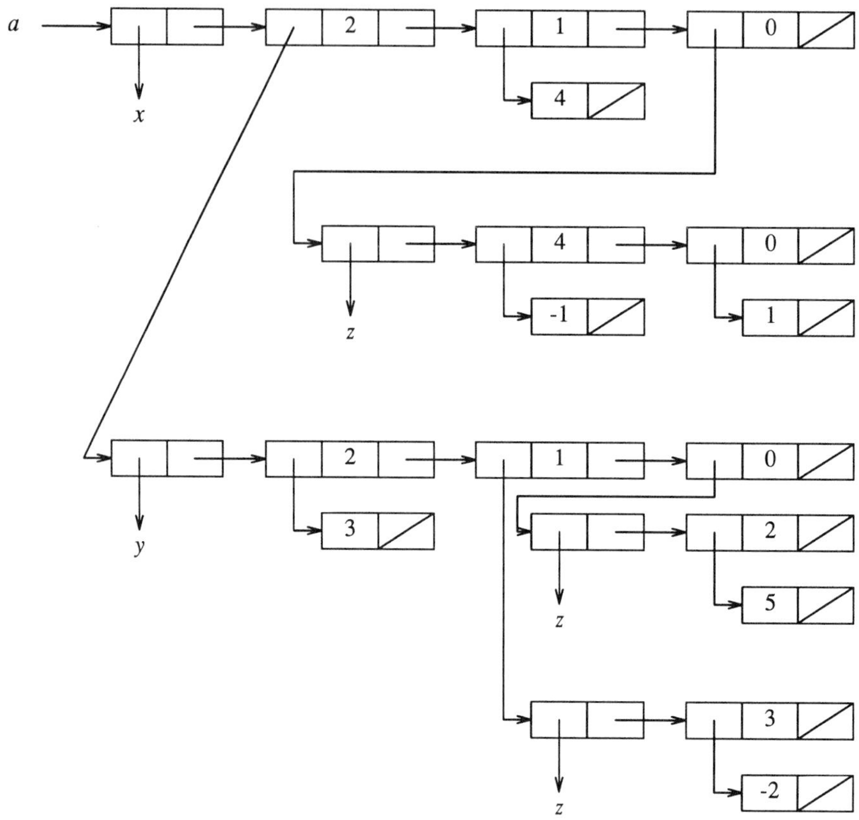

Figure 3.2. A linked list representation.

In Example 3.4 the elements in the coefficient domain **Z** are all represented as single-precision integers. Clearly the occurrence of a node representing an integer in this linked list structure could as well be a multiprecision integer in its linked list representation. More generally, the coefficient domain could be the field **Q** of rational numbers in which case rather than an integer node (or list of nodes) there would be a header node for a rational number in its linked list representation, pointing to a pair of multiprecision integers.

In a high-level list processing language using the linked list data structure presented here, it would be possible to distinguish the cases when a pointer is pointing to a polynomial, a multiprecision integer, or a rational number. A polynomial is distinguished by a header node, the first field of which points to the name of an indeterminate. A multiprecision integer is distinguished by the fact that the first field of its header node contains a single-precision integer rather than a pointer. A rational number is distinguished by a header node the first field of which points to a multiprecision integer.

A Dynamic Array Structure

When a list processing language, such as LISP, is used to implement a symbolic algebra system, linked data structures are typically used to represent algebraic objects. Alternate structures based on dynamic arrays have been used in some systems which are not based on list processors, for example MAPLE which is implemented using the C programming language. As with indefinite precision integers using dynamic array structures, these structures require the use of more sophisticated storage management tools. When a sufficiently powerful storage manager is available, dynamic array data structures can offer improved storage use by reducing the number of link fields. Such structures also offer improved execution speeds due to the efficiency of accessing sequential locations in computer memory.

A multivariate polynomial in distributive form can be represented by a dynamic array with a length field and with links to multiple terms having numeric coefficients:

TYPE/LENGTH	COEFF	TERM	...	COEFF	TERM

The SUM structure in MAPLE is a variation of this. Each term can be represented by a similar structure with links to multiple factors having integer exponents:

TYPE/LENGTH	EXPON	FACT	...	EXPON	FACT

The PRODUCT structure in MAPLE is a variation of this. As an example, the distributive form of the polynomial $a(x,y,z)$ from Example 3.4 can be represented by the structure shown in Figure 3.3. Note that the $4x$ term does not use a PRODUCT structure to hold the x, instead it points directly to x. This is one example of many simple optimizations which might be performed on such data structures.

A Descriptor Block Data Structure

The ALTRAN system offers an interesting historical perspective on the data structures used to represent polynomials. ALTRAN was developed in the mid-sixties for computer systems having very limited memory compared to modern machines. For this reason, the structures used by ALTRAN sacrifice simplicity and generality to minimize the storage needed for representing polynomials. The data structure used in ALTRAN is not purely sequential allocation; this would require the dense representation, which is not a practical alternative for multivariate polynomials (see Section 3.4). Instead it uses a *descriptor block* data structure which we now describe.

Using the distributive representation of multivariate polynomials in expanded canonical form, a polynomial $a(\mathbf{x}) \in D[\mathbf{x}]$ is viewed in the form

$$a(\mathbf{x}) = \sum_{\mathbf{e} \in \mathbf{N}^v} a_{\mathbf{e}} \mathbf{x}^{\mathbf{e}}$$

where $a_{\mathbf{e}} \in D$, $\mathbf{x} = (x_1, \ldots, x_v)$ is a vector of indeterminates and each $\mathbf{e} = (e_1, \ldots, e_v)$ is a corresponding vector of exponents. More explicitly, a term $a_{\mathbf{e}}\mathbf{x}^{\mathbf{e}}$ is of the form

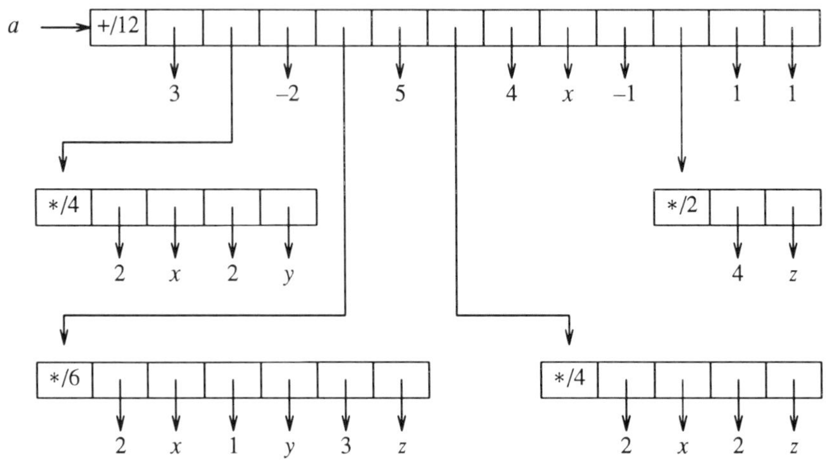

Figure 3.3. A dynamic array representation.

$$a_{\mathbf{e}} x_1^{e_1} x_2^{e_2} \cdots x_v^{e_v}.$$

With this point of view, the representation of a polynomial $a(\mathbf{x})$ can be accomplished by storing three *blocks* of information: (i) a *layout block* which records the names of the indeterminates $x_1, \ldots, x_v$; (ii) a *coefficient block* which records the list of all nonzero coefficients $a_{\mathbf{e}}$; and (iii) an *exponent block* which records the list of exponents vectors $(e_1, \ldots, e_v)$, one such v-vector corresponding to each coefficient $a_{\mathbf{e}}$ in the coefficient block. The order of the integers in the exponent vectors corresponds to the order of the indeterminates specified in the layout block.

Each block of information is stored as an array (or more specifically in ALTRAN, a block of sequential locations in a large array called the *workspace*). The use of sequential allocation of storage imposes the requirement that the precise ''width'' of each block of storage be pre-specified. Thus in the ALTRAN language each variable is associated with a declared layout which specifies the names of all indeterminates which may appear in expressions assigned to the variable and the maximum degree to which each variable may appear. The layout blocks are therefore specified by explicit declarations and the size of each exponent block to be allocated is also known from the declarations. The system then exploits the fact that the maximum size specified for each individual exponent e_i in an exponent vector $\mathbf{e}$ will generally be much smaller than the largest single-precision integer and hence several exponents can be *packed* into one computer word. The layout block is used to store detailed information about this packing of exponents into computer words.

The ''width'' of the coefficient block is determined by the range of values allowed for the coefficient domain D. In ALTRAN only multiprecision integers are allowed as coefficients. (Thus the domain $\mathbf{Q}[x]$ is not represented in this system but since $\mathbf{Z}(x)$, the quotient field of $\mathbf{Z}[x]$, will be represented there is no loss in generality.) Since ALTRAN uses the array representation of multiprecision integers (see Section 3.6) the length l (in computer words) of all multiprecision integers is a pre-specified constant. The coefficient block therefore consists of l computer words for each coefficient $a_{\mathbf{e}}$ to be represented. Finally, a polynomial $a(\mathbf{x})$ is represented by a descriptor block which is an array containing three pointers, pointing to the layout block, the coefficient block, and the exponent block for $a(\mathbf{x})$.

Example 3.5. Let $a(x,y,z) \in \mathbf{Z}[x,y,z]$ be the polynomial given in Example 3.4. Using the descriptor block data structure just described, suppose that the declared maximum degrees are degree 2 in x, degree 3 in y, and degree 4 in z. Suppose further that multiprecision integers are represented using base $\beta = 10^3$ and with pre-specified length $l = 2$. Then the polynomial $a(x,y,z)$ is represented as in Figure 3.4.

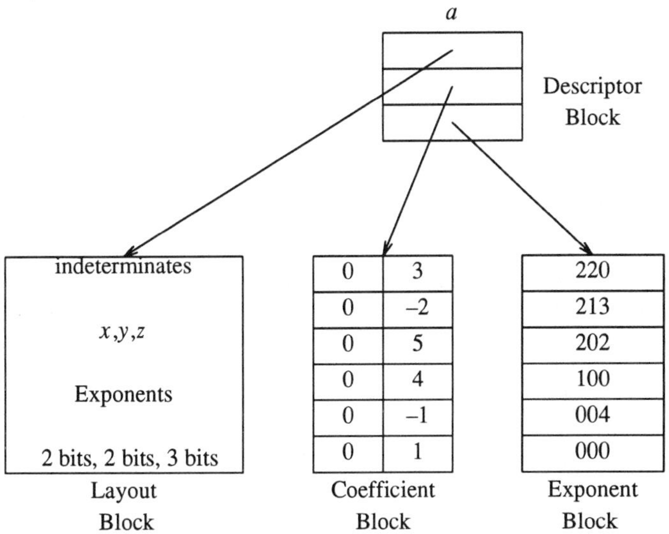

Figure 3.4. A descriptor block representation.

The layout block illustrated in this example indicates that the information stored in the actual layout block would include pointers to the names of the indeterminates and also a specification of the fact that each vector of three exponents is packed into one computer word, with the exponent of x occupying 2 bits, the exponent of y occupying 2 bits, and the exponent of z occupying 3 bits. In practice there is also a guard bit in front of each exponent (to facilitate performing arithmetic operations on the exponents) so this specification implies that the computer word consists of at least 10 bits. The coefficient block illustrated here reflects the specification of $l = 2$ words for each multiprecision integer although $l = 1$ would have sufficed in this particular example.

●

Implementing Factored Normal Form

The three basic data structures of this section have been described as they apply to the representation of multivariate polynomials in expanded canonical form. It is not difficult to use any of these basic data structures for the representation of polynomials in a non-canonical normal form or indeed in a non-normal form. The case of the factored normal form will be briefly examined here.

A polynomial p in factored normal form as defined in Definition 3.5 of Section 3.4 can be expressed as a product of factors

$$p = \prod_{i=1}^{k} f_i^{\alpha_i} \tag{3.23}$$

where α_i $(1 \le i \le k)$ is a positive integer, f_i $(1 \le i \le k)$ is a polynomial in expanded canonical form, and $f_i \ne f_j$ for $i \ne j$. Using a linked list data structure, the polynomial p in the product form (3.23) can be represented by the linked list

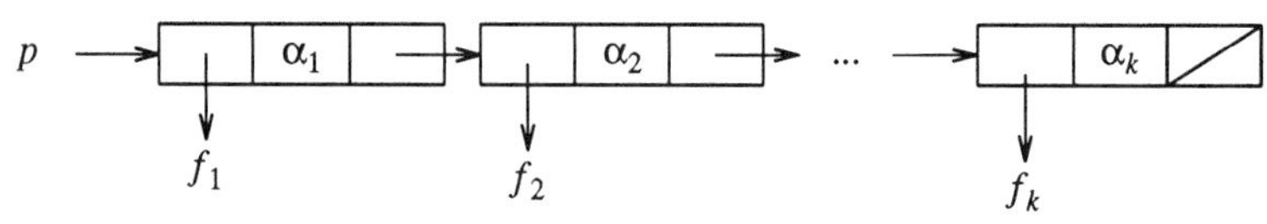

where each factor f_i $(1 \le i \le k)$ is represented by a linked list as previously described for polynomials in expanded canonical form. In a system using this scheme, all polynomials are assumed to be in product form and if a is a single polynomial factor in expanded canonical form then it is represented by

Thus we have simply introduced more structure on top of the original linked list data structure for representing multivariate polynomials.

A similar scheme can be used to represent the product form (3.23) based on a dynamic array data structure. The MAPLE system achieves this by allowing the pointers to factors in the PRODUCT structure to refer to SUM structures. This allows the representation of arbitrarily nested polynomial expressions.

The descriptor block data structure can also be modified to support factored normal form. Indeed the ALTRAN system uses the factored normal form as its basic polynomial form. Here the polynomial p in the product form (3.23) is represented by a *formal product block* which is an array containing one pointer to each factor f_i $(1 \le i \le k)$ and an additional pointer to a corresponding array of the powers α_i $(1 \le i \le k)$. Thus the representation of p is

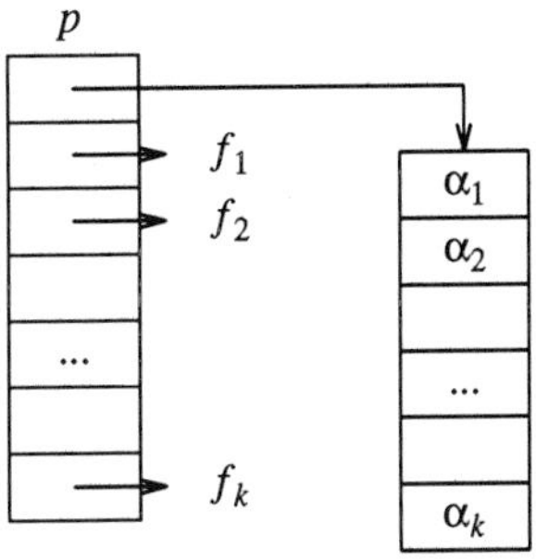

where each factor f_i $(1 \le i \le k)$ is represented by the descriptor block data structure as previously described for polynomials in expanded canonical form.

Rational Functions

A field $D(x_1, \ldots, x_\nu)$ of rational functions has a natural representation as the quotient field of a polynomial domain $D[x_1, \ldots, x_\nu]$. In this point of view a rational function is represented by a pair (*numerator*, *denominator*) where each of *numerator* and *denominator* is a representation of a polynomial lying in the domain $D[x_1, \ldots, x_\nu]$. Using either a linked list data structure or a descriptor block data structure for polynomials, a rational function is thus represented by a node

LINK1	LINK2

of pointers to the numerator and denominator polynomials. As discussed in Section 3.5 a rational function would usually be represented with numerator and denominator relatively prime and with unit normal denominator. In addition, a system would provide one or more (possibly independent) choices of normal forms for the numerator and for the denominator.

A slightly different representation for rational functions is obtained by a trivial generalization of the formal product representation described for polynomials. In the formal product

$$p = \prod_{i=1}^{k} f_i^{\alpha_i}$$

if we allow the powers α_i to be negative integers as well as positive integers then we immediately have a data structure for representing rational functions. No change is required in the data structures already described for formal product representation. This is the data structure used for rational functions in the MAPLE system. In this formal product representation, the numerator consists of all factors with positive powers and the denominator consists of all factors with negative powers.

Power Series

We are considering in this book univariate power series lying in a domain D[[x]] where the coefficient domain D is one of the domains previously discussed (i.e. integers, rational numbers, polynomials, or rational functions). A data structure for power series representation is therefore an extension of data structures previously discussed.

If the TPS representation of power series is used, the TPS

$$\sum_{k=0}^{t} a_k x^k + O(x^{t+1})$$

has a natural representation as a linear list $(a_0, a_1, \ldots, a_t)$. This linear list is easily implemented as either a linked list or an array of pointers to the coefficients a_k in their appropriate representation. Using linked list or dynamic arrays, the "sparse representation" would be natural (i.e. with only nonzero terms stored). With fixed-length arrays, the "dense representation" would be used. In the dense representation, the truncation degree t is implicitly specified by the fact that there are t+1 elements in the linear list, while in the sparse representation the value of t must be stored as an additional piece of information.

The non-truncated representations of power series can be implemented using a similar data structure. The power series

$$a(x) = \sum_{k=0}^{\infty} \mathrm{f}_a(k) x^k$$

can be represented as a linear list

$$(a_0, a_1, \ldots, a_{l-1}, \mathrm{f}_a(k))$$

where the number l of coefficients which have been explicitly computed may increase as a computation proceeds. Again this linear list can be implemented using either a linked list or an array of pointers, with all but the last element pointing to explicit representations of coefficients and with the last element pointing to a representation of the coefficient function $\mathrm{f}_a(k)$. We note that in general the representation of the coefficient function $\mathrm{f}_a(k)$ will involve expressions which are much more complicated than we have so far discussed.

Representations for extended power series are obtained by straightforward generalizations of the representations for ordinary power series. As noted at the end of Section 3.5, the representation of an extended power series $a(x)$ with coefficients lying in a field F can be viewed as the representation of an ordinary power series plus an additional piece of information specifying the power of x by which the ordinary power series is to be "divided". Thus if a particular data structure is chosen for ordinary power series, a data structure for extended power series is obtained by allowing for the representation of one additional integer.

Exercises

1. For each of the following expressions try to find the simplest equivalent expression. As a measure of "simplicity" one could count the number of characters used to write the expression but a higher level measure such as the number of "terms" in the expression would suffice.

(a) $a(x,y) = ((x^2 - xy + x) + (x^2 + 3)(x - y + 1)) \cdot$
$$((y^3 - 3y^2 - 9y - 5) + x^4(y^2 + 2y + 1));$$

(b) $b(x,y) = \dfrac{1}{x^9 + x^8y + x^7y^2 + x^6y^3 + x^5y^4 + x^4y^5 + x^3y^6 + x^2y^7 + xy^8 + y^9}$;

(c) $c(x,y) = \dfrac{(x-y)}{b(x,y)}$, where $b(x,y)$ is defined in part (b);

(d) $$d(x) = \frac{e^x \cos x + \cos x \sin^4 x + 2\cos^3 x \sin^2 x + \cos^5 x}{x^2 - x^2 e^{-2x}} - \frac{e^{-x} \cos x + \cos x \sin^2 x + \cos^3 x}{x^2 e^x - x^2 e^{-x}}.$$

2. Determine whether or not each of the following expressions is equivalent to zero.

(a) $$a(x,y) = \frac{2x}{3y^3(x^2 - y^2)} + \frac{x - y}{x^5 + x^4y + x^3y^2 + x^2y^3 + xy^4 + y^5} - \frac{x + y}{3y^3(x^2 - xy + y^2)} - \frac{x - y}{3y^3(x^2 + xy + y^2)} - \frac{x^2 + y^2}{x^6 - y^6};$$

(b) $$b(x,y) = \frac{x - y}{x^5 + x^4y + x^3y^2 + x^2y^3 + xy^4 + y^5} - \frac{x^2 - xy + y^2}{x^6 - y^6};$$

(c) $c(x) = 16 \sin^{-1}(x) + 2\cos^{-1}(2x) - 3\sinh^{-1}(\tan(\frac{1}{2}x))$;

(d) $$d(x) = 16\cos^3(x)\cosh(\tfrac{1}{2}x)\sinh(x) - 6\cos(x)\sinh(\tfrac{1}{2}x) - 6\cos(x)\sinh(\tfrac{3}{2}x) - \cos(3x)(e^{\frac{3}{2}x} + e^{\frac{1}{2}x})(1 - e^{-2x}).$$

3. In this problem you will show that, in a certain sense, if the zero equivalence problem can be solved for a given class of expressions then the general simplification problem can also be solved. Let E be a class of expressions and let f be a normal function defined on E. Suppose there is an algorithm A which will generate all of the syntactically valid expressions in the class E, in lexicographically increasing order. (That is, Algorithm A generates all syntactically valid expressions containing l characters, for $l = 1$, then $l = 2$, then $l = 3$, etc. and the expressions of a fixed length l are generated in increasing order with respect to some encoding of the characters.)

(a) Define a *simplification function* g on E in terms of normal function f and algorithm A such that g is a canonical function and moreover the canonical form g(a) of any expression $a \in$ E is the *shortest* expression equivalent to a.

(b) If E is a class of expressions obtained by performing the operations of addition and multiplication on the elements in a quotient field Q(D) of an integral domain D then the usual canonical function (i.e. "form a common denominator" and "reduce to lowest terms") is not a simplification in the sense of the function g of part (a). Illustrate this fact for the field **Q** of rational numbers by giving examples of expressions of the form $\frac{a_1}{b_1} + \frac{a_2}{b_2} + \frac{a_3}{b_3}$ (where $a_i, b_i \in \mathbf{Z}$ and $\frac{a_i}{b_i}$ is in lowest terms, for $i = 1,2,3$) such that the "reduced form" of the expression requires more characters for its representation than the original expression.

4. Consider the four forms for rational functions discussed in Section 3.5: *factored/factored, factored/expanded, expanded/factored, expanded/expanded,* with numerator and denominator relatively prime in each case.

(a) Put each of the expressions $a(x,y)$, $b(x,y)$, and $c(x,y)$ given in Exercise 1 into each of the above four forms. Similarly for the expressions $a(x,y)$ and $b(x,y)$ given in Exercise 2.

(b) Which (if any) of these four forms is useful for performing "simplification" as requested in Exercise 1? For determining "zero-equivalence" as requested in Exercise 2?

5. Consider the problem of computing the functions $f(x,y)$ and $g(x,y)$ defined by

$$f(x,y) = \frac{\partial a}{\partial x}\,\frac{\partial b}{\partial x}\,,$$

$$g(x,y) = \frac{\partial a}{\partial y}\,\frac{\partial b}{\partial y}$$

(where ∂ denotes partial differentiation) where a and b are the rational functions

$$a = \frac{(10x^2y^3 + 13x - 7)\,(3x^2 - 7y^2)^2}{(5x^2y^2 + 1)^2\,(x - y)^3\,(x + y)^2}\,,$$

$$b = \frac{13x^3y^3 + 75x^3y + 81xy - x + 19}{(5x^2y^2+1)\,(x - y)\,(x + y)^3}\,.$$

Perform this computation on a computer algebra system available to you using several different choices of normal (or non-normal) forms available in the system. Compare the results obtained using the various choices of form in terms of (i) processor time used, (ii) memory space required, and (iii) compactness (i.e. readability) of the output.

6. The TPS representation of a power series appears to be similar to a polynomial but must be distinguished from a polynomial. Consider the two power series defined by

$$a(x) = \sum_{k=0}^{\infty} (-1)^k x^k \qquad \left[= \frac{1}{1+x} \right],$$

$$b(x) = \sum_{k=0}^{\infty} \frac{1}{k!} x^k \qquad \left[= e^x \right].$$

(a) The TPS representations of $a(x)$ and $b(x)$ with truncation degree $t = 3$ are

$$\tilde{a}(x) = 1 - x + x^2 - x^3 + O(x^4),$$

$$\tilde{b}(x) = 1 + x + \frac{1}{2}x^2 + \frac{1}{6}x^3 + O(x^4)\,.$$

Let $p(x)$ and $q(x)$ be the corresponding polynomials defined by

$$p(x) = 1 - x + x^2 - x^3\,,$$

$$q(x) = 1 + x + \frac{1}{2}x^2 + \frac{1}{6}x^3.$$

What should be the result of performing the TPS multiplication $\tilde{a}(x)\,\tilde{b}(x)$? What is the result of performing the polynomial multiplication $p(x)\;q(x)$? What is the correct power series product $a(x)\cdot b(x)$ expressed as a TPS with truncation degree $t = 6$?

(b) Let $\tilde{a}(x)$ and $p(x)$ be as in part (a). The result of performing $\dfrac{1}{p(x)}$ is the rational function $\dfrac{1}{1 - x + x^2 - x^3}$. What is the result of performing the TPS division $\dfrac{1}{\tilde{a}(x)}$? What is the correct power series reciprocal $\dfrac{1}{a(x)}$?

7. Consider the problem of computing the power series solution of a linear ordinary differential equation with polynomial coefficients:

$$p_\nu(x)\,y^{(\nu)} + \cdots + p_1(x)\,y' + p_0(x)\,y = r(x)$$

where $p_i(x)$, $0 \le i \le \nu$, and $r(x)$ are polynomials in x and where y denotes the unknown function of x. Show that if this differential equation has a power series solution

$$y(x) = \sum_{k=0}^{\infty} y_k\,x^k$$

then the power series coefficients can be expressed, for $k \ge K$ for some K, as a finite linear recurrence:

$$y_k = u_1(k)\,y_{k-1} + u_2(k)\,y_{k-2} + \cdots + u_n(k)\,y_{k-n}$$

where $u_i(k)$, $1 \le i \le n$, are rational expressions in k. Thus an NTPS representation for the solution $y(x)$ is possible with the coefficient function $f_y(k)$ specified by a finite linear recurrence (and with the first k coefficients specified explicitly).

8. Show that a power series $a(x) = \sum_{k=0}^{\infty} a_k x^k$ has an NTPS representation in which the coefficient function can be expressed, for $k \geq$ K for some K, as a finite linear recurrence with *constant coefficients*:

$$a_k = u_1 a_{k-1} + u_2 a_{k-2} + \cdots + u_n a_{k-n}$$

if and only if $a(x)$ can be expressed as a rational function of x. (See Section 2.10.)

9. Generalize the results of Exercise 7 and of Exercise 8 into statements about the NTPS representation of extended power series rather than just ordinary power series.

10. Using a language in which linked list manipulation is convenient, implement algorithms for addition and multiplication of indefinite-precision integers (i.e. multiprecision integers in linked list representation). Base your algorithms on the methods you use to do integer arithmetic by hand.

11. Assuming that the algorithms of Exercise 10 are available, implement algorithms for addition and multiplication of multivariate polynomials in expanded canonical form with indefinite-precision integer coefficients. Use the linked list data structure described in Section 3.7. Base your algorithms on the methods you use to do polynomial arithmetic by hand.

12. Assuming that the algorithms of Exercise 10 and 11 are available, implement algorithms for addition and multiplication of multivariate rational functions in *expanded/expanded* form (i.e. in the expanded canonical form of Definition 3.6) with indefinite-precision integer coefficients. Use either of the linked list data structures for rational functions described in Section 3.7. You will need a recursive implementation of Algorithm 2.3 (or some other algorithm for GCD computation).

13. Assuming that the algorithms of Exercise 12 are available, implement algorithms for addition and multiplication of univariate power series with coefficients which are multivariate rational functions. Use the TPS representation implemented as a linked list.

14. Choose a specific representation for extended power series and implement algorithms for addition, multiplication, and division of extended power series with coefficients which are multivariate rational functions. You will need to have available the algorithms of Exercise 12 and you may wish to have available the algorithms of Exercise 13.

References

1. W.S. Brown, ''On Computing with Factored Rational Expressions,'' *ACM SIGSAM Bull.*, **8** pp. 26-34 (1974).
2. B.F. Caviness, ''On Canonical Forms and Simplification,'' *J. ACM*, **2** pp. 385-396 (1970).
3. B.W. Char, K.O. Geddes, G.H. Gonnet, B.L. Leong, M.B. Monagan, and S.M. Watt, *Maple V Language Reference Manual,*, Springer-Verlag (1991).
4. A.C. Hearn, ''Polynomial and Rational Function Representations,'' Tech. Report UCP-29, Univ. of Utah (1974).
5. E. Horowitz and S. Sahni, *Fundamentals of Computer Algorithms,* Computer Science Press, Maryland (1978).
6. A.D. Hall Jr., ''The Altran System for Rational Function Manipulation - A Survey,'' *Comm. ACM*, **14** pp. 517-521 (1971).
7. J. Moses, ''Algebraic Simplification: A Guide for the Perplexed,'' *Comm. ACM*, **14** pp. 527-537 (1971).
8. A.C. Norman, ''Computing with Formal Power Series,'' *ACM TOMS*, **1** pp. 346-356 (1975).
9. D. Richardson, ''Some Unsolvable Problems Involving Elementary Functions of a Real Variable,'' *J. Symbolic Logic*, **33** pp. 511-520 (1968).

CHAPTER 4

ARITHMETIC OF POLYNOMIALS,

RATIONAL FUNCTIONS,

AND POWER SERIES

4.1. INTRODUCTION

In Chapter 2 we introduced the basic algebraic domains which are of interest to computer algebra. This was followed by the representation problem, that is, the problem of how elements of these algebras are to be represented in a computer environment. Having described the types of objects along with the various representation issues, there follows the problem of implementing the various algebraic operations that define the algebras. In this chapter we describe the arithmetic operations of addition, subtraction, multiplication, and division for these domains. In particular, we describe these fundamental operations in the ring of integers modulo n, the ring of formal power series over a field, and the ring of polynomials over an integral domain along with their quotient fields. The latter includes the domain of multiprecision integers and rational numbers.

In addition to the fundamental algorithms, we will also describe algorithms which improve on the standard arithmetic algorithms. In particular, we describe fast algorithms for multiplication of multiprecision integers (Karatsuba's algorithm), for polynomial multiplication over certain fields (FFT) and for power series inversion (Newton's method). The interested reader should refer to Lipson [8] or Knuth[6] for additional algorithms in this area.

4.2. BASIC ARITHMETIC ALGORITHMS

Multiprecision Integers

As discussed in the previous chapter, we look at multiprecision integers as objects of the form

$$a = a_0 + a_1 \cdot B + \cdots + a_{m-1} \cdot B^{m-1}$$

where each a_i satisfies $a_i < B$. Here B is the base of the multiprecision number system. Choices for B vary from system to system. Clearly we wish that $B < W$ where W is the largest integer to fit into a single computer word. Implementation considerations may dictate that the base B be chosen so that it is less than one-half the computer word size. This is to allow for all component arithmetic operations to be performed in a single computer word. Still another consideration for choosing the base B is portability. A smaller choice of B allows a wider range of architectures for which multiprecision arithmetic can be implemented without alteration.

Addition and subtraction are carried out in the same manner as one would with base 10 arithmetic. One adds (or subtracts) two integers component-wise starting on the left. The only complication results from the need for a carry digit (either 0 or 1) as we proceed from left to right. It is an easy exercise to see that the addition of two multiprecision integers with m and n digits, respectively, has a complexity of $O(\max(m,n))$ operations (cf. Exercise 1(a)).

If a and b are two multiprecision integers

$$a = a_0 + a_1 B + \cdots + a_{m-1} B^{m-1},$$

$$b = b_0 + b_1 B + \cdots + b_{n-1} B^{n-1}$$

of m and n digits, respectively, then their product c is a multiprecision integer having at most $m + n$ digits. Multiplication of two multiprecision integers can be done in the same way that one multiplies two base 10 integers: for each i, multiply a by b_i carrying digits forward as necessary, and then add the accumulated partial products. However, this requires that all the partial products be stored, a significant drawback when the number of digits is very large. A slight modification to this algorithm, which accomplishes the calculation in place can be achieved by simply performing one addition after each partial product. If we let $b^{(0)} = 0$ and for each integer $1 \le k \le n$ let

$$b^{(k)} = b_0 + \cdots + b_{k-1} B^{k-1}$$

then, including carries, we have

$$a \cdot b^{(k)} = c_0^{(k)} + \cdots + c_{m+k-1}^{(k)} B^{m+k-1}$$

with $c = a \cdot b^{(n)}$. For each k, we have

$$a \cdot b^{(k+1)} = a \cdot (b^{(k)} + b_k B^k) = ab^{(k)} + ab_k B^k$$

and hence the first k digits of $a \cdot b^{(k+1)}$ remain unaltered. From this it is easy to see that multiplication of two multiprecision integers having m and n digits, respectively, requires $O(m \cdot n)$

operations (cf. Exercise 1(b)). A simple implementation of the multiplication algorithm is presented in Algorithm 4.1.

Algorithm 4.1. Multiprecision Integer Multiplication.

```
procedure BigIntegerMultiply(a, b, B)
  # Given two multiprecision integers a and b of
  # lengths m and n with base B, we determine
  # c = a·b = c_0 + c_1 B + ··· + c_{m+n-1} B^{m+n-1}

  for i from 0 to m-1 do c_i <- 0

  for k from 0 to n-1 do {
     carry <- 0
     for i from 0 to m-1 do {
        temp <- a_i·b_k + c_{i+k} + carry
        c_{i+k} <- rem(temp, B)
        carry <- quo(temp, B) }
     c_{k+m} <- carry }
  return(c_0 + c_1 B + ··· + c_{m+n-1} B^{m+n-1})
end
```

If a and b are two multiprecision integers of size m and n, then to determine integers q and r satisfying the division property

$$a = b \cdot q + r, \quad 0 \le r < b$$

also follows the usual grade school method. Assuming $m \ge n$, the division algorithm requires $m - n + 1$ steps, each step of which involves $O(n)$ operations. Hence this operation requires $O((m-n) \cdot n)$ operations.

Polynomial Arithmetic

In the ring of polynomials R[x], R a ring and x an indeterminate, the classical arithmetic operations are again those taught in grade school. If

$$a(x) = a_0 + a_1 x + \cdots + a_n x^n, \; b(x) = b_0 + b_1 x + \cdots + b_m x^m$$

then $a(x)$ and $b(x)$ are added component-wise

$$a(x) + b(x) = (a_0 + b_0) + (a_1 + b_1)x + \cdots + (a_k + b_k)x^k,$$

where $k = \max(m,n)$. Multiplication is given by

$$a(x) \cdot b(x) = h_0 + h_1 x + \cdots h_{m+n} x^{m+n}$$

where

$$h_i = a_i b_0 + \cdots + a_0 b_i .$$

As was the case with multiprecision arithmetic, these methods have complexity $O(\max(m,n))$ and $O(m \cdot n)$ ring operations, respectively. Similarly, when the coefficient domain is a field so that the division property holds in the polynomial domain, the grade school method determines the remainder and quotient with a complexity of $O((m-n) \cdot n)$ field operations.

Power Series Arithmetic

Polynomial-type arithmetic is also the basis for arithmetic in the ring of truncated power series over a field F. These truncated power series have their algebraic representation as

$$A(x) = a(x) + O(x^{n+1})$$

for some fixed integer n, where $a(x)$ is a polynomial of degree at most n. Arithmetic is essentially mod x^{n+1} arithmetic. Power series are added and subtracted component-wise and multiplied according to the Cauchy product rule

$$(a(x) + O(x^{n+1})) \cdot (b(x) + O(x^{n+1})) = c(x) + O(x^{n+1})$$

where

$$c_i = a_0 \cdot b_i + \cdots + a_i \cdot b_0, \quad i = 0, \ldots, n.$$

The quotient $A(x)/B(x)$, of two power series is well-defined when the constant term, b_0, of $B(x)$ is a unit in the coefficient domain. If we denote the quotient by $C(x)$, then its components are determined by

$$c_i = b_0^{-1} \cdot (a_i - c_0 \cdot b_i - \cdots - c_{i-1} \cdot b_1).$$

It is easily seen that both the product and quotient require $O(n^2)$ operations.

Up to now, the arithmetic operations on power series are identical to those found with polynomial arithmetic. However, power series domains allow for some additional arithmetic operations. For example, one can ask for the square root of a polynomial $a(x)$, that is, $b(x)$ such that $b(x)^2 = a(x)$. However such a $b(x)$ is usually a power series rather than a polynomial. Thus it is more natural to embed $a(x)$ into the domain of power series and ask to find the square root in this domain. Indeed, this is a natural domain for a *powering* operation. Another arithmetic operation important to power series domains is the operation of *reversion*. This is essentially the taking of inverse with respect to the composition operation. Such an operation is not available in polynomial domains.

Power series powering asks to calculate

$$B(x) = A(x)^p \tag{4.1}$$

for a given power series $A(x)$. Here p is some arbitrary real number. The arithmetic operations of taking reciprocals of power series (p=−1), square roots of power series (p=1/2) along with the obvious operations such as squaring and cubing are all covered in this operation. We will assume that the constant term a_0 is the unit 1 (the powering operation can always be reduced to this case; cf. Exercise 4.4) so that b_0 will also be the unit 1. The other coefficients can then be determined by the following procedure. Notice that taking derivatives on both sides of (4.1) gives

$$B'(x) = b_1 + 2b_2x + \cdots = p \cdot A(x)^{p-1} \cdot A'(x)$$

so

$$B'(x) \cdot A(x) = p \cdot B(x) \cdot A'(x). \tag{4.2}$$

Equating the $(i-1)$-th coefficients in (4.2) gives

$$b_1 a_{i-1} + 2b_2 a_{i-2} + \cdots + ib_i = p \cdot (a_1 \cdot b_{i-1} + 2a_2 \cdot b_{i-2} + \cdots + i \cdot a_i)$$

and so

$$b_i = (i \cdot p \cdot a_i + ((i-1) \cdot p - 1)a_{i-1}b_1 + \cdots + (p-(i-1))a_1 b_{i-1}) \,/\, i. \tag{4.3}$$

Example 4.1. Let

$$a(x) = 1 - x.$$

Then using the above recurrence we obtain

$$(1-x)^{-5} = 1 + 5x + 15x^2 + 35x^3 + \cdots .$$

●

Calculating $B(x)$ using (4.3) gives an algorithm that requires $O(n^2)$ operations to determine the first n terms.

Power series reversion can be stated as: given

$$t = x + a_2x^2 + a_3x^3 + \cdots = a(x) \in F[[x]]$$

find x in terms of t; that is, find

$$x = t + b_2t^2 + b_3t^3 + \cdots = b(t) \in F[[t]] \text{ with } a(b(t)) = t.$$

For example, we have

$$t = \sin(x) = x - \frac{1}{6}x^3 + \frac{1}{120}x^5 + O(x^7)$$

and from this would like to calculate the power series for $x = \arcsin(t)$. The classical method for solving the reversion operation is to use the inversion formula of Lagrange. This formula states that

$$x = t + \frac{c_1^{(2)}}{2}t^2 + \frac{c_2^{(3)}}{3}t^3 + \frac{c_3^{(4)}}{4}t^4 + \cdots$$

where

$$1 + c_1^{(i)}x + c_2^{(i)}x^2 + \cdots = (t/x)^{-i} = (1 + a_2x + a_3x^2 + \cdots)^{-i}$$

for $i = 1, 2, 3, \ldots$. Calculating the reversion of a power series using Lagrange's inversion formula leads to a computation requiring approximately $n^3/6$ multiplications to find the first n terms. Here we assume that the powering method given previously in this section is used to calculate the negative powers of t/x.

Example 4.2. If we use

$$t = \sin(x) = x - \frac{1}{6}x^3 + \frac{1}{120}x^5 + O(x^7)$$

then

$$\begin{aligned}
(1 - \frac{1}{6}x^2 + \frac{1}{120}x^4)^{-2} &= 1 + \frac{1}{3}x^2 + \cdots \\
(1 - \frac{1}{6}x^2 + \frac{1}{120}x^4)^{-3} &= 1 + \frac{1}{2}x^2 + \frac{17}{120}x^4 + \cdots \\
(1 - \frac{1}{6}x^2 + \frac{1}{120}x^4)^{-4} &= 1 + \frac{2}{3}x^2 + \frac{11}{45}x^4 + \cdots \\
(1 - \frac{1}{6}x^2 + \frac{1}{120}x^4)^{-5} &= 1 + \frac{5}{6}x^2 + \frac{3}{8}x^4 + \cdots ;
\end{aligned}$$

hence the first few coefficients of $\arcsin(t)$ are given by

$$s_1 = 0, \ s_2 = \frac{1/2}{3} = \frac{1}{6}, \ s_3 = 0, \ s_4 = \frac{3/8}{5} = \frac{3}{40}$$

(where $s_i = c_i^{(i+1)}/(i+1)$). Thus

$$\arcsin(t) = t + \frac{1}{6}t^3 + \frac{3}{40}t^5 + \cdots .$$

●

Integers mod n Arithmetic

The set $\mathbf{Z}_n$ can be represented in either a positive or symmetric representation. For example, the positive representation of $\mathbf{Z}_7$ is $\{0, 1, 2, 3, 4, 5, 6\}$ while its symmetric representation is $\{-3, -2, -1, 0, 1, 2, 3\}$. The integer 19 mod 7 is represented by 5 in the positive representation and -2 in the symmetric representation. The negative integer -8 mod 7 is 6 in the positive representation and -1 in the symmetric one. The modular representation of an integer is obtained by simple remainder operations in either form.

The operations of addition, subtraction and multiplication in $\mathbf{Z}_n$ are straightforward, regardless of the representation used. In each case the operation is performed on the representatives considered as integers and the results reduced modulo n.

When $n = p$, a prime, the set $\mathbf{Z}_p$ is a finite field, and hence it is possible to divide. The operation of inverting a nonzero element in this case, involves the use of the extended Euclidean algorithm (EEA) of Chapter 2. If an integer m is nonzero modulo the prime p, then it must be relatively prime to p. The EEA finds integers s and t such that

$$s \cdot m + t \cdot p = 1$$

so that

$$s \cdot m \equiv 1 \pmod{p}.$$

Thus, the representative of s in $\mathbf{Z}_p$ is the inverse of m. For example, the EEA applied to the integers 14 and 17 produces the equation

$$11 \cdot 14 - 9 \cdot 17 = 1$$

giving 11 as the inverse of 14 in the field $\mathbf{Z}_{17}$. The complexity of the EEA is $O(n^2)$ bit operations (cf. Exercise 4.2), hence the cost of division is the same as the cost of multiplication.

The last arithmetic operation of interest for integers mod p is the powering operation, i.e. determining

$$a^k \bmod p$$

for an integer k. This is efficiently accomplished by repeated squaring. Thus, if we write the integer k in binary as

$$k = \sum_{i=0}^{\lfloor \log_2 k \rfloor} b_i \cdot 2^i,$$

where each b_i is either a 0 or a 1, then a^k is computed by

$$a^k = \prod_{i=0}^{\lfloor \log_2 k \rfloor} a^{b_i 2^i}.$$

For example, to calculate 7^{11} in the field $\mathbf{Z}_{17}$ we first calculate

$$7^2 = 15, \; 7^4 = 15^2 = 4, \; 7^8 = 4^2 = 16$$

and then

$$7^{11} = 7 \cdot 7^2 \cdot 7^8 = 7 \cdot 15 \cdot 16 = 14.$$

The total cost complexity of this operation is $O(\log_2 k)$ multiplications (cf. Exercise 4.1(d)).

4.3. FAST ARITHMETIC ALGORITHMS: KARATSUBA'S ALGORITHM

In this section we describe an algorithm due to A. Karatsuba [5] which multiplies two polynomials with a complexity that is less than that of the classical grade school method. This algorithm, discovered in 1962 was the first algorithm to accomplish this multiplication in under $O(n^2)$ operations. Similar techniques have been used to obtain algorithms to speed up matrix multiplication (called Strassen's method – cf. Exercise 4.13).

We will describe Karatsuba's algorithm as a fast algorithm to multiply two multiprecision integers of size n digits. The modification to fast multiplication of two polynomials of degree n is straightforward and is left to the reader.

If the two integers x and y are of size n digits, with B the base, then x and y can be represented as

$x =$	a	b		$y =$	c	d

that is,

$$x = a \cdot B^{n/2} + b, \quad y = c \cdot B^{n/2} + d.$$

Therefore the classical form of the product is

$$x \cdot y = a \cdot c \cdot B^n + (a \cdot d + b \cdot c) \cdot B^{n/2} + b \cdot d. \tag{4.4}$$

One method of determining the complexity of the classical multiplication method is by the use of recurrence relations. If $\mathrm{T}(n)$ denotes the cost of multiplying two multiprecision integers of size n digits, then equation (4.4) shows that multiplying two multiprecision integers with n digits can be accomplished by four multiplications of integers having only $n/2$ digits and one addition of two $n/2$ digit integers. Thus we have the recurrence relation

$$\mathrm{T}(1) = 1, \quad \mathrm{T}(n) = 4 \cdot \mathrm{T}(n/2) + C \cdot n$$

for some constant C. For $n = 2^m$ this implies

$$\begin{aligned} \mathrm{T}(n) = \mathrm{T}(2^m) &= 4 \cdot (4 \cdot \mathrm{T}(2^{m-2}) + C \cdot 2^{m-1}) + C \cdot 2^m \\ &= 4^2 \cdot \mathrm{T}(2^{m-2}) + C \cdot 2^m \cdot (1 + 2) = \cdots \\ &= 4^m \cdot \mathrm{T}(1) + C \cdot 2^m \cdot (1 + 2 + \cdots + 2^{m-1}) \\ &= \hat{C} \cdot (2^m)^2 = \hat{C} \cdot n^2 \end{aligned}$$

giving a simple proof that the complexity of the grade school method is $O(n^2)$ operations.

Karatsuba's method depends on noticing that the product of x and y may also be written as

$$x \cdot y = ac\mathrm{B}^n + (ac + bd + (a-b)(d-c))\mathrm{B}^{n/2} + bd \tag{4.5}$$

(this formula may be verified by simple algebra). Although (4.5) appears far more complicated than (4.4), one sees that using (4.5) gives the product at a cost of four additions/subtractions but only three multiplications. Thus the cost function $\mathrm{T}(n)$ satisfies the recurrence relation

$$T(1) = 1, \quad T(n) = 3 \cdot T(n/2) + C \cdot n.$$

In this case, for $n = 2^m$ the recurrence relation implies

$$\begin{aligned} T(n) = T(2^m) &= 3 \cdot (3 \cdot T(2^{m-2}) + C \cdot 2^{m-1}) + C \cdot 2^m \\ &= 3^2 \cdot T(2^{m-2}) + C \cdot 2^m \cdot (1 + 3/2) = \cdots \\ &= 3^m \cdot T(1) + C \cdot 2^m \cdot (1 + 3/2 + \cdots + (3/2)^{m-1}) \\ &= 3^m + C \cdot 2^m \cdot \frac{(3/2)^m - 1}{(3/2) - 1} \approx \hat{C} \cdot 3^m = \hat{C} \cdot 2^{m \cdot \log_2 3} = \hat{C} \cdot n^{\log_2 3} \end{aligned}$$

showing that the complexity of Karatsuba's algorithm is $O(n^{1.58})$.

Algorithm 4.2. Karatsuba's Multiplication Algorithm.

procedure Karatsuba(a,b,n)

\# Given multiprecision integers a and b with n digits
\# and base B we compute their product $c=a \cdot b$.
\# The size n must be a power of 2.

if n=1 **then return**($a \cdot b$)
else {
  $c \leftarrow \text{sign}(a) \cdot \text{sign}(b)$
  $\hat{a} \leftarrow |a|$; $\hat{b} \leftarrow |b|$
  $a1 \leftarrow$ first $n/2$ digits of $\hat{a}$
  $a2 \leftarrow$ last $n/2$ digits of $\hat{a}$
  $b1 \leftarrow$ first $n/2$ digits of $\hat{b}$
  $b2 \leftarrow$ last $n/2$ digits of $\hat{b}$
  $m1 \leftarrow \text{Karatsuba}(a1, b1, n/2)$
  $m2 \leftarrow \text{Karatsuba}(a1 - a2, b2 - b1, n/2)$
  $m3 \leftarrow \text{Karatsuba}(a2, b2, n/2)$
  $c \leftarrow c \cdot (m1 \cdot B^n + (m1 + m2 + m3) \cdot B^{n/2} + m3)$
  return(c) }
end

We point out that Karatsuba's algorithm requires a much larger constant C than does the grade school method and so is only useful for large multiprecision integers (in practice at least 500 digits). In addition, the algorithm has one fundamental limitation: the large storage required for the intermediate calculations. To overcome this problem, one must implement the multiplication "in place", rather than making use of temporary local storage.

4.4. MODULAR REPRESENTATIONS

In this section we continue our quest to improve the cost of multiplication for integers and polynomials. In particular, we describe a new representation for integers and polynomials in which multiplication has the same complexity as addition or subtraction.

Multiprecision integers can be algebraically represented in ways other than the classical base B representation. In particular, multiprecision integers have a natural representation as vectors of modular numbers. This representation results in simple linear algorithms for both integer addition and multiplication, once the quantities are in modular form.

To formally define the concept of a modular representation, let $m_0, m_1, \ldots, m_n$, be a set of $n+1$ pairwise relatively prime integers, and set

$$m = m_0 \cdot m_1 \cdot \cdots \cdot m_n .$$

It is a classical result from commutative algebra (the Chinese remainder theorem, discussed in the next chapter) that there is a ring isomorphism

$$\phi : \mathbf{Z}_m \rightarrow \mathbf{Z}_{m_1} \times \cdots \times \mathbf{Z}_{m_n}$$

given by

$$\phi(x) = (x \bmod m_1, \ldots, x \bmod m_n).$$

Every positive multiprecision integer x less than m can be uniquely represented by a list

$$x = (x_0, \ldots, x_n)$$

where

$$x_i = x \bmod m_i .$$

Let y be a second positive multiprecision integer with

$$y = (y_0, \ldots, y_n)$$

as its modular representation. Then, as long as $x+y < m$, the sum can be uniquely represented by

$$x + y = (t_0, \ldots, t_n), \text{ where } t_i = x_i + y_i \bmod m_i .$$

Similarly, if $x \cdot y < m$, then the product can be uniquely represented by

$$x \cdot y = (t_0, \ldots, t_n), \text{ where } t_i = x_i \cdot y_i \bmod m_i .$$

In both cases, the complexity of the arithmetic operations is $O(n)$, a considerable improvement in the case of multiplication.

Modular representations also exist for polynomial domains $F[x]$, with F a field. If $x_0, \ldots, x_n$ are $n+1$ distinct points from the field F and

$$m_i(x) = (x - x_i), \text{ and } m(x) = m_0(x) \cdot \cdots \cdot m_n(x),$$

then the polynomial version of the Chinese remainder theorem (cf. Chapter 5) shows that there is a ring isomorphism

$$\phi : F[x] / <m(x)> \;\rightarrow\; F[x]/<m_1(x)> \times \cdots \times F[x]/<m_n(x)>$$

given by

$$\phi(a(x) \bmod m(x)) = (a(x) \bmod m_1(x), \ldots, a(x) \bmod m_n(x)).$$

The mod operation for polynomials is the same as remainder on division, hence, for any polynomial $a(x)$ of degree at most $n + 1$ we have

$$a(x) \bmod m(x) = a(x), \quad \text{and} \quad a(x) \bmod m_i(x) = a(x_i)$$

with each $F[x]/<m_i(x)>$ identified with F. For polynomials of degree at most n, ϕ can be viewed as the evaluation isomorphism

$$\phi(a(x)) = (a(x_0), \ldots, a(x_n)).$$

Thus, rather than represent a polynomial in its coefficient representation

$$a(x) \leftarrow (a_0, \ldots, a_n) \quad \text{where} \quad a(x) = \sum_{i=0}^{n} a_i x^i.$$

we can represent $a(x)$ in its evaluation representation

$$a(x) \leftarrow (\hat{a}_0, \ldots, \hat{a}_n) \quad \text{where} \quad \hat{a}_i = a(x_i).$$

As was the case with multiprecision integers, the resulting modular representation allows both addition/subtraction and multiplication to be implemented with $O(n)$ cost complexity. For example, if $a(x)$ and $b(x)$ are two polynomials each of degree at most $(n+1)/2$ then the modular representation of their product is given by

$$a(x) \cdot b(x) \leftarrow (\hat{a}_0 \cdot \hat{b}_0, \ldots, \hat{a}_n \cdot \hat{b}_n).$$

Clearly, polynomial multiplication in the modular representation is accomplished with a complexity $O(n)$, the same as the complexity of addition or subtraction.

Since multiplication is improved by one order of magnitude in a modular representation, it is natural to ask whether it is also possible to improve division in this representation. Thus, let $a(x)$ and $b(x)$ be polynomials of degree m and n, respectively, with say $m \geq n$. Let $x_0, \ldots, x_m$ be $n+1$ points, each of which satisfies $b(x_i) \neq 0$. Then $a(x)$ and $b(x)$ will have modular representations of

$$a(x) \leftarrow (\hat{a}_0, \ldots, \hat{a}_n), \; b(x) \leftarrow (\hat{b}_0, \ldots, \hat{b}_n)$$

where $\hat{a}_i = a(x_i)$ and $\hat{b}_i = b(x_i)$. If $b(x) \mid a(x)$ then their quotient, say $c(x)$, will have a modular representation

$$c(x) \leftarrow (\hat{a}_1 / \hat{b}_1, \ldots, \hat{a}_n / \hat{b}_n).$$

However, unlike multiplication, division is not always possible. Indeed, one usually wishes to determine if one polynomial divides a second and, if so, to obtain the quotient. In our case, if the quotient $r(x)$ does exist then we know that it must have degree $m-n$ when returned to its polynomial representation. Thus, we obtain an algorithm which in essence performs trial divisions in F[x] by using only divisions in F.

Algorithm 4.3. Polynomial Trial Division Algorithm.

procedure TrialDivision($a(x)$,$b(x)$,m,n)

\# Given two polynomials $a(x)$ and $b(x)$ with degrees m and n with
\# $m \geq n$, determine if $b(x)$ divides into $a(x)$ by trial division at the
\# points $x_0, \ldots, x_m$. If true then return the quotient $c(x)$.

for i **from** 0 **to** m **do** $\hat{c}_i = a(x_i)/b(x_i)$

$c(x) \leftarrow$ PolyInterp($\hat{c}_0, \ldots, \hat{c}_m$)

if deg($c(x)$) $= m-n$ **then return**($c(x)$)
else return(*does not divide*)

end

Hidden in all this is a significant drawback which limits the usefulness of multiplication and division using modular representations. The O(n) cost for multiplication is the cost given that the algebraic objects are already present in their modular representations. However, in most practical applications one must convert to the modular representation, do the necessary arithmetic in the modular domain and then convert the representation back. A problem arises because the conversion process (in either direction) is generally higher than the cost of classical methods of multiplication. Thus, as it is presented above, the use of modular representation for reducing the complexity of multiplication is impractical. However there are applications where one can convert the input values, do a significant amount of work in the modular representation, and then convert the final result back to standard form.

4.5. THE FAST FOURIER TRANSFORM (FFT)

The Forward Fourier Transform

In this section we will study the (forward) conversion process from a coefficient representation of a polynomial to a modular representation in more detail. In all cases, the constants come from a field F. We are given a set of n points $x_0, \ldots, x_{n-1}$ and wish to calculate a transformation of the type

$$T_{(x_0, \ldots, x_{n-1})}\ (a_0, \ldots, a_{n-1}) = (\hat{a}_0, \ldots, \hat{a}_{n-1}) \tag{4.6}$$

where, for each i

$$\hat{a}_i = a_0 + a_1 \cdot x_i + \cdots + a_{n-1} x_i^{n-1}.$$

By setting

$$a(x) = a_0 + a_1 x + \cdots + a_{n-1} x^{n-1} \tag{4.7}$$

we see that this is the same as the problem of evaluating polynomials of degree at most $n-1$ at the points $\{x_0, \ldots, x_{n-1}\}$.

In analyzing the conversion process in the polynomial case, note that the cost of the forward transform is the cost of polynomial evaluation. The normal cost of evaluating a polynomial of degree $n-1$ (using say Horner's method) at a single point is $O(n)$ operations. Thus, in general the forward transformation (4.6) from a coefficient representation to a modular representation requires $O(n^2)$ operations. Our goal is to reduce the cost of such a transformation. We accomplish this by picking a special set of evaluation points combined with a special divide-and-conquer evaluation technique that can be applied at these special points.

Consider the problem of polynomial evaluation. If $a(x)$ is given by (4.7) with n even, then one can rewrite $a(x)$ in the form

$$a(x) = b(x^2) + x \cdot c(x^2) \tag{4.8}$$

where

$$b(y) = a_0 + a_2 \cdot y + \cdots + a_{n-2} \cdot y^{n/2-1}$$

and

$$c(y) = a_1 + a_3 \cdot y + \cdots + a_{n-1} \cdot y^{n/2-1}.$$

Notice that both $b(y)$ and $c(y)$ have degree at most one-half the degree of $a(x)$.

Lemma 4.1. Let $\{x_0, \ldots, x_{n-1}\}$ be a set of n points satisfying the symmetry condition

$$x_{(n/2)+i} = -x_i \tag{4.9}$$

for $i \in \{0, 1, \ldots, n/2-1\}$. If $T(n)$ is the cost of evaluating a polynomial of degree $n-1$ at these n points, then

$$T(1) = 0, \text{ and } T(n) = 2 \cdot T(\frac{n}{2}) + c \cdot \frac{n}{2} \tag{4.10}$$

for some constant c.

Proof: Equation (4.9) implies that

$$x_0^2 = x_{n/2}^2, \; x_1^2 = x_{n/2+1}^2, \; \ldots, \; x_{n/2-1}^2 = x_{n-1}^2$$

and hence there are only $n/2$ distinct squares. A polynomial (4.7) of degree at most $n-1$ can be evaluated at the n points $\{x_0, \ldots, x_{n-1}\}$ by evaluating the polynomials $b(y)$ and $c(y)$ at the $n/2$ points

$$\{x_0^2, \ldots, x_{n/2-1}^2\} \tag{4.11}$$

and then using formula (4.8) to combine the results into the desired evaluation. The overhead cost of such a process is $n/2$ multiplications to obtain the squares, and $n/2$ multiplications, additions and subtractions, respectively, to combine the smaller evaluations. Thus, the cost of evaluation satisfies the relation (4.10).

●

The *fast Fourier transform (FFT)* exploits Lemma 4.1 in a recursive manner. However, for the recursion to work we need the symmetry property to also hold for the $n/2$ points (4.11) and so on down the line. For this to work we need the symmetry of n-th roots of unity. We will restrict ourselves to working over fields having primitive n-th roots of unity.

Definition 4.1. An element ω of the field F is a primitive n-th root of unity if

$$\omega^n = 1 \;, \text{ but } \omega^k \neq 1 \text{ for } 0 < k < n.$$

When ω is a primitive n-th root of unity, the set of n points

$$\{1, \omega, \omega^2, \ldots, \omega^{n-1}\} \tag{4.12}$$

are called *Fourier* points. The evaluation transformation at the Fourier points

$$T_{(1, \omega, \ldots, \omega^{n-1})} \tag{4.13}$$

given by (4.6) is called the *discrete Fourier transform (DFT)*.

●

Example 4.3. Let F = **C**, the field of complex numbers, and let $n = 8$. Then

$$\omega = e^{\pi \cdot i/4} = \frac{(1+i)}{\sqrt{2}}$$

is a primitive 8-th root of unity, while

$$\omega = e^{\pi \cdot i/2} = i$$

satisfies $\omega^8 = 1$, but also $\omega^4 = 1$ and hence is an 8-th root of unity which is not primitive.

●

Example 4.4. In $\mathbf{Z}_{17}$, 4 is a 4-th root of unity since $4^4 = 256 = 1 \bmod 17$. It is also primitive since

$$4^2 = 16\ , \text{ and } 4^3 = 13.$$

The corresponding set of Fourier points are

$$\{1, 4, 4^2, 4^3\} = \{1, 4, 16, 13\}.$$

The associated DFT, $T_{(1,4,16,13)}$, is a linear transformation from the vector space $(\mathbf{Z}_{17})^4$ to itself. Its matrix in the standard basis is given by

$$\begin{bmatrix} 1 & 1 & 1 & 1 \\ 1 & 4 & 16 & 13 \\ 1 & 16 & 1 & 16 \\ 1 & 13 & 16 & 4 \end{bmatrix}. \tag{4.14}$$

●

Lemma 4.2. If ω is a primitive n-th root of unity, then the n Fourier points satisfy the symmetry condition (4.9).

Proof: Since ω is a primitive n-th root of unity, we have

$$(\omega^{n/2+j})^2 = \omega^n \cdot (\omega^j)^2 = (\omega^j)^2$$

hence

$$(\omega^{n/2+j} - \omega^j) \cdot (\omega^{n/2+j} + \omega^j) = ((\omega^{n/2+j})^2 - (\omega^j)^2) = 0.$$

If

$$\omega^{n/2+j} - \omega^j = 0$$

then

$$\omega^{n/2} = 1,$$

contradicting the assumption that ω is a primitive n-th root of unity. Therefore

$$\omega^{n/2+j} + \omega^j = 0$$

which is equivalent to equation (4.9) in the case of the Fourier points.

●

Lemma 4.3. Let ω be a primitive n-th root of unity with n even. Then

(a) ω^2 is a primitive $n/2$-th root of unity and

(b) the $n/2$ squares

$$\{1, w^2, w^4, \ldots, w^n\}$$

satisfy the symmetry condition (4.9).

Proof: That ω^2 is an $n/2$-th root of unity follows from

$$(\omega^2)^{n/2} = \omega^n = 1.$$

To see that it is also primitive, suppose $k < n/2$ and

$$(\omega^2)^k = 1.$$

Then

$$\omega^{2 \cdot k} = 1 \quad \text{with} \quad 2 \cdot k < n$$

contradicting that ω is a primitive n-th root of unity. Hence ω^2 is a primitive $n/2$ -th root of unity. The second statement of Lemma 4.3 follows directly from Lemma 4.2.

●

Lemma 4.3 implies that when ω is a primitive n-th root of unity, the set of Fourier points generated by ω provides a set of points that allow equation (4.8) to be evaluated recursively.

Example 4.5. In $\mathbf{Z}_{41}$, 14 is a primitive 8-th root of unity. The corresponding set of Fourier points is

$$\{ 1, 14, -9, -3, -1, -14, 9, 3 \}$$

which clearly satisfies (4.9). Also, $14^2 = -9$ is a primitive 4-th root of unity with the set of Fourier points given by

$$\{ 1, -9, -1, 9 \}.$$

These points also clearly satisfy (4.9). Finally, $(-9)^2 = -1$ is a primitive 2-nd root of unity with the Fourier points given by

$$\{ 1, -1 \}.$$

Again the symmetry condition (4.9) holds.

●

Theorem 4.1. Let ω be a primitive n-th root of unity. Then the DFT defined from the n Fourier points can be calculated in $O(n \cdot \log n)$ operations.

Proof: We will prove Theorem 4.1 when $n = 2^m$ for some integer m. Lemma 4.3 implies that the cost function $T(n)$ satisfies the recursion

$$T(1) = 0, \quad T(2^k) = 2 \cdot T(2^{k-1}) + c \cdot 2^{k-1} \quad \text{for} \quad k \geq 1.$$

Therefore the cost function simplifies to

$$\begin{aligned}T(n) &= T(2^m) = 2 \cdot T(2^{m-1}) + c \cdot 2^{m-1} \\ &= 2^2 \cdot T(2^{m-2}) + c \cdot 2^{m-1} \cdot 2 \\ &= 2^3 \cdot T(2^{m-3}) + c \cdot 2^{m-1} \cdot 3 = \cdots = 2^m \cdot T(1) + c \cdot 2^{m-1} \cdot m \\ &= c \cdot 2^{m-1} \cdot m = c \cdot \frac{n}{2} \cdot \log n,\end{aligned}$$

proving our result. •

Example 4.6. Let $a(x)$ be the polynomial

$$a(x) = 5x^6 + x^5 + 3x^3 + x^2 - 4x + 1,$$

a polynomial in $\mathbf{Z}_{41}[x]$. Then $a(x)$ is the same as

$$a(x) = b(y) + xc(y)$$

where $y = x^2$ and

$$b(y) = 5y^3 + y + 1, \;\; c(y) = y^2 + 3y - 4.$$

Thus, if we wish to evaluate $a(x)$ at the 8 Fourier points

$$\{ 1, 14, -9, -3, -1, -14, 9, 3 \}$$

then this would be the same as evaluating $b(y)$ and $c(y)$ at the 4 Fourier points

$$\{1, -9, -1, 9 \}.$$

Writing $b(y)$ as

$$b(y) = d(z) + ye(z)$$

where $z = y^2$ and

$$d(z) = 1, \;\; e(z) = 5z + 1,$$

we see that evaluating $b(y)$ at the 4 Fourier points is the same as evaluating $d(z)$ and $e(z)$ at the two points

$$\{ 1, -1 \}.$$

A similar operation will be done when evaluating $c(y)$ at the 4 Fourier points. As a result, we obtain

$$d(1) = 1, e(1) = 6, \;\; \Rightarrow \;\; b(1) = 7, b(-1) = -5,$$

$$d(-1) = 1, e(-1) = -4, \;\; \Rightarrow \;\; b(-9) = -4, b(9) = 6.$$

In a similar manner we deduce that

$$c(1) = 0, \;\; c(-1) = -2, \;\; c(-9) = 9, c(9) = -19.$$

Thus we may now determine the components of the FFT of $a(x)$. For example, we have

$$a(3) = b(9) + 3 \cdot c(9) = -10, \text{ and } a(-3) = b(9) - 3 \cdot c(9) = -19.$$

Calculating the other 6 components gives

$$A \leftarrow \text{FFT}(8, 14, a(x)) = (7, -1, 8, -19, 7, -7, -18, -10).$$

●

Thus, as long as there exist n-th roots of unity in the coefficient field, we can transform the coefficient domain to the modular domain in $O(n \cdot \log n)$ operations, rather than the $O(n^2)$ operations required before.

Algorithm 4.4. Fast Fourier Transform (FFT).

procedure FFT($N, \omega, a(x)$)

\# Given N, a power of 2, ω a primitive N-th root of
\# unity and $a(x)$ a polynomial of degree $\le N-1$, we calculate
\# the N components of the Fourier transform of $a(x)$.

if N=1 **then** $A_0 \leftarrow a_0$

else {

$b(x) \leftarrow \sum_{i=0}^{N/2-1} a_{2i} \cdot x^i$; $c(x) \leftarrow \sum_{i=0}^{N/2-1} a_{2i+1} \cdot x^i$

$B \leftarrow \text{FFT}(N/2, \omega^2, b(x))$; $C \leftarrow \text{FFT}(N/2, \omega^2, c(x))$

for i **from** 0 **to** $N/2 - 1$ **do** {

$A_i \leftarrow B_i + \omega^i \cdot C_i$

$A_{N/2+i} \leftarrow B_i - \omega^i \cdot C_i$ } }

return($(A_0, A_1, \ldots, A_{N-1})$)

end

4.6. THE INVERSE FOURIER TRANSFORM

Consider now the problem of transforming from the modular domain back to a coefficient domain. Thus, for a set of points $\{x_0, \ldots, x_{n-1}\}$ we are looking for

$$T_{(x_0, \ldots, x_{n-1})}^{-1}.$$

Since the matrix of the linear transformation $T_{(x_0, \ldots, x_{n-1})}$ with respect to the standard basis is the Vandermonde matrix

$$V(x_0, \ldots, x_{n-1}) = \begin{bmatrix} 1 & x_0 & \cdot & (x_0)^{n-1} \\ 1 & x_1 & \cdot & (x_1)^{n-1} \\ \cdot & \cdot & \cdot & \cdot \\ \cdot & \cdot & \cdot & \cdot \\ 1 & x_{n-1} & \cdot & (x_{n-1})^{n-1} \end{bmatrix}$$

the problem is the same as finding the inverse of this $n \times n$ matrix. Using Gaussian elimination such an inverse can be determined in $O(n^3)$ operations.

However, the Vandermonde matrix is a highly structured matrix, and hence it comes as no surprise that such a matrix can be inverted in less than $O(n^3)$ operations. Indeed, the problem of transforming from the modular domain to the coefficient domain is really a problem of polynomial interpolation. That is, we are given a set of n points $(q_0, \ldots, q_{n-1})$ and we are looking for a polynomial $a(x)$ of degree at most $n-1$ such that

$$\hat{a}_i = a(x_i) = q_i, \quad \text{for} \quad i = 0, 1, \ldots, n-1.$$

Thus, this problem can be solved using either Lagrange interpolation or Newton interpolation (cf. Chapter 5) at the reduced cost of $O(n^2)$ operations, an improvement of one order of magnitude. However, as mentioned in Section 4.4, converting from a coefficient domain to a modular domain to take advantage of the efficient multiplication in a modular domain, and then converting back again, is only useful when both transforms can be done with less than $O(n^2)$ operations. Hence we seek a faster inverse transform. Again primitive roots of unity and the corresponding Fourier points provide the correct mechanism via the DFT.

We note that transforming problems from one domain to a second (simpler) domain and then back again is a common technique in mathematics and its related disciplines. Indeed, a standard technique in engineering, low level image processing, makes use of the Fourier transform to convert from the time domain to the frequency domain and back again. The advantage of such a transformation is that the frequency domain represents a more natural environment for formulating and solving problems.

To solve the problem of inverting our DFT, consider what happens in the case of inverting the *continuous* Fourier transform used in engineering. The analytic Fourier transform of a continuous function $f(x)$ is defined by

$$F(s) = \int_{-\infty}^{\infty} f(x) \cdot e^{2\pi \cdot i \cdot s \cdot x} dx. \tag{4.15}$$

Notice that, if the function $f(x)$ is defined by a discrete (i.e. finite) set of samples, $f_0, \ldots, f_{n-1}$, rather than a continuous sample, then the discrete version of (4.15) becomes

$$F_s = \sum_{k=0}^{n-1} f_k \cdot e^{2\pi \cdot i \cdot s \cdot k/n} = \sum_{k=0}^{n-1} f_k (\omega^s)^k \tag{4.16}$$

where $\omega = e^{2\pi \cdot i/n}$. Since ω is a primitive n-th root of unity over the complex numbers, the discrete version of (4.15) given by (4.16) is precisely the forward DFT of the preceding section. This explains how our discrete Fourier transform gets its name. It will also help us in inverting our transform.

In the continuous analytic case, the inverse Fourier transform problem is similar to the Fourier transform problem. If $F(s)$ given in (4.15) is the Fourier transform of $f(x)$, then the inverse Fourier transform is given by

$$f(x) = \frac{1}{2\pi} \cdot \int_{-\infty}^{\infty} F(s) \cdot e^{-2\pi \cdot i \cdot s \cdot x} ds. \tag{4.17}$$

The discrete transform corresponding to (4.17) is given by

$$f_j = \frac{1}{n} \cdot \sum_{k=0}^{n-1} F_k \cdot (e^{-2\pi \cdot i \cdot j / n})^k = 1/n \cdot \sum_{k=0}^{n-1} F_k \cdot (\omega^{-j})^k$$

where $\omega = e^{2\pi \cdot i / n}$ is a primitive n-th root of unity over the field of complex numbers.

Definition 4.2. The *inverse discrete Fourier transform (IDFT)* for a Fourier set of points (4.12) is

$$S_{(1,\, \omega,\, \ldots,\, \omega^{n-1})}(q_0, \ldots, q_{n-1}) = (\bar{q}_0, \ldots, \bar{q}_{n-1})$$

where

$$\bar{q}_j = n^{-1} \cdot \sum_{k=0}^{n-1} q_k \cdot (\omega^{-j})^k,$$

and ω is a primitive n-th root of unity.

●

Theorem 4.2. The DFT and the IDFT transform are inverses of each other.

Proof: Let $0 < p < n$. Then

$$(\omega^p)^n = (\omega^n)^p = 1, \quad \text{and} \quad (\omega^p) \neq 1,$$

since ω is a primitive n-th root of unity. Since

$$(x^n - 1) = (x - 1) \cdot (x^{n-1} + x^{n-2} + \cdots + x + 1)$$

ω^p must be a root of the second factor, that is

$$0 = (\omega^p)^{n-1} + (\omega^p)^{n-2} + \cdots + (\omega^p) + 1. \tag{4.18}$$

By multiplying through by $\omega^{-p \cdot (n-1)}$, it is also possible to see that equation (4.18) is true for $-n < p < 0$. Of course, when $p = 0$ the right hand side of equation (4.18) will be n rather than 0.

The simple observations of the previous paragraph provide the tools necessary for a proof of Theorem 4.2. Suppose that

$$T_{(1,\omega,\, \ldots,\, \omega^{n-1})} \cdot (a_0, \ldots, a_{n-1}) = (\hat{a}_0, \ldots, \hat{a}_{n-1})$$

that is,

$$\hat{a}_i = \sum_{j=0}^{n-1} a_j (\omega^i)^j, \text{ for } i = 0, 1, \ldots, n-1.$$

For any integer k we have

$$\begin{aligned} n^{-1} \cdot \sum_{i=0}^{n-1} \hat{a}_i\, \omega^{-ki} &= n^{-1} \cdot \sum_{i=0}^{n-1} \sum_{j=0}^{n-1} a_j\, \omega^{ij} \cdot \omega^{-ki} \\ &= n^{-1} \cdot \sum_{j=0}^{n-1} a_j \sum_{i=0}^{n-1} \omega^{ij} \cdot \omega^{-ki} \\ &= n^{-1} \cdot \sum_{j=0}^{n-1} a_j \left(\sum_{i=0}^{n-1} \omega^{(j-k)i} \right) = a_k \end{aligned}$$

since the inside summation works out to 0 in the cases when $j - k \neq 0$, and n in the case where $j = k$. Using Definition 4.2, we obtain

$$S_{(1,\, \omega,\, \ldots,\, \omega^{n-1})} \cdot (\hat{a}_0, \ldots, \hat{a}_{n-1}) = (a_0, \ldots, a_{n-1})$$

showing that $S_{(1,\, \omega,\, \ldots,\, \omega^{n-1})}$ is the inverse of $T_{(1,\, \omega,\, \ldots,\, \omega^{n-1})}$. This proves Theorem 4.2.

●

In terms of our matrix interpretation for the interpolation, Theorem 4.2 gives the inverse of the Vandermonde matrix as

$$V(1, \omega, \ldots, \omega^{n-1})^{-1} = n^{-1} \cdot \begin{bmatrix} 1 & 1 & \cdot & 1 \\ 1 & \omega^{-1} & \cdot & \omega^{-(n-1)} \\ \cdot & \cdot & \cdot & \cdot \\ \cdot & \cdot & \cdot & \cdot \\ 1 & \omega^{-(n-1)} & \cdot & \omega^{-(n-1)^2} \end{bmatrix}$$

$$= n^{-1} \cdot V(1, \omega^{-1}, \ldots, \omega^{-(n-1)}).$$

Example 4.7. Let the field be $\mathbf{Z}_{17}$ with $\omega = 4$, a primitive 4-th root of unity. Then the inverse transform $S_{(1,4,16,13)}$ is a linear transformation from $(\mathbf{Z}_{17})^4$ to itself having as its matrix with respect to the standard basis

$$4^{-1} \begin{bmatrix} 1 & 1 & 1 & 1 \\ 1 & 13 & 16 & 4 \\ 1 & 16 & 1 & 16 \\ 1 & 4 & 16 & 4 \end{bmatrix} = \begin{bmatrix} 13 & 13 & 13 & 13 \\ 13 & 16 & 4 & 1 \\ 13 & 4 & 13 & 4 \\ 13 & 1 & 4 & 16 \end{bmatrix}.$$

This is easily checked to be the inverse of the matrix (4.14) of Example 4.4.

●

The significance of Theorem 4.2 is that both the forward and inverse transformations are Fourier transforms. Thus the transformation from coefficient domain to modular domain is via the calculation of

$$T_{(1,\omega,\ \dots,\ \omega^{n-1})}\cdot\mathbf{p} = \mathbf{q}$$

while the transformation from the modular domain to the coefficient domain is via the calculation of

$$n^{-1}\cdot T_{(1,\omega^{-1},\ \dots,\ \omega^{-(n-1)})}\cdot\mathbf{q} = \mathbf{p}\ .$$

The complexity in either case is $O(n \cdot \log n)$, which is faster than $O(n^2)$.

4.7. FAST POLYNOMIAL MULTIPLICATION

Let us now return to our original goal, that of multiplying two polynomials $a(x)$ and $b(x)$. If our polynomials are of degree m and n, and our coefficient field F has a required N-th root of unity where N is the first power of 2 greater than the sum of the two degrees, then a fast multiplication algorithm is given by Algorithm 4.5.

Algorithm 4.5. Fast Fourier Polynomial Multiplication.

procedure FFT_Multiply($a(x), b(x), m, n$)

    # Given polynomials $a(x)$ and $b(x)$ of degree m and n
    # calculate $c(x) = a(x) \cdot b(x)$ using FFT's.

    $N \leftarrow$ first power of 2 greater than $m+n$
    $\omega \leftarrow$ primitive N–th root of unity
    $A \leftarrow \mathrm{FFT}(N, \omega, a(x))$
    $B \leftarrow \mathrm{FFT}(N, \omega, b(x))$

    for i **from** 0 **to** $N-1$ **do** $C_i = A_i \cdot B_i$

    $c \leftarrow N^{-1} \cdot \mathrm{FFT}(N, \omega^{-1}, C(x))$

    $c(x) \leftarrow \sum_{i=0}^{N-1} c_i\, x^i$

    return($c(x)$)

end

The fast multiplication algorithm above has a complexity of $O((m+n) \cdot \log\,(m+n))$ compared with $O(m \cdot n)$, the complexity of the classical method, a considerable improvement. Of course, this does not take into consideration the constants of proportionality of each method. In practice, depending on the method of recursion used, the fast method is better than the classical method approximately when $m + n \geq 600$ (cf. Moenck [9]).

Example 4.8. Multiply the two polynomials

$$a(x) = 3x^3 + x^2 - 4x + 1, \;\; b(x) = x^3 + 2x^2 + 5x - 3$$

in the field $\mathbf{Z}_{41}$. Use 14 as a primitive 8-th root of unity for this problem. From the previous section we have

$$A \leftarrow \mathrm{FFT}(8, 14, a(x)) = (1, 9, -19, -18, 3, 16, 19, -3)$$

and

$$B \leftarrow \mathrm{FFT}(8, 14, b(x)) = (5, 5, 0, 14, -7, -6, -10, 16).$$

Multiplying these together gives

$$C \leftarrow (5, 4, 0, -6, 20, -14, 15, -7)$$

which is the FFT of $c(x) = a(x) \cdot b(x)$. To obtain $c(x)$ from C we do

$$c \leftarrow 8^{-1} \cdot \mathrm{FFT}(8, 3, -7x^7 + 15x^6 - 14x^5 + 20x^4 - 6x^3 + 4x + 5)$$

$$= (-3, 17, 20, -11, 13, 7, 3, 0)\,;$$

hence the product is

$$c(x) = 3x^6 + 7x^5 + 13x^4 - 11x^3 + 20x^2 + 17x - 3.$$

●

4.8. COMPUTING PRIMITIVE N-th ROOTS OF UNITY

When applying the FFT for polynomial multiplication over a given field F, we are faced with the problem of finding a primitive n-th root of unity in F. When F is the field of complex numbers there is no problem with determining a primitive n-th root of unity for a given integer n. A simple example of such a primitive root is

$$\omega = e^{2\pi i/n}.$$

For example,

$$e^{\pi \cdot i/6} = (\sqrt{3} + i)/2$$

is a primitive 12-th root of unity over the complex numbers. When dealing with other fields, for example, the finite fields $\mathbf{Z}_p$, the situation is not as simple. In this section we discuss the practical problem of finding primitive n-th roots of unity for finite fields.

Theorem 4.3. The finite field $\mathbf{Z}_p$ has a primitive n-th root of unity if and only if n divides $p - 1$.

Proof: If ω is a primitive n-th root of unity in $\mathbf{Z}_p$, then the set of Fourier points

$$\{1, \omega, \ldots, \omega^{n-1}\}$$

forms a (cyclic) subgroup of the multiplicative group of $\mathbf{Z}_p$. Since this multiplicative group has $p-1$ elements while the subgroup has n elements, Lagrange's theorem from group theory (cf. Herstein [4]) implies that n must divide $p - 1$.

Conversely, we know from finite field theory that the multiplicative group of the field $\mathbf{Z}_p$ is a cyclic group (this is true in general for all finite fields of order p^k for some integer k). Let α be any generator of this multiplicative group, that is

$$\mathbf{Z}_p = \{ 1, \alpha, \alpha^2, \ldots, \alpha^{p-2} \} \quad \text{with} \quad \alpha^{p-1} = 1.$$

Let n be an integer which divides $p - 1$. If we set

$$\omega = \alpha^{(p-1)/n}, \tag{4.19}$$

then

$$\omega^n = \alpha^{p-1} = 1;$$

so, ω is an n-th root of unity. For $0 < k < n$, we have the inequality $(p-1) \cdot k/n < (p-1)$, so

$$\omega^k = \alpha^{(p-1) \cdot k/n} \neq 1. \tag{4.20}$$

Equation (4.20) implies that ω is a primitive n-th root of unity, completing our proof.

●

Example 4.9. In $\mathbf{Z}_{41}$ we have $8 \mid (41 - 1)$ so that there are primitive 8-th roots of unity in $\mathbf{Z}_{41}$. Indeed, the element 14 is a primitive 8-th root of unity in $\mathbf{Z}_{41}$.

●

The problem we now face is that, although we may know when a primitive n-th root of unity exists in $\mathbf{Z}_p$, we have no way of determining what that primitive n-th root might be. One approach is to check each ω in $\mathbf{Z}_p$ and stop when there is one satisfying $\omega^n = 1$ and $\omega^k \neq 1$ for $k < n$. This is a very costly procedure however (as could well be imagined). A better approach would be to find the particular α that generates the multiplicative group and then use equation (4.19) to define ω.

In our case, we are actually interested in primes p and integers n of the form $n = 2^r$, for which there is a primitive n-th root of unity in $\mathbf{Z}_p$. By Theorem 4.3, $\mathbf{Z}_p$ has such a root if and only if 2^r divides $p-1$, that is, if and only if p is a prime of the form

$$p = 2^r \cdot k + 1$$

for some odd integer k. Such primes are called *Fourier primes*. In the case of Fourier primes, the brute force method mentioned in the last paragraph does have some merit. This is because there are a large number of primitive elements in $\mathbf{Z}_p$ in this case. To see how many, we quote a fundamental result from analytic number theory.

Theorem 4.4. Let a and b be two relatively prime integers. Then the number of primes $\leq x$ in the arithmetic progression

$$a \cdot k + b, \quad k = 1, 2, \ldots$$

is approximately $\frac{x}{\log x \cdot \phi(a)}$ where $\phi(a)$ is the Euler phi function (i.e. the number of integers less than and relatively prime to a).

●

Since all odd integers less than 2^r are relatively prime to 2^r, and these account for approximately half the total number of integers, we have $\phi(2^r) \approx 2^{r-1}$. Theorem 4.4 then tells us that there are approximately

$$\frac{x}{\log x \cdot 2^{r-1}}$$

Fourier primes less than a given integer x.

Example 4.10. Let $x = 2^{31}$, which represents the usual size required for single precision integers in most computers. When $r = 20$ there are approximately

$$\frac{2^{31}}{\log(2^{31}) \cdot \phi(2^{19})} \approx 130$$

primes of the form $2^e \cdot k + 1$ with $e \geq 20$ in the interval 2^{20} to 2^{31}. Any such Fourier prime could be used to compute FFT's of size 2^{20}.

●

Even though we know that we can use brute force to find our primitive generators of the multiplicative group of $\mathbf{Z}_p$ (and correspondingly our primitive roots of unity), we still need a simple way to recognize that a given element is a primitive generator. For this we have

Theorem 4.5. An element a is a generator of the multiplicative group of $\mathbf{Z}_p$ iff

$$a^{(p-1)/q} \neq 1 \mod p$$

for every prime factor, q, of the integer $p - 1$.

Proof: This is a simple consequence of Lagrange's theorem from group theory.

●

Theorem 4.5 allows for a probabilistic algorithm to determine a generating element of $\mathbf{Z}_p$. First factor $p - 1$. Note that this is tractable for $p \approx 2^{31}$, and besides this only needs to be done a single time with the factors stored in a table in the program. Choose an integer a at random from $2, \ldots, p-1$. Then for every prime factor q, of $p-1$, calculate

$$a^{\frac{p-1}{q}}.$$

If this quantity is not 1, then it is a generating element.

Example 4.11. Since $41 - 1 = 40 = 2^3 \cdot 5$, an element of $\mathbf{Z}_{41}$ is a generator of the multiplicative group if and only if it is not the identity when taken to either the 8-th or 20-th powers.

Choose a random element of $\mathbf{Z}_{41}$, say 15. Then, modulo 41 we have

$$15^8 = 18 \neq 1, \quad \text{and} \quad 15^{20} = -1 \neq 1\,;$$

hence 15 is a primitive 40-th root of unity.

●

How lucky do we need to be for this probabilistic procedure to work? From finite group theory the number of primitive generators in $\mathbf{Z}_p$ is just $\phi(p-1)$ and so the percentage of primitive elements will be $\phi(p-1)/(p-1)$. Number theory tells us that this quantity is, on average, $3/\pi^2$. Thus we would be expected on average to find a primitive generator with probability ≈ 0.3, i.e. a success rate of about one in every three attempts, an acceptable ratio.

4.9. NEWTON'S METHOD FOR POWER SERIES DIVISION

In the previous sections we presented an algorithm that (over certain coefficient fields) multiplied two polynomials of degrees m and n, respectively, with an asymptotic complexity of $O((m+n)\cdot\log(m+n))$, rather than $O(m\cdot n)$, the complexity of the classical polynomial multiplication. It is a natural question to ask whether the same speedup may be applied to division of two polynomials. That is, given two polynomials $a(x)$ and $b(x)$ is it possible to find polynomials $q(x)$ and $r(x)$ such that

$$\frac{a(x)}{b(x)} = q(x) + \frac{r(x)}{b(x)} \quad \text{with} \quad \deg(r(x)) < \deg(b(x)) \tag{4.21}$$

and such that the number of operations is less than the number of operations in the classical method?

Let

$$a^*(x) = x^m a(1/x), \; b^*(x) = x^n b(1/x) \tag{4.22}$$

where m and n are the degrees of $a(x)$ and $b(x)$, respectively. The $a^*(x)$ and $b^*(x)$ defined as in (4.22) are called reciprocal polynomials of $a(x)$ and $b(x)$, respectively. They are the same polynomials as $a(x)$ and $b(x)$, except with the coefficients in reversed order. Thus, for example, if

$$a(x) = 3 - x^2 + 4x^3$$

then

$$a^*(x) = x^3\cdot(3 - (\frac{1}{x})^2 + 4(\frac{1}{x})^3) = 3x^3 - x + 4.$$

Equation (4.21) can be rewritten as

$$\frac{a^*(x)}{b^*(x)} = q^*(x) + x^{m-n+\lambda}.\frac{r^*(x)}{b^*(x)} \quad \text{with} \quad \lambda \geq 1,$$

and hence to find the quotient $q(x)$ in (4.21) it is sufficient to find the first $m-n+1$ terms of the quotient power series defined by $a^*(x)/b^*(x)$. This formulation of the problem can in turn be stated as finding the first $m-n+1$ terms of the power series for $b^*(x)^{-1}$ and multiplying the result by $a^*(x)$. Thus to find a fast division algorithm, we look for a fast algorithm for calculating the truncated power series for $1/p(x)$, with $p(x)$ a polynomial.

A fast technique for calculating the terms in a power series expansion is based on the well known Newton's iteration scheme for solving a nonlinear equation

$$f(x) = 0. \tag{4.23}$$

Newton's method assumes the existence of a point x_0, "near" the correct answer. It then approximates the graph of the function about this point by the tangent line to the graph at the point $(x_0, f(x_0))$, that is

$$f(x) \approx T(x) = f(x_0) + f'(x_0) \cdot (x - x_0).$$

Equation (4.23) is therefore approximated by the linear equation

$$T(x) = 0. \tag{4.24}$$

The solution to equation (4.24) is the point where the tangent line crosses the x-axis. This point, x_1, becomes the next approximation to the root. Solving for x_1 in equation (4.24) gives

$$x_1 = x_0 - \frac{f(x_0)}{f'(x_0)}. \tag{4.25}$$

The process is then repeated with x_1 representing the new point "near" the correct solution of (4.23). With a good initial point x_0 and a suitably well-behaved function $f(x)$, the iteration does indeed converge to the correct root of the nonlinear equation. Furthermore, at least in the case where the solution is a simple root, the convergence is quadratic. Let e_k be the error at the k-th step of Newton's iteration. Then quadratic convergence means that there exists a constant C with

$$\| e_{n+1} \| \le C \cdot \| e_n \|^2$$

for all n. In the context of nonlinear equations this is roughly the statement that there is an integer N such that the number of correct digits in the approximations doubles every N iterations. The fact that Newton's method converges quadratically in the case of a simple root follows from considering the Taylor series expansion of f from which the linear term has been cancelled by the choice of x_{n+1}.

Newton's method for solving a nonlinear equation has been used previously to determine approximations to "reciprocals". In the early days of computers, division was often not implemented in hardware. Rather Newton's method was used to find floating-point representations of reciprocals of integers. For example, to find the floating-point representation of 1/7 is the same as finding the solution to the nonlinear equation

$$f(x) = 7 - \frac{1}{x} = 0.$$

In this case, Newton's iteration formula simplifies to

$$x_{n+1} = x_n - \frac{7-1/x_n}{1/x_n{}^2} = 2 \cdot x_n - 7 \cdot x_n{}^2$$

which requires no division. If we use as our initial point $x_0 = 0.1$, then the subsequent approximations are

$$x_1 = 0.13, \ x_2 = 0.1417, \ x_3 = 0.14284777, \ \ x_4 = 0.1428571423, \ldots .$$

The last approximation is correct to 9 decimal places. Notice that from x_1 to x_2 there is 1 extra correct digit, from x_2 to x_3 there are 2 more correct digits, and from x_3 to x_4 there are at least 4 more correct digits. This provides a good example of quadratic convergence.

It is the fact that Newton's method is primarily an algebraic algorithm, rather than an analytic algorithm, that enables us to use it to solve problems in algebraic computation. However, to use such an algorithm, we must make clear the notions of convergence in the domain of power series F[[x]].

Definition 4.3. A power series $\bar{y}(x)$ is an *order n approximation* of $y(x)$ if

$$\bar{y}(x) = y(x) + \mathrm{O}(x^n). \tag{4.26}$$

●

Thus $\bar{y}(x)$ is an order n approximation to $y(x)$ if the two power series have the same first n terms. A sequence of power series will then converge to a given power series if the sequence approximates the power series to higher and higher terms. In this context, quadratic convergence will mean that there are twice as many terms that are correct than at a previous step as we iterate.

In a polynomial or power series domain F[x] or F[[x]], we can do "mod x^k" arithmetic. A polynomial or power series modulo x^k is just the remainder after dividing by the polynomial x^k. In this form equation (4.26) becomes

$$\bar{y}(x) = y(x) \mod x^{n+1},$$

providing an alternate description of an order n approximation.

Consider now the special case of finding the power series of the reciprocal of a polynomial $a(x)$ with $a(0) \neq 0$. One can of course solve this directly as in Chapter 2 by

$$a(x)^{-1} = \hat{a}_0 + \hat{a}_1 x + \hat{a}_2 x^2 + \cdots$$

where $\hat{a}_0 = \dfrac{1}{a_0}$ and

$$\hat{a}_i = \hat{a}_0 \cdot (\hat{a}_{i-1} \cdot a_1 + \cdots + \hat{a}_0 \cdot a_i) \ \text{ for } i \geq 1.$$

Unfortunately, the cost of determining the first n terms of the reciprocal satisfies the recurrence relation

$$\mathrm{T}(1) = 0, \ \text{ and } \ \mathrm{T}(n) = \mathrm{T}(n-1) + c \cdot (n-1)$$

for some constant c. Solving this recurrence in the same way as earlier in the chapter gives

$$T(n) = C \cdot n^2,$$

that is, the cost of determining the first n terms of the power series is $O(n^2)$.

Suppose that we use Newton's method instead, in a manner similar to the case of computing the floating-point representation of 1/7. Finding the reciprocal power series is the same as solving the equation

$$f(y) = a(x) - \frac{1}{y} = 0.$$

Newton's iteration scheme in this case will be stated as

$$y_{n+1} = y_n - \{ \frac{f(y_n)}{f'(y_n)} \mod x^{2^{n+1}} \}. \tag{4.27}$$

The mod operation is there because we will be finding only the first 2^{n+1} terms of our reciprocal power series per iteration. In this case, as was the case where we worked out 1/7, Newton's iteration simplifies to

$$y_{n+1} = y_n - \frac{a(x) - 1/y_n}{1/y_n{}^2} \mod x^{2^{n+1}} = y_n \cdot (2 - y_n \cdot a(x)) \mod x^{2^{n+1}}.$$

Since

$$f(\frac{1}{a_0}) = a(x) - a_0 = 0 \mod x \tag{4.28}$$

we can let

$$y_0 = \frac{1}{a_0}$$

and so obtain a starting value. Equation (4.28) can be thought of as representing a solution that has the first "decimal" of the answer correct.

Example 4.12. Let

$$a(x) = 1 - 2x + 3x^2 + x^4 - x^5 + 2x^6 + x^7 + \cdots \quad \in \mathbf{Z}_7[[x]].$$

Then, using the iteration scheme above gives the approximations

$$a(x)^{-1} \bmod x = 1,$$

$$a(x)^{-1} \bmod x^2 = 1 + 2x,$$

$$a(x)^{-1} \bmod x^4 = 1 + 2x + x^2 + 3x^3,$$

$$a(x)^{-1} \bmod x^8 = 1 + 2x + x^2 + 3x^3 + 2x^4 + x^5 + 2x^6 + x^7.$$

As before, it is easy to see the quadratic convergence to the reciprocal in this case in the context of *power series convergence*.

•

Algorithm 4.6. Newton's Method for Power Series Inversion.

```
procedure FastNewtonInversion(a(x),n)
    # Given a power series a(x) in x with a(0) = a_0 ≠ 0,
    # find the first 2^n terms of the power series 1/a(x).
    y ← 1/a_0
    for k from 0 to n−1 do
      y ← y·(2 − y·a(x)) mod x^(2^(k+1))
    return(y)
end
```

There are some observations that are useful when considering the cost of Newton's method for finding reciprocals. Notice that

$$2 - y_k \cdot a(x) = 1 + O(x^{2^k})$$

which, combined with the knowledge that y_k is a polynomial of degree at most 2^k, means that the cost of determining y_{k+1}, given that y_k is known, is the same as the cost of multiplying two polynomials of degree 2^k. Thus, using Newton's method to invert a power series gives a recurrence relation

$$T(2^{k+1}) = T(2^k) + c \cdot M(2^k) \tag{4.29}$$

where $M(2^k)$ is the cost of multiplying two polynomials of degree 2^k.

Theorem 4.6. Let F be a field. If F supports fast Fourier multiplication then the cost of finding the first n terms of the power series of the reciprocal of a polynomial is $O(n \cdot \log n)$. Otherwise the cost is $O(n^2)$.

Proof: Let $n = 2^m$. From equation (4.29) we obtain

$$T(2^m) = T(2^{m-1}) + c \cdot 2^{m-1} \cdot (m-1) \tag{4.30}$$

since $M(2^{m-1}) = 2^{m-1} \cdot \log(2^{m-1}) = 2^{m-1} \cdot (m-1)$. We will show by induction that

$$T(2^m) \le c \cdot 2^m \cdot m \tag{4.31}$$

for all m. This will prove Theorem 4.6.

Certainly (4.31) holds for $m = 0$, hence we may assume that $m > 0$ and that the result is true for $m-1$. Since the result holds for $m-1$, equation (4.30) implies

$$T(2^m) \le c \cdot 2^{m-1} \cdot (m-1) + c \cdot 2^{m-1} \cdot (m-1) = c \cdot 2^m \cdot (m-1) \le c \cdot 2^m \cdot m.$$

Thus equation (4.31) holds for all m by the induction hypothesis.

●

Returning to the initial query from the beginning of this section that motivated this search, we have

Corollary 4.7. Let F be a field that supports fast Fourier multiplication. Let $a(x)$ and $b(x)$ be two polynomials in F[x] of degrees m and n, respectively, with $m \geq n$. Then the quotient and remainder of $a(x)/b(x)$ can be calculated in O($n \cdot \log n$) operations when $n \approx m/2$.

Proof: Corollary 4.7 follows directly from Theorem 4.6 and the discussion at the start of this section.

●

Generalizing Newton's Method For Solving P(y) = 0

Just as Newton's method in the real setting is developed in a setting more general than just finding floating-point approximations of reciprocals, we will formulate Newton's method for power series as a general problem of solving algebraic equations involving power series. To this end, recall that for a field F, the set F<x> denotes the field of extended power series over F with indeterminate x. Then

Definition 4.4. An extended power series $y(x)$ in F<x> is called an *algebraic function* over the field F<x> if it satisfies a polynomial equation

$$P(y) = 0 \tag{4.32}$$

where $P(y)$ is a polynomial in y with coefficients from the field F<x>.

●

Example 4.13. The Legendre polynomials

$$L_0(t) = 1,\ \ L_1(t) = t,\ L_2(t) = \frac{3}{2}t^2 - \frac{1}{2},\ L_3(t) = \frac{5}{2}t^3 - \frac{3}{2}t, \ldots$$

have a generating function

$$G(t,x) = (1 - 2tx + x^2)^{-1/2},$$

that is,

$$G(t,x) = \sum_{i=0}^{\infty} L_i(t)x^i.$$

To find all the Legendre polynomials is therefore the same problem as finding the power series expansion in x of a function y satisfying

$$P(y) = (1 - 2t \cdot x + x^2) \cdot y^2 - 1 = 0.$$

Notice that $P(y)$ is a polynomial in y with coefficients from the field F<x>. F in turn is $\mathbf{Q}(t)$, the quotient field of $\mathbf{Q}[t]$, the ring of polynomials over the rationals with indeterminate t. The solution, $y(x)$, will also have coefficients which are not just rationals, but polynomials in $\mathbf{Q}[t]$. Thus $y(x)$ is in $\mathbf{Q}[t][[x]]$ which in turn is a subset of $\mathbf{Q}(t)$<x>. Here $\mathbf{Q}[t][[x]]$ is the domain of power series in x with coefficients that are polynomials with rational coefficients, while $\mathbf{Q}(t)<x>$ is the domain of extended power series in x having rational functions over $\mathbf{Q}$ as the coefficients.

●

The notion of convergence in power series domains has been presented in the previous section. To complete our generalization of Newton's iteration procedure to algebraic functions over power series domains, we need to describe a starting point for the iteration process. Thus, we must obtain an initial approximation y_0. This initial approximation is obtained by solving the simpler equation

$$P(y) = 0 \mod x. \tag{4.33}$$

At the k-th step of the iteration we wish to have the first 2^k terms of the power series; hence we need only calculate our iteration up to x^{2^k}. Therefore the power series generalization of equation (4.25) is given by

$$y_{n+1} = y_n - \left(\frac{P(y_n)}{P'(y_n)} \mod x^{2^{n+1}} \right). \tag{4.34}$$

Example 4.14. Let us solve the first two steps of Newton's method as applied to the previous example. The initial value y_0 is determined by solving the equation (4.33), which in our example is just the equation

$$y^2 - 1 = 0.$$

One solution is given by $y_0 = 1$, hence

$$P(y_0) = (1 - 2t \cdot x + x^2) \cdot y_0^2 - 1 = 0 \mod x$$

and so is an order one approximation to the exact power series answer. This gives a starting point for the iteration.

Equation (4.34) with $n = 0$ and $y_0 = 1$ is

$$\begin{aligned} y_1 &= 1 - \left(\frac{P(1)}{P'(1)} \mod x^2 \right) \\ &= 1 - \left(\frac{(-2tx + x^2)}{2(1 - 2tx + x^2)} \mod x^2 \right) \\ &= 1 - \left(-tx + \frac{1}{2}(1 - 4t^2) \cdot x^2 + \cdots \mod x^2 \right) = 1 + tx. \end{aligned}$$

Simple algebra verifies that $P(y_1) = 0 \mod x^2$.

Equation (4.34) with $n = 1$ and $y_1 = 1 + tx$ is

$$y_2 = (1 + tx) - \left(\frac{P(1 + tx)}{P'(1 + tx)} \mod x^4 \right)$$

$$= (1 + tx) - \left(\frac{(1-3t^2)x^2 + (2t-2t^3)x^3 + t^2x^4}{2 - 2tx + (1-2t^2)x^2 + tx^3} \mod x^4 \right)$$

$$= (1 + tx) - \left((\tfrac{3}{2}t^2 - \tfrac{1}{2})\cdot x^2 + (\tfrac{5}{2}t^3 - \tfrac{3}{2}t)\cdot x^3 + \cdots \mod x^4 \right)$$

$$= 1 + t\cdot x + (\tfrac{3}{2}t^2 - \tfrac{1}{2})\cdot x^2 + (\tfrac{5}{2}t^3 - \tfrac{3}{2}t)\cdot x^3.$$

Again, a simple check shows that $P(y_2) = 0 \mod x^4$.

Hence our approximations are

$$y \mod x = 1,$$

$$y \mod x^2 = 1 + tx,$$

$$y \mod x^4 = 1 + tx + (\tfrac{3}{2}t^2 - \tfrac{1}{2})x^2 + (\tfrac{5}{2}t^3 - \tfrac{3}{2}t)x^3$$

which demonstrates the notion of quadratic convergence of our approximations. •

Example 4.15. Newton's method gives a very good method for determining n-th roots of a power series. For example, to calculate the square root of

$$a(x) = 4 + x + 2x^2 + 3x^3 + \cdots$$

means finding a power series y in x satisfying $y^2 = a(x)$, that is

$$P(y) = y^2 - a(x) = 0.$$

Newton's iteration scheme becomes

$$y_{n+1} = y_n - \left(\frac{y_n{}^2 - a(x)}{2y_n} \mod x^{2^{n+1}} \right). \tag{4.35}$$

The initial point of the iteration process is the solution of

$$0 = P(y) \mod x = y^2 - 4$$

which in this case gives one solution as $y_0 = 2$. To determine y_1, we use equation (4.35) to obtain

$$y_1 = 2 - \frac{4 - a(x)}{4} \mod x^2 = 2 + \frac{1}{4}x.$$

Notice that $(2 + \frac{1}{4}x)^2 \mod x^2 = (4 + x)$.

To determine y_2 we calculate

$$y_2 = 2 + \frac{1}{4}x - \left(\frac{(4 + x + 1/16x^2) - a(x)}{4 + 1/2x} \mod x^4 \right)$$

$$= 2 + \frac{1}{4}x + \left(\frac{31}{64}x^2 + \frac{353}{512}x^3 + \cdots \mod x^4 \right)$$

$$= 2 + \frac{1}{4}x + \frac{31}{64}x^2 + \frac{353}{512}x^3.$$

Notice that

$$(2 + \frac{1}{4}x + \frac{31}{64}x^2 + \frac{353}{512}x^3)^2 \bmod x^4 = (4 + x + 2x^2 + 3x^3).$$

Continuing in this manner, we obtain solutions to our algebraic equation modulo $x^8, x^{16}, \ldots$. ●

Algorithm 4.7. Newton's Method for Solving $P(y) = 0$.

```
procedure NewtonSolve(P(y), y_0, n)
  # Given P(y) ∈ F[[x]][y], and a point y_0 satisfying
  # P(y_0)=0 mod x and P'(y_0) ≠ 0, we determine the first
  # 2^n terms of a solution to P(y)=0 via Newton's method.

  y ← y_0
  for k from 0 to n−1 do
    y ← y − ( P(y)·P'(y)^(−1) mod x^(2^(k+1)) )
  return(y)
end
```

Theorem 4.8. Let $P(y) \in F[[x]][y]$ be a polynomial with power series coefficients. Let y_0 be an $O(x)$ approximation to the correct solution y. If

$$P(y_0) = 0 \bmod x, \text{ and } P'(y_0) \neq 0 \bmod x$$

then the iteration given by (4.34) converges quadratically, that is, if $\hat{y}$ is the exact solution of (4.32) then

$$y_k = \hat{y} + O(x^{2^k}).$$

Proof: We prove Theorem 4.8 by induction on k.

Clearly, by our choice of y_0, the theorem is true for $k = 0$. Thus assume that our result is true for $k = n$, and we wish to show that it is true for $k = n+1$. From Theorem 2.8 (with $x = y_n$ and $y = \hat{y} - y_n$) we can expand $P(\hat{y})$ in its Taylor expansion as

$$P(\hat{y}) = P(y_n) + P'(y_n)\cdot(\hat{y} - y_n) + Q(y_n, \hat{y} - y_n)\cdot(\hat{y} - y_n)^2 \qquad (4.36)$$

where Q is a polynomial in y_n and $\hat{y} - y_n$. Since

$$P'(y_n) \bmod x = P'(y_n \bmod x) = P'(y_0 \bmod x) = P'(y_0) \bmod x$$

we obtain $P'(y_n) \neq 0$. Using this, along with the fact that $\hat{y}$ solves (4.32) exactly, we may transform equation (4.36) into

$$0 = \frac{P(y_n)}{P'(y_n)} + (\hat{y} - y_n) + \frac{Q(y_n,\hat{y}-y_n)}{P'(y_n)} \cdot (\hat{y} - y_n)^2.$$

This in turn can be rewritten as

$$\hat{y} = y_n - \frac{P(y_n)}{P'(y_n)} - \frac{Q(y_n,\hat{y}-y_n)}{P'(y_n)} \cdot (\hat{y} - y_n)^2. \tag{4.37}$$

By definition we have

$$y_{n+1} = y_n - \left(\frac{P(y_n)}{P'(y_n)} \mod x^{2^n} \right)$$

while our induction hypothesis gives

$$\hat{y} = y_n \mod x^{2^n},$$

hence

$$(\hat{y} - y_n)^2 = 0 \mod x^{2^{n+1}},$$

and equation (4.37) becomes

$$\hat{y} = y_{n+1} \mod x^{2^{n+1}}.$$

Since the result is true for $k = n+1$, Theorem 4.8 follows by induction. ●

Exercises

1. Determine recurrence relations for the cost of the following arithmetic operations. Then determine the complexity of the operations in each case.
   (a) The arithmetic operation of addition of two multiprecision integers of length m and n.
   (b) The arithmetic operation of multiplication of two multiprecision integers of length m and n.
   (c) The arithmetic operation of division with remainder of two polynomials of degrees m and n.
   (d) The arithmetic operation of binary powering in $\mathbf{Z}_n$.

2. Show that the extended Euclidean algorithm applied to two n-bit integers has a complexity of $O(n^2)$. Show also that the worst case occurs when taking the GCD of two consecutive Fibonacci numbers (cf. Chapter 2).

3. Modify the power series powering algorithm of this chapter to determine $e^{a(x)}$ for a given power series satisfying $a(0) = 0$.

4. Modify the power series powering operation $B(x) = A(x)^p$ to allow for the case where $A(0) = 0$.

5. Determine a recurrence relation for the cost of power series powering. What is the complexity of calculating the first n terms of the power series for $A(x)^p$ when $A(0) = 1$?

6. Apply the reversion algorithm of this chapter twice to see that the process does indeed return to its original power series.

7. Apply the reversion algorithm to the polynomial $a(x) = x - x^3$.

8. Determine a recurrence relation for the cost of reversion using Lagrange's inversion formula and our power series powering algorithm. What is the complexity of calculating the first n terms of the power series for this case?

9. Let $a(x)$ and $b(x)$ be the polynomials

$$a(x) = x^4 - 2x^2 + 3x - 3, \;\; b(x) = x^2 + 2x + 3$$

from the domain $\mathbf{Z}_{11}[x]$.

(a) Multiply the two polynomials by using the modular algorithm.

(b) Determine if $b(x)$ divides $a(x)$ by using the trial division algorithm.

(c) Repeat part (b) with $b(x) = x^2 + x + 3$.

10. Develop a trial division algorithm for division of integers by performing divisions in $\mathbf{Z}_p$ for primes p.

11. What are the storage requirements of Karatsuba's algorithm for multiplying two multiprecision integers of length n? How do the space requirements compare to those required by the grade school method?

12. Develop a Karatsuba algorithm for the fast multiplication of two polynomials.

13. The following divide-and-conquer approach to matrix multiplication is due to V. Strassen [13]. Note the similarity of ideas with Karatsuba's algorithm. To compute the matrix product

$$\begin{bmatrix} c_{11} & c_{12} \\ c_{21} & c_{22} \end{bmatrix} = \begin{bmatrix} a_{11} & a_{12} \\ a_{21} & a_{22} \end{bmatrix} \cdot \begin{bmatrix} b_{11} & b_{12} \\ b_{21} & b_{22} \end{bmatrix}$$

first compute the following products:

$$m_1 = (a_{12}-a_{22})(b_{21}+b_{22}), \quad m_2 = (a_{11}+a_{22})(b_{11}+b_{22}),$$

$$m_3 = (a_{11}-a_{21})(b_{11}+b_{12}), \quad m_4 = (a_{11}+a_{12})b_{22},$$

$$m_5 = a_{11}(b_{12}-b_{22}), \quad m_6 = a_{22}(b_{21}-b_{11}),$$

$$m_7 = (a_{21}+a_{22})b_{11}.$$

Then compute the c_{ij} by

$$c_{11} = m_1 + m_2 - m_4 + m_6, \quad c_{12} = m_4 + m_5,$$

$$c_{21} = m_6 + m_7, \quad c_{22} = m_2 - m_3 + m_5 - m_7.$$

Analyze the performance of this algorithm. We mention in passing that Strassen's algorithm was the first known method of multiplying two $n \times n$ matrices with a complexity less than $O(n^3)$. The result was very surprising and has spawned numerous similar approaches to reduce the complexity of matrix multiplication even further (cf. Pan [10] or Coppersmith and Winograd [3]). The last named reference gives the lowest known exponent (2.375477) at the time of writing.

14. Let

$$a(x) = -x^3 + 3x + 1, \; b(x) = 2x^4 - 3x^3 - 2x^2 + x + 1$$

be polynomials from $\mathbf{Z}_{17}[x]$. Determine the FFT of both $a(x)$ and $b(x)$. Use this to calculate the product of the two polynomials.

15. Let $a(x)$ be a polynomial of degree 3^n-1.
    (a) Show that $a(x)$ can be decomposed into

    $$a(x) = b(x^3) + x \cdot c(x^3) + x^2 \cdot d(x^3)$$

    where b, c, and d are polynomials of degree at most $3^{n-1} - 1$.
    (b) Determine a symmetry condition similar to that found in Lemma 4.1 which will allow $a(x)$ to be evaluated at 3^n points, by evaluating 3 polynomials at 3^{n-1} points.
    (c) Show that if ω is a primitive 3^n-th root of unity in a field F, then the points $1, \omega, \ldots, \omega^{3^n-1}$ satisfy the symmetry condition determined in part (b).
    (d) What is the cost of evaluating $a(x)$ at the 3^n points given in part (c)?
    (e) Use parts (a) to (d) develop a 3-ary FFT algorithm. Notice that a similar approach can be used to develop a b-ary FFT algorithm for any positive integer b.

16. Use Newton's method to find the first 8 terms of the reciprocal of

$$a(x) = 2 - x^2 + x^3 + 4x^4 - 5x^5 + x^7 + \cdots .$$

17. Calculate the first 8 terms of

$$y(x) = \left(\frac{1}{(1-x)} \right)^{\frac{1}{3}}$$

using Newton's iteration on the equation

$$P(y) = (1-x)\cdot y^3 - 1 = 0.$$

18. Determine an algorithm to find the p-th root of a power series $a(x)$ based on Newton's method.

19. This question determines the cost of Newton's iteration

$$y_{n+1} = y_n - (\, P(y_n)\cdot P'(y_n)^{-1} \bmod x^{2^{n+1}} \,)$$

in the case where $P(y)$ is an arbitrary polynomial over F[[x]] of degree m.

(a) For a polynomial $P(y)$ and a point $\hat{y}$, show that there are polynomials $Q(y)$ and $R(y)$ such that

$$P(y) = (y - \hat{y})\cdot Q(y) + P(\hat{y}),$$

$$Q(y) = (y - \hat{y})\cdot R(y) + P'(\hat{y}).$$

(b) Using part (a), show that $P(\hat{y})$ and $P'(\hat{y})$ can be calculated with $2m - 1$ multiplications.

(c) Show that Newton's iteration requires $O(m\cdot M(n))$ operations to compute the first n terms of the power series for a solution of $P(y)=0$ where $M(n)$ is the cost of multiplying two polynomials in F[[x]] of degree n.

(d) Deduce that if F is a field supporting the FFT, then Newton's method applied to an algebraic equation of degree m requires $O(m\cdot n\cdot \log n)$ operations to determine the first n terms of the power series expansion of the root.

20. Determine an algorithm for power series reversion using Newton's iteration procedure. Use this algorithm to determine the first 8 terms of arctan(x).

References

1. R.P. Brent and H.T. Kung, "Fast Algorithms for Composition and Reversion of Power Series," pp. 217-225 in *Algorithms and Complexity*, ed. J.F. Traub, (1976).

2. J.W. Cooley and J.W. Tuckey, "An Algorithm for the Machine Calculation of Complex Fourier Series," *Math. Comp.*, **19** pp. 297-301 (1965).

3. D. Coppersmith and S. Winograd, "Matrix Multiplication via Arithmetic Progressions," *J. Symbolic Comp.*, **8**(3)(1990).

4. I.N. Herstein, *Topics in Algebra,* Blaisdell (1964).

5. A. Karatsuba, ''Multiplication of Multidigit Numbers on Automata,'' *Soviet Physics - Doklady*, **7** pp. 595-596 (1963).

6. D.E. Knuth, *The Art of Computer Programming, Volume 2: Seminumerical Algorithms (second edition),* Addison-Wesley (1981).

7. J.D. Lipson, ''Newton's Method: A Great Algebraic Algorithm,'' pp. 260-270 in *Proc. SYMSAC '76*, ed. R.D. Jenks, ACM Press (1976).

8. J.D. Lipson, *Elements of Algebra and Algebraic Computing,* Addision-Wesley (1981).

9. R.T. Moenck, ''Practical Fast Polynomial Multiplication,'' pp. 136-145 in *Proc. SYMSAC '76*, ed. R.D. Jenks, ACM Press (1976).

10. V. Pan, ''Strassen's Algorithm is not Optimal,'' pp. 166-176 in *Proc. of 19-th IEEE Symp. on Foundations of Computer Science*, (1978).

11. J.M. Pollard, ''The Fast Fourier Transform in a Finite Field,'' *Math. Comp.*, **25** pp. 365-374 (1971).

12. A. Schonhage and V. Strassen, ''Schnelle Multiplikation Grosser Zahlen,'' *Computing*, **7** pp. 281-292 (1971).

13. V. Strassen, ''Gaussian Elimination is not Optimal,'' *Numerische Mathematik*, **13** pp. 354-356 (1969).

14. J.F. Traub and H.T. Kung, ''All Algebraic Functions Can Be Computed Fast,'' *J. ACM*, pp. 245-260 (1978).

CHAPTER 5

HOMOMORPHISMS AND

CHINESE REMAINDER ALGORITHMS

5.1. INTRODUCTION

In the previous three chapters we have introduced the general mathematical framework for computer algebra systems. In Chapter 2 we discussed the algebraic domains which we will be working with. In Chapter 3 we concerned ourselves with the representations of these algebraic domains in a computer environment. In Chapter 4 we discussed algorithms for performing the basic arithmetic operations in these algebraic domains.

In this and subsequent chapters we will be concerned with various algorithms which are fundamental for computer algebra systems. These include factorization algorithms, GCD algorithms, and algorithms for solving equations. We will see that there are fundamental differences between algorithms required for computer algebra systems versus those required for systems supporting only floating-point arithmetic. We must deal with the problem of intermediate expression swell, where it is not uncommon for the coefficients involved in intermediate steps to grow exponentially in size. In addition, we must concern ourselves with the special properties of the computer environment in which we are working. Thus we realize that computing with coefficients which are single-precision integers (that is, integers which fit inside one computer word) is considerably more efficient than computing with multiprecision coefficients. At the same time, we realize that computer algebra systems require exact rather than approximate answers. Thus having multiprecision coefficients in a solution to a given problem will be the norm, rather than the exception.

5.2. INTERMEDIATE EXPRESSION SWELL: AN EXAMPLE

Consider the simple problem of solving a system of linear equations with integer coefficients, for example the system

$$\begin{aligned} 22x + 44y + 74z &= 1, \\ 15x + 14y - 10z &= -2, \\ -25x - 28y + 20z &= 34. \end{aligned}$$

If we wish to solve this system using only integer arithmetic, then a variation of Gaussian elimination may be applied. Thus, to eliminate the x term from the second equation we simply multiply equation one by 15, equation two by 22 and take the difference. The system of equations, without the x term in equations two and three, will then be

$$\begin{aligned} 22x + 44y + 74z &= 1\,, \\ -352y - 1330z &= -59\,, \\ 484y + 2290z &= 773\,. \end{aligned}$$

Continuing in this manner until we isolate x, y and z gives the "reduced" system

$$\begin{aligned} 1257315840x \qquad\qquad &= 7543895040\,, \\ -57150720y \qquad &= 314328960\,, \\ 162360z &= 243540\,, \end{aligned}$$

which gives our solutions as

$$x = \frac{7543895040}{1257315840} = 6\,,$$

$$y = \frac{314328960}{-57150720} = \frac{-11}{2}\,,$$

$$z = \frac{243540}{162360} = \frac{3}{2}\,.$$

Notice the growth in the size of the coefficients. Indeed it is easy to see that if each coefficient in a linear system having n equations and n unknowns requires ω computer words of storage, then each coefficient of the reduced linear system may well need $2^{n-1}\omega$ computer words of storage. That is, this method suffers from exponential growth in the size of the coefficients. Notice that the same method may also be applied to linear systems having polynomials, rather than integers, as coefficients. We would then have both exponential growth in the size of the numerical coefficients of the polynomials, and exponential growth in the degrees of the polynomials themselves.

Although our method in the previous example may suffer from exponential growth, our final answer will always be of reasonable size. For, by Cramer's rule we know that each component of the solution to such a linear system is a ratio of two determinants, each of which requires approximately $n \cdot \omega$ computer words. Thus the simplification found in the final form of the answer is the norm rather than the exception.

Of course we can always solve the system by reverting to rational arithmetic, but this also involves a hidden cost. Each rational arithmetic operation requires GCD computations, which are expensive when compared to ordinary operations such as addition or multiplication. (While this may not be so apparent with linear systems having integer coefficients, one only need consider linear systems having polynomial coefficients.) In Chapter 9 we will give a fraction-free method that closely resembles the method just described, but which has a considerable improvement in efficiency while still avoiding quotient field computation. We remark that the problem of exponential intermediate coefficient growth is not unique to linear systems. It appears in numerous other calculations, most notably in polynomial GCD calculations (cf. Chapter 7).

In this chapter we describe one approach, commonly called the modular or multiple homomorphism approach, to solving problems such as our previous example. This method involves breaking down a problem posed in one algebraic domain into a similar problem in a number of (much simpler) algebraic domains. Thus in our example, rather than solving our system over $\mathbf{Z}$ we solve the linear system over $\mathbf{Z}_p$, for a number of primes p. These domains are simpler in that they are fields, and so allow for the normal use of the Gaussian elimination method. In addition, since the coefficient domains are finite fields, the coefficients that arise during the solution are all of fixed size. Thus we avoid the problem of exponential coefficient growth that plagues us in numerous applications. In some sense we may think of ourselves as working in a number of fixed-precision domains.

Of course, we still must know how to piece together the solutions from these ''homomorphic images'' of our original problem. This is the subject of the Chinese remainder algorithm (in both integer and polynomial forms) which provides the central focus of this chapter. Along the way we must also deal with such problems as determining which domains are most suitable for our reductions, and deciding how many such homomorphic reductions are needed to piece together a final solution.

5.3. RING MORPHISMS

In this section we introduce the concept of mapping an algebraic system onto a simpler ''model'' of itself and, alternatively, the concept of embedding an algebraic system into a larger algebraic system. A related concept is that of an isomorphism between two algebraic systems which we have already encountered in Chapter 2. It is convenient to adopt the terminology of universal algebra when discussing these concepts.

Subalgebras

By an *algebra* (or *algebraic system*) we understand a set S together with a collection of operations defined on S. Specifically for our purposes, by an algebra we shall mean any one of the following types of *rings:*

commutative ring
integral domain
unique factorization domain (UFD)
Euclidean domain
field .

Thus the operations in the algebras we are considering are the *binary* operations of addition and multiplication, the *unary* operation of negation, and the *nullary* operations of "select 0" and "select 1".[6] In addition, if the algebra is a field then there is also the unary operation of inversion which maps each element (except 0) onto its multiplicative inverse. When the collection of operations is implied by context, we often refer to the algebra S when what we mean is the algebra consisting of the set S together with the operations defined on S.

Definition 5.1. Let S be an algebra. A subset S′ of the set S is called a *subalgebra* if S′ is closed under the operations defined on S.

●

We use the following terminology for subalgebras of the specific algebras listed above. If S is a (commutative) ring then a subalgebra of S is called a *subring*. If S is an integral domain, UFD, or a Euclidean domain then a subalgebra of S is called a *subdomain*. If S is a field then a subalgebra of S is called a *subfield*. For any algebra S, it is clear that if a subset S′ is closed under all of the operations defined on S then all of the axioms which hold in S are automatically inherited by the subalgebra S′. In particular, a subring of a commutative ring is itself a commutative ring, a subdomain of an integral domain (UFD, Euclidean domain) is itself an integral domain (UFD, Euclidean domain), and a subfield of a field is itself a field.

Example 5.1. In Figure 2.2 of Chapter 2, if two domains S and R are related by the notation $S \rightarrow R$ then S is a subdomain of R. For example, $D[x]$ and $D[[x]]$ are subdomains of $F_D((x))$. Also, $D(x)$ and $D((x))$ are subfields of $F_D((x))$.

●

Morphisms

In discussing mappings between two rings R and R′ we will adopt the convention of using the same notation to denote the operations in R and in R′ Thus + will denote addition in R or in R′, depending on context, multiplication in both R and R′ will be denoted by juxtaposition without any operator symbol, and 0 and 1 will denote (respectively) the additive and multiplicative identities in R or R′, depending on context. This convention is particularly appropriate in the common situation where one of R, R′ is a subring of the other.

6. A binary operation takes two operands and a unary operation takes one operand. Similarly, a nullary operation takes no operands and is simply a *selection function* – in our case, the operations of selecting the additive identity 0 and the multiplicative identity 1.

Definition 5.2. Let R and R′ be two rings. Then a mapping $\phi : R \to R'$ is called a *ring morphism* if

(i) $\phi(a+b) = \phi(a) + \phi(b)$ for all $a,b \in R$;

(ii) $\phi(ab) = \phi(a)\,\phi(b)$ for all $a,b \in R$;

(iii) $\phi(1) = 1$.

•

The general (universal algebra) concept of a *morphism* between two algebras is that of a mapping which preserves *all* of the operations defined on the algebras. In Definition 5.2, note that properties (i) - (iii) ensure that three of the ring operations are preserved but that no mention has been made of the unary operation of negation and the nullary operation "select 0". This is because the two additional properties:

$$\phi(0) = 0;$$

$$\phi(-a) = -\phi(a) \quad \text{for all } a \in R$$

are simple consequences of the ring axioms and properties (i) - (iii). Similarly, if R and R′ are fields with the additional unary operation of inversion then the ring morphism of Definition 5.2 is in fact a *field morphism* because the additional property:

$$\phi(a^{-1}) = [\phi(a)]^{-1} \quad \text{for all } a \in R-\{0\}$$

is a consequence of the field axioms and properties (i) - (iii). Therefore in the sequel when we refer to a morphism it will be understood that we are referring to a ring morphism as defined in Definition 5.2.

Morphisms are classified according to their properties as functions. If $\phi : R \to R'$ is a morphism then it is called a *monomorphism* if the function ϕ is injective (i.e. one-to-one), an *epimorphism* if the function ϕ is surjective (i.e. onto), and an *isomorphism* if the function ϕ is bijective (i.e. one-to-one and onto). The classical term *homomorphism* in its most general usage is simply a synonym for the more modern term "morphism" used in the context of universal algebra. However in common usage the term "homomorphism" is most often identified with an epimorphism and in particular if $\phi : R \to R'$ is an epimorphism then R′ is called a *homomorphic image* of R.

A monomorphism $\phi : R \to R'$ is called an *embedding* of R into R′ since clearly the mapping $\phi : R \to \phi(R)$ onto the image set

$$\phi(R) = \{r' \in R' \colon \ \phi(r) = r' \quad \text{for some } r \in R\}$$

is an isomorphism – i.e. the ring R′ contains R (more correctly, an isomorphic copy of R) as a subring. An epimorphism $\phi : R \to R'$ is called a *projection* of R onto the homomorphic image R′. In this terminology, it is clear that for any morphism $\phi : R \to R'$ the image set $\phi(R)$ is a homomorphic image of R. An important property of morphisms is that a homomorphic image of a (commutative) ring is itself a (commutative) ring. However, a homomorphic image of an integral domain is not necessarily an integral domain (see Example 5.4).

Example 5.2. Several instances of isomorphic algebras were encountered in Chapter 2. For any commutative ring R, the polynomial domains $R[x,y]$, $R[x][y]$, and $R[y][x]$ are isomorphic; for example, the natural mapping

$$\phi : R[x,y] \rightarrow R[x][y]$$

defined by

$$\phi(\sum_{i=0}^{m} \sum_{j=0}^{n} a_{ij} x^i y^j) = \sum_{j=0}^{n} (\sum_{i=0}^{m} a_{ij} x^i) y^j$$

is an isomorphism. Similarly, for any integral domain D with quotient field F_D the fields of rational functions $D(\mathbf{x})$ and $F_D(\mathbf{x})$ are isomorphic with a natural mapping between them. Also, for any field F the fields $F((x))$ and $F\langle x \rangle$ are isomorphic with a natural mapping from the canonical form of a power series rational function in $F((x))$ onto an extended power series in $F\langle x \rangle$.

●

Example 5.3. Let D be an integral domain and let F_D be its quotient field. The mapping

$$\phi : D \rightarrow F_D$$

defined by

$$\phi(a) = [a/1] \quad \text{for all } a \in D$$

is a monomorphism. Thus ϕ is an embedding of D into F_D and we call F_D an *extension* of D (the smallest extension of the integral domain D into a field).

●

Example 5.4. Let $\mathbf{Z}$ be the integers and let $\mathbf{Z}_6$ be the set of integers modulo 6. Let $\phi : \mathbf{Z} \rightarrow \mathbf{Z}_6$ be the mapping defined by

$$\phi(a) = \mathrm{rem}(a,6) \quad \text{for all } a \in \mathbf{Z}$$

where the remainder function ‘‘rem’’ is as defined in Chapter 2. Then ϕ is an epimorphism and thus ϕ is a projection of $\mathbf{Z}$ onto the homomorphic image $\mathbf{Z}_6$. $\mathbf{Z}_6$ is a commutative ring because $\mathbf{Z}$ is a commutative ring. $\mathbf{Z}_6$ is not an integral domain (see Exercise 2.3) even though $\mathbf{Z}$ is an integral domain.

●

Example 5.5. Let R and R′ be commutative rings with R a subring of R′. Let $R[x]$ be the commutative ring of univariate polynomials over R and let

$$\phi : R[x] \rightarrow R'$$

be the mapping defined by

$$\phi(a(x)) = a(\alpha)$$

for some fixed element $\alpha \in R'$ (i.e. the image of $a(x)$ is obtained by evaluating $a(x)$ at the value $x = \alpha$). Then ϕ is a morphism of rings.

●

Modular and Evaluation Homomorphisms

In the sequel the morphisms of interest to us will be projections of a ring R onto (simpler) homomorphic images of R. In keeping with common usage we will use the term "homomorphism" for such projections. We now consider two particular classes of homomorphisms which have many practical applications in algorithms for symbolic computation.

The first homomorphism of interest is a generalization of the projection of the integers considered in Example 5.4. Formally, a *modular homomorphism*

$$\phi_m : \mathbf{Z}[x_1, \ldots, x_v] \rightarrow \mathbf{Z}_m[x_1, \ldots, x_v]$$

is a homomorphism defined for a fixed integer $m \in \mathbf{Z}$ by the rules:

$$\phi_m(x_i) = x_i, \quad \text{for } 1 \le i \le v;$$

$$\phi_m(a) = \mathrm{rem}(a, m), \quad \text{for all coefficients } a \in \mathbf{Z}.$$

In other words, a modular homomorphism ϕ_m is a projection of $\mathbf{Z}[x_1, \ldots, x_v]$ onto $\mathbf{Z}_m[x_1, \ldots, x_v]$ obtained by simply replacing every coefficient of a polynomial $a(\mathbf{x}) \in \mathbf{Z}[x_1, \ldots, x_v]$ by its "modulo m" representation. Of course ϕ_m remains well-defined in the case $v = 0$ in which case it is simply a projection of $\mathbf{Z}$ onto $\mathbf{Z}_m$.

Example 5.6. In $\mathbf{Z}[x,y]$ let $a(x,y)$ and $b(x,y)$ be the polynomials

$$a(x,y) = 3x^2y^2 - x^2y + 5x^2 + xy^2 - 3xy; \tag{5.1}$$

$$b(x,y) = 2xy + 7x + y^2 - 2. \tag{5.2}$$

The modular homomorphism ϕ_5 maps these two polynomials onto the following polynomials in the domain $\mathbf{Z}_5[x,y]$:

$$\phi_5(a(x,y)) = 3x^2y^2 + 4x^2y + xy^2 + 2xy;$$

$$\phi_5(b(x,y)) = 2xy + 2x + y^2 + 3.$$

Similarly, the modular homomorphism ϕ_7 maps (5.1) - (5.2) onto the following polynomials in the domain $\mathbf{Z}_7[x,y]$:

$$\phi_7(a(x,y)) = 3x^2y^2 + 6x^2y + 5x^2 + xy^2 + 4xy;$$

$$\phi_7(b(x,y)) = 2xy + y^2 + 5.$$

●

The second homomorphism of interest is a special case of the ring morphism considered in Example 5.5 applied in the context of a multivariate polynomial domain $D[x_1, \ldots, x_\nu]$. In the notation of Example 5.5, we identify x with a *particular* indeterminate x_i and we choose

$$R = R' = D[x_1, \ldots, x_{i-1}, x_{i+1}, \ldots, x_\nu]$$

so that

$$R[x] = D[x_1, \ldots, x_\nu].$$

Formally, an *evaluation homomorphism*

$$\phi_{x_i-\alpha} : D[x_1, \ldots, x_\nu] \to D[x_1, \ldots, x_{i-1}, x_{i+1}, \ldots, x_\nu]$$

is a homomorphism defined for a particular indeterminate x_i and a fixed element $\alpha \in D$ such that for any polynomial $a(x_1, \ldots, x_\nu) \in D[x_1, \ldots, x_\nu]$,

$$\phi_{x_i-\alpha}(a(x_1, \ldots, x_\nu)) = a(x_1, \ldots, x_{i-1}, \alpha, x_{i+1}, \ldots, x_\nu).$$

In other words, an evaluation homomorphism $\phi_{x_i-\alpha}$ is a projection of $D[x_1, \ldots, x_\nu]$ onto $D[x_1, \ldots, x_{i-1}, x_{i+1}, \ldots, x_\nu]$ obtained by simply substituting the value $\alpha \in D$ for the i-th indeterminate x_i. Thus the notation $\phi_{x_i-\alpha}$ can be read "substitute α for x_i". (The particular choice of notation $\phi_{x_i-\alpha}$ for an evaluation homomorphism is such that the subscript $x_i-\alpha$ corresponds to the subscript m in the notation ϕ_m for a modular homomorphism. The reason for this correspondence of notation will become clear in a later section.)

Compositions of modular and evaluation homomorphisms will be used frequently in later chapters for projecting the multivariate polynomial domain $\mathbf{Z}[x_1, \ldots, x_\nu]$ onto simpler homomorphic images of itself. In most such applications a modular homomorphism ϕ_p, where p is a positive prime integer, will be chosen to project $\mathbf{Z}[x_1, \ldots, x_\nu]$ onto $\mathbf{Z}_p[x_1, \ldots, x_\nu]$ where the coefficient domain Z_p is now a field. A sequence of evaluation homomorphisms (one for each indeterminate) can then be applied to project the multivariate polynomial domain $\mathbf{Z}_p[x_1, \ldots, x_\nu]$ onto a homomorphic image of the form $\mathbf{Z}_p[x_1]$ (a Euclidean domain) or, if desired, onto a homomorphic image of the form $\mathbf{Z}_p$ (a field). It will be seen in later chapters that for the problem of GCD computation in $\mathbf{Z}[x_1, \ldots, x_\nu]$, and also for the problem of polynomial factorization in $\mathbf{Z}[x_1, \ldots, x_\nu]$, very efficient algorithms can be obtained by projecting to homomorphic images of the form $\mathbf{Z}_p[x_1]$ where the ordinary Euclidean algorithm applies. The following example considers the more elementary problem of polynomial multiplication in which case projections onto fields $\mathbf{Z}_p$ are appropriate.

Example 5.7. In the domain $\mathbf{Z}[x]$ let

$$a(x) = 7x + 5 \text{ and } b(x) = 2x - 3.$$

Suppose we wish to determine the product polynomial

$$c(x) = a(x)\, b(x).$$

Rather than directly multiplying these polynomials in the domain $\mathbf{Z}[x]$ we could choose to project $\mathbf{Z}[x]$ onto homomorphic images $\mathbf{Z}_p$ and perform the (simpler) multiplications in the fields $\mathbf{Z}_p$. For example, the composite homomorphism

$$\phi_{x-0}\, \phi_5 : \mathbf{Z}[x] \to \mathbf{Z}_5$$

maps $a(x)$ and $b(x)$ as follows:

$$a(x) \xrightarrow{\phi_5} 2x \xrightarrow{\phi_{x-0}} 0\,;$$
$$b(x) \xrightarrow{\phi_5} 2x + 2 \xrightarrow{\phi_{x-0}} 2\,.$$

Thus the product in this particular homomorphic image $\mathbf{Z}_5$ is $0 \cdot 2 = 0$. Using standard congruence notation for "mod p" arithmetic we represent this as follows:

$$c(0) \equiv 0 \pmod 5.$$

Similarly, applying the composite homomorphism $\phi_{x-1}\, \phi_5$ yields:

$$a(x) \xrightarrow{\phi_5} 2x \xrightarrow{\phi_{x-1}} 2\,;$$
$$b(x) \xrightarrow{\phi_5} 2x + 2 \xrightarrow{\phi_{x-1}} 4\,.$$

This times the product in $\mathbf{Z}_5$ is $2 \cdot 4 = 3$. Thus,

$$c(1) \equiv 3 \pmod 5.$$

Similarly, we find

$$c(2) \equiv 4 \pmod 5)$$

by applying the composite homomorphism $\phi_{x-2}\, \phi_5$. If in addition we apply the triple of composite homomorphisms:

$$\phi_{x-0}\, \phi_7\,;\ \phi_{x-1}\, \phi_7\,;\ \phi_{x-2}\, \phi_7$$

each of which projects $\mathbf{Z}[x]$ onto $\mathbf{Z}_7$, we get

$$c(0) \equiv 6 \pmod 7;$$
$$c(1) \equiv 2 \pmod 7;$$
$$c(2) \equiv 5 \pmod 7).$$

The above process is only useful if we can 'invert' the homomorphisms to reconstruct the polynomial $c(x) \in \mathbf{Z}[x]$, given information about the images of $c(x)$ in fields $\mathbf{Z}_p$. The inverse process involves the concepts of interpolation and Chinese remaindering which will be discussed later in this chapter. Briefly, since we know that $\deg(c(x)) = \deg(a(x)) + \deg(b(x)) = 2$, the polynomial $c(x)$ is completely specified by its values at 3 points. Using the above information, we obtain by interpolation:

$$c(x) \equiv 4x^2 + 4x \pmod{5};$$

$$c(x) \equiv 3x + 6 \pmod{7}.$$

Thus we know the images of $c(x)$ in $\mathbf{Z}_5[x]$ and in $\mathbf{Z}_7[x]$. Finally, we can determine

$$c(x) = c_2x^2 + c_1x + c_0 \in \mathbf{Z}[x]$$

by a process known as Chinese remaindering. For example, since we know that

$$c_2 \equiv 4 \pmod{5} \quad \text{and} \quad c_2 \equiv 0 \pmod{7}$$

we can determine that

$$c_2 \equiv 14 \pmod{35}$$

(where $35 = 5 \cdot 7$). We eventually get:

$$c(x) = 14x^2 - 11x - 15 \in \mathbf{Z}[x].$$

●

5.4. CHARACTERIZATION OF MORPHISMS

Ideals

A ring morphism $\phi : \mathrm{R} \rightarrow \mathrm{R}'$ can be conveniently characterized in terms of its action on particular subsets of R known as ideals.

Definition 5.3. Let R be a commutative ring. A nonempty subset I of R is called an *ideal* if

(i) $a - b \in \mathrm{I}$ for all $a, b \in \mathrm{I}$;

(ii) $ar \in \mathrm{I}$ for all $a \in \mathrm{I}$ and for all $r \in \mathrm{R}$.

●

Two very special ideals in any commutative ring R are the subsets $\{0\}$ and R since properties (i) and (ii) of Definition 5.3 are clearly satisfied by these two subsets. We call $\{0\}$ the *zero ideal* and R the *universal ideal*. By a *proper ideal* we mean any ideal I such that $\mathrm{I} \neq \{0\}$ and $\mathrm{I} \neq \mathrm{R}$. Note that the subset $\{0\}$ is not a subring of R according to Definition 5.1 since it is not closed under the nullary operation "select 1" defined on R (i.e. $\{0\}$ does not contain the multiplicative identity of R). This is a characteristic property of ideals which we formulate as the following theorem.

Theorem 5.1. Every proper ideal I in a commutative ring R is closed under all of the ring operations defined on R except that I is not closed under the nullary operation "select 1" (i.e. $1 \notin \mathrm{I}$).

Proof: It is easy to verify that property (i) of Definition 5.3 guarantees that I is closed under the operations + (binary), – (unary), and "select 0" (nullary). (Indeed property (i) is used precisely because it is sufficient to guarantee closure with respect to these three "group" operations.) It is also trivial to see that property (ii) guarantees that I is closed under multiplication. As for the nullary operation "select 1", if $1 \in$ I then by property (ii) $r \in$ I for all $r \in$ R – i.e. I = R so that I is not a proper ideal.

●

The crucial property of an ideal I, apart from the closure properties of Theorem 5.1, is the "extended closure" property (ii) of Definition 5.3 which guarantees that I is closed under multiplication *by any element of the ring* R.

Example 5.8. In the integral domain **Z** of integers, the subset

$$<m> = \{ mr : r = 0, \pm 1, \pm 2, \ldots \}$$

for some fixed integer $m \in \mathbf{Z}$ is an ideal called the *ideal generated by m*. For example, the ideal <4> is the set

$$<4> = \{0, \pm 4, \pm 8, \pm 12, \ldots \}.$$

●

Example 5.9. In the polynomial domain **Q**[x], the subset

$$<p(x)> = \{ p(x) \cdot a(x) : a(x) \in \mathbf{Q}[x] \}$$

for some fixed polynomial $p(x) \in \mathbf{Q}[x]$ is an ideal called the *ideal generated by* $p(x)$. For example, the ideal $<x - \alpha>$ for some fixed $\alpha \in \mathbf{Q}$ is the set of all polynomials over **Q** which have $x - \alpha$ as a factor (i.e. polynomials $a(x)$ such that $a(\alpha) = 0$).

●

Example 5.10. In the bivariate polynomial domain **Z**[x,y], the subset

$$\{ p_1(x,y)\, a_1(x,y) + p_2(x,y)\, a_2(x,y) : \; a_1(x,y), a_2(x,y) \in \mathbf{Z}[x,y] \}$$

for some fixed polynomials $p_1(x,y), p_2(x,y) \in \mathbf{Z}[x,y]$ is an ideal called the *ideal generated by* $p_1(x,y)$ *and* $p_2(x,y)$. We use the notation $<p_1(x,y), p_2(x,y)>$ for this ideal. For example, the ideal $<x,y>$ is the set of all bivariate polynomials over **Z** with constant term zero. Also note that $<y - \alpha>$ for some fixed $\alpha \in \mathbf{Z}$ is an ideal in $\mathbf{Z}[x,y]$ consisting of all bivariate polynomials over **Z** which have $y - \alpha$ as a factor (i.e. polynomials $a(x,y)$ such that $a(x,\alpha) = 0$). Similarly $<m>$ for some fixed integer $m \in \mathbf{Z}$ is an ideal in $\mathbf{Z}[x,y]$ consisting of all bivariate polynomials whose integer coefficients are multiples of m.

●

The fact that an ideal I in a commutative ring R is closed under addition and is closed under multiplication by any element of R, implies that if I contains the n elements $a_1, \ldots, a_n$ then it must contain the set of all linear combinations of these elements, defined by:

$$< a_1, \ldots, a_n > = \{a_1 r_1 + \cdots + a_n r_n : r_i \in \mathrm{R}\}.$$

On the other hand, it is easy to verify that for any given elements $a_1, \ldots, a_n \in \mathrm{R}$, the set $< a_1, \ldots, a_n >$ of all linear combinations of these elements is an ideal in R. The ideal $< a_1, \ldots, a_n >$ is called the ideal with *basis* $a_1, \ldots, a_n$.

Definition 5.4. An ideal I in a commutative ring R is called an *ideal with finite basis* if I can be expressed as the set $< a_1, \ldots, a_n >$ of all linear combinations of a finite number n of elements $a_1, \ldots, a_n \in \mathrm{R}$.

●

Definition 5.5. An ideal I in a commutative ring R is called a *principal ideal* if I can be expressed as the set $< a >$ of all multiples of a single element $a \in \mathrm{R}$.

●

Domains with Special Ideals

Definition 5.6. An integral domain D is called a *Noetherian integral domain* if every ideal in D is an ideal with finite basis.

●

Definition 5.7. An integral domain D is called a *principal ideal domain* if every ideal in D is a principal ideal.

●

It can be proved that every Euclidean domain is a principal ideal domain and therefore the domains **Z** and **Q**$[x]$ considered in Examples 5.8 and 5.9 are principal ideal domains. The polynomial domain **Z**$[x,y]$ considered in Example 5.10 is an example of an integral domain that is not a principal ideal domain since it is not possible to generate the ideal $< x,y >$, for example, by a single element. However it can be proved that if D is a Noetherian integral domain then so is the domain D$[x]$, which implies by induction that **Z**$[x,y]$ and indeed any multivariate polynomial domain over **Z** or over a field is a Noetherian integral domain.

In the hierarchy of domains given in Table 2.3 of Chapter 2, the principal ideal domain lies between the unique factorization domain (UFD) and the Euclidean domain (i.e. every Euclidean domain is a principal ideal domain and every principal ideal domain is a UFD). However the multivariate polynomial domains considered in this book are Noetherian integral domains but are not principal ideal domains. The abstract concept of a Noetherian integral domain, unlike a principal ideal domain, is not simply a UFD which satisfies additional axioms. (For example, the integral domain

$$S = \{a + b\sqrt{-5} : a, b \in \mathbf{Z}\}$$

considered in Exercises 2.10 and 2.11 is a Noetherian integral domain but is not a UFD.)

In the sequel we will require the concepts of the sum and product of two ideals and also the concept of an integral power of an ideal. These concepts are defined in the following definition in the context of an arbitrary Noetherian integral domain. Before proceeding to the definition let us note the following generalization of our notation for specifying ideals in a Noetherian integral domain D. If I and J are two ideals in D then by the notation $<\mathrm{I},\mathrm{J}>$ we understand the ideal $< a_1, \ldots, a_n, b_1, \ldots, b_m>$ where $a_1, \ldots, a_n \in$ D forms a basis for I and where $b_1, \ldots, b_m \in$ D forms a basis for J (that is, $\mathrm{I} = < a_1, \ldots, a_n>$ and $\mathrm{J} = < b_1, \ldots, b_m>$). The notation $<\mathrm{I}, b>$ or $<b, \mathrm{I}>$ where $b \in$ D and I is an ideal in D is similarly defined – i.e. $<\mathrm{I}, b> = <\mathrm{I}, <b>>$.

Definition 5.8. Let I and J be two ideals in a Noetherian integral domain D and suppose $\mathrm{I} = < a_1, \ldots, a_n>$, $\mathrm{J} = < b_1, \ldots, b_m>$ for elements $a_i \in$ D $(1 \le i \le n)$, $b_j \in$ D $(1 \le j \le m)$.

(i) The *sum* of the ideals I and J in D is the ideal defined by $<\mathrm{I}, \mathrm{J}> = < a_1, \ldots, a_n, b_1, \ldots, b_m>$. Note that the ideal $<\mathrm{I}, \mathrm{J}>$ consists of all possible sums $a + b$ where $a \in$ I and $b \in$ J.

(ii) The *product* $\mathrm{I} \cdot \mathrm{J}$ of the ideals I and J in D is the ideal generated by all elements $a_i b_j$ such that a_i is a basis element for I and b_j is a basis element for J. Thus the product of I and J can be expressed as

$$\mathrm{I} \cdot \mathrm{J} = < a_1 b_1, \ldots, a_1 b_m, a_2 b_1, \ldots, a_2 b_m, \ldots, a_n b_1, \ldots, a_n b_m>.$$

(iii) The i-th *power* of the ideal I in D (for i a positive integer) is defined recursively in terms of products of ideals as follows:

$$\mathrm{I}^1 = \mathrm{I};$$

$$\mathrm{I}^i = \mathrm{I} \cdot \mathrm{I}^{i-1} \quad \text{for } i \ge 2.$$

●

The application of Definition 5.8 to the case of principal ideals should be noted in particular. For the product of two principal ideals $< a>$ and $< b>$ in D it follows from Definition 5.8 that

$$< a> \cdot < b> = < ab>.$$

Similarly for the i-th power of the ideal $< a>$ in D we have

$$< a>^i = < a^i> \quad \text{for } i \ge 1.$$

The sum of the ideals $< a>$ and $< b>$ in D is simply the ideal $< a, b>$ which may not be a principal ideal. However if D is a principal ideal domain then the sum $< a, b>$ must be a principal ideal. It can be proved that in any principal ideal domain,

$$<a,b> = <\mathrm{GCD}(a,b)>.$$

(Note that since D is a principal ideal domain it is also a UFD and therefore the GCD exists by Theorem 2.1.)

The Characterization Theorem

Definition 5.9. Let R and R′ be commutative rings and let $\phi : R \rightarrow R'$ be a morphism. The *kernel* K of the morphism ϕ is the set defined by:

$$K = \phi^{-1}(0) = \{a : a \in R \text{ and } \phi(a) = 0\}.$$

●

Theorem 5.2. Let R and R′ be commutative rings. The kernel K of a morphism $\phi : R \rightarrow R'$ is an ideal in R.

Proof: The set K is not empty since $\phi(0) = 0$. If $a, b \in K$ then

$$\phi(a-b) = \phi(a) - \phi(b) = 0 - 0 = 0$$

so that $a - b \in K$, proving property (i) of Definition 5.3. Similarly property (ii) holds because if $a \in K$ and $r \in R$ then

$$\phi(ar) = \phi(a)\,\phi(r) = 0 \cdot \phi(r) = 0$$

so that $ar \in K$.

●

There is a direct connection between the homomorphic images of a commutative ring R and the set of ideals in R. Recall that every morphism $\phi : R \rightarrow R'$ determines a homomorphic image $\phi(R)$ of the ring R. We see from Theorem 5.2 that to each morphism $\phi : R \rightarrow R'$ there corresponds an ideal in R which is the kernel K of ϕ. Conversely, we shall see in the next section that to each ideal I in R there corresponds a homomorphic image R′ of R such that I is the kernel of the corresponding morphism $\phi : R \rightarrow R'$. We first prove that a homomorphic image of R is completely determined (up to isomorphism) by the ideal of elements mapped onto zero.

Theorem 5.3 (Characterization Theorem). Let R be a commutative ring and let K be an ideal in R. If $\phi_1 : R \rightarrow R'$ and $\phi_2 : R \rightarrow R''$ are two morphisms both having kernel K then the correspondence between the two homomorphic images $\phi_1(R)$ and $\phi_2(R)$ defined by

$$\phi_1(a) \leftrightarrow \phi_2(a)$$

is an isomorphism.

Proof: Suppose ϕ_1 and ϕ_2 have kernel K. The correspondence mentioned above can be formally specified as follows. For any element $\alpha \in \phi_1(R)$ the set of pre-images of α is the set

$$\phi_1^{-1}(\alpha) = \{a \in \mathrm{R}:\ \phi_1(a) = \alpha\}.$$

We define the mapping

$$\psi : \phi_1(\mathrm{R}) \rightarrow \phi_2(\mathrm{R})$$

by

$$\psi(\alpha) = \phi_2(\phi_1^{-1}(\alpha)) \text{ for all } \alpha \in \phi_1(\mathrm{R}) \tag{5.3}$$

where we claim first that the image under ϕ_2 of the set $\phi_1^{-1}(\alpha)$ is a *single* element in $\phi_2(\mathrm{R})$. To see this, note that if $a, b \in \mathrm{R}$ are two elements in the set $\phi_1^{-1}(\alpha)$ then $a - b \in \mathrm{K}$ (the kernel of ϕ_1) since

$$\phi_1(a - b) = \phi_1(a) - \phi_1(b) = \alpha - \alpha = 0.$$

Hence,

$$\phi_2(a - b) = \phi_2(a) - \phi_2(b) = 0$$

(because K is also the kernel of ϕ_2) yielding

$$\phi_2(a) = \phi_2(b).$$

Thus (5.3) defines a valid mapping of $\phi_1(\mathrm{R})$ into $\phi_2(\mathrm{R})$ and clearly ψ specifies the correspondence mentioned above. We may calculate (5.3) by letting $a \in \phi_1^{-1}(\alpha)$ be *any* particular pre-image of α and setting $\psi(\alpha) = \phi_2(a)$.

We now claim that ψ is an isomorphism. Properties (i) - (iii) of Definition 5.2 are satisfied by ψ because they are satisfied by the morphisms ϕ_1 and ϕ_2. To see this, for any $\alpha, \beta \in \phi_1(\mathrm{R})$ let $a \in \phi_1^{-1}(\alpha)$ and $b \in \phi_1^{-1}(\beta)$ be particular pre-images of α and β, respectively. Then a particular pre-image of $\alpha + \beta \in \phi_1(\mathrm{R})$ is the element $a + b \in \mathrm{R}$ since

$$\phi_1(a + b) = \phi_1(a) + \phi_1(b) = \alpha + \beta;$$

similarly, $ab \in \mathrm{R}$ is a particular pre-image of $\alpha\beta \in \phi_1(\mathrm{R})$. Thus,

$$\psi(\alpha + \beta) = \phi_2(\phi_1^{-1}(\alpha + \beta)) = \phi_2(a+b) = \phi_2(a) + \phi_2(b) = \psi(\alpha) + \psi(\beta)$$

and

$$\psi(\alpha\beta) = \phi_2(\phi_1^{-1}(\alpha\beta)) = \phi_2(ab) = \phi_2(a)\,\phi_2(b) = \psi(\alpha)\,\psi(\beta)$$

verifying properties (i) and (ii). To verify property (iii), note that $1 \in \mathrm{R}$ is a particular pre-image of $1 \in \phi_1(\mathrm{R})$ because $\phi_1(1) = 1$ (i.e. ϕ_1 is a morphism) and therefore

$$\psi(1) = \phi_2(\phi_1^{-1}(1)) = \phi_2(1) = 1$$

(because ϕ_2 is a morphism). We have thus proved that ψ is a morphism. It is easy to see that the mapping ψ is surjective since the mappings

$$\phi_1 : R \to \phi_1(R) \text{ and } \phi_2 : R \to \phi_2(R)$$

are surjective. To see that ψ is injective, let $\alpha, \beta \in \phi_1(R)$ have particular pre-images $a, b \in R$ (i.e. $\alpha = \phi_1(a)$ and $\beta = \phi_1(b)$) and suppose that $\psi(\alpha) = \psi(\beta)$. Then we have

$$\phi_2(\phi_1^{-1}(\alpha)) = \phi_2(\phi_1^{-1}(\beta))$$

$$\Rightarrow \phi_2(a) = \phi_2(b)$$

$$\Rightarrow a - b \in K \text{ (the kernel of } \phi_2)$$

$$\Rightarrow \phi_1(a) = \phi_1(b) \text{ (because K is the kernel of } \phi_1)$$

$$\Rightarrow \alpha = \beta.$$

Hence the mapping ψ is injective and ψ defines an isomorphism between $\phi_1(R)$ and $\phi_2(R)$.

●

Corollary to Theorem 5.3.

Let $\phi : R \to R'$ be a morphism between commutative rings R and R'. If K denotes the kernel of ϕ then:

(i) $K = \{0\}$ if and only if ϕ is injective (i.e. $\phi(R) = R$ in the sense of isomorphism);

(ii) $K = R$ if and only if $\phi(R) = \{0\}$.

Proof:

(i) If ϕ is injective then $K = \{0\}$ because $\phi(0) = 0$. In the other direction, suppose $K = \{0\}$. Then since the identity mapping $\Phi : R \to R$ is also a morphism with kernel $\{0\}$, we have from Theorem 5.3 that the mapping $\phi : R \to \phi(R)$ is an isomorphism; i.e. ϕ is injective.

(ii) By definition of the kernel K, if $\phi(R) = \{0\}$ then $K = R$ and if $K = R$ then $\phi(R) = \{0\}$.

●

By Theorem 5.3, we can specify a homomorphic image of a commutative ring R by simply specifying the ideal of elements which is mapped onto zero. The above corollary specifies the two "degenerate" cases corresponding to the two choices of ideals which are not proper ideals. By a *proper homomorphic image* of a commutative ring R we mean a homomorphic image specified by a morphism ϕ whose kernel is a proper ideal in R.

5.5. HOMOMORPHIC IMAGES

Quotient Rings

If R is a commutative ring and if I is any ideal in R, we now show how to construct a homomorphic image $\phi(R)$ such that I is the kernel of the morphism ϕ. Note that if $\phi : R \to R'$ is to be a morphism with kernel I then we must have

$$\phi(a) = \phi(b) \text{ if and only if } a - b \in I.$$

We therefore define the following *congruence relation* on R:

$$a \equiv b \text{ if and only if } a - b \in I. \qquad (5.4)$$

It is readily verified that the congruence relation $\equiv$ is an equivalence relation on R and it therefore divides R into equivalence classes, called *residue classes*. For any element $a \in R$, it is easy to prove that every element in the set

$$a + I = \{a + c : c \in I\}$$

belongs to the same residue class with respect to the congruence relation $\equiv$, that $a \in a + I$, and moreover that if b is in the same residue class as a (i.e. if $b \equiv a$) then $b \in a + I$. Thus the residue class containing a is precisely the set $a + I$.

The set of all residue classes with respect to the congruence relation $\equiv$ defined by (5.4) is called a *quotient set*, denoted by

$$R/I = \{a+I : a \in R\}$$

(read "R modulo the ideal I"). Note that if a and b are in the same residue class (i.e. if $a \equiv b$) then $a + I$ and $b + I$ are two representatives for the same element in the quotient set R/I. We define the operations of addition and multiplication on the quotient set R/I, in terms of the operations defined on R, as follows:

$$(a + I) + (b + I) = (a + b) + I; \qquad (5.5)$$

$$(a + I)\,(b + I) = (ab) + I. \qquad (5.6)$$

Using the fact that I is an ideal, it can be verified that the operations of addition and multiplication on residue classes in R/I are well-defined by (5.5) - (5.6) in the sense that the definitions are independent of the particular representatives used for the residue classes. (Note that the terminology being used here is very similar to the terminology used in Chapter 2 for defining the quotient field of an integral domain.) The following theorem proves that the quotient set R/I with the operations (5.5) - (5.6) is a commutative ring, and R/I is called the *quotient ring* of R modulo the ideal I. Moreover, the theorem specifies a "natural" homomorphism $\phi : R \to R/I$ such that I is the kernel of ϕ and the quotient ring R/I is the desired homomorphic image of R.

Theorem 5.4. Let R be a commutative ring and let I be an ideal in R. The quotient set R/I is a commutative ring under the operations (5.5) - (5.6) and the mapping $\phi : \mathrm{R} \to \mathrm{R/I}$ defined by

$$\phi(a) = a + \mathrm{I} \ \text{ for all } a \in \mathrm{R}$$

is an epimorphism with kernel I.

Proof: First note that the residue classes 0 + I and 1 + I act as the zero and identity (respectively) in R/I since from (5.5) - (5.6) we have:

$$(a + \mathrm{I}) + (0 + \mathrm{I}) = a + \mathrm{I} \ \text{ for any } a + \mathrm{I} \in \mathrm{R/I};$$

$$(a + \mathrm{I})\,(1 + \mathrm{I}) = a + \mathrm{I} \ \text{ for any } a + \mathrm{I} \in \mathrm{R/I}.$$

Now consider the mapping $\phi : \mathrm{R} \to \mathrm{R/I}$ defined by

$$\phi(a) = a + \mathrm{I} \ \text{ for all } a \in \mathrm{R}.$$

It follows immediately from (5.5) - (5.6) that for any $a, b \in \mathrm{R}$,

$$\phi(a + b) = (a + b) + \mathrm{I} = (a + \mathrm{I}) + (b + \mathrm{I}) = \phi(a) + \phi(b)$$

and

$$\phi(ab) = (ab) + \mathrm{I} = (a + \mathrm{I})\,(b + \mathrm{I}) = \phi(a)\,\phi(b).$$

Also,

$$\phi(1) = 1 + \mathrm{I}$$

by definition of ϕ. Thus ϕ is a morphism according to Definition 5.2. But ϕ is surjective by the definition of R/I, so ϕ is an epimorphism. The fact that R/I is a homomorphic image of R implies that R/I is a commutative ring. Finally, we can prove that the kernel of ϕ is precisely I as follows:

$$a \in \mathrm{I} \ \Rightarrow \ \phi(a) = a + \mathrm{I} = 0 + \mathrm{I}$$

and

$$\phi(a) = 0 + \mathrm{I} \ \Rightarrow \ a + \mathrm{I} = 0 + \mathrm{I} \ \Rightarrow \ a - 0 \in \mathrm{I} \ \Rightarrow \ a \in \mathrm{I}.$$

●

Example 5.11. In the integral domain **Z** of integers, we noted in Example 5.8 that $<m>$ is an ideal, for some fixed $m \in \mathbf{Z}$. Thus the quotient ring $\mathbf{Z}/<m>$ is a homomorphic image of **Z** and $<m>$ is the kernel of the natural homomorphism $\phi : \mathbf{Z} \to \mathbf{Z}/<m>$. Assuming that m is positive, the elements of $\mathbf{Z}/<m>$ are given by:

$$\mathbf{Z}/<m> = \{0+<m>, 1+<m>, \ldots, m-1+<m>\}.$$

We usually denote $\mathbf{Z}/<m>$ by $\mathbf{Z}_m$ (the ring of integers modulo m) and we may denote its elements simply by $\{0, 1, \ldots, m-1\}$. The natural homomorphism is precisely the modular homomorphism $\phi_m : \mathbf{Z} \to \mathbf{Z}_m$ defined in Section 5.3.

●

Example 5.12. In the polynomial domain **Q**$[x]$, we noted in Example 5.9 that $<p(x)>$ is an ideal for a fixed polynomial $p(x) \in$ **Q**$[x]$. Thus the quotient ring **Q**$[x]/<p(x)>$ is a homomorphic image of **Q**$[x]$ and $<p(x)>$ is the kernel of the natural homomorphism $\phi : \mathbf{Q}[x] \to \mathbf{Q}[x]/<p(x)>$. Two polynomials $a(x), b(x) \in \mathbf{Q}[x]$ are in the same residue class if they have the same remainder after division by $p(x)$. In particular if $p(x) = x - \alpha$ for some constant $\alpha \in \mathbf{Q}$ then

$$\mathbf{Q}[x]/<x - \alpha> = \{r + <x - \alpha> : \ r \in \mathbf{Q}\}.$$

In this case we may identify $\mathbf{Q}[x]/<x - \alpha>$ with **Q** and the natural homomorphism is precisely the evaluation homomorphism $\phi_{x-\alpha} : \mathbf{Q}[x] \to \mathbf{Q}$ defined in Section 5.3. (See Exercise 5.14.)

•

Example 5.13. In the bivariate polynomial domain $\mathbf{Z}[x,y]$, we noted in Example 5.10 that $<m>$ is an ideal for some fixed integer $m \in \mathbf{Z}$. The quotient ring $\mathbf{Z}[x,y]/<m>$ can be identified with the ring $\mathbf{Z}_m[x,y]$ and the natural homomorphism is precisely the modular homomorphism $\phi_m : \mathbf{Z}[x,y] \to \mathbf{Z}_m[x,y]$ defined in Section 5.3. We also noted in Example 5.10 that $<y - \alpha>$ is an ideal in $\mathbf{Z}[x,y]$, for some fixed $\alpha \in \mathbf{Z}$. The quotient ring $\mathbf{Z}[x,y]/<y - \alpha>$ can be identified with the ring $\mathbf{Z}[x]$ (see Example 5.12) and the natural homomorphism is the evaluation homomorphism $\phi_{y-\alpha} : \mathbf{Z}[x,y] \to \mathbf{Z}[x]$ defined in Section 5.3. (See Exercise 5.15.)

•

Ideal Notation for Homomorphisms

The choice of notation used for the modular and evaluation homomorphisms defined in Section 5.3 and used in the above examples can now be justified. In general if R is a commutative ring then any ideal I in R determines a homomorphic image R/I and we use the notation ϕ_I to denote the corresponding natural homomorphism from R to R/I. Thus if $R = \mathbf{Z}[x_1, \ldots, x_v]$ and if $I = <m>$ for some fixed integer $m \in \mathbf{Z}$ then $\phi_{<m>}$ (or simply ϕ_m) denotes the modular homomorphism which projects $\mathbf{Z}[x_1, \ldots, x_v]$ onto the quotient ring $\mathbf{Z}[x_1, \ldots, x_v]/<m> = \mathbf{Z}_m[x_1, \ldots, x_v]$. Similarly for the evaluation homomorphism we have $R = D[x_1, \ldots, x_v]$ for some coefficient domain D (usually D will be a field $\mathbf{Z}_p$ in practical applications), and if $I = <x_i - \alpha>$ for some fixed $\alpha \in D$ then $\phi_{<x_i - \alpha>}$ (or simply $\phi_{x_i - \alpha}$) denotes the evaluation homomorphism which projects $D[x_1, \ldots, x_v]$ onto $D[x_1, \ldots, x_v]/<x_i - \alpha> = D[x_i, \ldots, x_{i-1}, x_{i+1}, \ldots, x_v]$.

As we have noted previously, modular and evaluation homomorphisms will be used in practice to project the multivariate polynomial domain $\mathbf{Z}[x_1, \ldots, x_v]$ onto a Euclidean domain $\mathbf{Z}_p[x_1]$ or else onto a field $\mathbf{Z}_p$. For example the projection of a polynomial domain $D[x_1, \ldots, x_v]$ onto its coefficient domain D can be accomplished by a composite homomorphism of the form

$$\phi_{x_1-\alpha_1}\,\phi_{x_2-\alpha_2}\,\cdots\,\phi_{x_v-\alpha_v}$$

where $\alpha_i \in D$, $1 \leq i \leq v$. It is convenient to express such a composite homomorphism as a single homomorphism ϕ_I but in order to do so we must specify the kernel I of the composite homomorphism. The following theorem proves that under special conditions (which are satisfied in the cases of interest here) the kernel of a composite homomorphism is simply the sum of the individual kernels, where the "sum" of two ideals was defined in Definition 5.8.

Theorem 5.5. Let $D[x_1, \ldots, x_v]$ be a polynomial domain over a UFD D. Let $\phi_{x_i-\alpha_i}$ be an evaluation homomorphism defined on $D[x_1, \ldots, x_v]$ with kernel $<x_i-\alpha_i>$ and let ϕ_I be another homomorphism defined on $D[x_1, \ldots, x_v]$ with kernel I. Suppose that the homomorphism ϕ_I is *independent of* the homomorphism $\phi_{x_i-\alpha_i}$ in the sense that the composite mappings $\phi_{x_i-\alpha_i}\,\phi_I$ and $\phi_I\,\phi_{x_i-\alpha_i}$ are valid homomorphisms defined on $D[x_1, \ldots, x_v]$ and moreover the composition of these two homomorphisms is commutative (i.e. $\phi_{x_i-\alpha_i}\,\phi_I = \phi_I\,\phi_{x_i-\alpha_i}$). Then the kernel of the composite homomorphism is the sum $<x_i-\alpha_i, I>$ of the two kernels. Notationally,

$$\phi_{x_i-\alpha_i}\,\phi_I = \phi_{<x_i-\alpha_i,\,I>}.$$

Proof: We must prove that for any polynomial $a \in D[x_1, \ldots, x_v]$, $\phi_{x_i-\alpha_i}\,\phi_I(a) = 0$ if and only if $a \in <x_i-\alpha_i, I>$.

"if":

Suppose $a \in <x_i-\alpha_i, I>$. Then $a = p + r$ for some polynomials $p \in <x_i-\alpha_i>$ and $r \in I$. Hence

$$\phi_{x_i-\alpha_i}\,\phi_I(a) = \phi_{x_i-\alpha_i}\,\phi_I(p) + \phi_{x_i-\alpha_i}\,\phi_I(r)$$

$$= \phi_{x_i-\alpha_i}\,\phi_I(p) \ \text{ because } r \in I$$

$$= \phi_I\,\phi_{x_i-\alpha_i}(p) \ \text{ by commutativity}$$

$$= 0 \ \text{ because } p \in <x_i-\alpha_i>.$$

"only if":

Suppose $\phi_{x_i-\alpha_i}\,\phi_I(a) = 0$ for a polynomial $a \in D[x_1, \ldots, x_v]$. Consider the polynomial domain $D[x_1, \ldots, x_v]$ as the univariate domain $C[x_i]$ over the coefficient domain $C = D[x_1, \ldots, x_{i-1}, x_{i+1}, \ldots, x_v]$. Then since $C[x_i]$ is a UFD the pseudo-division property holds and applying it to the polynomials a and $(x_i - \alpha_i)$, we can write

$$a = (x_i - \alpha_i)\, q + r \tag{5.7}$$

for some polynomials $q, r \in C[x_i]$ with either $r = 0$ or $\deg_i(r) < \deg_i(x_i - \alpha_i) = 1$. (Note that in applying the pseudo-division property to obtain (5.7) we have used the fact that the leading coefficient of the "divisor" $x_i - \alpha_i$ is 1.) Hence a can be expressed as the sum (5.7) where the first term of the sum is clearly a member of the ideal $<x_i - \alpha_i>$ and it remains only to prove that $r \in$ I. We will then have the desired result that $a \in <x_i - \alpha_i, \mathrm{I}>$.

To prove that $r \in$ I, apply the composite homomorphism $\phi_{x_i - \alpha_i}\, \phi_{\mathrm{I}}$ to equation (5.7). Then since by supposition $\phi_{x_i - \alpha_i}\, \phi_{\mathrm{I}}\,(a) = 0$ we get

$$\begin{aligned} 0 &= \phi_{x_i - \alpha_i}\, \phi_{\mathrm{I}}\,((x_i - \alpha_i)\, q) + \phi_{x_i - \alpha_i}\, \phi_{\mathrm{I}}\,(r) \\ &= \phi_{\mathrm{I}}\, \phi_{x_i - \alpha_i}\,(r) \text{ by commutativity} \\ &= \phi_{\mathrm{I}}\,(r) \text{ because either } r = 0 \text{ or } \deg_i(r) = 0 \end{aligned}$$

(where in the last step we have used the fact that the evaluation homomorphism $\phi_{x_i - \alpha_i}$ clearly acts as the identity mapping on any polynomial r which is independent of x_i). But $\phi_{\mathrm{I}}\,(r) = 0$ implies that $r \in$ I.

●

From Theorem 5.5 we see that if $\phi_{x_i - \alpha_i}$ and $\phi_{x_j - \alpha_j}$ $(j \neq i)$ are two distinct evaluation homomorphisms defined on a polynomial domain $D[x_1, \ldots, x_v]$ (where D is a UFD) then

$$\phi_{x_i - \alpha_i}\, \phi_{x_j - \alpha_j} = \phi_{<x_i - \alpha_i,\, x_j - \alpha_j>}.$$

By repeated application of Theorem 5.5 we have the more general result that for any n distinct evaluation homomorphisms $\phi_{x_1 - \alpha_1}, \ldots, \phi_{x_n - \alpha_n}$ defined on $D[x_1, \ldots, x_v]$, where $1 \le n \le v$,

$$\phi_{x_1 - \alpha_1}\, \phi_{x_2 - \alpha_2} \cdots \phi_{x_n - \alpha_n} = \phi_{<x_1 - \alpha_1, \ldots,\, x_n - \alpha_n>}.$$

Thus the notation $\phi_{<x_1 - \alpha_1, \ldots,\, x_n - \alpha_n>}$ can be read "substitute α_i for x_i, $1 \le i \le n$" and we call this a *multivariate evaluation homomorphism.* (Note that the order in which the substitutions are performed is irrelevant.)

It also follows from Theorem 5.5 that if

$$\phi_p : \mathbf{Z}[x_1, \ldots, x_v] \to \mathbf{Z}_p[x_1, \ldots, x_v]$$

is a modular homomorphism (with p a prime integer) and if

$$\phi_{x_i - \alpha_i} : \mathbf{Z}_p[x_1, \ldots, x_v] \to \mathbf{Z}_p[x_1, \ldots, x_{i-1}, x_{i+1}, \ldots, x_v]$$

is an evaluation homomorphism (with $\alpha_i \in \mathbf{Z}_p$) then

$$\phi_{x_i - \alpha_i} \phi_p = \phi_{<x_i - \alpha_i, p>}.$$

Again by repeated application of Theorem 5.5 we can generalize this result to show that

$$\phi_{\rm I} \phi_p = \phi_{<{\rm I},p>}$$

if I is the kernel of a multivariate evaluation homomorphism. In practical applications the most commonly used homomorphisms will be of the form

$$\phi_{<{\rm I},p>} : \mathbf{Z}[x_1, \ldots, x_v] \to \mathbf{Z}_p[x_1]$$

where $\mathrm{I} = <x_2 - \alpha_2, \ldots, x_v - \alpha_v>$ with $\alpha_i \in \mathbf{Z}_p$ $(2 \le i \le v)$. For implementation purposes a composite homomorphism $\phi_{<{\rm I},p>}$ where p is a prime integer and I is the kernel of a multivariate evaluation homomorphism will be viewed as the composition of precisely two mappings, namely a modular homomorphism

$$\phi_p : \mathbf{Z}[x_1, \ldots, x_v] \to \mathbf{Z}_p[x_1, \ldots, x_v]$$

followed by a multivariate evaluation homomorphism

$$\phi_{\rm I} : \mathbf{Z}_p[x_1, \ldots, x_v] \to \mathbf{Z}_p[x_1, \ldots, x_v] / \mathrm{I}.$$

The notation $\phi_{<{\rm I},p>}$ will be freely used for this pair of mappings but for computational efficiency it will be important that the order of application of the mappings is as specified above, namely $\phi_{<{\rm I},p>} = \phi_{\rm I} \phi_p$.

Congruence Arithmetic

It is useful to formally specify a *congruence notation* that is used when performing arithmetic on residue classes in a homomorphic image R/I of a ring R. Recall that if I is an ideal in a commutative ring R then the residue classes (i.e. equivalence classes) which form the quotient ring R/I are determined by the congruence relation $\equiv$ defined on R by

$$a \equiv b \text{ if and only if } a - b \in \mathrm{I}.$$

We read this relation as "a is congruent to b modulo I" and we write

$$a \equiv b \pmod{\mathrm{I}}.$$

In the particular case where I is a principal ideal $<q>$ for some fixed element $q \in \mathrm{R}$, we write (mod q) rather than (mod $<q>$). (This notation was already seen briefly in Example 5.7 for the particular case of "modulo p" arithmetic in the quotient ring $\mathbf{Z} / <p> = \mathbf{Z}_p$.)

We will have occasion to solve certain equations involving the congruence relation $\equiv$, so let us note some useful properties of $\equiv$ in addition to the standard properties of an equivalence relation. For any commutative ring R and I an ideal in R we have the following relationships. For any $a,b,c,d \in \mathrm{R}$, if $a \equiv b \pmod{\mathrm{I}}$ and $c \equiv d \pmod{\mathrm{I}}$ then

$$a + c \equiv b + d \pmod{\mathrm{I}}; \tag{5.8}$$

$$a - c \equiv b - d \pmod{\mathrm{I}}; \tag{5.9}$$

$$ac \equiv bd \pmod{\mathrm{I}}. \tag{5.10}$$

For (5.8) and (5.9) it is easy to see that

$$(a \pm c) - (b \pm d) = (a - b) \pm (c - d) \in \mathrm{I}.$$

The proof of (5.10) is only slightly less obvious, namely

$$ac - bd = c\,(a - b) + b\,(c - d) \in \mathrm{I}.$$

We will need another property which will allow us to solve a congruence equation of the form

$$ax \equiv b \pmod{\mathrm{I}}$$

for x if a and b are given. Clearly if there is an element, say a^{-1}, such that $aa^{-1} \equiv 1 \pmod{\mathrm{I}}$ then by (5.10) it follows that $aa^{-1}b \equiv b \pmod{\mathrm{I}}$ so that choosing $x = a^{-1}b$ yields a solution to the given congruence equation. Since an element in an arbitrary commutative ring does not necessarily have a multiplicative inverse, the property which will allow us to solve congruence equations in the above sense will be less general than properties (5.8) - (5.10).

In order to obtain the desired property we will restrict attention to the case where the ring R is a Euclidean domain D. As we noted in Section 5.4, every ideal I in a Euclidean domain D is a principal ideal so $\mathrm{I} = <q>$ for some fixed element $q \in \mathrm{D}$. The following theorem states a condition under which an element $a \in \mathrm{D}$ has an inverse modulo $<q>$. The proof of the theorem is constructive – i.e. it gives an algorithm for computing inverses modulo $<q>$.

Theorem 5.6. Let $<q>$ be an ideal in a Euclidean domain D and let $a \in \mathrm{D}$ be relatively prime to q (i.e. $\mathrm{GCD}(a,q) = 1$). Then there exists an element $a^{-1} \in \mathrm{D}$ such that

$$aa^{-1} \equiv 1 \pmod{q}.$$

This is equivalent to saying that in the homomorphic image $\mathrm{D}/<q>$ the element $\phi_q(a)$ has a multiplicative inverse.

Proof: Since D is a Euclidean domain we can apply the extended Euclidean algorithm (Algorithm 2.2) to $a,q \in \mathrm{D}$ yielding elements $s,t \in \mathrm{D}$ such that

$$sa + tq = 1,$$

where we have used the fact that $\mathrm{GCD}(a,q) = 1$. Then $sa - 1 \in <q>$, or $sa \equiv 1 \pmod{q}$. Thus $a^{-1} = s$ is the desired inverse.

To show the equivalence of the last statement in the theorem, first suppose that $aa^{-1} \equiv 1 \pmod{q}$. Then $aa^{-1} - 1 \in <q>$, so $\phi_q(aa^{-1} - 1) = 0$ which yields $\phi_q(a)\,\phi_q(a^{-1}) = 1$ – i.e. $\phi_q(a^{-1})$ is the multiplicative inverse of $\phi_q(a)$ in $\mathrm{D}/<q>$. In the other direction, suppose $\phi_q(a)$ has a multiplicative inverse $\bar{b} \in \mathrm{D}/<q>$. Then there is an element $b \in \mathrm{D}$ such that $\phi_q(b) = \tilde{b}$. We have $\phi_q(a)\,\phi_q(b) = 1$ which implies that $\phi_q(ab-1) = 0$, or $ab - 1 \in \langle q \rangle$, or $ab \equiv 1 \pmod{q}$.

●

Finally we are able to state the property of congruence relations that we have been seeking. For any Euclidean domain D and $\langle q \rangle$ the ideal generated by a fixed element $q \in$ D the following property holds:

For any $a,b \in$ D with a relatively prime to q there is an element $a^{-1} \in$ D which is the inverse (mod q) of a and any element $x \in$ D such that

$$x \equiv a^{-1} b \pmod{q} \tag{5.11}$$

is a solution of the congruence equation

$$ax \equiv b \pmod{q}.$$

5.6. THE INTEGER CHINESE REMAINDER ALGORITHM

We now turn to the development of algorithms for inverting homomorphisms. The basic tenet of these "inversion" algorithms is that under appropriate conditions an element a in a ring R can be reconstructed if its images $\phi_{I_i}(a)$, $i = 1, 2, \ldots$ are known in an "appropriate number" of homomorphic images R/I_i of R.

The Chinese Remainder Problem

Recall that for any fixed integer $m \in \mathbf{Z}$ the modular homomorphism $\phi_m : \mathbf{Z} \to \mathbf{Z}_m$ which projects the ring $\mathbf{Z}$ of integers onto the finite ring $\mathbf{Z}_m$ of "integers modulo m" is specified by

$$\phi_m(a) = \text{rem}(a,m) \text{ for all } a \in \mathbf{Z}. \tag{5.12}$$

Using congruence notation, if $a \in \mathbf{Z}$ and if $\phi_m(a) = \tilde{a} \in \mathbf{Z}_m$ then we write

$$a \equiv \tilde{a} \pmod{m}.$$

The classical mathematical problem known as the *Chinese remainder problem* can be stated as follows:

Given *moduli* $m_0, m_1, \ldots, m_n \in \mathbf{Z}$ and given corresponding *residues* $u_i \in \mathbf{Z}_{m_i}$, $0 \le i \le n$, find an integer $u \in \mathbf{Z}$ such that

$$u \equiv u_i \pmod{m_i}, \quad 0 \le i \le n.$$

(This problem, in a less general form, was considered by the ancient Chinese and by the ancient Greeks about 2000 years ago.) Note that an algorithm for solving the Chinese remainder problem will be an algorithm for "inverting" the modular homomorphism, since if we know the images (residues) $u_i = \phi_{m_i}(u)$ of an integer u, for several modular homomorphisms ϕ_{m_i}, then such an algorithm will reconstruct the integer u. (More correctly, the latter statement will be true once we have determined conditions such that there exists a *unique* integer u which solves the problem.) The following theorem specifies conditions under which there exists a unique solution to the Chinese remainder problem.

Theorem 5.7 (Chinese Remainder Theorem). Let $m_0, m_1, \ldots, m_n \in \mathbf{Z}$ be integers which are pairwise relatively prime – i.e.

$$\mathrm{GCD}(m_i, m_j) = 1 \ \text{ for } \ i \neq j,$$

and let $u_i \in \mathbf{Z}_{m_i}$, $i = 0, 1, \ldots, n$ be $n + 1$ specified residues. For any fixed integer $a \in \mathbf{Z}$ there exists a unique integer $u \in \mathbf{Z}$ which satisfies the following conditions:

$$a \le u < a + m, \text{ where } m = \prod_{i=0}^{n} m_i; \tag{5.13}$$

$$u \equiv u_i \pmod{m_i}, \ 0 \le i \le n. \tag{5.14}$$

Proof:

Uniqueness:

Let $u, v \in \mathbf{Z}$ be two integers satisfying conditions (5.13) and (5.14). Then using the fact that $\equiv$ is an equivalence relation, it follows from condition (5.14) that

$$u \equiv v \pmod{m_i}, \text{ for } i = 0, 1, \ldots, n$$

$$\Rightarrow \ u - v \in \langle m_i \rangle, \text{ for } i = 0, 1, \ldots, n$$

$$\Rightarrow \ u - v \in \langle m \rangle \text{ where } m = \prod_{i=0}^{n} m_i$$

where in the last step we have used the fact that since the moduli $m_0, m_1, \ldots, m_n$ are pairwise relatively prime, an integer which is a multiple of each m_i must also be a multiple of the product m. But from condition (5.13) it follows that

$$| \, u - v \, | < m$$

and hence $u - v = 0$ since 0 is the only element of the ideal $\langle m \rangle$ which has absolute value less than m. Thus $u = v$.

Existence:

Let u run through the m distinct integer values in the range specified by condition (5.13) and consider the corresponding $(n + 1)$-tuples $(\phi_{m_0}(u), \phi_{m_1}(u), \ldots, \phi_{m_n}(u))$, where ϕ_{m_i} is the modular homomorphism defined by (5.12). By the uniqueness proof above, no two of these $(n + 1)$-tuples can be identical and hence the $(n + 1)$-tuples also take on m distinct values. But since the finite ring $\mathbf{Z}_{m_i}$ contains precisely m_i elements there are exactly $m = \prod_{i=0}^{n} m_i$ distinct $(n + 1)$-tuples $(v_0, v_1, \ldots, v_n)$ such that $v_i \in \mathbf{Z}_{m_i}$. Hence each possible $(n+1)$-tuple occurs exactly once and therefore there must be one value of u in the given range such that

$$(\phi_{m_0}(u), \phi_{m_1}(u), \ldots, \phi_{m_n}(u)) = (u_0, u_1, \ldots, u_n).$$

●

It is important to note the sense in which the solution to the Chinese remainder problem is unique. If we are given $n+1$ residues $u_i \in \mathbf{Z}_{m_i}$ $(0 \le i \le n)$ corresponding to $n+1$ moduli m_i $(0 \le i \le n)$ (assumed to be pairwise relatively prime) then the Chinese remainder problem has an infinite set of integer solutions, but by property (5.13) of Theorem 5.7 (choosing $a = 0$) we see that the solution is unique if we restrict it to the range $0 \le u < m$. Thus we say that the solution is *unique modulo m*. In other words, given $u_i \in \mathbf{Z}_{m_i}$ $(0 \le i \le n)$ the system of congruences (5.14) does not have a unique solution in the ring $\mathbf{Z}$ but it does have a unique solution in the ring $\mathbf{Z}_m$, where $m = \prod_{i=0}^{n} m_i$.

Different choices of values for the arbitrary integer a in Theorem 5.7 correspond to different representations for the ring $\mathbf{Z}_m$. The choice $a = 0$ corresponds to the familiar *positive representation* of $\mathbf{Z}_m$ as

$$\mathbf{Z}_m = \{0, 1, \ldots, m-1\}$$

(where we are assuming that m is positive). In practical applications all of the moduli $m_0, m_1, \ldots, m_n$ and m will be *odd positive* integers and another useful representation will be the *symmetric representation* of $\mathbf{Z}_m$ as

$$\mathbf{Z}_m = \{-\frac{m-1}{2}, \ldots, -1, 0, 1, \ldots, \frac{m-1}{2}\}.$$

The choice of value for the integer a in Theorem 5.7 which corresponds to the symmetric representation of $\mathbf{Z}_m$ is clearly

$$a = -\frac{m-1}{2}.$$

The proof given above for Theorem 5.7 is not a constructive proof since it would be highly impractical to determine the solution u by simply trying each element of the ring $\mathbf{Z}_m$ when m is a large integer. We will now proceed to develop an efficient algorithm for solving the Chinese remainder problem.

Garner's Algorithm

The algorithm which is generally used to solve the Chinese remainder problem is named after H. L. Garner who developed a version of the algorithm in the late 1950's (cf.[2]). Given positive moduli $m_i \in Z$ $(0 \le i \le n)$ which are pairwise relatively prime and given corresponding residues $u_i \in \mathbf{Z}_{m_i}$ $(0 \le i \le n)$, we wish to compute the unique $u \in \mathbf{Z}_m$ (where $m = \prod_{i=0}^{n} m_i$) which satisfies the system of congruences (5.14). The key to Garner's algorithm is to express the solution $u \in \mathbf{Z}_m$ in the *mixed radix representation*

$$u = v_0 + v_1(m_0) + v_2(m_0 m_1) + \cdots + v_n(\prod_{i=0}^{n-1} m_i) \tag{5.15}$$

where $v_k \in \mathbf{Z}_{m_k}$ for $k = 0, 1, \ldots, n$.

The mixed radix representation (5.15) is not meaningful in the full generality stated above since the addition and multiplication operations appearing in (5.15) are to be performed in the ring $\mathbf{Z}_m$ but each *mixed radix coefficient* v_k lies in a different ring $\mathbf{Z}_{m_k}$. In order to make (5.15) meaningful, we will require that the rings $\mathbf{Z}_{m_k}$ $(0 \le k \le n)$ and $\mathbf{Z}_m$ be represented in one of the following two *consistent representations:*

(i) Each ring $\mathbf{Z}_{m_k}$ $(0 \le k \le n)$ and $\mathbf{Z}_m$ is represented in its positive representation; or

(ii) Each ring $\mathbf{Z}_{m_k}$ $(0 \le k \le n)$ and $\mathbf{Z}_m$ is represented in its symmetric representation (where we assume that each m_k is odd).

Then the natural identification of elements in a ring $\mathbf{Z}_{m_k}$ with elements in the larger ring $\mathbf{Z}_m$ gives the desired interpretation of (5.15). It can be proved that any $u \in \mathbf{Z}_m$ can be represented in the form (5.15) and if one of the consistent representations (i) or (ii) is used then the coefficients v_k $(0 \le k \le n)$ are uniquely determined. It should be noted that in the case when the positive consistent representation (i) is used, (5.15) is a straightforward generalization of the familiar fact the any integer u in the range $0 \le u < \beta^{n+1}$ (i.e. $u \in \mathbf{Z}_{\beta^{n+1}}$), for a positive integer $\beta > 1$, can be uniquely represented in the *radix* β *representation:*

$$u = v_0 + v_1\beta + v_2\beta^2 + \cdots + v_n\beta^n$$

where $0 \le v_k < \beta$ (i.e. $v_k \in \mathbf{Z}_\beta$).

Example 5.14. Let $m_0 = 3, m_1 = 5$, and $m = m_0 m_1 = 15$. Using the positive consistent representation, the integer $u = 11 \in \mathbf{Z}_{15}$ has the unique mixed radix representation

$$11 = v_0 + v_1(3)$$

with $v_0 = 2 \in \mathbf{Z}_3$ and $v_1 = 3 \in \mathbf{Z}_5$. Using the symmetric consistent representation, the integer $\tilde{u} = -4 \in \mathbf{Z}_{15}$ has the unique mixed radix representation

$$-4 = \tilde{v_0} + \tilde{v_1}(3)$$

with $\tilde{v_0} = -1 \in \mathbf{Z}_3$ and $\tilde{v_1} = -1 \in \mathbf{Z}_5$. Note that $u = 11$ and $\tilde{u} = -4$ are simply two different representations for the same element in $\mathbf{Z}_{15}$ but that the corresponding coefficients v_1 and $\tilde{v_1}$ are *not* simply two different representations for the same element in $\mathbf{Z}_5$.

●

Writing the solution u of the system of congruences (5.14) in the mixed radix representation (5.15), it is easy to determine formulas for the coefficients v_k $(0 \le k \le n)$ appearing in (5.15). It is obvious from (5.15) that

$$u \equiv v_0 \,(\mathrm{mod}\ m_0)$$

and therefore the case $i = 0$ of the system of congruences (5.14) will be satisfied if v_0 is chosen such that

$$v_0 \equiv u_0 \,(\mathrm{mod}\ m_0). \tag{5.16}$$

In general for $k \ge 1$, if coefficients $v_0, v_1, \ldots, v_{k-1}$ have been determined then noting from (5.15) that

$$u \equiv v_0 + v_1(m_0) + \cdots + v_k \,(\prod_{i=0}^{k-1} m_i) \,(\mathrm{mod}\ m_k),$$

we can satisfy the case $i = k$ of the system of congruences (5.14) by choosing v_k such that

$$v_0 + v_1(m_0) + \cdots + v_k \,(\prod_{i=0}^{k-1} m_i) \equiv u_k \,(\mathrm{mod}\ m_k).$$

Using properties (5.8) - (5.11) to solve this congruence equation for v_k we get for $k \ge 1$:

$$v_k \equiv \left(u_k - [v_0 + v_1(m_0) + \cdots + v_{k-1}(\prod_{i=0}^{k-2} m_i)] \right) \left(\prod_{i=0}^{k-1} m_i \right)^{-1} (\mathrm{mod}\ m_k) \tag{5.17}$$

where the inverse appearing here is valid because $\prod_{i=0}^{k-1} m_i$ is relatively prime to m_k. Finally we note that once a consistent representation has been chosen, there is a *unique* integer $v_0 \in \mathbf{Z}_{m_0}$ satisfying (5.16) (namely $v_0 = u_0 \in \mathbf{Z}_{m_0}$) and similarly for $k = 1, 2, \ldots, n$ there is a *unique* integer $v_k \in \mathbf{Z}_{m_k}$ satisfying (5.17).

Implementation Details for Garner's Algorithm

Garner's algorithm is presented formally as Algorithm 5.1. Some details about the implementation of this algorithm need further discussion. It is important to note that in the usual applications of Garner's algorithm the moduli m_i $(0 \le i \le n)$ are single-precision integers (typically, large single-precision integers) and therefore the residues u_i $(0 \le i \le n)$ are also single-precision integers. The integer u being computed will be a multiprecision integer and indeed the list of residues $(u_0, u_1, \ldots, u_n)$ can be viewed simply as a different representation for the multiprecision integer u (see Chapter 4). Algorithm 5.1 is organized so that in this typical situation operations on multiprecision integers are completely avoided until the last step. In particular we use the notation ϕ_{m_k} in Algorithm 5.1 in a manner that is consistent with its mathematical meaning as a modular homomorphism but we give it the following more precise algorithmic specification:

$$\phi_{m_k}(expression) \text{ means ``evaluate } expression \text{ in the ring } \mathbf{Z}_{m_k}\text{''}.$$

More specifically, it means that when *expression* is decomposed into a sequence of binary operations, the intermediate result of *each* binary operation is to be reduced modulo m_k before proceeding with the evaluation of *expression*. In this way we are guaranteed that every variable (except of course u) appearing in Algorithm 5.1 is a single-precision variable and moreover that every operation appearing in step 1 and step 2 is an operation on single-precision integers. (Note however that if a and b are single-precision integers then the operation $\phi_{m_k}(a \cdot b)$, for example, is usually performed by an ordinary integer multiplication $a \cdot b$ yielding a double-precision integer, say c, followed by an integer division operation to compute $\mathrm{rem}(c, m_k)$.)

For $k = 1, 2, \ldots, n$ the integer v_k satisfying (5.17) is computed in step 2 of Algorithm 5.1 by evaluating the right hand side of (5.17) in the ring $\mathbf{Z}_{m_k}$. The inverses appearing in (5.17):

$$\gamma_k = (\prod_{i=0}^{k-1} m_i)^{-1} \pmod{m_k}, \text{ for } k = 1, 2, \ldots, n$$

are all computed in step 1. Note that a method for implementing the procedure

$$\mathrm{reciprocal}(a, q)$$

to compute $a^{-1} \pmod q$ for relatively prime a and q, is given in the proof of Theorem 5.6; namely, apply the extended Euclidean algorithm (Algorithm 2.2) to $a, q \in \mathbf{Z}$ yielding integers s and t such that

$$sa + tq = 1$$

and then $\phi_q(s) = \mathrm{rem}(s, q)$ is the desired inverse in the ring $\mathbf{Z}_q$. The computation of the inverses $\{\gamma_k\}$ was purposely separated from the rest of the computation in Algorithm 5.1 because $\{\gamma_k\}$ depend only on the moduli $\{m_i\}$. For typical applications of Garner's algorithm in a system using the modular representation for multiprecision integers, the moduli $\{m_i\}$ would be fixed so that step 1 would be removed from Algorithm 5.1 and the inverses $\{\gamma_k\}$ would be given to the algorithm as precomputed constants. It is also worth noting that there are situations when both step 1 and step 3 would be removed from Algorithm 5.1. For example, in the above-mentioned setting if it is desired to compare two multiprecision integers a and b represented in their modular representations then it is sufficient to compute their (single-precision) mixed radix coefficients and compare them (cf. Knuth [3]).

Finally, step 3 needs some justification. We have stated that if consistent representations are used for $\mathbf{Z}_{m_k}$ $(0 \le k \le n)$ and $\mathbf{Z}_m$ then the mixed radix representation (5.15) for $u \in \mathbf{Z}_m$ is unique. However we have not shown that if the operations in (5.15) are performed in the ring $\mathbf{Z}$ rather than in the ring $\mathbf{Z}_m$, we will still obtain the unique $u \in \mathbf{Z}_m$ as desired – i.e. in step 3 of Algorithm 5.1 there is no need to write the **for**-loop statement as

Algorithm 5.1. Garner's Chinese Remainder Algorithm.

```
procedure IntegerCRA((m_0, ..., m_n),(u_0, ..., u_n))
    # Given positive moduli m_i ∈ Z (0 ≤ i ≤ n) which are pairwise
    # relatively prime and given corresponding residues u_i ∈ Z_{m_i},
    # compute the unique integer u ∈ Z_m (where m = ∏m_i) such that
    # u ≡ u_i (mod m_i), i = 0, 1, ..., n.

    # Step 1: Compute the required inverses using a procedure
    #         reciprocal(a, q) which computes a^{-1} (mod q).

    for k from 1 to n do {
        product ← φ_{m_k}(m_0)
        for i from 1 to k − 1 do
            product ← φ_{m_k}(product · m_i)
        γ_k ← reciprocal(product, m_k)  }

    # Step 2: Compute the mixed radix coeffs {v_k}.

    v_0 ← u_0
    for k from 1 to n do {
        temp ← v_{k−1}
        for j from k − 2 to 0 by −1 do
            temp ← φ_{m_k}(temp · m_j + v_j)

        v_k ← φ_{m_k}((u_k − temp) · γ_k)  }

    # Step 3: Convert from mixed radix representation
    #         to standard representation.

    u ← v_n
    for k from n − 1 to 0 by −1 do
        u ← u · m_k + v_k
    return(u)
end
```

$$u \leftarrow \phi_m(u \cdot m_k + v_k).$$

To justify this, note from (5.15) that if $|v_k| \le (m_k - 1)/2$ for $k = 0, 1, \ldots, n$ (i.e. if the symmetric consistent representation is used) then

$$|u| \le \frac{m_0 - 1}{2} + \frac{m_1 - 1}{2}(m_0) + \cdots + \frac{m_n - 1}{2}(\prod_{i=0}^{n-1} m_i)$$

$$\le \frac{1}{2}[(\prod_{i=0}^{n} m_i) - 1]$$

proving that u lies in the correct range. Similarly if $0 \le v_k \le m_k - 1$ for $k = 0, 1, \ldots, n$ (i.e. if the positive consistent representation is used) then clearly $u \ge 0$ and, proceeding as above,

$$u \le (\prod_{i=0}^{n} m_i) - 1$$

proving again that u lies in the correct range. Finally, step 3 performs the evaluation of (5.15) using the method of nested multiplication:

$$u = v_0 + m_0(v_1 + m_1(v_2 + \cdots + m_{n-2}(v_{n-1} + m_{n-1}(v_n)) \cdots)).$$

Example 5.15. Suppose that the single-precision integers on a particular computer are restricted to the range $-100 < a < 100$ (i.e. two-digit integers). Consider as moduli the three largest single-precision integers which are odd and pairwise relatively prime:

$$m_0 = 99; \; m_1 = 97; \; m_2 = 95.$$

Then $m = m_0 m_1 m_2 = 912285$. Using the symmetric consistent representation, the range of integers in $\mathbf{Z}_{912285}$ is $-456142 \le u \le 456142$.

Now consider the problem of determining u given that:

$u \equiv 49 \pmod{99}$;

$u \equiv -21 \pmod{97}$;.

$u \equiv -30 \pmod{95}$

Applying Algorithm 5.1, we compute in step 1 the following inverses:

$$\gamma_1 = m_0^{-1} \pmod{m_1} = 99^{-1} \pmod{97} = 2^{-1} \pmod{97} = -48;$$

$$\gamma_2 = (m_0 m_1)^{-1} \pmod{m_2} = 8^{-1} \pmod{95} = 12.$$

Carrying out the computation of step 2, we get the following mixed radix coefficients for u:

$$v_0 = 49; \; v_1 = -35; \; v_2 = -28.$$

At this point we have the following mixed radix representation for u:

$$u = 49 - 35\,(99) - 28\,(99)\,(97).$$

Finally, carrying out the conversion of step 3 using ''multiprecision'' arithmetic we find

$$u = -272300.$$

●

Let us return to the example in the introduction of this chapter. We may look at our linear system over the domains $\mathbf{Z}_p$, for various primes p. However, piecing together the solutions using the CRA will give us integer values for x, y and z, which we know happens infrequently. Rather, if we let

$$x_1 = \det \begin{bmatrix} 1 & 44 & 74 \\ -2 & 14 & -10 \\ 34 & -28 & 20 \end{bmatrix}, \quad y_1 = \det \begin{bmatrix} 22 & 1 & 74 \\ 15 & -2 & -10 \\ -25 & 34 & 20 \end{bmatrix}$$

$$z_1 = \det \begin{bmatrix} 22 & 44 & 1 \\ 15 & 14 & -2 \\ -25 & -28 & 34 \end{bmatrix}, \quad d = \det \begin{bmatrix} 22 & 44 & 74 \\ 15 & 14 & -10 \\ -25 & -28 & 20 \end{bmatrix}$$

then we know that x_1, y_1, z_1 and d will be integers and that

$$x = \frac{x_1}{d}, \qquad y = \frac{y_1}{d}, \qquad z = \frac{z_1}{d}.$$

However, for a given domain $\mathbf{Z}_p$ we need not calculate these determinants. Rather we find the modular solutions

$$x \pmod p), \ y \pmod p), \ z \pmod p), \text{ and } d \pmod p),$$

via the usual efficient Gaussian elimination method, and use

$$x_1 \equiv x \cdot d \pmod p, \quad y_1 \equiv y \cdot d \pmod p, \quad z_1 \equiv z \cdot d \pmod p.$$

In this way we obtain modular representations for x_1, y_1, z_1, and d. Using the integer CRA gives integer representations for these four quantities, and hence rational number answers for x_1, y_1, and z.

For example, working over $\mathbf{Z}_7$ the system becomes

$$\begin{aligned} x + 2y - 3z &= 1, \\ x \qquad - 3z &= -2, \\ 3x \qquad - z &= -1. \end{aligned}$$

Gaussian elimination gives

$$x \equiv -1 \pmod 7, \ y \equiv -2 \pmod 7, \ z \equiv -2 \pmod 7 \text{ and } d \equiv -2 \pmod 7.$$

Similarly, working over the domains $\mathbf{Z}_{11}$, $\mathbf{Z}_{13}$, $\mathbf{Z}_{17}$ and $\mathbf{Z}_{19}$ gives

$$x_1 \equiv -5 \pmod{11},\ \ y_1 \equiv 0 \pmod{11},\ \ z_1 \equiv -4 \pmod{11},\ \ d \equiv 1 \pmod{11},$$
$$x_1 \equiv -2 \pmod{13},\ \ y_1 \equiv 4 \pmod{13},\ \ z_1 \equiv 6 \pmod{13},\ \ d \equiv 4 \pmod{13},$$
$$x_1 \equiv 5 \pmod{17},\ \ y_1 \equiv -6 \pmod{17},\ \ z_1 \equiv -3 \pmod{17},\ \ d \equiv -2 \pmod{17},$$
$$x_1 \equiv 9 \pmod{19},\ \ y_1 \equiv 6 \pmod{19},\ \ z_1 \equiv 7 \pmod{19},\ \ d \equiv -8 \pmod{19}.$$

Thus, for example the "modular representations" for x_1 and d are

$$x_1 = (2, -5, -2, 5, 9) \ \text{ and } \ d \equiv (-2, 1, 4, -2, -8)$$

with respect to the moduli 7, 11, 13, 17 and 19.

Using Garner's algorithm, we find that corresponding integer representations are then

$$x_1 = -44280 \ \text{ and } \ d = -7380$$

giving

$$x = \frac{-44280}{-7380} = 6\,.$$

Similarly, we obtain

$$y = \frac{40590}{-7380} = \frac{-11}{2} \ \text{ and } \ z = \frac{-11070}{-7380} = \frac{3}{2}\ .$$

We will return to the topic of modular methods for solving linear systems in Chapter 9.

5.7. THE POLYNOMIAL INTERPOLATION ALGORITHM

We now consider the corresponding inversion process for evaluation homomorphisms. Recall that we are primarily interested in homomorphisms $\phi_{I,p}$ which project the multivariate polynomial domain $\mathbf{Z}[x_1, \ldots, x_v]$ onto the Euclidean domain $\mathbf{Z}_p[x_1]$ (or perhaps onto the field $\mathbf{Z}_p$). In the notation $\phi_{<I,p>}$, p denotes a prime integer, I denotes the kernel of a multivariate evaluation homomorphism, and $\phi_{<I,p>}$ denotes the composite homomorphism $\phi_I \phi_p$ with domains of definition indicated by:

$$\phi_p : \mathbf{Z}[x_1, \ldots, x_v] \rightarrow \mathbf{Z}_p[x_1, \ldots, x_v] \tag{5.18}$$

and

$$\phi_I : \mathbf{Z}_p[x_1, \ldots, x_v] \rightarrow \mathbf{Z}_p[x_1] \tag{5.19}$$

(or the homomorphic image in (5.19) could as well be $\mathbf{Z}_p$). The inversion process for homomorphisms of the form (5.18) is the Chinese remainder algorithm of the preceding section. (Note that Garner's Chinese remainder algorithm can be applied coefficient-by-coefficient in the polynomial case, with the polynomials expressed in expanded canonical form.) The inversion process for homomorphisms of the form (5.19) is the problem of polynomial interpolation.

The Polynomial Interpolation Problem

The inversion of multivariate evaluation homomorphisms of the form (5.19) will be accomplished one indeterminate at a time, viewing ϕ_I in the natural way as a composition of univariate evaluation homomorphisms. Therefore it is sufficient to consider the inversion of univariate evaluation homomorphisms of the form

$$\phi_{x-\alpha_i} : D[x] \to D$$

where D is (in general) a multivariate polynomial domain over a field $\mathbf{Z}_p$ and where $\alpha_i \in \mathbf{Z}_p$. It will be important computationally that α_i lies in the field $\mathbf{Z}_p$.

The development of an algorithm for polynomial interpolation will directly parallel the development of Garner's algorithm for the integer Chinese remainder problem. Indeed it should become clear that the two processes are identical if one takes an appropriately abstract (ring-theoretic) point of view. In particular, by paraphrasing the statement of the integer Chinese remainder problem we get the following statement of the *polynomial interpolation problem:*

> Let D be a domain of polynomials (in zero or more indeterminates other than x) over a coefficient field $\mathbf{Z}_p$. Given *moduli* $x-\alpha_0, x-\alpha_1, \ldots, x-\alpha_n$ where $\alpha_i \in \mathbf{Z}_p, 0 \le i \le n$, and given corresponding *residues* $u_i \in D, 0 \le i \le n$, find a polynomial $u(x) \in D[x]$ such that
>
> $$u(x) \equiv u_i \ (\mathrm{mod}\ x-\alpha_i),\ 0 \le i \le n. \qquad (5.20)$$

Note that in this case the congruences (5.20) are usually stated in the following equivalent form:

$$u(\alpha_i) = u_i,\ 0 \le i \le n \qquad (5.21)$$

and the elements $\alpha_i \in \mathbf{Z}_p$ $(0 \le i \le n)$ are usually called *evaluation points* or *interpolation points*. As in the case of the integer Chinese remainder problem, in order to guarantee that a solution exists we must impose the additional condition that the moduli $\{x-\alpha_i\}$ be pairwise relatively prime. But clearly

$$\mathrm{GCD}(x-\alpha_i, x-\alpha_j) = 1 \text{ if and only if } \alpha_i \ne \alpha_j$$

so the additional condition reduces to the rather obvious condition that the moduli $\{x-\alpha_i\}$ must be *distinct* (i.e. the evaluation points $\{\alpha_i\}$ must be distinct). Also as in the integer Chinese remainder problem, the solution to the polynomial interpolation problem is only unique modulo $\prod_{i=0}^{n}(x-\alpha_i)$, which is to say that the solution is unique if we restrict it to be of degree less than $n+1$.

The following theorem proves the above existence and uniqueness results in a more general setting where the domain D is an arbitrary integral domain and the evaluation points $\{\alpha_i\}$ are arbitrary distinct points in D. However this theorem allows the solution $u(x)$ to lie in $F_D[x]$ rather than in $D[x]$, where F_D denotes the quotient field of the integral domain D.

We will then proceed to develop an efficient algorithm for solving the polynomial interpolation problem and it will be obvious that in the particular setting presented above, the solution $u(x)$ will lie in D[x] because the only divisions required will be divisions in the coefficient field $\mathbf{Z}_p$.

Theorem 5.8. Let D be an arbitrary integral domain, let $\alpha_i \in$ D, $i = 0, 1, \ldots, n$ be n+1 distinct elements in D, and let $u_i \in$ D, $i = 0, 1, \ldots, n$ be n+1 specified values in D. There exists a unique polynomial $u(x) \in F_D[x]$ (where F_D is the quotient field of D) which satisfies the following conditions:

(i) $\deg(u(x)) \le n$;

(ii) $u(\alpha_i) = u_i, \; 0 \le i \le n$.

Proof: By condition (i) we may write $u(x)$ in the form

$$u(x) = a_0 + a_1 x + \cdots + a_n x^n$$

where the coefficients $a_i \in F_D$ $(0 \le i \le n)$ are to be determined. Condition (ii) then becomes the following linear system of order (n+1):

$$\mathrm{V}\mathbf{a} = \mathbf{u}$$

where V is the *Vandermonde matrix* with (i,j)-th entry α_i^j $(i,j = 0,1, \ldots, n)$,$\mathbf{u}$ is the vector with i-th entry u_i $(i = 0, 1, \ldots, n)$, and $\mathbf{a}$ is the vector of unknowns with i-th entry a_i $(i = 0, 1, \ldots, n)$. From elementary linear algebra, this linear system can be solved in the field F_D and the solution is unique if $\det(\mathrm{V}) \ne 0$. Employing the classical formula for the Vandermonde determinant:

$$\det(\mathrm{V}) = \prod_{0 \le i < k \le n} (\alpha_k - \alpha_i),$$

we see that $\det(\mathrm{V}) \ne 0$ because the elements $\alpha_0, \alpha_1, \ldots, \alpha_n \in$ D are distinct.

•

The Newton Interpolation Algorithm

The proof of Theorem 5.8 is a constructive proof since we can solve linear equations over the quotient field of an integral domain (see Chapter 9). However the solution to the interpolation problem can be computed by an algorithm requiring much less work than solving a system of linear equations (cf. Exercise 5.22). The algorithm we will develop for polynomial interpolation dates back to Newton in the 17th century. As with Garner's algorithm, the key to the development is to express the solution $u(x) \in F_D[x]$ in the following mixed radix representation (sometimes called the *Newton form* or the *divided-difference form*):

$$u(x) = v_0 + v_1(x - \alpha_0) + v_2(x - \alpha_0)(x - \alpha_1) + \cdots + v_n \prod_{i=0}^{n-1}(x - \alpha_i) \tag{5.22}$$

where the *Newton coefficients* $v_k \in F_D$ $(0 \le k \le n)$ are to be determined. The justification for this mixed radix representation is the fact from elementary linear algebra that *any* set of polynomials $m_k(x) \in F_D[x]$, $k = 0, 1, \ldots, n$ with $\deg(m_k(x)) = k$ forms a valid *basis* for polynomials of degree n in x over the field F_D; in this case we are choosing $m_0(x) = 1$, $m_k(x) = \prod_{i=0}^{k-1}(x - \alpha_i)$ for $k = 1,2, \ldots, n$.

The Newton interpolation algorithm can be developed for the general setting of Theorem 5.8 in which case the Newton coefficients $\{v_k\}$ in (5.22) will be quotients of elements in D (called *divided-differences*). However we will develop the algorithm for the case of practical interest to us, namely the setting indicated in the preamble to (5.20). In this case no quotients of elements (polynomials) in D will arise since the only divisions which arise will be divisions (i.e. multiplications by inverses) in the coefficient field $\mathbf{Z}_p$ of the polynomial domain D.

Writing the solution $u(x)$ in the Newton form (5.22) we apply the conditions (5.21) to obtain formulas for the Newton coefficients v_k $(0 \le k \le n)$. It is obvious from (5.22) that

$$u(\alpha_0) = v_0$$

and therefore the case i=0 of the conditions (5.21) will be satisfied if v_0 is chosen to be

$$v_0 = u_0 \,. \tag{5.23}$$

In general for $k \ge 1$, if the Newton coefficients $v_0, v_1, \ldots, v_{k-1}$ have been determined then noting from (5.22) that

$$u(\alpha_k) = v_0 + v_1(\alpha_k - \alpha_0) + \cdots + v_k \prod_{i=0}^{k-1}(\alpha_k - \alpha_i),$$

the case i=k of the conditions (5.21) will be satisfied if v_k is chosen such that

$$v_0 + v_1(\alpha_k - \alpha_0) + \cdots + v_k \prod_{i=0}^{k-1}(\alpha_k - \alpha_i) = u_k.$$

Now since $\alpha_i \in \mathbf{Z}_p$ $(0 \le i \le n)$ we can compute in the field $\mathbf{Z}_p$ the inverse of the nonzero element $\prod_{i=0}^{k-1}(\alpha_k - \alpha_i) \in \mathbf{Z}_p$, using once again the extended Euclidean algorithm since any nonzero integer in $\mathbf{Z}_p$ is relatively prime (in $\mathbf{Z}$) to the prime integer p. Solving for v_k, we get for $k \ge 1$:

$$v_k = \left(u_k - [v_0 + \cdots + v_{k-1}\prod_{i=0}^{k-2}(\alpha_k - \alpha_i)] \right) \left(\prod_{i=0}^{k-1}(\alpha_k - \alpha_i) \right)^{-1}. \tag{5.24}$$

It is important to note that u_k and v_k $(0 \le k \le n)$ will be, in general, multivariate polynomials in a domain D with coefficients lying in a field $\mathbf{Z}_p$ and all coefficient arithmetic arising in equation (5.24) will be performed in the field $\mathbf{Z}_p$.

The Newton interpolation algorithm is presented formally as Algorithm 5.2. Comparing with Algorithm 5.1 it can be seen that the two algorithms are statement-by-statement identical except for computational details. As with Garner's algorithm, the Newton interpolation algorithm is divided into three steps. Step 1 again could be removed and precomputed if the evaluation points $\{\alpha_i\}$ are fixed, although in the multivariate case the computational cost of step 1 will be insignificant compared with step 2 (because step 1 involves only operations on integers in $\mathbf{Z}_p$ while step 2 involves operations on polynomials in $\mathbf{Z}_p[\mathbf{y}]$). The notation ϕ_p has an algorithmic specification as before:

> $\phi_p(expression)$ means "evaluate *expression* with all operations on integers being performed modulo p".

Note that in Algorithm 5.2 *all* coefficient arithmetic is to be performed modulo p (i.e. in the field $\mathbf{Z}_p$).

Example 5.16. Let us determine the polynomial $u(x,y) \in \mathbf{Z}_{97}[x,y]$ of maximum degree 2 in x and maximum degree 1 in y specified by the following values in the field $\mathbf{Z}_{97}$:

$$u(0,0) = -21; \quad u(0,1) = -30;$$

$$u(1,0) = 20; \quad u(1,1) = 17;$$

$$u(2,0) = -36; \quad u(2,1) = -31.$$

Let us first reconstruct the image of $u(x,y)$ in $\mathbf{Z}_{97}[x,y]/<x-0>$ (i.e. the case $x = 0$). In the notation of Algorithm 5.2 we have $D = \mathbf{Z}_{97}$, $\alpha_0 = 0, \alpha_1 = 1, u_0 = -21, u_1 = -30$, and we are computing a polynomial $u(0,y) \in \mathbf{Z}_{97}[y]$ (i.e. the indeterminate x in Algorithm 5.2 is y for now). Step 1 is in this case trivial:

$$\gamma_1 = (\alpha_1 - \alpha_0)^{-1} \pmod{97} = 1^{-1} \pmod{97} = 1.$$

Step 2 computes the following Newton coefficients for $u(0,y)$:

$$v_0 = -21; \quad v_1 = -9;$$

and therefore in step 3 we find

$$u(0,y) = -21 - 9(y - 0) = -9y - 21.$$

Similarly, reconstructing the images of $u(x,y)$ in $\mathbf{Z}_{97}[x,y]/<x-1>$ and $\mathbf{Z}_{97}[x,y]/<x-2>$ we find

$$u(1,y) = -3y + 20;$$

$$u(2,y) = 5y - 36.$$

Now for the multivariate step, we apply Algorithm 5.2 with $D = \mathbf{Z}_{97}[y]$, $\alpha_0 = 0, \alpha_1 = 1, \alpha_2 = 2$, $u_0 = u(0,y), u_1 = u(1,y), u_2 = u(2,y)$, and we compute the polynomial $u(x,y) \in D[x] = \mathbf{Z}_{97}[y][x]$. Step 1 in this case computes the following inverses:

Algorithm 5.2. Newton Interpolation Algorithm.

procedure NewtonInterp$((\alpha_0, \ldots, \alpha_\nu), (u_0, \ldots, u_\nu))$

\# Let D = $\mathbf{Z}_p[\mathbf{y}]$ denote a domain of polynomials in $v \geq 0$
\# indeterminates $\mathbf{y} = (y_1, \ldots, y_v)$ over a finite field $\mathbf{Z}_p$
\# (D = $\mathbf{Z}_p$ in case $v = 0$). Given distinct evaluation points $\alpha_i \in \mathbf{Z}_p$
\# $(0 \leq i \leq n)$ and given corresponding values $u_i \in D$ $(0 \leq i \leq n)$,
\# compute the unique polynomial $u(x) \in$ D[x] such that $\deg(u(x)) \leq n$
\# and $u(\alpha_i) = u_i$, $i = 0, 1, \ldots, n$.

\# Step 1: Compute the required inverses using a procedure
\# reciprocal(a, q) which computes a^{-1} (mod q)

for k **from** 1 **to** n **do** {
  $product \leftarrow \phi_p(\alpha_k - \alpha_0)$
  for i **from** 1 **to** $k - 1$ **do**
    $product \leftarrow \phi_p(product \cdot (\alpha_k - \alpha_i))$
  $\gamma_k \leftarrow$ reciprocal$(product, p)$ }

\# Step 2: Compute the Newton coefficients $\{v_k\}$

$v_0 \leftarrow u_0$
for k **from** 1 **to** n **do** {
  $temp \leftarrow v_{k-1}$
  for j **from** $k - 2$ **to** 0 **by** -1 **do**
    $temp \leftarrow \phi_p(temp \cdot (\alpha_k - \alpha_j) + v_j)$
  $v_k \leftarrow \phi_p((u_k - temp) \cdot \gamma_k)$ }

\# Step 3: Convert from Newton form to standard form

$u \leftarrow v_n$
for k **from** $n - 1$ **to** 0 **by** -1 **do**
  $u \leftarrow \phi_p(u \cdot (x - \alpha_k) + v_k)$
return$(u(x))$

end

$$\gamma_1 = (\alpha_1 - \alpha_0)^{-1} \pmod{97} = 1^{-1} \pmod{97} = 1;$$

$$\gamma_2 = [(\alpha_2 - \alpha_0)(\alpha_2 - \alpha_1)]^{-1} \pmod{97} = 2^{-1} \pmod{97} = -48.$$

Step 2 computes the following Newton coefficients:

$$v_0 = -9y - 21;$$

$$v_1 = 6y + 41;$$

$$v_2 = y.$$

Finally in step 3 we find

$$u(x,y) = (-9y - 21) + (6y + 41)(x - 0) + y(x - 0)(x - 1)$$
$$= x^2y + 5xy + 41x - 9y - 21$$

which is the desired polynomial in the domain $\mathbf{Z}_{97}[x,y]$.

●

5.8. FURTHER DISCUSSION OF THE TWO ALGORITHMS

Integer and Polynomial Representations

It is important in some circumstances to recognize that in each of Algorithms 5.1 and 5.2, three different representations arise for the same object. In the polynomial case (Algorithm 5.2), the polynomial $u(x) \in D[x]$ is initially represented uniquely by its $n + 1$ *values (residues)* $\{u_0, u_1, \ldots, u_n\}$ corresponding to the $n + 1$ distinct evaluation points $\{\alpha_0, \alpha_1, \ldots, \alpha_n\}$. At the end of step 2, the polynomial $u(x)$ is represented uniquely in *Newton form* by its $n + 1$ Newton coefficients $\{v_0, v_1, \ldots, v_n\}$ with respect to the basis polynomials

$$1, (x - \alpha_0), (x - \alpha_0)(x - \alpha_1), \ldots, \prod_{i=0}^{n-1}(x - \alpha_i).$$

In step 3 the Newton form of $u(x)$ is converted to *standard polynomial form*, which can be characterized as uniquely representing $u(x)$ by its $n + 1$ coefficients $\{a_0, a_1, \ldots, a_n\}$ with respect to the standard basis polynomials $1, x, x^2, \ldots, x^n$. Similarly in the integer Chinese remainder case (Algorithm 5.1), the initial representation for the integer u is by its $n + 1$ *residues* $\{u_0, u_1, \ldots, u_n\}$ with respect to the $n + 1$ moduli $\{m_0, m_1, \ldots, m_n\}$. The second representation is the *mixed radix representation* $\{v_0, v_1, \ldots, v_n\}$ with respect to the mixed radices

$$1, m_0, m_0, m_1, \ldots, \prod_{i=0}^{n-1} m_i.$$

The final step converts the mixed radix representation to the more familiar *radix* β *representation* where the base β depends on the representation being used for multiprecision integers (see Chapter 3).

Residue representations of integers and of polynomials arise (by the application of homomorphisms) because some operations are easier to perform in this representation. For example, multiplication of integers or of polynomials is a simpler operation when residue representations are used than when standard representations are used. The conversion processes of Algorithms 5.1 and 5.2 are required not only because the human computer user will generally want to see his answers presented in the more standard representations of objects but also because some required operations cannot be performed directly on the residue representations. For example, the result of the comparison "Is $u < v$?" where u and v are integers cannot be determined directly from knowledge of the residue representations of u and v, but as previously noted the result of such a comparison can be directly determined by comparing the mixed radix coefficients of u and v. As another example, if a polynomial $u(x)$ is to be evaluated for arbitrary values of x then the residue representation of $u(x)$ is not appropriate, but $u(x)$ can be evaluated in the Newton form as well as in standard polynomial form. These two examples indicate circumstances where conversions from residue representations are required but where step 3 of the algorithms may be considered to be unnecessary extra computations. However in the context of applying Algorithms 5.1 and 5.2 to the inversion of composite modular/evaluation homomorphisms on the polynomial domain $\mathbf{Z}[x_1, \ldots, x_v]$ (which is the context of primary interest in this book), step 3 of the algorithms will always be applied. The reason for this in the polynomial case (i.e. Algorithm 5.2) will be explained shortly. In the integer case (i.e. Algorithm 5.1) the reason is simply that subsequent operations on the integer coefficients (whether output operations or arithmetic operations) will require the standard integer representation. For output this is a user requirement, while for arithmetic operations there is no practical advantage in requiring a system to support arithmetic operations on integers in more than one representation. (Of course it is conceivable to have a system in which the "standard" integer representation is not a radix β representation but all of the present-day systems of interest to us use a radix β representation for integers.)

Another issue which arises in the practical application of modular and evaluation homomorphisms and their corresponding inversion algorithms is to determine the number of moduli (evaluation points) needed to uniquely represent an unknown integer (polynomial). In the polynomial case, the information needed is an upper bound D for the degree of the result since then $D+1$ moduli (evaluation points) are sufficient to uniquely represent the polynomial result. Similarly in the integer case, if an upper bound M for the magnitude of the integer result is known then by choosing enough moduli $\{m_i\}$ such that

$$m = \prod_{i=0}^{n} m_i > 2M,$$

we are guaranteed that the ring $\mathbf{Z}_m$ is large enough to represent the integer result. In other words, the result determined by Algorithm 5.1 lies in the ring $\mathbf{Z}_m$ and this result will be the same (when expressed in the symmetric representation) as the desired result in $\mathbf{Z}$, since

$$\mathbf{Z}_m = \left\{-\frac{m-1}{2}, \ldots, -1, 0, 1, \ldots, \frac{m-1}{2}\right\}$$

with $\frac{m-1}{2} \geq M$. Such polynomial degree bounds and integer magnitude bounds can usually be calculated with little effort for practical problems, although it is quite common for easily-calculated bounds to be very pessimistic (i.e. much larger than necessary). An alternate computational approach is available in situations where it is easy to verify the correctness of the result. This alternate approach is based on the observation that the mixed radix coefficients (or Newton coefficients) $\{v_k\}$ will be zero for $k > K$ if the moduli $m_0, m_1, \ldots, m_K$ (or $x - \alpha_0, x - \alpha_1, \ldots, x - \alpha_K$) are sufficient to uniquely represent the result. The computation therefore can be halted once $v_{K+1} = 0$ for some K (on the assumption that $v_k = 0$ for all $k > K$) as long as the result is later verified to be correct.

A Generalization of Garner's Algorithm

There is a slight generalization of Garner's Chinese remainder algorithm which is useful in the applications of interest to us. Recall that we wish to invert composite homomorphisms $\phi_{<I_{ij}, p_i>} = \phi_{I_{ij}} \phi_{p_i}$ where

$$\phi_{p_i} : \mathbf{Z}[x_1, \ldots, x_v] \rightarrow \mathbf{Z}_{p_i}[x_1, \ldots, x_v],\ i = 0, 1, \ldots, n \tag{5.25}$$

is a sequence of modular homomorphisms for some chosen prime moduli $p_0, p_1, \ldots, p_n$, and for *each* i there is a corresponding sequence of some N multivariate evaluation homomorphisms

$$\phi_{I_{ij}} : \mathbf{Z}_{p_i}[x_i, \ldots, x_v] \rightarrow \mathbf{Z}_{p_i}[x_1],\ j = 1, 2, \ldots, N \tag{5.26}$$

with kernels of the form $I_{ij} = <x_2 - \alpha_{2ij}, \ldots, x_v - \alpha_{vij}>$. In this notation, for a fixed i the evaluation points $\alpha_{2ij}, \ldots, \alpha_{vij}$ all lie in the field $\mathbf{Z}_{p_i}$ and the number N of different kernels I_{ij} is determined by the degree of the solution in each indeterminate. Now suppose that Algorithm 5.2 is applied (as in Example 5.16) to invert the evaluation homomorphisms (5.26) and suppose that the $n+1$ polynomials which arise are $u_i(\mathbf{x}) \in \mathbf{Z}_{p_i}[x_1, \ldots, x_v]$, for $i = 0, 1, \ldots, n$. If the polynomials $u_i(\mathbf{x})$ are all expressed in expanded canonical form then Algorithm 5.1 can be applied coefficient-by-coefficient to reconstruct the coefficients of the desired solution $u(\mathbf{x}) \in \mathbf{Z}[x_1, \ldots, x_v]$ (i.e. to invert the modular homomorphisms (5.25)).

The desired generalization of Garner's algorithm is obtained by simply noting that Algorithm 5.1 can be applied directly to the polynomials $u_i(\mathbf{x})$ to reconstruct $u(\mathbf{x})$, rather than being applied many times separately for each coefficient of the polynomial $u(\mathbf{x})$. To see this, suppose $u(\mathbf{x})$ is the polynomial

$$u(\mathbf{x}) = \sum_{\mathbf{e}} u_{\mathbf{e}} \mathbf{x}^{\mathbf{e}} \in \mathbf{Z}[\mathbf{x}]$$

with images

$$u_i(\mathbf{x}) = \sum_{\mathbf{e}} u_{\mathbf{e},i}\mathbf{x}^{\mathbf{e}} \in \mathbf{Z}_{p_i}[\mathbf{x}],\ i = 0, 1, \ldots, n$$

where $u_{\mathbf{e},i} = \phi_{p_i}(u_{\mathbf{e}})$. If Algorithm 5.1 is applied separately for each coefficient $u_{\mathbf{e}}$, it calculates (in step 2) each integer $u_{\mathbf{e}}$ in its mixed radix representation

$$u_{\mathbf{e}} = \sum_{k=0}^{n} v_{\mathbf{e},k} \left(\prod_{j=0}^{k-1} p_j \right)$$

where $v_{\mathbf{e},k} \in \mathbf{Z}_{p_k}$, $0 \le k \le n$. But since the same mixed radices appear in the mixed radix representations for each different coefficient $u_{\mathbf{e}}$, we may express the polynomial $u(\mathbf{x})$ as follows:

$$u(\mathbf{x}) = \sum_{\mathbf{e}} \left(\sum_{k=0}^{n} v_{\mathbf{e},k} \left(\prod_{j=0}^{k-1} p_j \right) \right) \mathbf{x}^{\mathbf{e}}$$

$$= \sum_{k=0}^{n} \left(\sum_{\mathbf{e}} v_{\mathbf{e},k}\mathbf{x}^{\mathbf{e}} \right) \left(\prod_{j=0}^{k-1} p_j \right)$$

The latter expression for the polynomial $u(\mathbf{x})$ is called a *polynomial mixed radix representation* with respect to the mixed radices $1, p_0, p_0 p_1, \ldots, \prod_{j=0}^{n-1} p_j$, and its general form is

$$u(\mathbf{x}) = v_0(\mathbf{x}) + v_1(\mathbf{x})(p_0) + v_2(\mathbf{x})(p_0 p_1) + \cdots + v_n(\mathbf{x}) \left(\prod_{j=0}^{n-1} p_j \right)$$

where $v_k(\mathbf{x}) \in \mathbf{Z}_{p_k}[\mathbf{x}]$ for $k = 0, 1, \ldots, n$. It can be seen that step 2 of Algorithm 5.1 will directly generate the polynomial $u(\mathbf{x})$ in its polynomial mixed radix representation if we simply change the specification of Algorithm 5.1 to allow the residues to be polynomials $u_i(\mathbf{x}) \in \mathbf{Z}_{p_i}[\mathbf{x}]$ $(0 \le i \le n)$. Note that step 3 of Algorithm 5.1 also remains valid to convert the polynomial to its standard representation as a polynomial with integer coefficients. The validity of this *generalized Garner's algorithm* follows immediately from the fact that the operations of multiplying a polynomial by a constant and of adding two polynomials are by definition coefficient-by-coefficient operations. This generalization can be viewed simply as a method for computing "in parallel" the separate Chinese remainder processes for each coefficient of the polynomial solution $u(\mathbf{x})$.

The generalized Garner's algorithm is only valid if all of the polynomial residues $u_i(\mathbf{x})$, $0 \le i \le n$ are expressed in expanded canonical form for only then can we be assured that the operations in the algorithm are the correct coefficient-by-coefficient operations. Since the given polynomial residues $u_i(\mathbf{x}) \in \mathbf{Z}_{p_i}[\mathbf{x}]$ will usually result from a previous interpolation step, it is worth noting in particular why the polynomials cannot be left in Newton form. The reason is that the basis polynomials for the Newton form of one polynomial residue $u_i(\mathbf{x}) \in \mathbf{Z}_{p_i}[\mathbf{x}]$ involve evaluation points lying in the field $\mathbf{Z}_{p_i}$ while the basis polynomials for the Newton form of a different polynomial residue $u_j(\mathbf{x}) \in \mathbf{Z}_{p_j}[\mathbf{x}]$ involve evaluation points lying in the different field $\mathbf{Z}_{p_j}$. There is in general no consistent interpretation of these various polynomial residues as images of the solution $u(\mathbf{x})$ unless each polynomial residue $u_i(\mathbf{x}) \in \mathbf{Z}_{p_i}[\mathbf{x}]$ is first converted to expanded canonical form in its own domain $\mathbf{Z}_{p_i}[\mathbf{x}]$. The basis polynomials for the expanded canonical form are independent of the evaluation points. This explains why step 3 of Algorithm 5.2 is an essential step of the Newton interpolation algorithm in the context of inverting composite modular/evaluation homomorphisms.

Example 5.17. Let us complete the details of the process of inverting the homomorphisms used in Example 5.7 at the end of Section 5.3. The problem was to determine the product polynomial

$$c(x) = a(x)\,b(x) = (7x + 5)(2x - 3) \in \mathbf{Z}[x].$$

To determine the number of evaluation homomorphisms to use, note that $\deg(c(x)) = \deg(a(x)) + \deg(b(x)) = 2$ so that three evaluation points will be sufficient. For a bound on the magnitudes of the integers in the product $c(x)$, it is easy to see that the product of two linear polynomials $a(x)$ and $b(x)$ will have coefficients bounded in magnitude by

$$M = 2\|a\|_\infty \cdot \|b\|_\infty$$

where $\|a\|_\infty$ and $\|b\|_\infty$ are the magnitudes of the largest coefficients in $a(x)$ and $b(x)$ respectively. Thus $M = 42$ in this example so it will be sufficient to use moduli such that

$$m = \prod_{i=0}^{n} m_i > 84.$$

In Example 5.7 it was seen that the three composite homomorphisms

$$\phi_{x-\alpha_i}\phi_5 : \mathbf{Z}[x] \to \mathbf{Z}_5, \text{ where } \alpha_0 = 0,\ \alpha_1 = 1,\ \alpha_2 = 2$$

yield the following images in the field $\mathbf{Z}_5$ (when converted to the symmetric representation):

$$c(0) = 0,\ c(1) = -2,\ c(2) = -1.$$

Applying Algorithm 5.2 to this interpolation problem yields

$$c(x) = -x^2 - x \in \mathbf{Z}_5[x].$$

Next the three composite homomorphisms

$$\phi_{x-\alpha_i}\phi_7 : \mathbf{Z}[x] \to \mathbf{Z}_7, \text{ where } \alpha_0 = 0,\ \alpha_1 = 1,\ \alpha_2 = 2$$

led to the following images in the field $\mathbf{Z}_7$:

$$c(0)=-1;\ c(1)=2;\ c(2)=-2.$$

Applying Algorithm 5.2 to this interpolation problem yields

$$c(x) = 3x - 1 \in \mathbf{Z}_7[x].$$

Now since the moduli $p_0 = 5$ and $p_1 = 7$ do not satisfy $p_0 p_1 > 84$, let us choose also $p_2 = 3$. Then $p_0 p_1 p_2 = 105 > 84$ so these moduli will be sufficient. The three composite homomorphisms

$$\phi_{x-\alpha_i}\phi_3 : \mathbf{Z}[x] \to \mathbf{Z}_3, \text{ where } \alpha_0 = 0,\ \alpha_1 = 1,\ \alpha_2 = -1$$

yield the following images in the field $\mathbf{Z}_3$:

$$c(0)=0;\ c(1)=0;\ c(-1)=1.$$

Applying Algorithm 5.2 to this interpolation problem yields

$$c(x) = -x^2 + x \in \mathbf{Z}_3[x].$$

Now let us apply the generalized Garner's algorithm to invert the three modular homomorphisms:

$$\phi_{p_i} : \mathbf{Z}[x] \to \mathbf{Z}_{p_i}[x], \text{ where } p_0 = 5,\ p_1 = 7,\ p_2 = 3.$$

The given residues are

$$u_0(x) = -x^2 - x;\ u_1(x) = 3x - 1;\ u_2(x) = -x^2 + x.$$

The inverses computed in step 1 are:

$$\gamma_1 = p_0^{-1} (\text{mod } p_1) = 5^{-1} (\text{mod } 7) = 3;$$

$$\gamma_2 = (p_0 p_1)^{-1} (\text{mod } p_2) = (-1)^{-1} (\text{mod } 3) = -1.$$

In step 2 the following polynomial mixed radix coefficients are computed:

$$v_0(x) = -x^2 - x;\ v_1(x) = 3x^2 - 2x - 3;\ v_2(x) = 0.$$

Finally in step 3 we find

$$u(x) = (-x^2 - x) + (3x^2 - 2x - 3)(5) + (0)(5)(7)$$
$$= 14x^2 - 11x - 15 \in \mathbf{Z}_{105}[x].$$

Note that the last polynomial mixed radix coefficient $v_2(x)$ is zero which implies that the two moduli $p_0 = 5$ and $p_1 = 7$ would have been sufficient for this problem. In other words, the bound $M = 42$ on the magnitudes of the integer coefficients in the result is a large overestimate. In any case, we are guaranteed that $u(x)$ is the desired result – i.e.

$$c(x) = 14x^2 - 11x - 15 \in \mathbf{Z}[x].$$

●

A Homomorphism Diagram

Finally in this chapter, Figure 5.1 presents a *homomorphism diagram* which is a convenient way to visualize the computational "route" to the solution of a problem when homomorphism methods are used. The particular homomorphism diagram expressed in Figure 5.1 is for the case of applying composite modular/evaluation homomorphisms as in (5.25) - (5.26) to project the multivariate polynomial domain $\mathbf{Z}[x_1, \ldots, x_v]$ onto Euclidean domains $\mathbf{Z}_{p_i}[x_1]$. Of course, the same diagram is valid if $\mathbf{Z}_{p_i}[x_1]$ is replaced by $\mathbf{Z}_{p_i}$ which would express the case of Example 5.17. Note that for the particular problem considered in Example 5.17 the homomorphism method in fact requires much more work than a "direct method" (i.e. ordinary polynomial multiplication), which can be expressed in the diagram of Figure 5.1 by drawing an arrow from the "Given problem" box directly to the "Desired solution" box. However for problems such as multivariate GCD computation and multivariate polynomial factorization, the "long route" of homomorphism methods can yield substantial decreases in total computational cost in many cases as we shall see in later chapters.

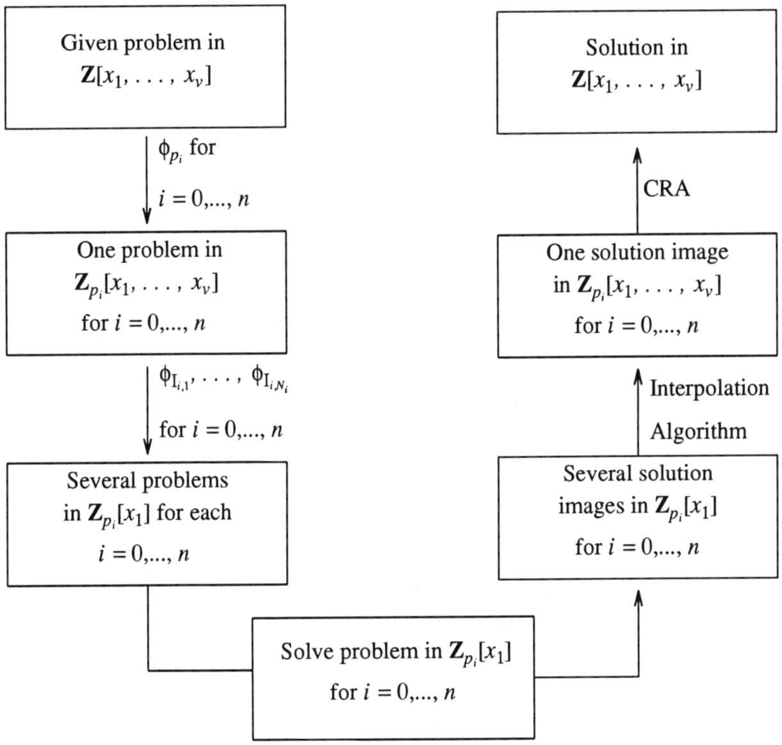

Figure 5.1. Homomorphism diagram for Chinese remainder and interpolation algorithms.

Exercises

1. (a) Let R and R′ be two rings and let $\phi : R \rightarrow R'$ be a ring morphism as defined by Definition 5.2. Properties (i) - (iii) of Definition 5.2 guarantee that ϕ preserves three of the ring operations. Using properties (i) - (iii) and the fact that R and R′ are rings, prove that the other two ring operations are also preserved – i.e. prove that

$$\phi(0) = 0;$$

$$\phi(-a) = -\phi(a) \quad \text{for all } a \in R.$$

   (b) Suppose that the rings R and R′ in part (a) are fields. Prove that the operation of inversion is also preserved by any ring morphism $\phi : R \rightarrow R'$ – i.e. prove that

$$\phi(a^{-1}) = [\phi(a)]^{-1} \quad \text{for all } a \in R - \{0\}.$$

   (c) Suppose that the ring R is commutative. Prove that if R′ is a homomorphic image of R then R′ is also commutative.

2. (a) In the integral domain **Z**, give a complete description of each of the following ideals: $<3>$, $<-3>$, $<4, 6>$, $<4, 7>$.

   (b) In the polynomial domain **Q**[x] let $\alpha \in \mathbf{Q}$ be a fixed constant. The subset $I = \{a(x) : a(\alpha) = 0\}$ is an ideal in **Q**[x] (see Example 5.9). Consider the subset $J = \{a(x) : a(\alpha) = 1\}$. Prove or disprove that J is an ideal in **Q**[x].

3. In any integral domain D, prove that $<a> = <b>$ if and only if a and b are associates in D.

4. In the bivariate polynomial domain **Z**[x, y] consider the ideals $I = <x, y>$ and $J = <x>$. The subset relationships between I, J and **Z**[x, y] can be specified as follows:

$$J \subset I \subset \mathbf{Z}[x, y].$$

   The ideal I can be described as the set of all bivariate polynomials over **Z** with no constant term and the ideal J can be described as the set of all polynomials in I which have no constant term when expressed as univariate polynomials in x – i.e. when expressed as elements of the domain **Z**[y][x].

   (a) Express in the usual notation for ideals the following three ideals: the sum $<I, J>$, the product $I \cdot J$, and the power I^2.

   (b) Specify the subset relationships between $<I, J>$, $I \cdot J$, and I^2.

   (c) Given a description (in the sense of the descriptions of I and J given above) of each of the ideals $<I, J>$, $I \cdot J$, and I^2.

5. Let D be a Euclidean domain and let $a, b \in D$ be any two elements. Use Theorem 2.2 (i.e. the extended Euclidean algorithm) to prove that the ideal $<a, b>$ generated by these two elements is a principal ideal. More specifically, prove that

$$< a,b > = < g >$$

where $g = \text{GCD}(a,b)$. (*Remark:* A proof that every Euclidean domain is a principal ideal domain can be based on this result.)

6. (a) Let D be a principal ideal domain and let $a,b \in$ D be any two elements. Then the ideal $< a,b >$ generated by these two elements must be a principal ideal, say

$$< a,b > = < g >$$

for some element $g \in$ D. Prove that g is a greatest common divisor of a and b.

(b) Use the result of part (a) to prove that the extended Euclidean property of Theorem 2.2 holds in any principal ideal domain.

7. Show that the domain $\mathbf{Z}[x]$ is not a principal ideal domain by exhibiting an ideal in $\mathbf{Z}[x]$ which is not a principal ideal.

8. (a) Determine all of the ideals in $\mathbf{Z}_5$. Thus determine all of the homomorphic images of $\mathbf{Z}_5$.

(b) Prove that a field has no proper homomorphic images.

9. (a) Determine all of the ideals in $\mathbf{Z}_6$. Thus determine all of the homomorphic images of $\mathbf{Z}_6$.

(b) Determine all of the ideals in $\mathbf{Z}_m$, for every integer m. Thus determine all of the homomorphic images of $\mathbf{Z}_m$.

10. (a) Prove that the only proper homomorphic images of the ring $\mathbf{Z}$ are rings of the form $\mathbf{Z}_m$. (*Hint:* $\mathbf{Z}$ is a principal ideal domain.)

(b) Prove that the quotient ring $\mathbf{Z}_p = \mathbf{Z}/< p >$ of the ring $\mathbf{Z}$ is an integral domain if and only if p is a prime integer. (*Hint:* A fundamental step in the proof is to deduce that if p is a prime integer then

$$ab \in < p > \Rightarrow a \in < p > \text{ or } b \in < p >.)$$

(c) Prove that if p is a prime integer then the integral domain $\mathbf{Z}_p$ of part (b) is in fact a field. (*Hint:* Use Theorem 5.6.)

11. Generalize Exercise 10 (a), (b), and (c) to the case where the Euclidean domain $\mathbf{Z}$ is replaced by the Euclidean domain F$[x]$ of univariate polynomials over a field F.

12. (a) In the integral domain $\mathbf{Q}[[x]]$ of power series over $\mathbf{Q}$, describe the ideal $< x^e >$ where e is a fixed positive integer.

(b) Consider the natural homomorphism

$$\phi_{< x^e >} : \mathbf{Q}[[x]] \rightarrow \mathbf{Q}[[x]]/< x^e >.$$

Describe the elements in the homomorphic image $\mathbf{Q}[[x]]/< x^e >$. Describe a practical representation for the elements in this homomorphic image. (cf. Chapter 3.)

13. For extended power series in $\mathbf{Q}<x>$, what is the ideal $<x^e>$ where e is a fixed integer? Does $\mathbf{Q}<x>$ have a homomorphic image comparable to the case of ordinary power series considered in the preceding problem?

14. (a) Let $p(x) \in \mathbf{Q}[x]$ be a fixed polynomial and consider the quotient ring $\mathbf{Q}[x]/<p(x)>$. Prove that two polynomials $a(x), b(x) \in \mathbf{Q}[x]$ lie in the same residue class in this quotient ring if and only if

$$\mathrm{rem}(a(x), p(x)) = \mathrm{rem}(b(x), p(x)).$$

Thus deduce a practical representation for the elements in the homomorphic image $\mathbf{Q}[x]/<p(x)>$.

(b) Let the polynomial $p(x)$ in part (a) be the linear polynomial

$$p(x) = x - \alpha \quad \text{for some fixed } \alpha \in \mathbf{Q}.$$

Prove that the evaluation homomorphism

$$\phi_{x-\alpha} : \mathbf{Q}[x] \to \mathbf{Q}$$

as defined in Section 5.3 can be defined equivalently by

$$\phi_{x-\alpha}(a(x)) = \mathrm{rem}(a(x), x - \alpha) \quad \text{for all } a(x) \in \mathbf{Q}[x].$$

Thus deduce that the evaluation homomorphism $\phi_{x-\alpha}$ is indeed the natural homomorphism with kernel $<x - \alpha>$ which projects $\mathbf{Q}[x]$ onto the homomorphic image $\mathbf{Q}[x]/<x - \alpha>$.

15. Generalize the preceding problem to the case of a multivariate polynomial domain $D[x_1, \ldots, x_v]$ over an arbitrary UFD D, as follows.

(a) Let $p(x_i) \in D[x_i]$ be a *monic* univariate polynomial over D in the particular indeterminate x_i and consider the quotient ring $D[x_1, \ldots, x_v]/<p(x_i)>$. Prove that two polynomials $a(x_1, \ldots, x_v), b(x_1, \ldots, x_v) \in D[x_1, \ldots, x_v]$ lie in the same residue class in this quotient ring if and only if

$$\mathrm{prem}(a(x_1, \ldots, x_v),\ p(x_i)) = \mathrm{prem}(b(x_1, \ldots, x_v),\ p(x_i)),$$

where the prem operation is performed in the (univariate) polynomial domain $D[x_1, \ldots, x_{i-1}, x_{i+1}, \ldots, x_v][x_i]$. (Note that since $p(x_i)$ is assumed to be monic, the operation of pseudo-division is in fact just ordinary polynomial division.) Thus deduce a practical representation for the elements in the homomorphic image $D[x_1, \ldots, x_v]/<p(x_i)>$.

(b) Let the polynomial $p(x_i)$ in part (a) be the linear polynomial

$$p(x_i) = x_i - \alpha \quad \text{for some fixed } \alpha \in D.$$

Prove that the evaluation homomorphism

$$\phi_{x_i - \alpha} : D[x_1, \ldots, x_v] \rightarrow D[x_1, \ldots, x_{i-1}, x_{i+1}, \ldots, x_v]$$

as defined in Section 5.3 can be defined equivalently by

$$\phi_{x_i - \alpha}(a(x_1, \ldots, x_v)) = \text{prem}(a(x_1, \ldots, x_v), x_i - \alpha)$$

$$\text{for all } a(x_1, \ldots, x_v) \in D[x_1, \ldots, x_v].$$

Thus deduce that the evaluation homomorphism $\phi_{x_i - \alpha}$ is indeed the natural homomorphism with kernel $< x_i - \alpha >$ which projects $D[x_1, \ldots, x_v]$ onto the homomorphic image $D[x_1, \ldots, x_v]/< x_i - \alpha >$.

16. (a) Describe a practical representation for the elements in the quotient ring $\mathbf{Q}[x]/< x^2 + 1 >$. (cf. Exercise 14(a).) Prove that this quotient ring is a field. (cf. Exercise 11.)

(b) In part (a) suppose that the coefficient field **Q** is changed to **R** (the real numbers). What is the field $\mathbf{R}[x]/< x^2 + 1 >$?

(c) Is the quotient ring $\mathbf{Z}[x]/< x^2 + 1 >$ a field? Is it an integral domain? What is the relationship between this quotient ring and the domain G of Gaussian integers defined in Exercise 2.9?

17. (a) In the congruence notation defined in Section 5.5, what is the meaning of

$$a \equiv b \pmod 0?$$

(b) In the integral domain **Z**, compute the inverse of 173 modulo 945 – i.e. solve the following congruence equation for $x \in \mathbf{Z}$:

$$173\,x \equiv 1 \pmod{945}.$$

Use the method given in the proof of Theorem 5.6.

(c) In the polynomial domain $\mathbf{Q}[x]$, solve the following congruence equation for $u(x)$:

$$(x + 1)(x + 2)\,u(x) \equiv x \pmod{x(x - 1)(x - 2)}.$$

Be sure to reduce the solution modulo $x(x - 1)(x - 2)$.

18. (a) Suppose that the single-precision integers on a particular computer are restricted to the range $-100 < a < 100$ (i.e. two-digit integers). Determine the ten largest such single-precision integers which are odd and pairwise relatively prime. (Note that Example 5.15 uses the three largest such integers.)

(b) What range of integers can be represented in a modular representation using as moduli the ten integers determined in part (a)? In particular, how many decimal digits long must an integer be in order that it not be representable?

19. (a) Using the positive consistent representation, express the integer u = 102156 in mixed radix representation with respect to the moduli $m_0 = 99$, $m_1 = 97$, and $m_2 = 95$.

(b) Repeat part (a) using the symmetric consistent representation.

20. Apply Algorithm 5.1 by hand to solve the following Chinese remainder problems for the integer u:

(a) $u \equiv 1 \pmod 5$; $u \equiv -3 \pmod 7$; $u \equiv -2 \pmod 9$.

(b) $u \equiv 1 \pmod 5$; $u \equiv -2 \pmod 7$; $u \equiv -4 \pmod 9$.

21. (a) Step 2 of Algorithm 5.1 is based on formulas (5.16) - (5.17) with the computation of the required inverses performed in step 1. Show that an alternate method to compute the same mixed radix coefficient v_k $(0 \le k \le n)$ can be based on formula (5.16) and the following rearrangement of formula (5.17) for $k \ge 1$:

$$v_k \equiv (\;\cdots\; ((u_k - v_0)m_0^{-1} - v_1)m_1^{-1} - \;\cdots\; - v_{k-1})m_{k-1}^{-1} \pmod{m_k}.$$

Note that the inverses appearing in this formula are inverses modulo m_k.

(b) If step 2 of Algorithm 5.1 were based on the alternate formula of part (a), what set of inverses would have to be computed in step 1? In particular, how many inverses are now required?

(c) Compare the computational efficiency of Algorithm 5.1 with the alternate algorithm proposed above. Consider the case where the set $\{m_i\}$ of moduli is fixed (i.e. the case where the computation of inverses in step 1 would be removed from the algorithm and pre-computed) and also consider the case where no pre-computation is possible.

22. (a) The proof of Theorem 5.8 outlines a method for solving the polynomial interpolation problem by solving a system of linear equations. For the specific problem described in the preamble of Algorithm 5.2 (in particular, D $= \mathbf{Z}_p[\mathbf{y}]$ and $\alpha_i \in \mathbf{Z}_p$, $0 \le i \le n$) the linear system of Theorem 5.8 can be solved by an algorithm based on the familiar Gaussian elimination method and the only divisions required are divisions in the field $\mathbf{Z}_p$. In this case the solution $u(x)$ will lie in the domain D[x] and not in the larger domain $F_D[x]$ of Theorem 5.8. Give an algorithmic description of such a method for solving the polynomial interpolation problem.

(b) Compare the computational cost of the algorithm of part (a) with the cost of Algorithm 5.2.

23. Use Algorithm 5.2 (by hand) to determine the polynomial $u(x,y,z) \in \mathbf{Z}_5[x,y,z]$ with maximum degrees 2 in x, 1 in y, and 1 in z specified by the following values in the field $\mathbf{Z}_5$:

$$\begin{aligned} u(0,0,0) &= 1; & u(0,0,1) &= 2; \\ u(0,1,0) &= -1; & u(0,1,1) &= 0; \\ u(1,0,0) &= 0; & u(1,0,1) &= 2; \\ u(1,1,0) &= 2; & u(1,1,1) &= -2; \\ u(2,0,0) &= 1; & u(2,0,1) &= 2; \\ u(2,1,0) &= 0; & u(2,1,1) &= 0. \end{aligned}$$

Express the result in expanded canonical form.

24. Consider the problem of inverting composite modular/evaluation homomorphisms of the form

$$\phi_{x-\alpha_i}\phi_{p_i} : \mathbf{Z}[x] \rightarrow \mathbf{Z}_{p_i}.$$

Suppose that the polynomial $u(x) \in \mathbf{Z}[x]$ to be determined is known to be of degree 3 with coefficients not exceeding 17 in magnitude, and suppose that the following images of $u(x)$ have been determined.

p_i	α_i	$\phi_{x-\alpha_i}\phi_{p_i}(u(x))$
5	0	1
5	1	1
5	2	2
5	–1	0
7	0	0
7	1	–2
7	2	0
7	–1	2

(a) Verify that the image of $u(x)$ in $\mathbf{Z}_5[x]$ is

$$u_0(x) = 1 - 2x(x-1) + 2x(x-1)(x-2) \in \mathbf{Z}_5[x]$$

and that the image of $u(x)$ in $\mathbf{Z}_7[x]$ is

$$u_1(x) = -2x + 2x(x-1) + 3x(x-1)(x-2) \in \mathbf{Z}_7[x].$$

Note that these interpolating polynomials have been left in Newton (mixed radix) form.

(b) To complete the inversion process, we must solve the following Chinese remainder problem:

$$u(x) \equiv u_0(x) \pmod 5;$$

$$u(x) \equiv u_1(x) \pmod 7.$$

Suppose that this Chinese remainder problem is solved by leaving the polynomials $u_0(x)$ and $u_1(x)$ of part (a) in Newton form, yielding the result in the Newton form

$$u(x) = c_0 + c_1 x + c_2 x(x-1) + c_3 x(x-1)(x-2).$$

Calculate the values of c_0, c_1, c_2, and c_3 that would result.

(c) Determine the polynomial $u(x) \in \mathbf{Z}[x]$ by solving the Chinese remainder problem of part (b) after expressing the polynomials $u_0(x)$ and $u_1(x)$ of part (a) in expanded canonical form in their respective domains. Is the result the same as the result in part (b)? Is there any relationship between the two results?

(d) In the problem considered above, the set of evaluation points is the same for each modulus $p_0 = 5$ and $p_1 = 7$. In many practical problems the set of evaluation points will be different for different moduli. Does this affect the possibility of avoiding the conversion from Newton form to expanded canonical form as contemplated in part (b)?

25. There are other well-known techniques for solving the interpolating polynomial problem that differ from Newton's method. Some of these methods provide other constructive proofs of the Chinese remainder theorem. In particular consider the following method.

(a) Let $m_0, \ldots, m_k$ be pairwise relatively prime moduli. For each integer $0 \le i \le k$ construct integers M_i such that

$$M_i = \begin{cases} 1 & \text{mod } m_i \\ 0 & \text{mod } m_j \text{ if } i \ne j \,. \end{cases}$$

(b) Using the integers from part (a) let

$$u = u_0 \cdot M_0 + \cdots + u_k \cdot M_k .$$

Show that u satisfies the requirements of the integer Chinese remainder theorem in the special case when $a = 0$. What do we do for other values of a?

(c) Use the above method to give a second solution for Example 5.15. Is there an obvious drawback to this method compared to Newton's method?

(d) The corresponding technique for polynomial interpolation is called Lagrange's method. Give the parallels to parts (a) and (b) in this case.

(e) Give a Lagrange construction in the case of the Chinese remainder theorem when working over an arbitrary Euclidean domain.

References

1. W.S. Brown, ''On Euclid's Algorithm and the Computation of Polynomial Greatest Divisors,'' *J. ACM*, **18** pp. 476-504 (1971).

2. H. Garner, ''The Residue Number System,'' *IRE Transactions, EC-8*, pp. 140-147 (1959).

3. D.E. Knuth, *The Art of Computer Programming, Volume 2: Seminumerical Algorithms (second edition)*, Addison-Wesley (1981).

4. M. Lauer, ''Computing by Homomorphic Images,'' pp. 139-168 in *Computer Algebra - Symbolic and Algebraic Computation*, ed. B. Buchberger, G.E. Collins and R. Loos, Springer-Verlag (1982).

5. J.D. Lipson, ''Chinese Remainder and Interpolation Algorithms,'' pp. 372-391 in *Proc. SYMSAM '71*, ed. S.R. Petrick, ACM Press (1971).

6. J.D. Lipson, *Elements of Algebra and Algebraic Computing*, Addision-Wesley (1981).

7. B.L. van der Waerden, *Modern Algebra (Vols. I and II)*, Ungar (1970).

CHAPTER 6

NEWTON'S ITERATION AND

THE HENSEL CONSTRUCTION

6.1. INTRODUCTION

In this chapter we continue our discussion of techniques for inverting modular and evaluation homomorphisms defined on the domain $\mathbf{Z}[x_1, \ldots, x_v]$. The particular methods developed in this chapter are based on Newton's iteration for solving a polynomial equation. Unlike the integer and polynomial Chinese remainder algorithms of the preceding chapter, algorithms based on Newton's iteration generally require only *one* image of the solution in a domain of the form $\mathbf{Z}_p[x_1]$ from which to reconstruct the desired solution in the larger domain $\mathbf{Z}[x_1, \ldots, x_v]$. A particularly important case of Newton's iteration to be discussed here is the *Hensel construction*. It will be seen that multivariate polynomial computations (such as GCD computation and factorization) can be performed much more efficiently (in most cases) by methods based on the Hensel construction than by methods based on the Chinese remainder algorithms of the preceding chapter.

6.2. P-ADIC AND IDEAL-ADIC REPRESENTATIONS

The reason for introducing homomorphism techniques is that a gain in efficiency can be realized for many practical problems by solving several *image problems* in simpler domains (such as $\mathbf{Z}_{p_i}[x_1]$ or $\mathbf{Z}_{p_i}$) rather than directly solving a given problem in a more complicated domain (such as $\mathbf{Z}[x_1, \ldots, x_v]$) (cf. Figure 5.1). However the particular homomorphism techniques discussed in Chapter 5 have the potentially serious drawback that the *number* of image problems that have to be solved grows exponentially with the size of the solution (where by the "size" of the solution we mean some measure that takes into account the magnitudes of the integer coefficients, the number of indeterminates, and the degree of the solution in each indeterminate). This fact can be seen by referring to the homomorphism diagram of Figure 5.1 in which we see that the number of image problems that have to be solved is $(n+1)N$, where $n+1$ is the number of moduli required for the residue representation of the integer coefficients that can appear in the solution and where N is the number of multivariate evaluation homomorphisms required. Noting that if the degree of the solution in each of ν indeterminates is d then $N = (d+1)^{\nu-1}$ (because $d+1$ evaluation points are required for each of the $\nu-1$ indeterminates being eliminated), we see that the number of image problems is given by

$$(n+1)\,(d+1)^{\nu-1}$$

which grows exponentially as the number of indeterminates increases.

There is a homomorphism technique of a rather different nature in which only *one* image problem is constructed and solved and then this image solution is "lifted" to the solution in the original domain by solving some nonlinear equations associated with the problem. Because this method requires the solution of nonlinear equations, it can only be used for specific problems where such equations are solvable. The specific problems where it is applicable, however, include such important areas as GCD calculations and polynomial factorization. In addition, nothing really comes for free and this new method can be viewed as trading off a sharp decrease in the computational cost of solving image problems with a sharp increase in the computational cost of "lifting" the image solution to the larger domain. In other words, the new algorithm will be rather more complicated than the interpolation and Chinese remainder algorithms which perform the corresponding "lifting" process in the diagram of Figure 5.1. However this new approach has been found to be significantly more efficient for many practical problems. (Actually the efficiency of the new approach lies mainly in its ability to take advantage of sparseness in the polynomial solution, and as we noted in Chapter 3 polynomials which arise in practical problems involving several indeterminates will invariably be sparse.)

P-adic Representation and Approximation

Consider the problem of inverting the modular homomorphism $\phi_p : \mathbf{Z}[x] \to \mathbf{Z}_p[x]$. The starting point in the development of the new algorithm is to consider yet another representation for integers and polynomials. Recall that, in applying Garner's algorithm to solve the Chinese remainder problem, the integer solution u is developed (in step 2 of Algorithm 5.1) in its mixed radix representation:

$$u = v_0 + v_1(m_0) + v_2(m_0 m_1) + \cdots + v_n(\prod_{i=0}^{n-1} m_i)$$

where m_i $(0 \le i \le n)$ are odd positive moduli such that $\prod_{i=0}^{n} m_i > 2|u|$ and $v_k \in \mathbf{Z}_{m_k}$ $(0 \le k \le n)$. The new approach is based on developing an integer solution u in its *p-adic representation*:

$$u = u_0 + u_1 p + u_2 p^2 + \cdots + u_n p^n \qquad (6.1)$$

where p is an odd positive prime integer, n is such that $p^{n+1} > 2|u|$, and $u_i \in \mathbf{Z}_p$ $(0 \le i \le n)$. As in the case of the mixed radix representation, the p-adic representation can be developed using either the positive or the symmetric representation of $\mathbf{Z}_p$. Obviously if the positive representation is used then (6.1) is simply the familiar *radix p representation* of the nonnegative integer u (and it is sufficient for n to be such that $p^{n+1} > u$). However as we have seen, the symmetric representation is more useful in practice because then the integer u is allowed to be negative.

There is a simple procedure for developing the p-adic representation for a given integer u. Firstly we see from equation (6.1) that $u \equiv u_0 \pmod p$, so using the modular mapping $\phi_p(a) = \mathrm{rem}(a, p)$ we have

$$u_0 = \phi_p(u). \tag{6.2}$$

For the next p-adic coefficient u_1, note that $u - u_0$ must be divisible by p and from equation (6.1) it follows that

$$\frac{u - u_0}{p} = u_1 + u_2 p + \cdots + u_n p^{n-1}.$$

Hence as before, we have

$$u_1 = \phi_p(\frac{u - u_0}{p}).$$

Continuing in this manner, we get

$$u_i = \phi_p(\frac{u - [u_0 + u_1 p + \cdots + u_{i-1} p^{i-1}]}{p^i}), \; i = 1, \ldots, n \tag{6.3}$$

where the division by p^i is guaranteed to be an exact integer division. In formula (6.3) it is important to note that the calculation is to be performed in the domain **Z** and then the modular mapping ϕ_p is applied (unlike the "algorithmic specification" of the ϕ_p notation previously used).

Example 6.1. Let $u = -272300$ be the integer which arose as the solution in Example 5.15 where a mixed radix representation of u was developed. Let us develop the p-adic representation of u choosing p to be the largest two-digit prime, namely $p = 97$. The p-adic coefficients are

$$u_0 = \phi_p(u) = -21,$$

$$u_1 = \phi_p(\frac{u - u_0}{p}) = 6,$$

$$u_2 = \phi_p(\frac{u - [u_0 + u_1 p]}{p^2}) = -29.$$

If we try to compute another coefficient u_3 we find that $u - [u_0 + u_1 p + u_2 p^2] = 0$ so we are finished. Thus the p-adic representation of $u = -272300$ when $p = 97$ is:

$$-272300 = -21 + 6\,(97) - 29\,(97)^2.$$

•

As in the case of a mixed radix representation, the concept of a p-adic representation can be readily extended to polynomials. Consider the polynomial

$$u(x) = \sum_e u_e x^e \in \mathbf{Z}[x]$$

and let p and n be chosen such that $p^{n+1} > 2\,u_{\max}$, where $u_{\max} = \max_e |\,u_e\,|$. If each integer coefficient u_e is expressed in its p-adic representation

$$u_e = \sum_{i=0}^{n} u_{e,i} p^i \text{ with } u_{e,i} \in \mathbf{Z}_p$$

then the polynomial $u(x)$ can be expressed as

$$u(x) = \sum_e \left(\sum_{i=0}^{n} u_{e,i} p^i \right) x^e = \sum_{i=0}^{n} \left(\sum_e u_{e,i} x^e \right) p^i .$$

The latter expression for the polynomial $u(x)$ is called a *polynomial p-adic representation* and its general form is

$$u(x) = u_0(x) + u_1(x)p + u_2(x)p^2 + \cdots + u_n(x)p^n \tag{6.4}$$

where $u_i(x) \in \mathbf{Z}_p[x]$ for $i = 0, 1, \ldots, n$. Formulas (6.2) and (6.3) remain valid when u and u_i $(0 \le i \le n)$ are polynomials.

Example 6.2. Let $u(x) = 14x^2 - 11x - 15 \in \mathbf{Z}[x]$ be the polynomial which arose as the solution in Example 5.17 where a polynomial mixed radix representation of $u(x)$ was developed. Let us develop the polynomial p-adic representation of $u(x)$ choosing $p = 5$. The polynomial p-adic coefficients are:

$$u_0(x) = \phi_p(u(x)) = -x^2 - x,$$

$$u_1(x) = \phi_p\left(\frac{u(x) - u_0(x)}{p}\right) = -2x^2 - 2x + 2,$$

$$u_2(x) = \phi_p\left(\frac{u(x) - [u_0(x) + u_1(x)p]}{p^2}\right) = x^2 - 1.$$

If we try to compute another coefficient $u_3(x)$ we find that $u(x) - [u_0(x) + u_1(x)p + u_2(x)p^2]$ $= 0$ so we are finished. Thus the polynomial p-adic representation of the given polynomial $u(x) \in \mathbf{Z}[x]$ when $p = 5$ is:

$$u(x) = (-x^2 - x) + (-2x^2 - 2x + 2)\,5 + (x^2 - 1)\,5^2.$$

●

It is useful to introduce a concept of *approximation* which is associated with a polynomial p-adic representation. Recall that the congruence relation

$$a(x) \equiv b(x) \pmod{<q>}$$

defined on the domain $\mathbf{Z}[x]$ with respect to a principal ideal $<q>$ in $\mathbf{Z}[x]$ has the meaning:

$$a(x) - b(x) \in \; <q>$$

(i.e. $a(x) - b(x)$ is a multiple of q). Using this congruence notation, it is readily seen that the following relations hold for the polynomials appearing in the polynomial p-adic representation (6.4):

$$u(x) \equiv u_0(x) \pmod{p}$$

and more generally

$$u(x) \equiv u_0(x) + u_1(x)p + \cdots + u_{k-1}(x)p^{k-1} \pmod{p^k},$$

for $1 \le k \le n+1$. We thus have a finite sequence of approximations to the polynomial $u(x)$ in the sense of the following definition.

Definition 6.1. Let $a(x) \in \mathbf{Z}[x]$ be a given polynomial. A polynomial $b(x) \in \mathbf{Z}[x]$ is called an *order n p-adic approximation* to $a(x)$ if

$$a(x) \equiv b(x) \pmod{p^n}.$$

The *error* in approximating $a(x)$ by $b(x)$ is $a(x) - b(x) \in \mathbf{Z}[x]$.

●

Note the similarity of Definition 6.1 with order n approximations of power series used in Chapter 4.

Multivariate Taylor Series Representation

We now consider a generalization of the p-adic representation which will lead to a new technique for inverting a multivariate evaluation homomorphism

$$\phi_I : \mathbf{Z}_p[x_1, \ldots, x_v] \to \mathbf{Z}_p[x_1] \tag{6.5}$$

with kernel $I = \; <x_2 - \alpha_2, \ldots, x_v - \alpha_v>$ for some specified values $\alpha_i \in \mathbf{Z}_p$ $(2 \le i \le v)$. As before, the key to the development of the new algorithm is to choose an appropriate representation for the solution. In this case the "solution" is a multivariate polynomial $\tilde{u} = u(x_1, \ldots, x_v) \in \mathbf{Z}_p[x_1, \ldots, x_v]$ and the "first term" of $\tilde{u}$ is a univariate polynomial $u^{(1)} \in \mathbf{Z}_p[x_1]$, where

$$u^{(1)} = \phi_I(\tilde{u}). \tag{6.6}$$

Note that

$$u^{(1)} = u(x_1, \alpha_2, \ldots, \alpha_v).$$

Corresponding to the previous representation, suppose that we choose a representation for the solution $\tilde{u}$ of the form

$$\tilde{u} = u^{(1)} + \Delta u^{(1)} + \Delta u^{(2)} + \Delta u^{(3)} + \cdots \tag{6.7}$$

with the first term given by (6.6). In order to determine the remaining terms, consider the "error" $e^{(1)} = \tilde{u} - u^{(1)}$ and note that from (6.6) we clearly have $\phi_I(e^{(1)}) = 0$ whence

$$e^{(1)} \in I. \tag{6.8}$$

Now any element of the ideal I can be expressed as a linear combination of the basis elements of I, so (6.8) can be expressed as

$$e^{(1)} = \sum_{i=2}^{v} c_i(x_i - \alpha_i), \text{ where } c_i \in \mathbf{Z}_p[x_1, \ldots, x_v]. \tag{6.9}$$

For the first "correction term" $\Delta u^{(1)}$ in the representation (6.7), we choose the linear terms in the error expression (6.9) defined by

$$\Delta u^{(1)} = \sum_{i=2}^{v} u_i(x_1)\,(x_i - \alpha_i) \tag{6.10}$$

where the coefficients $u_i(x_1) \in \mathbf{Z}_p[x_1]$ are given by

$$u_i(x_1) = \phi_I(c_i),\ 2 \le i \le v. \tag{6.11}$$

Note that $\Delta u^{(1)} \in I$. At this point we have the "approximation" to $\tilde{u}$

$$u^{(2)} = u^{(1)} + \Delta u^{(1)}$$

defined by (6.6) and (6.10). Consider the new error term

$$e^{(2)} = \tilde{u} - u^{(2)} = e^{(1)} - \Delta u^{(1)}.$$

Applying (6.9) and (6.10) we have

$$e^{(2)} = \sum_{i=2}^{v} (c_i - u_i(x_1))\,(x_i - \alpha_i).$$

Now

$$c_i - u_i(x_1) \in I,\ 2 \le i \le v$$

because from (6.11) clearly $\phi_I(c_i - u_i(x_1)) = 0$, which implies that

$$e^{(2)} \in I^2. \tag{6.12}$$

In order to understand the statement (6.12) (and similar statements in the sequel) let us recall from Chapter 5 the definition of the i-th power of an ideal I with finite basis. Specifically, I^2 is the ideal generated by all pairs of products of basis elements of I, I^3 is the ideal generated by all triples of products of basis elements of I, and so on. In our particular case since the basis elements of $I = \langle x_2 - \alpha_2, \ldots, x_v - \alpha_v \rangle$ are linear terms, the basis elements of I^2 will be multivariate terms of total degree 2 and, in general, the basis elements of I^i will be multivariate terms of total degree i. As a clarification, consider the particular case where $v = 3$ in which case we have

$$\mathrm{I} = <x_2 - \alpha_2, x_3 - \alpha_3>;$$

$$\mathrm{I}^2 = <(x_2-\alpha_2)^2, (x_2-\alpha_2)(x_3-\alpha_3), (x_3-\alpha_3)^2>;$$

$$\mathrm{I}^3 = <(x_2-\alpha_2)^3, (x_2-\alpha_2)^2(x_3-\alpha_3), (x_2-\alpha_2)(x_3-\alpha_3)^2, (x_3-\alpha_3)^3>;$$

$$\vdots$$

$$\mathrm{I}^i = <(x_2-\alpha_2)^i, (x_2-\alpha_2)^{i-1}(x_3-\alpha_3), \ldots, (x_3-\alpha_3)^i>.$$

The result (6.12) should now be evident. Expressing $e^{(2)} \in \mathrm{I}^2$ as a linear combination of the basis elements of I^2 yields

$$e^{(2)} = \sum_{i=2}^{v} \sum_{j=i}^{v} c_{ij}\,(x_i - \alpha_i)\,(x_j - \alpha_j), \text{ where } c_{ij} \in \mathbf{Z}_p[x_1, \ldots, x_v].$$

The next correction term in the representation (6.7) is the term $\Delta u^{(2)} \in \mathrm{I}^2$ defined by

$$\Delta u^{(2)} = \sum_{i=2}^{v} \sum_{j=i}^{v} u_{ij}(x_1)\,(x_i - \alpha_i)\,(x_j - \alpha_j) \tag{6.13}$$

where the coefficients $u_{ij}(x_1) \in \mathbf{Z}_p[x_1]$ are given by

$$u_{ij}(x_1) = \phi_{\mathrm{I}}(c_{ij}),\ 2 \le i \le j \le v.$$

We then have the "approximation" to $\tilde{u}$

$$u^{(3)} = u^{(2)} + \Delta u^{(2)} = u^{(1)} + \Delta u^{(1)} + \Delta u^{(2)}$$

defined by (6.6), (6.10), and (6.13). Continuing in this manner, we can show that

$$e^{(3)} \in \mathrm{I}^3$$

where $e^{(3)} = \tilde{u} - u^{(3)}$ and we can proceed to define the next correction term $\Delta u^{(3)} \in \mathrm{I}^3$ in the form

$$\Delta u^{(3)} = \sum_{i=2}^{v} \sum_{j=i}^{v} \sum_{k=j}^{v} u_{ijk}(x_1)\,(x_i - \alpha_i)\,(x_j - \alpha_j)\,(x_k - \alpha_k)$$

for some coefficients $u_{ijk}(x_1) \in \mathbf{Z}_p[x_1]$. This process will eventually terminate because the solution $\tilde{u}$ is a polynomial. Specifically, if d denotes the *total degree* of $\tilde{u}$ as an element of the domain $\mathbf{Z}_p[x_1][x_2, \ldots, x_v]$ (i.e. as a polynomial in the indeterminates $x_2, \ldots, x_v$) then with

$$u^{(d+1)} = u^{(1)} + \Delta u^{(1)} + \cdots + \Delta u^{(d)}$$

we will have $e^{(d+1)} = \tilde{u} - u^{(d+1)} = 0$ so that $u^{(d+1)}$ is the desired polynomial. This must be so because each correction term $\Delta u^{(k)} \in \mathrm{I}^k$ is of total degree k (with respect to $x_2, \ldots, x_v$).

The representation (6.7) which we have just developed for a polynomial $\tilde{u} = u(x_1, \ldots, x_v) \in \mathbf{Z}_p[x_1, \ldots, x_v]$ is called the *multivariate Taylor series representation* with respect to the ideal $\mathrm{I} = <x_2 - \alpha_2, \ldots, x_v - \alpha_v>$ and its general form is

$$u(x_1, \ldots, x_v) = u(x_1, \alpha_2, \ldots, \alpha_v) + \sum_{i=2}^{v} u_i(x_1)(x_i - \alpha_i)$$
$$+ \sum_{i=2}^{v}\sum_{j=i}^{v} u_{ij}(x_1)(x_i - \alpha_i)(x_j - \alpha_j)$$
$$+ \sum_{i=2}^{v}\sum_{j=i}^{v}\sum_{k=j}^{v} u_{ijk}(x_1)(x_i - \alpha_i)(x_j - \alpha_j)(x_k - \alpha_k)$$
$$+ \cdots . \qquad (6.14)$$

The number of terms here will be finite with the last term containing d nested summations, where d is the total degree of $u(x_1, \ldots, x_v)$ with respect to the indeterminates $x_2, \ldots, x_v$.

Ideal-adic Representation and Approximation

The multivariate Taylor series representation (6.14) for a polynomial $u(\mathbf{x}) \in \mathbf{Z}_p[\mathbf{x}]$ can be viewed as a direct generalization of a polynomial p-adic representation. Recall that the polynomial p-adic representation of a polynomial $\tilde{u} = u(\mathbf{x}) \in \mathbf{Z}[\mathbf{x}]$ can be expressed in the form

$$\tilde{u} = u^{(1)} + \Delta u^{(1)} + \Delta u^{(2)} + \cdots + \Delta u^{(n)}$$

where

$$u^{(1)} = u_0(\mathbf{x}) \in \mathbf{Z}[\mathbf{x}] / \langle p \rangle;$$
$$\Delta u^{(k)} = u_k(\mathbf{x})\, p^k \in \langle p \rangle^k, \text{ for } k = 1, 2, \ldots, n.$$

Note here that $\mathbf{Z}[\mathbf{x}] / \langle p \rangle = \mathbf{Z}_p[\mathbf{x}]$ and that $\langle p \rangle^k = \langle p^k \rangle$. We also have the property that the coefficient $u_k(\mathbf{x})$ in the expression for $\Delta u^{(k)}$ as a multiple of the basis element of the ideal in which it lies satisfies

$$u_k(\mathbf{x}) \in \mathbf{Z}[\mathbf{x}] / \langle p \rangle, \; 1 \le k \le n.$$

In the p-adic case, we may define a sequence of order k+1 p-adic approximations

$$u^{(k+1)} \in \mathbf{Z}[\mathbf{x}] / \langle p \rangle^{k+1}, \text{ for } k = 1, 2, \ldots, n$$

where

$$u^{(k+1)} = u^{(1)} + \Delta u^{(1)} + \cdots + \Delta u^{(k)}.$$

In defining the k-th element of this sequence, we have an approximation $u^{(k)} \in \mathbf{Z}[\mathbf{x}] / \langle p \rangle^k$ and we define the new approximation $u^{(k+1)} \in \mathbf{Z}[\mathbf{x}] / \langle p \rangle^{k+1}$ by adding the term $\Delta u^{(k)} \in \langle p \rangle^k$. The addition

$$u^{(k+1)} = u^{(k)} + \Delta u^{(k)}$$

is an addition in the larger domain $\mathbf{Z}[\mathbf{x}] / \langle p \rangle^{k+1}$ and is made valid by assuming the natural embedding of the domain $\mathbf{Z}[\mathbf{x}] / \langle p \rangle^k$ into the larger domain $\mathbf{Z}[\mathbf{x}] / \langle p \rangle^{k+1}$. Thus the successive p-adic approximations $u^{(1)}, u^{(2)}, u^{(3)}, \ldots$ to $\tilde{u} \in \mathbf{Z}[\mathbf{x}]$ lie in a sequence of subdomains of $\mathbf{Z}[\mathbf{x}]$ of increasing size indicated by

$$\mathbf{Z}[\mathbf{x}]/<p> \subset \mathbf{Z}[\mathbf{x}]/<p>^2 \subset \mathbf{Z}[\mathbf{x}]/<p>^3 \subset \cdots \subset \mathbf{Z}[\mathbf{x}].$$

Noting that a polynomial $\tilde{u} \in \mathbf{Z}[\mathbf{x}]$ has a finite polynomial p-adic representation, it is clear that for some $k = n$ the subdomain $\mathbf{Z}[\mathbf{x}]/<p>^{n+1}$ will be large enough to contain the polynomial $\tilde{u}$.

The multivariate Taylor series representation (6.14) for a polynomial $\tilde{u} = u(\mathbf{x}) \in \mathbf{Z}_p[\mathbf{x}]$ can be viewed in an abstractly equivalent manner with the ideal I taking the place of the ideal $<p>$ above. The polynomial $\tilde{u}$ was developed in the form

$$\tilde{u} = u^{(1)} + \Delta u^{(1)} + \Delta u^{(2)} + \cdots + \Delta u^{(d)}$$

where

$$u^{(1)} = u(x_1, \alpha_2, \ldots, \alpha_v) \in \mathbf{Z}_p[\mathbf{x}]/\mathrm{I};$$

$$\Delta u^{(k)} \in \mathrm{I}^k, \text{ for } k = 1, 2, \ldots, d.$$

Here $\mathbf{x} = (x_1, \ldots, x_v)$, $\mathrm{I} = <x_2 - \alpha_2, \ldots, x_v - \alpha_v>$, and note that $\mathbf{Z}_p[\mathbf{x}]/\mathrm{I} = \mathbf{Z}_p[x_1]$. Corresponding to the p-adic case, we have the additional property that for each k the coefficients in the expression for $\Delta u^{(k)}$ as a linear combination of the basis elements of the ideal I^k all lie in the domain $\mathbf{Z}_p[\mathbf{x}]/\mathrm{I}$. (For example,

$$\Delta u^{(2)} = \sum_{i=2}^{v} \sum_{j=i}^{v} u_{ij}(x_1)(x_i - \alpha_i)(x_j - \alpha_j)$$

with $u_{ij}(x_1) \in \mathbf{Z}_p[\mathbf{x}]/\mathrm{I}$, $2 \le i \le j \le v$.) It is therefore appropriate to speak of a sequence of approximations (see Definition 6.2) to $\tilde{u}$ defined by

$$u^{(k+1)} \in \mathbf{Z}_p[\mathbf{x}]/\mathrm{I}^{k+1}, \text{ for } k = 1, 2, \ldots, d$$

where

$$u^{(k+1)} = u^{(1)} + \Delta u^{(1)} + \cdots + \Delta u^{(k)}.$$

Again we must assume a natural embedding of domains and the sequence of approximations $u^{(1)}, u^{(2)}, u^{(3)}, \ldots$ to $\tilde{u} \in \mathbf{Z}_p[\mathbf{x}]$ lie in the following sequence of subdomains of $\mathbf{Z}_p[\mathbf{x}]$ of increasing size:

$$\mathbf{Z}_p[\mathbf{x}]/\mathrm{I} \subset \mathbf{Z}_p[\mathbf{x}]/\mathrm{I}^2 \subset \mathbf{Z}_p[\mathbf{x}]/\mathrm{I}^3 \subset \cdots \subset \mathbf{Z}_p[\mathbf{x}].$$

As in the p-adic case, since the multivariate Taylor series representation for $\tilde{u}$ is finite there is an index $k = d$ such that the subdomain $\mathbf{Z}_p[\mathbf{x}]/\mathrm{I}^{d+1}$ is large enough to contain the polynomial $\tilde{u}$.

In view of this close correspondence with the p-adic representation of a polynomial, the multivariate Taylor series representation (6.14) of a polynomial $u(\mathbf{x}) \in \mathbf{Z}_p[\mathbf{x}]$ is also called the *ideal-adic representation* of $u(\mathbf{x})$ with respect to the ideal $\mathrm{I} = <x_2 - \alpha_2, \ldots, x_v - \alpha_v>$. The concept of approximation mentioned above is made precise by the following definition which is an obvious abstraction of Definition 6.1.

Definition 6.2. Let D be a Noetherian integral domain and let I be an ideal in D. For a given element $a \in$ D, the element $b \in$ D is called an *order n ideal-adic approximation* to a with respect to the ideal I if

$$a \equiv b \mod \mathrm{I}^n.$$

The *error* in approximating a by b is the element $a - b \in \mathrm{I}^n$.

●

Recalling that $a \equiv b \mod \mathrm{I}^n$ means that $a - b \in \mathrm{I}^n$, it is clear from the development of the ideal-adic representation (multivariate Taylor series representation) (6.14) for $u(x_1, \ldots, x_v) \in \mathbf{Z}_p[x_1, \ldots, x_v]$ that $u^{(k)}$ is an order k ideal-adic approximation to $u(x_1, \ldots, x_v)$ with respect to the ideal $\mathrm{I} = <x_2 - \alpha_2, \ldots, x_v - \alpha_v>$, where

$$u^{(1)} = u(x_1, \alpha_2, \ldots, \alpha_v);$$

$$u^{(k+1)} = u^{(k)} + \Delta u^{(k)}, \text{ for } k = 1, 2, \ldots, d;$$

with $\Delta u^{(k)}$ defined to be the term in (6.14) of total degree k with respect to I (i.e. the term represented by k nested summations). In connection with the concept of ideal-adic approximation it is useful to note the following computational definition of the homomorphism ϕ_{I^n} defined on the domain $\mathbf{Z}_p[\mathbf{x}]$, where $\mathrm{I} = <x_2 - \alpha_2, \ldots, x_v - \alpha_v>$. Since

$$\phi_{\mathrm{I}^n} : \mathbf{Z}_p[\mathbf{x}] \to \mathbf{Z}_p[\mathbf{x}] / \mathrm{I}^n$$

denotes the homomorphism with kernel I^n, if the polynomial $a(\mathbf{x}) \in \mathbf{Z}_p[\mathbf{x}]$ is represented in its ideal-adic representation with respect to the ideal I then $\phi_{\mathrm{I}^n}(a(\mathbf{x}))$ is precisely the order n ideal-adic approximation to $a(\mathbf{x})$ obtained by dropping all terms in the ideal-adic representation of $a(\mathbf{x})$ which have total degree equal to or exceeding n (with respect to I).

6.3. NEWTON'S ITERATION FOR F(u) = 0

Linear p-adic Iteration

We wish to develop a method corresponding to the Chinese remainder algorithm for inverting the modular homomorphism $\phi_p : \mathbf{Z}[\mathbf{x}] \to \mathbf{Z}_p[\mathbf{x}]$. In the new approach we assume that we use only one prime p and that we know the image $u_0(\mathbf{x}) \in \mathbf{Z}_p[\mathbf{x}]$ of the desired solution $u(\mathbf{x}) \in \mathbf{Z}[\mathbf{x}]$. In the terminology of the preceding section, $u_0(\mathbf{x})$ is an order 1 (or first-order) p-adic approximation to $u(\mathbf{x})$ and it is also the first term in the polynomial p-adic representation of $u(\mathbf{x})$. We will develop a method to compute successively the order k approximation

$$u_0(\mathbf{x}) + u_1(\mathbf{x})p + \cdots + u_{k-1}(\mathbf{x})p^{k-1} \in \mathbf{Z}_{p^k}[\mathbf{x}],$$

for $k = 1, 2, \ldots, n+1$. Then the order $n+1$ approximation which lies in the domain $\mathbf{Z}_{p^{n+1}}[\mathbf{x}]$ is the desired solution $u(\mathbf{x}) \in \mathbf{Z}[\mathbf{x}]$ (assuming that n was chosen large enough). This general process is called *lifting* the image $u_0(\mathbf{x}) \in \mathbf{Z}_p[\mathbf{x}]$ to the solution $u(\mathbf{x})$ in the larger domain $\mathbf{Z}[\mathbf{x}]$.

The lifting process clearly requires more information about the solution $u(\mathbf{x})$ than simply the single image $u_0(\mathbf{x})$. We will assume that the additional information can be specified in the form of one or more equations (usually nonlinear) which $u(\mathbf{x})$ must satisfy. For now, let us assume that the solution $u = u(\mathbf{x})$ is known to satisfy

$$F(u) = 0 \tag{6.15}$$

where $F(u) \in \mathbf{Z}[\mathbf{x}]\,[u]$ – i.e. $F(u)$ is some polynomial expression in u with coefficients lying in the domain $\mathbf{Z}[\mathbf{x}]$. The basic idea of the new approach is to have an iterative method which will improve the given first-order p-adic approximation $u_0(\mathbf{x})$ into successively higher-order p-adic approximations to the solution $u(\mathbf{x})$ of (6.15). The iterative process will be finite if (6.15) has a polynomial solution $u(\mathbf{x})$ since, in the above notation, the order $n+1$ p-adic approximation to $u(\mathbf{x})$ will be $u(\mathbf{x})$ itself.

Recall again as in Chapter 4 the classical Newton's iteration for solving a nonlinear equation of the form (6.15) in the traditional analytic setting where $F(u)$ is a differentiable real-valued function of a real variable u. Letting $u^{(k)}$ denote an approximation to a solution $\tilde{u}$ and expanding the function $F(u)$ in a Taylor series about the point $u^{(k)}$, we have

$$F(u) = F(u^{(k)}) + F'(u^{(k)})\,(u - u^{(k)}) + \frac{1}{2}F''(u^{(k)})\,(u - u^{(k)})^2 + \cdots .$$

Setting $u = \tilde{u}$, the left hand side becomes zero and retaining only linear terms in the Taylor series we have the approximate equality

$$0 \approx F(u^{(k)}) + F'(u^{(k)})\,(\tilde{u} - u^{(k)}).$$

Solving for $\tilde{u}$ and calling it the new approximation $u^{(k+1)}$, we have Newton's iterative formula

$$u^{(k+1)} = u^{(k)} - \frac{F(u^{(k)})}{F'(u^{(k)})}$$

(where we need the assumption that $F'(u^{(k)}) \neq 0$). The iteration must be started with an initial guess $u^{(1)}$ and using techniques of real analysis it can be proved that if $u^{(1)}$ is "close enough" to a solution $\tilde{u}$ of $F(u) = 0$ and if $F'(\tilde{u}) \neq 0$ then the infinite iteration specified above will converge (quadratically) to the solution $\tilde{u}$. We will develop a similar iterative formula for our polynomial setting and it will have two significant computational advantages over the traditional analytic case: (i) the first-order p-adic approximation will be sufficient to give *guaranteed convergence*, and (ii) the iteration will be *finite*.

We wish to solve the polynomial equation *assuming that it has a polynomial solution* $\tilde{u} = u(\mathbf{x}) \in \mathbf{Z}[\mathbf{x}]$, given the first-order p-adic approximation $u_0(\mathbf{x}) \in \mathbf{Z}_p[\mathbf{x}]$ to $\tilde{u}$. (Note that an arbitrary polynomial equation of the form (6.15) would not in general have a polynomial solution but we are assuming a context in which a polynomial solution is known to exist.) Writing the solution in its polynomial p-adic representation

$$\tilde{u} = u_0(\mathbf{x}) + u_1(\mathbf{x})p + \cdots + u_n(\mathbf{x})p^n \tag{6.16}$$

we wish to determine the polynomial p-adic coefficients $u_i(\mathbf{x}) \in \mathbf{Z}_p[\mathbf{x}]$ for $i = 1, 2, \ldots, n$ ($u_0(\mathbf{x})$ is given). Let us denote by $u^{(k)}$ the order k p-adic approximation to $\tilde{u}$ given by the first k terms of (6.16). Thus $u^{(1)} = u_0(\mathbf{x})$ and in general

$$u^{(k)} = u_0(\mathbf{x}) + u_1(\mathbf{x})p + \cdots + u_{k-1}(\mathbf{x})p^{k-1}, \; 1 \le k \le n+1.$$

We would like an iteration formula which at step k is given the order k approximation $u^{(k)}$ and which computes the polynomial p-adic coefficient $u_k(\mathbf{x}) \in \mathbf{Z}_p[\mathbf{x}]$ yielding the order $k+1$ approximation

$$u^{(k+1)} = u^{(k)} + u_k(\mathbf{x})p^k, \; 1 \le k \le n. \tag{6.17}$$

By Theorem 2.8 of Chapter 2 applied to the polynomial $F(u) \in \mathrm{D}[u]$ where $\mathrm{D} = \mathbf{Z}[\mathbf{x}]$, we have the following "Taylor series expansion":

$$F(u^{(k)} + u_k(\mathbf{x})p^k) = F(u^{(k)}) + F'(u^{(k)})\, u_k(\mathbf{x})p^k + G(u^{(k)}, u_k(\mathbf{x})p^k)\, [u_k(\mathbf{x})]^2 p^{2k} \tag{6.18}$$

for some polynomial $G(u,w) \in \mathrm{D}[u,w]$.

At this point we need to use a property of congruences. Recall the congruence properties developed in Chapter 5 which show that congruences can be added, subtracted, and multiplied. As a direct consequence of these properties, it follows that if I is any ideal in a commutative ring R and if $h(x) \in \mathrm{R}[x]$ is any polynomial expression over R then for $a, b \in \mathrm{R}$

$$a \equiv b \pmod{\mathrm{I}} \implies h(a) \equiv h(b) \pmod{\mathrm{I}}. \tag{6.19}$$

Now since $u^{(k)} \equiv \tilde{u} \pmod{p^k}$, applying property (6.19) and the fact that $F(\tilde{u}) = 0$ yields

$$F(u^{(k)}) \equiv 0 \pmod{p^k}.$$

Similarly,

$$F(u^{(k)} + u_k(\mathbf{x})p^k) \equiv 0 \pmod{p^{k+1}}$$

if (6.17) is to define the order $k+1$ approximation $u^{(k+1)}$. Therefore we can divide by p^k in (6.18) yielding

$$\frac{F(u^{(k)} + u_k(\mathbf{x})p^k)}{p^k} = \frac{F(u^{(k)})}{p^k} + F'(u^{(k)})u_k(\mathbf{x}) + G(u^{(k)}, u_k(\mathbf{x})p^k)[u_k(\mathbf{x})]^2 p^k.$$

Now applying the modular homomorphism ϕ_p and noting that the left hand side is still a multiple of p, we find that the desired polynomial p-adic coefficient $u_k(\mathbf{x}) \in \mathbf{Z}_p[\mathbf{x}]$ must satisfy

$$0 = \phi_p\left(\frac{F(u^{(k)})}{p^k}\right) + \phi_p(F'(u^{(k)}))u_k(\mathbf{x}) \in \mathbf{Z}_p[\mathbf{x}].$$

Finally since $u^{(k)} \equiv u^{(1)} \pmod{p}$ for all $k \ge 1$, it follows from property (6.19) that

$$F'(u^{(k)}) \equiv F'(u^{(1)}) \pmod{p}.$$

Therefore if the given first-order approximation $u^{(1)}$ satisfies the condition

$$F'(u^{(1)}) \not\equiv 0 \pmod{p}$$

then the desired polynomial p-adic coefficient is given by

$$u_k(\mathbf{x}) = -\frac{\phi_p\left(\dfrac{F(u^{(k)})}{p^k}\right)}{\phi_p(F'(u^{(1)}))} \in \mathbf{Z}_p[\mathbf{x}]. \tag{6.20}$$

The division appearing in (6.20) must be an exact division in the polynomial domain $\mathbf{Z}_p[\mathbf{x}]$ if equation (6.15) has a polynomial solution. The iteration formula (6.17) together with the *linear update formula* (6.20) is known as the *linear p-adic Newton's iteration.* Note that in formula (6.20) the calculation of $F(u^{(k)})$ must be performed in the domain $\mathbf{Z}[\mathbf{x}]$, followed by an exact division by p^k in $\mathbf{Z}[\mathbf{x}]$, before the modular homomorphism ϕ_p is applied.

Example 6.3. Consider the problem of determining a polynomial $u(x) \in \mathbf{Z}[x]$ which is a square root of the polynomial

$$a(x) = 36x^4 - 180x^3 + 93x^2 + 330x + 121 \in \mathbf{Z}[x]$$

(assuming that $a(x)$ is a perfect square). Then $u(x)$ can be expressed as the solution of the polynomial equation

$$F(u) = a(x) - u^2 = 0.$$

Choosing $p = 5$, the first-order p-adic approximation $u^{(1)} = u_0(x) \in \mathbf{Z}_5[x]$ must be a square root of $\phi_5(a(x))$ in $\mathbf{Z}_5[x]$. Now

$$\phi_5(a(x)) = x^4 - 2x^2 + 1$$

which clearly has the square root

$$u^{(1)} = u_0(x) = x^2 - 1 \in \mathbf{Z}_5[x].$$

Now to apply the linear p-adic Newton's iteration, first note that

$$\phi_5(F'(u^{(1)})) = \phi_5(-2u^{(1)}) = -2x^2 + 2.$$

Then

$$u_1(x) = -\frac{\phi_5(\frac{F(u^{(1)})}{5})}{(-2x^2+2)} = -\frac{\phi_5(\frac{35x^4 - 180x^3 + 95x^2 + 330x + 120}{5})}{(-2x^2+2)}$$

$$= -\frac{(2x^4 - x^3 - x^2 + x - 1)}{(-2x^2+2)} = x^2 + 2x - 2 \in \mathbf{Z}_5[x]$$

yielding

$$u^{(2)} = (x^2 - 1) + (x^2 + 2x - 2)5 \in \mathbf{Z}_{25}[x].$$

Similarly we get

$$u_2(x) = -\frac{(-2x^3 + 2x)}{(-2x^2 + 2)} = -x \in \mathbf{Z}_5[x]$$

yielding

$$u^{(3)} = (x^2 - 1) + (x^2 + 2x - 2)5 + (-x)5^2 \in \mathbf{Z}_{125}[x].$$

If we proceed to calculate another polynomial p-adic coefficient $u_3(x)$ we find that $F(u^{(3)}) = 0$ (in the domain $\mathbf{Z}[x]$) so we are finished. The desired square root of $a(x)$ is therefore

$$u(x) = u^{(3)} = 6x^2 - 15x - 11 \in \mathbf{Z}[x].$$

•

Quadratic p-adic Iteration

Newton's iteration as specified by (6.17) and (6.20) increases the order of approximation by one per iteration. However it is possible to develop Newton's iteration in such a way that the order of approximation *doubles* per iteration and this corresponds to the concept of *quadratic* convergence familiar in the analytic applications of Newton's iteration. In the quadratic version, at step k we have the order $n_k = 2^{k-1}$ approximation

$$u^{(k)} = u_0(\mathbf{x}) + u_1(\mathbf{x})p + \cdots + u_{n_k-1}(\mathbf{x})p^{n_k-1}$$

to a solution $\tilde{u}$ of $F(u) = 0$ and we compute an update $\Delta u^{(k)}$ such that

$$u^{(k+1)} = u^{(k)} + \Delta u^{(k)} \tag{6.21}$$

is an order $2n_k = 2^k$ approximation, namely

$$\Delta u^{(k)} = u_{n_k}(\mathbf{x})p^{n_k} + \cdots + u_{2n_k-1}(\mathbf{x})p^{2n_k-1}$$
$$= p^{n_k}\left[u_{n_k}(\mathbf{x}) + \cdots + u_{2n_k-1}(\mathbf{x})p^{n_k-1}\right].$$

Corresponding to formula (6.18) we have from Theorem 2.8 of Chapter 2

$$F(u^{(k)} + \Delta u^{(k)}) = F(u^{(k)}) + F'(u^{(k)})\,\Delta u^{(k)} + G(u^{(k)},\Delta u^{(k)})\,[\Delta u^{(k)}]^2$$

for some polynomial $G(u,w)$. Noting from above that $\Delta u^{(k)}$ can be divided by p^{n_k} and using arguments similar to the linear case, we get the following formula which must be satisfied by the update $\Delta u^{(k)}$:

$$0 = \phi_{p^{n_k}}\left(\frac{F(u^{(k)})}{p^{n_k}}\right) + \phi_{p^{n_k}}(F'(u^{(k)}))\,\frac{\Delta u^{(k)}}{p^{n_k}} \in \mathbf{Z}_{p^{n_k}}[\mathbf{x}] \tag{6.22}$$

where $n_k = 2^{k-1}$. As before we have the result that for all $k \ge 1$,

$$F'(u^{(k)}) \equiv F'(u^{(1)}) \pmod{p}.$$

This time this result does not yield a simplification of the derivative since the modular homomorphism being applied to the derivative is now $\phi_{p^{n_k}}$ rather than ϕ_p. However we again wish to divide by the derivative term in (6.22) and the condition needed to guarantee that it is nonzero is precisely as in the linear update formula

$$F'(u^{(1)}) \not\equiv 0 \pmod{p},$$

since from above this guarantees that $F'(u^{(k)}) \not\equiv 0 \pmod{p}$ whence $F'(u^{(k)}) \not\equiv 0 \pmod{p^{n_k}}$. (In other words, if the derivative term in (6.22) is nonzero for $k = 1$ then it is nonzero for all $k \geq 1$). Finally, solving for the update term in (6.22) yields the *quadratic update formula*

$$\frac{\Delta u^{(k)}}{p^{n_k}} = -\frac{\phi_{p^{n_k}}\left(\dfrac{F(u^{(k)})}{p^{n_k}}\right)}{\phi_{p^{n_k}}(F'(u^{(k)}))} \in \mathbf{Z}_{p^{n_k}}[\mathbf{x}]. \tag{6.23}$$

As in the case of the linear update formula, the division in (6.23) must be an exact division in the polynomial domain $\mathbf{Z}_{p^{n_k}}[\mathbf{x}]$ if there exists a polynomial solution to the original problem.

Theorem 6.1 formally proves the quadratic convergence property of the p-adic Newton's iteration (6.21) with the quadratic update formula (6.23). There are cases where the quadratic method has a significant advantage over the linear method (cf. Loos [3]). However, it is also the case that in many practical problems requiring a p-adic Newton's iteration, the linear iteration is used rather than the quadratic iteration. The quadratic iteration does require fewer iteration steps but the cost of each iteration step beyond the first is significantly higher than in the linear iteration because the domain $\mathbf{Z}_{p^{n_k}}[\mathbf{x}]$ in which the update formula (6.23) must be computed becomes larger as k increases. Moreover, the derivative appearing in the divisor in (6.23) must be recomputed at each iteration step while the divisor in the linear update formula (6.20) is fixed for all iteration steps. For these reasons the linear iteration is sometimes preferable to the quadratic iteration in terms of overall efficiency.

Theorem 6.1. Let $F(u) \in \mathbf{Z}[\mathbf{x}][u]$ be such that the polynomial equation $F(u) = 0$ has a polynomial solution $\tilde{u} = u(\mathbf{x}) \in \mathbf{Z}[\mathbf{x}]$. Let $u_0(\mathbf{x}) \in \mathbf{Z}_p[\mathbf{x}]$ be a first-order p-adic approximation to the solution $\tilde{u}$ so that

$$F(u_0(\mathbf{x})) \equiv 0 \pmod{p}.$$

Further suppose that $u_0(\mathbf{x})$ satisfies

$$F'(u_0(\mathbf{x})) \not\equiv 0 \pmod{p}.$$

Then the sequence of iterates defined by

$$u^{(1)} = u_0(\mathbf{x});$$

$$u^{(k+1)} = u^{(k)} + \Delta u^{(k)}, k = 1, 2, 3, \ldots$$

where $\Delta u^{(k)}$ is defined by the quadratic update formula (6.23), is such that $u^{(k+1)}$ is an order 2^k p-adic approximation to the solution $\tilde{u}$.

Proof: The proof is by induction on k. The basis holds trivially: $u^{(1)}$ is an order 1 p-adic approximation to $\tilde{u}$.

For the induction step, assume for $k \geq 1$ that $u^{(k)}$ is an order $n_k = 2^{k-1}$ p-adic approximation to $\tilde{u}$. This means that

$$\tilde{u} \equiv u^{(k)} \ (\mathrm{mod}\ p^{n_k})$$

or, defining the error $e^{(k)} = \tilde{u} - u^{(k)}$ we have

$$e^{(k)} \equiv 0 \ (\mathrm{mod}\ p^{n_k}).$$

Applying Theorem 2.8 of Chapter 2 yields

$$F(u^{(k)} + e^{(k)}) = F(u^{(k)}) + F'(u^{(k)})\, e^{(k)} + G(u^{(k)}, e^{(k)})\, [e^{(k)}]^2$$

for some polynomial $G(u,w)$. Now $u^{(k)} + e^{(k)} = \tilde{u}$ so the left hand side becomes zero and, since $F(u^{(k)})$ and $e^{(k)}$ are multiples of p^{n_k}, we have

$$0 = \frac{F(u^{(k)})}{p^{n_k}} + F'(u^{(k)})\frac{e^{(k)}}{p^{n_k}} + G(u^{(k)}, e^{(k)})\frac{e^{(k)}}{p^{n_k}} e^{(k)}.$$

Applying the modular homomorphism $\phi_{p^{n_k}}$ then yields

$$0 = \phi_{p^{n_k}}\left(\frac{F(u^{(k)})}{p^{n_k}}\right) + \phi_{p^{n_k}}(F'(u^{(k)}))\, \phi_{p^{n_k}}\left(\frac{e^{(k)}}{p^{n_k}}\right)$$

where we note that the last term vanishes because $\phi_{p^{n_k}}(e^{(k)}) = 0$. Now applying the definition of the quadratic update formula (6.23), this becomes

$$0 = -\frac{\Delta u^{(k)}}{p^{n_k}} + \phi_{p^{n_k}}\left(\frac{e^{(k)}}{p^{n_k}}\right) \in \mathbf{Z}_{p^{n_k}}[\mathbf{x}].$$

Hence

$$\frac{e^{(k)} - \Delta u^{(k)}}{p^{n_k}} \equiv 0 \ \left(\mathrm{mod}\ p^{n_k}\right)$$

or

$$e^{(k)} - \Delta u^{(k)} \equiv 0 \ \left(\mathrm{mod}\ p^{2n_k}\right).$$

Finally, since $e^{(k)} = \tilde{u} - u^{(k)}$ we have

$$\tilde{u} - (u^{(k)} + \Delta u^{(k)}) \equiv 0 \ \left(\mathrm{mod}\ p^{2n_k}\right)$$

or

$$\tilde{u} \equiv u^{(k+1)} \left(\text{mod } p^{2n_k}\right)$$

which proves that $u^{(k+1)}$ is an order $2n_k = 2^k$ p-adic approximation to $\tilde{u}$.

•

Ideal-adic Iteration

We now turn to the problem of inverting a multivariate evaluation homomorphism

$$\phi_I : \mathbf{Z}_p[x_1, \ldots, x_v] \to \mathbf{Z}_p[x_1]$$

with kernel $I = < x_2 - \alpha_2, \ldots, x_v - \alpha_v >$ for some specified values $\alpha_i \in \mathbf{Z}_p$ $(2 \le i \le v)$. The inversion process will be accomplished by an ideal-adic version of Newton's iteration. We are given the order 1 ideal-adic approximation

$$u^{(1)} = \phi_I(\tilde{u}) \in \mathbf{Z}_p[x_1] = \mathbf{Z}_p[\mathbf{x}] \,/\, I$$

to the solution $\tilde{u} \in \mathbf{Z}_p[\mathbf{x}]$ and, as before, let us assume for now that the additional information about the solution $\tilde{u}$ is that it satisfies a polynomial equation

$$F(u) = 0$$

where $F(u) \in \mathbf{Z}_p[\mathbf{x}][u]$. We wish to define an iteration formula such that at step k the order k ideal-adic approximation $u^{(k)}$ is updated to the order k+1 ideal-adic approximation $u^{(k+1)}$ by the addition of the correction term $\Delta u^{(k)} \in I^k$. By Theorem 2.8 of Chapter 2 applied to the polynomial $F(u) \in \mathbf{Z}_p[\mathbf{x}][u]$, we have the following "Taylor series expansion"

$$F(u^{(k)} + \Delta u^{(k)}) = F(u^{(k)}) + F'(u^{(k)})\Delta u^{(k)} + G(u^{(k)}, \Delta u^{(k)})\,[\Delta u^{(k)}]^2 \tag{6.24}$$

for some polynomial $G(u,w)$. Now if $u^{(k)} + \Delta u^{(k)}$ is to be the order $k + 1$ ideal-adic approximation $u^{(k+1)}$ then using property (6.19) we deduce that

$$F(u^{(k)} + \Delta u^{(k)}) \in I^{k+1}.$$

Also since $\Delta u^{(k)} \in I^k$ it follows that

$$[\Delta u^{(k)}]^2 \in I^{2k}.$$

Hence applying the homomorphism $\phi_{I^{k+1}}$ to (6.24) yields the equation

$$0 = \phi_{I^{k+1}}(F(u^{(k)})) + \phi_{I^{k+1}}(F'(u^{(k)}))\Delta u^{(k)} \in \mathbf{Z}_p[\mathbf{x}] \,/\, I^{k+1} \tag{6.25}$$

which must be satisfied by the correction term $\Delta u^{(k)} \in I^k$.

Consider iteration step $k = 1$. In this case the correction term $\Delta u^{(1)} \in I$ takes the form

$$\Delta u^{(1)} = \sum_{i=2}^{v} u_i(x_1)\,(x_i - \alpha_i) \tag{6.26}$$

where the coefficients $u_i(x_1) \in \mathbf{Z}_p[x_1]$ are to be determined. Using property (6.19) and the fact that $u^{(1)} \equiv \tilde{u}$ (mod I), we deduce that $F(u^{(1)}) \in I$ and therefore we can write

$$F(u^{(1)}) = \sum_{i=2}^{v} c_i (x_i - \alpha_i) \tag{6.27}$$

for some coefficients $c_i \in \mathbf{Z}_p[\mathbf{x}]$, $2 \le i \le v$. Now the homomorphism being applied in equation (6.25) is ϕ_{I^2} when $k = 1$ and since the effect of ϕ_{I^2} is to drop the ideal-adic terms of total degree equal to or exceeding 2, it follows from (6.27) that

$$\phi_{I^2}(F(u^{(1)})) = \sum_{i=2}^{v} c_i(x_1)(x_i - \alpha_i)$$

where the coefficients $c_i(x_1) \in \mathbf{Z}_p[x_1]$ are defined from the coefficients $c_i \in \mathbf{Z}_p[\mathbf{x}]$ appearing in (6.27) by

$$c_i(x_1) = \phi_I(c_i),\ 2 \le i \le v.$$

Equation (6.25) is now

$$0 = \sum_{i=2}^{v} c_i(x_1)(x_i - \alpha_i) + \phi_{I^2}(F'(u^{(1)})) \left[\sum_{i=2}^{v} u_i(x_1)(x_i - \alpha_i) \right] \in \mathbf{Z}_p[\mathbf{x}] \,/\, I^2. \tag{6.28}$$

Now the ideal-adic representation of $\phi_{I^2}(F'(u^{(1)}))$ can be written in the form

$$\phi_{I^2}(F'(u^{(1)})) = \phi_I(F'(u^{(1)})) + \sum_{i=2}^{v} d_i(x_1)(x_i - \alpha_i)$$

for some coefficients $d_i(x_1) \in \mathbf{Z}_p[x_1]$, $2 \le i \le v$. Putting this form into equation (6.28) yields

$$0 = \sum_{i=2}^{v} c_i(x_1)(x_i - \alpha_i) + \phi_I(F'(u^{(1)})) \left[\sum_{i=2}^{v} u_i(x_1)(x_i - \alpha_i) \right] \in \mathbf{Z}_p[\mathbf{x}] \,/\, I^2 \tag{6.29}$$

where we have noted that

$$\left[\sum_{i=2}^{v} d_i(x_1)(x_i - \alpha_i) \right] \left[\sum_{i=2}^{v} u_i(x_1)(x_i - \alpha_i) \right] \in I^2.$$

Equating coefficients on the left and right in equation (6.29) yields finally

$$u_i(x_1) = -\frac{c_i(x_1)}{\phi_I(F'(u^{(1)}))} \in \mathbf{Z}_p[x_1],\ 2 \le i \le v. \tag{6.30}$$

Equation (6.30) is the desired update formula which defines the correction term (6.26) and the division appearing in (6.30) must be an exact division in the univariate polynomial domain $\mathbf{Z}_p[x_1]$ if the given equation $F(u) = 0$ has a polynomial solution. Note that the coefficients $c_i(x_1)$ appearing in (6.30) are simply the coefficients of the linear terms in the ideal-adic representation of $F(u^{(1)})$.

Turning now to the general iteration step, the k-th correction term $\Delta u^{(k)} \in I^k$ is the term of total degree k in the ideal-adic representation of the solution $\tilde{u} = u(x_1, \ldots, x_v)$ and its general form consists of k nested summations as follows

$$\Delta u^{(k)} = \sum_{i_1=2}^{v} \cdots \sum_{i_k=i_{k-1}}^{v} u_{\mathbf{i}}(x_1) \prod_{j=1}^{k} (x_{i_j} - \alpha_{i_j}) \tag{6.31}$$

where the subscript $\mathbf{i}$ denotes the vector of indices $\mathbf{i} = (i_1, \ldots, i_k)$. The coefficients $u_{\mathbf{i}}(x_1) \in \mathbf{Z}_p[x_1]$ are to be determined. We are given the order k ideal-adic approximation $u^{(k)}$ and the correction term $\Delta u^{(k)}$ must satisfy equation (6.25). As before, we deduce that $F(u^{(k)}) \in \mathrm{I}^k$ from which it follows that

$$\phi_{\mathrm{I}^{k+1}}(F(u^{(k)})) = \sum_{i_1=2}^{v} \cdots \sum_{i_k=i_{k-1}}^{v} c_{\mathbf{i}}(x_1) \prod_{j=1}^{k} (x_{i_j} - \alpha_{i_j})$$

for some coefficients $c_{\mathbf{i}}(x_1) \in \mathbf{Z}_p[x_1]$. Also, the term $\phi_{\mathrm{I}^{k+1}}(F'(u^{(k)}))$ in equation (6.25) can be replaced by $\phi_{\mathrm{I}}(F'(u^{(k)}))$ because just as in the case $k = 1$, the terms of order greater than 1 in the ideal-adic representation of $\phi_{\mathrm{I}^{k+1}}(F'(u^{(k)}))$ disappear when multiplied by $\Delta u^{(k)} \in \mathrm{I}^k$ (since the multiplication is in the domain $\mathbf{Z}_p[\mathbf{x}] \,/\, \mathrm{I}^{k+1}$). But for all $k \geq 1$, $u^{(k)} \equiv u^{(1)} \pmod{\mathrm{I}}$ which implies by property (6.19) that $F'(u^{(k)}) \equiv F'(u^{(1)}) \pmod{\mathrm{I}}$; i.e.

$$\phi_{\mathrm{I}}(F'(u^{(k)})) = \phi_{\mathrm{I}}(F'(u^{(1)})) \text{ for all } k \geq 1.$$

Equation (6.25) therefore becomes

$$0 = \sum_{i_1=2}^{v} \cdots \sum_{i_k=i_{k-1}}^{v} c_{\mathbf{i}}(x_1) \prod_{j=1}^{k} (x_{i_j} - \alpha_{i_j})$$
$$+ \phi_{\mathrm{I}}(F'(u^{(1)})) \left[\sum_{i_1=2}^{v} \cdots \sum_{i_k=i_{k-1}}^{v} u_{\mathbf{i}}(x_1) \prod_{j=1}^{k} (x_{i_j} - \alpha_{i_j}) \right] \in \mathbf{Z}_p[\mathbf{x}] \,/\, \mathrm{I}^{k+1}. \tag{6.32}$$

Finally, if the given first-order approximation $u^{(1)}$ satisfies the condition

$$F'(u^{(1)}) \not\equiv 0 \pmod{\mathrm{I}} \tag{6.33}$$

then by equating coefficients on the left and right in equation (6.32) we get the *linear ideal-adic Newton's iteration:*

$$u^{(k+1)} = u^{(k)} + \Delta u^{(k)} \tag{6.34}$$

where $\Delta u^{(k)}$ is the correction term (6.31) with coefficients defined by

$$u_{\mathbf{i}}(x_1) = -\frac{c_{\mathbf{i}}(x_1)}{\phi_{\mathrm{I}}(F'(u^{(1)}))} \in \mathbf{Z}_p[x_1]. \tag{6.35}$$

Once again, the division appearing in (6.35) must be an exact division in the univariate polynomial domain $\mathbf{Z}_p[x_1]$ if the given equation $F(u) = 0$ has a polynomial solution. Note that the coefficients $c_{\mathbf{i}}(x_1)$ appearing in (6.35) are simply the coefficients of the terms of total degree k in the ideal-adic representation of $F(u^{(k)})$ and note further that $F(u^{(k)})$ has no terms of total degree less than k (with respect to I).

The linear ideal-adic Newton's iteration (6.34) and (6.35) proceeds by computing in iteration step k *all* ideal-adic terms in the solution $\tilde{u}$ which have total degree k (with respect to I). It is possible to define a *quadratic* ideal-adic Newton's iteration just as in the p-adic case. Such an iteration would produce an order 2^k ideal-adic approximation $u^{(k+1)}$ in iteration step k. In other words, the quadratic iteration would compute in iteration step k all ideal-adic terms in the solution $\tilde{u}$ which have total degrees $2^{k-1}, 2^{k-1}+1, \ldots, 2^k - 1$. However as was noted in the p-adic case, the quadratic iteration entails a cost per iteration which is higher than that of linear iteration, so much so that in terms of overall efficiency the linear iteration has been found to been superior in many practical problems.

Example 6.4. Consider the problem of determining a polynomial $u(x,y,z) \in \mathbf{Z}_5[x,y,z]$ which is a square root of the polynomial

$$a(x,y,z) = x^4 + x^3y^2 - x^2y^4 + x^2yz + 2x^2z - 2x^2 - 2xy^3z + xy^2z$$
$$- xy^2 - y^2z^2 + yz^2 - yz + z^2 - 2z + 1 \in \mathbf{Z}_5[x,y,z]$$

(assuming that $a(x,y,z)$ is a perfect square). Then $u(x,y,z)$ can be expressed as the solution of the polynomial equation

$$F(u) = a(x,y,z) - u^2 = 0.$$

Choosing the ideal $\mathrm{I} = \langle y,z \rangle$ (i.e. choosing the evaluation points $y = 0$ and $z = 0$), the first-order ideal-adic approximation $u^{(1)} = u(x,0,0) \in \mathbf{Z}_5[x]$ must be a square root of $a(x,0,0)$ in $\mathbf{Z}_5[x]$. Now

$$a(x,0,0) = x^4 - 2x^2 + 1$$

which clearly has the square root

$$u^{(1)} = u(x,0,0) = x^2 - 1 \in \mathbf{Z}_5[x].$$

To apply the linear ideal-adic Newton's iteration, first note that

$$\phi_{\mathrm{I}}(F'(u^{(1)})) = \phi_{\mathrm{I}}(-2u^{(1)}) = -2x^2 + 2.$$

It is convenient to express $a(x,y,z)$ in its ideal-adic representation with respect to I, which is

$$a(x,y,z) = [(x^4-2x^2+1)] + [(2x^2-2)z] + [(x^3-x)y^2 + (x^2-1)yz + z^2]$$
$$+ [(x)y^2z + yz^2] + [(-x^2)y^4 + (-2x)y^3z - y^2z^2].$$

Now

$$\phi_{\mathrm{I}^2}(F(u^{(1)})) = \phi_{\mathrm{I}^2}(a(x,y,z) - (x^2-1)^2) = (2x^2-2)z \in \mathbf{Z}_5[x,y,z]\,/\,\mathrm{I}^2.$$

The first correction term is

$$\Delta u^{(1)} = u_2(x)y + u_3(x)z$$

where $u_2(x) = 0$ (because the corresponding term in $\phi_{\mathrm{I}^2}(F(u^{(1)}))$ is zero) and where

$$u_3(x) = -\frac{c_3(x)}{(-2x^2+2)} = -\frac{(2x^2-2)}{(-2x^2+2)} = 1 \in \mathbf{Z}_5[x].$$

Hence

$$u^{(2)} = u^{(1)} + \Delta u^{(1)} = (x^2-1) + z \in \mathbf{Z}_5[x,y,z] \,/\, \mathrm{I}^2.$$

For the next iteration, we have

$$\phi_{\mathrm{I}^3}(F(u^{(2)})) = \phi_{\mathrm{I}^3}(a(x,y,z) - [(x^2-1)+z]^2) = (x^3-x)y^2 + (x^2-1)yz$$

which lies in $\mathbf{Z}_5[x,y,z] \,/\, \mathrm{I}^3$. The new correction term is

$$\Delta u^{(2)} = u_{22}(x)y^2 + u_{23}(x)yz + u_{33}(x)z^2$$

where $u_{33}(x) = 0$ (because the corresponding term in $\phi_{\mathrm{I}^3}(F(u^{(2)}))$ is zero) and where

$$u_{22}(x) = -\frac{c_{22}(x)}{(-2x^2+2)} = -\frac{(x^3-x)}{(-2x^2+2)} = -2x \in \mathbf{Z}_5[x];$$

$$u_{23}(x) = -\frac{c_{23}(x)}{(-2x^2+2)} = -\frac{(x^2-1)}{(-2x^2+2)} = -2 \in \mathbf{Z}_5[x].$$

Hence

$$u^{(3)} = u^{(2)} + \Delta u^{(2)} = (x^2-1) + z + (-2x)y^2 + (-2)yz$$

a member of $\mathbf{Z}_5[x,y,z] \,/\, \mathrm{I}^3$. If we proceed to the next iteration we find that $F(u^{(3)}) = 0$ (in the domain $\mathbf{Z}_5[x,y,z]$) so we are finished. The desired square root of $a(x,y,z)$ is therefore

$$u(x,y,z) = u^{(3)} = x^2 - 2xy^2 - 2yz + z - 1 \in \mathbf{Z}_5[x,y,z].$$

•

A Homomorphism Diagram

Finally in this section, Figure 6.1 shows a homomorphism diagram for the case of solving a multivariate polynomial problem using the p-adic and ideal-adic Newton's iterations. This diagram should be compared with the diagram of Figure 5.1 where many image problems had to be constructed and solved rather than just one image problem. Note that in order to apply Newton's iteration it is assumed that the desired polynomial can be expressed as a solution of a polynomial equation $F(u) = 0$.

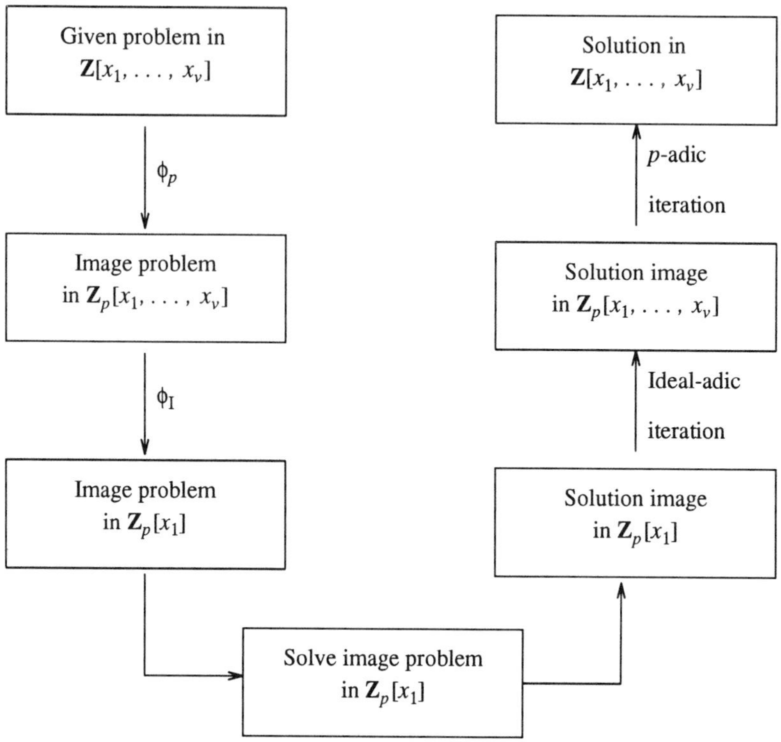

Figure 6.1. Homomorphism diagram for p-adic and ideal-adic Newton's iterations.

6.4. HENSEL'S LEMMA

Bivariate Newton's Iteration

In the preceding discussion of Newton's iteration for lifting an image polynomial $\phi_{I,p}(u) \in \mathbf{Z}_p[x_1]$ up to a desired polynomial $u \in \mathbf{Z}[x_1, \ldots, x_v]$ – i.e. for inverting a composite homomorphism

$$\phi_{I,p} : \mathbf{Z}[x_1, \ldots, x_v] \to \mathbf{Z}_p[x_1], \tag{6.36}$$

it was assumed that the polynomial u could be expressed as the solution of a polynomial equation

$$F(u) = 0 \tag{6.37}$$

for some $F(u) \in \mathbf{Z}[x_1, \ldots, x_v][u]$. However the most common applications of Newton's iteration for such a lifting process involve problems which cannot generally be expressed in the form (6.37), but rather can be expressed in the form

$$F(u,w) = 0 \tag{6.38}$$

for some bivariate polynomial $F(u,w) \in \mathbf{Z}[x_1, \ldots, x_v][u,w]$. An equation such as (6.38) will have a pair of solutions u and w so we will in fact be lifting two polynomials, not just one.

The fundamental problem which can be expressed in the form (6.38) is the polynomial factorization problem. Suppose we wish to find factors in the domain $\mathbf{Z}[x_1, \ldots, x_v]$ of a polynomial $a(x_1, \ldots, x_v) \in \mathbf{Z}[x_1, \ldots, x_v]$. By applying a composite homomorphism of the form (6.36), the factorization problem is reduced to a problem of factoring a univariate polynomial over the field $\mathbf{Z}_p$ (which as we see in Chapter 8 is a comparatively simple problem). Let $a_0(x_1)$ denote the image of $a(x_1, \ldots, x_v)$ in $\mathbf{Z}_p[x_1]$ and suppose we discover that $u_0(x_1)$ is a factor of $a_0(x_1)$ in the domain $\mathbf{Z}_p[x_1]$. Then we have the following relationship in the domain $\mathbf{Z}_p[x_1]$:

$$a_0(x_1) = u_0(x_1)\, w_0(x_1) \ \text{ where } \ w_0(x_1) = \frac{a_0(x_1)}{u_0(x_1)} \in \mathbf{Z}_p[x_1].$$

We therefore pose the problem of finding multivariate polynomials $u(x_1, \ldots, x_v)$, $w(x_1, \ldots, x_v) \in \mathbf{Z}[x_1, \ldots, x_v]$ which satisfy the bivariate polynomial equation

$$F(u,w) = a(x_1, \ldots, x_v) - u \cdot w = 0 \tag{6.39}$$

such that

$$\begin{cases} u(x_1, \ldots, x_v) \equiv u_0(x_1) \ (\text{mod} < \mathrm{I}, p>), \\ w(x_1, \ldots, x_v) \equiv w_0(x_1) \ (\text{mod} < \mathrm{I}, p>). \end{cases} \tag{6.40}$$

In other words, we wish to lift the factors $u_0(x_1), w_0(x_1) \in \mathbf{Z}_p[x_1]$ to factors $u(x_1, \ldots, x_v), w(x_1, \ldots, x_v) \in \mathbf{Z}[x_1, \ldots, x_v]$ by applying a form of Newton's iteration to the nonlinear equation (6.39). (Note that this process could be applied recursively to further factor the polynomials $u(x_1, \ldots, x_v)$ and $w(x_1, \ldots, x_v)$ in order to ultimately obtain the complete factorization of $a(x_1, \ldots, x_v)$ in the domain $\mathbf{Z}[x_1, \ldots, x_v]$.) Sufficient conditions for such a lifting process to be possible will be determined shortly. A detailed discussion of the polynomial factorization problem is given in Chapter 8.

Another problem which can be posed in the form (6.39) is the problem of computing the GCD of multivariate polynomials $a(x_1, \ldots, x_v)$, $b(x_1, \ldots, x_v) \in \mathbf{Z}[x_1, \ldots, x_v]$. Applying a composite homomorphism of the form (6.36) the problem is reduced to computing $\mathrm{GCD}(a_0(x_1), b_0(x_1))$ in the Euclidean domain $\mathbf{Z}_p[x_1]$, which can be easily accomplished by the basic Euclidean algorithm (Algorithm 2.1). Then if $u_0(x_1) = \mathrm{GCD}(a_0(x_1), b_0(x_1))$, we define the *cofactor* $w_0(x_1) = \dfrac{a_0(x_1)}{u_0(x_1)}$ and pose the problem of lifting the image polynomials $u_0(x_1), w_0(x_1) \in \mathbf{Z}_p[x_1]$ to multivariate polynomials

$$u(x_1, \ldots, x_v), w(x_1, \ldots, x_v) \in \mathbf{Z}[x_1, \ldots, x_v]$$

which satisfy (6.39) and (6.40). (Note that the polynomial $b(x_1, \ldots, x_v)$ could as well play the role of $a(x_1, \ldots, x_v)$ in this lifting process.) The problem of computing the GCD of polynomials by this method (and other methods) is discussed in more detail in Chapter 7.

In this section we discuss how, and under what conditions, Newton's iteration can be applied to solve the problem (6.39) and (6.40). Noting that (6.39) is a single nonlinear equation in two unknowns, we would expect from general mathematical principles that it would not have a unique solution without imposing additional conditions. Rather than imposing the additional conditions explicitly as a second equation of the form $G(u,w) = 0$, the additional conditions will appear more indirectly in the following development.

The general form of Newton's iteration for the bivariate polynomial equation

$$F(u,w) = 0$$

can be determined by applying Theorem 2.9. Suppose that we have a pair of approximations $u^{(k)}, w^{(k)}$ to the solution pair $\bar{u}, \bar{w}$ and that we wish to compute a pair of correction terms $\Delta u^{(k)}, \Delta w^{(k)}$. Theorem 2.9 yields the equation

$$F(u^{(k)} + \Delta u^{(k)}, w^{(k)} + \Delta w^{(k)}) = F(u^{(k)}, w^{(k)}) + F_u(u^{(k)}, w^{(k)})\, \Delta u^{(k)}$$
$$+ F_w(u^{(k)}, w^{(k)})\, \Delta w^{(k)} + E$$

where the term E involves higher-order expressions with respect to $\Delta u^{(k)}, \Delta w^{(k)}$. By arguments which can be formalized as before (or loosely speaking, setting the left hand side to zero and ignoring the higher-order term E), we find that the correction terms should be chosen to satisfy the following equation (modulo some ideal):

$$F_u(u^{(k)}, w^{(k)})\, \Delta u^{(k)} + F_w(u^{(k)}, w^{(k)})\, \Delta w^{(k)} = -F(u^{(k)}, w^{(k)}). \tag{6.41}$$

Thus we see that the basic computation to be performed in applying a step of Newton's iteration will be to solve the polynomial diophantine equation (6.41) which takes the form

$$A^{(k)} \Delta u^{(k)} + B^{(k)} \Delta w^{(k)} = C^{(k)}$$

where $A^{(k)}, B^{(k)}, C^{(k)}$ are given polynomials and $\Delta u^{(k)}, \Delta w^{(k)}$ are the unknown polynomials to be determined. Equation (6.41) will in general have either no solution or else a whole family of solutions. However Theorem 2.6 of Chapter 2 shows that under certain conditions the polynomial diophantine equation (6.41) has a unique solution.

From now on, we will specialize the development of Newton's iteration to the particular bivariate polynomial equation (6.39). As we have seen, the problem of polynomial factorization and also the problem of polynomial GCD computation can be posed in the particular form (6.39). Other problems may lead to different bivariate polynomial equations $F(u,w) = 0$ but the validity of Newton's iteration will depend on the particular problem. This is because of the need to introduce additional conditions which ensure the existence and uniqueness of a solution to the polynomial diophantine equation (6.41) which must be solved in each step of Newton's iteration.

Hensel's Lemma

Let us consider the univariate case of solving the problem (6.39) and (6.40). Thus we are given a polynomial $a(x) \in \mathbf{Z}[x]$ and a pair of factors $u_0(x), w_0(x) \in \mathbf{Z}_p[x]$ such that

$$a(x) \equiv u_0(x)\, w_0(x) \pmod p$$

and we wish to lift the factors $u_0(x), w_0(x)$ from the image domain $\mathbf{Z}_p[x]$ up to a pair of factors $u(x), w(x) \in \mathbf{Z}[x]$. In other words, we wish to invert the modular homomorphism

$$\phi_p : \mathbf{Z}[x] \to \mathbf{Z}_p[x]$$

by applying Newton's iteration to compute the solution $\bar{u} = u(x), \bar{w} = w(x)$ in $\mathbf{Z}[x]$ of the nonlinear equation

$$F(u, w) = a(x) - uw = 0 \tag{6.42}$$

such that

$$u(x) \equiv u_0(x) \pmod p, \quad w(x) \equiv w_0(x) \pmod p. \tag{6.43}$$

Writing the solution polynomials $\bar{u}$ and $\bar{w}$ in their polynomial p-adic representations

$$\begin{aligned} \bar{u} &= u_0(x) + u_1(x)p + \cdots + u_n(x)p^n\,; \\ \bar{w} &= w_0(x) + w_1(x)p + \cdots + w_n(x)p^n \end{aligned} \tag{6.44}$$

(where n must be large enough so that $\frac{1}{2}p^{n+1}$ bounds the magnitudes of all integer coefficients appearing in $a(x)$ and its factors $\bar{u}$ and $\bar{w}$), we wish to determine the polynomial p-adic coefficients $u_i(x), w_i(x) \in \mathbf{Z}_p[x]$ for $i = 1, 2, \ldots, n$. Let $u^{(k)}, w^{(k)}$ denote the order k p-adic approximations to $\bar{u}, \bar{w}$ given by the first k terms in (6.44) and let $\Delta u^{(k)} = u_k(x)p^k, \Delta w^{(k)} = w_k(x)p^k$. Note that $u^{(1)} = u_0(x)$ and $w^{(1)} = w_0(x)$. We find that the correction terms must satisfy the polynomial diophantine equation (6.41) modulo p^{k+1}, which for the particular nonlinear equation (6.42) takes the form

$$-w^{(k)}\Delta u^{(k)} - u^{(k)}\Delta w^{(k)} \equiv -\left(a(x) - u^{(k)}w^{(k)}\right) \pmod{p^{k+1}}.$$

Since $u^{(k)}w^{(k)}$ must be an order k p-adic approximation to $a(x)$ we can divide through by p^k, and also removing the negative signs we get

$$w^{(k)}u_k(x) + u^{(k)}w_k(x) \equiv \frac{a(x) - u^{(k)}w^{(k)}}{p^k} \pmod p.$$

Now we may apply the modular homomorphism ϕ_p to the left and right (because this is a congruence modulo p) and, noting that $\phi_p(w^{(k)}) = w_0(x)$ and $\phi_p(u^{(k)}) = u_0(x)$, we get the following polynomial diophantine equation to solve in the domain $\mathbf{Z}_p[x]$:

$$w_0(x)\, u_k(x) + u_0(x) w_k(x) = \phi_p\left(\frac{a(x) - u^{(k)}w^{(k)}}{p^k}\right).$$

Since $\mathbf{Z}_p[x]$ is a Euclidean domain (we choose p to be a prime integer), Theorem 2.6 shows that if $u_0(x), w_0(x) \in \mathbf{Z}_p[x]$ are relatively prime then we can find unique polynomials

$\sigma(x), \tau(x) \in \mathbf{Z}_p[x]$ such that

$$\sigma(x)u_0(x) + \tau(x)w_0(x) = \phi_p\left(\frac{a(x) - u^{(k)}w^{(k)}}{p^k}\right)$$

and

$$\deg(\sigma(x)) < \deg(w_0(x)).$$

We then define $u^{(k+1)} = u^{(k)} + \tau(x)p^k$, $w^{(k+1)} = w^{(k)} + \sigma(x)p^k$ and we claim that these are order k+1 p-adic approximations to the solutions $\bar{u}, \bar{w}$ respectively.

The following theorem formally proves the validity of the above method. This theorem is a standard result in algebra known as Hensel's lemma and it dates back to the early 1900's. The proof of Hensel's lemma is a constructive proof which follows naturally from the above development and this process is referred to as the *Hensel construction*.

Theorem 6.2 (Hensel's Lemma). Let p be a prime in $\mathbf{Z}$ and let $a(x) \in \mathbf{Z}[x]$ be a given polynomial over the integers. Let $u^{(1)}(x), w^{(1)}(x) \in \mathbf{Z}_p[x]$ be two relatively prime polynomials over the field $\mathbf{Z}_p$ such that

$$a(x) \equiv u^{(1)}(x)w^{(1)}(x) \pmod{p}.$$

Then for any integer $k \geq 1$ there exist polynomials $u^{(k)}(x), w^{(k)}(x) \in \mathbf{Z}_{p^k}[x]$ such that

$$a(x) \equiv u^{(k)}(x)w^{(k)}(x) \pmod{p^k} \tag{6.45}$$

and

$$u^{(k)}(x) \equiv u^{(1)}(x) \pmod{p}, \quad w^{(k)}(x) \equiv w^{(1)}(x) \pmod{p}. \tag{6.46}$$

Proof: The proof is by induction on k. The case $k = 1$ is given. Assume for $k \geq 1$ that we have $u^{(k)}(x), w^{(k)}(x) \in \mathbf{Z}_{p^k}[x]$ satisfying (6.45) and (6.46). Define

$$c^{(k)}(x) = \phi_p\left(\frac{a(x) - u^{(k)}(x)w^{(k)}(x)}{p^k}\right) \tag{6.47}$$

where all operations are performed in the domain $\mathbf{Z}_{p^{k+1}}[x]$ before applying ϕ_p. Since $u^{(1)}(x), w^{(1)}(x) \in \mathbf{Z}_p[x]$ are relatively prime, by Theorem 2.6 we can find unique polynomials $\sigma^{(k)}(x), \tau^{(k)}(x) \in \mathbf{Z}_p[x]$ such that

$$\sigma^{(k)}(x)u^{(1)}(x) + \tau^{(k)}(x)w^{(1)}(x) \equiv c^{(k)}(x) \pmod{p} \tag{6.48}$$

and

$$\deg(\sigma^{(k)}(x)) < \deg(w^{(1)}(x)). \tag{6.49}$$

Then by defining

$$u^{(k+1)}(x) = u^{(k)}(x) + \tau^{(k)}(x)p^k, \quad w^{(k+1)}(x) = w^{(k)}(x) + \sigma^{(k)}(x)p^k \tag{6.50}$$

we have by performing multiplication modulo p^{k+1}:

$$u^{(k+1)}(x)w^{(k+1)}(x) \equiv u^{(k)}(x)w^{(k)}(x) + (\sigma^{(k)}(x)u^{(1)}(x)+\tau^{(k)}(x)w^{(1)}(x))p^k \pmod{p^{k+1}}$$

$$\equiv u^{(k)}(x)w^{(k)}(x) + c^{(k)}(x)p^k \pmod{p^{k+1}}, \qquad \text{by (6.48)}$$

$$\equiv a(x) \pmod{p^{k+1}}, \qquad \text{by (6.47)}.$$

Thus (6.45) holds for k+1. Also, from (6.50) it is clear that

$$u^{(k+1)}(x) \equiv u^{(k)}(x) \pmod{p}, \quad w^{(k+1)}(x) \equiv w^{(k)}(x) \pmod{p}$$

and therefore since (6.46) holds for k it also holds for k+1.

●

Corollary (Uniqueness of the Hensel Construction). In Theorem 6.2, if the given polynomial $a(x) \in \mathbf{Z}[x]$ is monic and correspondingly if the relatively prime factors $u^{(1)}(x), w^{(1)}(x) \in \mathbf{Z}_p[x]$ are chosen to be monic, then for any integer $k \geq 1$ conditions (6.45) and (6.46) uniquely determine the monic polynomial factors $u^{(k)}(x), w^{(k)}(x) \in \mathbf{Z}_{p^k}[x]$.

Proof: The proof is again by induction on k. For the case k=1, the given polynomials $u^{(1)}(x), w^{(1)}(x)$ are clearly the unique monic polynomials in $\mathbf{Z}_p[x]$ which satisfy conditions (6.45) and (6.46). For the induction assumption, assume for some $k \geq 1$ that the uniqueness of the monic polynomials $u^{(k)}(x), w^{(k)}(x) \in \mathbf{Z}_{p^k}[x]$ satisfying (6.45) and (6.46) has been determined. Then we must prove the uniqueness of the monic polynomials $u^{(k+1)}(x), w^{(k+1)}(x) \in \mathbf{Z}_{p^{k+1}}[x]$ satisfying the conditions

$$a(x) \equiv u^{(k+1)}(x)w^{(k+1)}(x) \pmod{p^{k+1}} \tag{6.51}$$

and

$$u^{(k+1)}(x) \equiv u^{(1)}(x) \pmod{p}, \quad w^{(k+1)}(x) \equiv w^{(1)}(x) \pmod{p}. \tag{6.52}$$

Condition (6.51) implies, in particular, that

$$a(x) \equiv u^{(k+1)}(x)w^{(k+1)}(x) \pmod{p^k}$$

which together with (6.52) yields, by the induction assumption,

$$u^{(k+1)}(x) \equiv u^{(k)}(x) \pmod{p^k}; \quad w^{(k+1)}(x) \equiv w^{(k)}(x) \pmod{p^k}.$$

We may therefore write

$$u^{(k+1)}(x) = u^{(k)}(x) + \tau(x)p^k, \; w^{(k+1)}(x) = w^{(k)}(x) + \sigma(x)p^k \tag{6.53}$$

for some polynomials $\sigma(x), \tau(x) \in \mathbf{Z}_p[x]$ and it remains to prove the uniqueness of $\sigma(x)$ and $\tau(x)$.

Since $a(x), u^{(1)}(x)$, and $w^{(1)}(x)$ are given to be monic, it follows that for any $k \geq 1$ the polynomials $\sigma(x)$ and $\tau(x)$ appearing in (6.53) must satisfy

$$\deg(\sigma(x)) < \deg(w^{(1)}(x)) \text{ and } \deg(\tau(x)) < \deg(u^{(1)}(x)) \tag{6.54}$$

(i.e. $u^{(k+1)}(x)$ and $w^{(k+1)}(x)$ must always have the same leading terms as $u^{(1)}(x)$ and $w^{(1)}(x)$, respectively). Now by multiplying out the two polynomials expressed in (6.53) and using (6.51), we get (performing the multiplication modulo p^{k+1}):

$$a(x) \equiv u^{(k)}(x)w^{(k)}(x) + (\sigma(x)u^{(1)}(x) + \tau(x)w^{(1)}(x))p^k \pmod{p^{k+1}}$$

which can be expressed in the form

$$\sigma(x)u^{(1)}(x) + \tau(x)w^{(1)}(x) \equiv \frac{a(x) - u^{(k)}(x)w^{(k)}(x)}{p^k} \pmod{p}. \tag{6.55}$$

By Theorem 2.6, the polynomials $\sigma(x), \tau(x) \in \mathbf{Z}_p[x]$ satisfying (6.55) and (6.54) are unique. ●

6.5. THE UNIVARIATE HENSEL LIFTING ALGORITHM

Description of the Algorithm

The Hensel construction of Theorem 6.2 is based on a linear p-adic Newton's iteration. Zassenhaus [9] in 1969 was the first to propose the application of Hensel's lemma to the problem of polynomial factorization over the integers and he proposed the use of a quadratic p-adic Newton's iteration. This quadratic iteration is usually referred to as the *Zassenhaus construction* and it computes a sequence of factors modulo p^{2^k}, for $k = 1, 2, 3, \ldots$. However as we noted in Section 6.3, a quadratic iteration is not necessarily more efficient than a linear iteration because the added complexity of each iteration step in the quadratic iteration may outweigh the advantage of fewer iteration steps. For example, in each iteration step of the quadratic Zassenhaus construction one must solve a polynomial diophantine equation of the form

$$\sigma^{(k)}(x)u^{(k)}(x) + \tau^{(k)}(x)w^{(k)}(x) \equiv c^{(k)}(x) \pmod{p^{2^{k-1}}} \tag{6.56}$$

for $\sigma^{(k)}(x), \tau^{(k)}(x) \in \mathbf{Z}_{p^{2^{k-1}}}[x]$. The corresponding computation in the linear Hensel construction is to solve the same polynomial diophantine equation modulo p for $\sigma^{(k)}(x), \tau^{(k)}(x) \in \mathbf{Z}_p[x]$. The latter computation is simpler because it is performed in the smaller domain $\mathbf{Z}_p[x]$ and another level of efficiency arises because the $u^{(k)}(x)$ and $w^{(k)}(x)$ in (6.56) can be replaced by the fixed polynomials $u^{(1)}(x)$ and $w^{(1)}(x)$ in the linear Hensel case. A detailed comparison of these two p-adic constructions was carried out by Miola and Yun [5] in 1974 and their analysis showed that the computational cost of the quadratic Zassenhaus construction is higher than that of the linear Hensel construction for achieving the same p-adic order of approximation. Therefore we will not present the details of the quadratic Zassenhaus construction, and instead we leave it for the exercises.

The basic algorithm for lifting a factorization in $\mathbf{Z}_p[x]$ up to a factorization in $\mathbf{Z}[x]$ is presented as Algorithm 6.1. In the monic case Algorithm 6.1 corresponds precisely to the Hensel construction presented in the proof of Hensel's lemma, since by the corollary to Theorem 6.2 the factors at each step of the lifting process are uniquely determined in the monic case. However in the non-monic case the nonuniqueness of the factors modulo p^k leads to the "leading coefficient problem" to be discussed shortly, and as we shall see this accounts for the additional conditions and the additional operations appearing in Algorithm 6.1. For the moment, Algorithm 6.1 may be understood for the monic case if we simply ignore the stated conditions (other than the conditions appearing in Hensel's lemma), ignore step 1, ignore the "replace_lc" operation in step 3 (using instead the initialization $u(x) \leftarrow u^{(1)}(x)$; $w(x) \leftarrow w^{(1)}(x)$), and note that no adjustment of $u(x)$ and $w(x)$ is required in step 5 for the monic case.

Example 6.5. Consider the problem of factoring the following monic polynomial over the integers:

$$a(x) = x^3 + 10x^2 - 432x + 5040 \in \mathbf{Z}[x].$$

Choosing $p = 5$ and applying the modular homomorphism ϕ_5 to $a(x)$ yields

$$\phi_5(a(x)) = x^3 - 2x \in \mathbf{Z}_5[x].$$

Algorithm 6.1. Univariate Hensel Lifting Algorithm.

procedure UnivariateHensel($a, p, u^{(1)}, w^{(1)}, B, \gamma$)

```
# INPUT:
#  (1) A primitive polynomial a(x) ∈ Z[x].
#  (2) A prime integer p which does not divide lcoeff(a(x)).
#  (3) Two relatively prime polynomials u^(1)(x), w^(1)(x) ∈ Z_p[x] such that
#        a(x) ≡ u^(1)(x) w^(1)(x) (mod p).
#  (4) An integer B which bounds the magnitudes of all integer coefficients
#      appearing in a(x) and in any of its possible factors with degrees
#      not exceeding max{deg(u^(1)(x)), deg(w^(1)(x))}.
#  (5) Optionally, an integer γ ∈ Z which is known to be a multiple of
#      lcoeff(u(x)), where u(x) (see OUTPUT below) is one of the factors of
#      a(x) in Z[x] to be computed.

# OUTPUT:
#   (1) If there exist polynomials u(x), w(x) ∈ Z[x] such that
#         a(x) = u(x)w(x) ∈ Z[x]
#       and
#         n(u(x)) ≡ n(u^(1)(x)) (mod p), n(w(x)) ≡ n(w^(1)(x)) (mod p)
#       where n denotes the normalization "make the polynomial monic as an
#       element of the domain Z_p[x]", then u(x) and w(x) will be computed.
#   (2) Otherwise, the value returned will signal "no such factorization".
```

Algorithm 6.1 (continued). Univariate Hensel Lifting Algorithm.

```
# 1. Define new polynomial and its modulo p factors
α ← lcoeff(a(x))
if γ is undefined then γ ← α
a(x) ← γ · a(x)
u^(1)(x) ← φ_p(γ · n(u^(1)(x)));  w^(1)(x) ← φ_p(α · n(w^(1)(x)))

# 2. Apply extended Euclidean algorithm to u^(1)(x), w^(1)(x) ∈ Z_p[x]
s(x), t(x) ← polynomials in Z_p[x] computed by Algorithm 2.2 such that
        s(x) u^(1)(x) + t(x) w^(1)(x) ≡ 1 (mod p)

# 3. Initialization for the iteration
u(x) ← replace_lc(u^(1)(x), γ);  w(x) ← replace_lc(w^(1)(x), α)
e(x) ← a(x) − u(x) · w(x); modulus ← p

# 4. Iterate until either the factorization in Z[x] is obtained or
#    else the bound on modulus is reached
while e(x) ≠ 0 and modulus < 2 · B · γ do {
    # 4.1. Solve in the domain Z_p[x] the polynomial equation
    #      σ(x) u^(1)(x) + τ(x) w^(1)(x) ≡ c(x) (mod p)
    #      where c(x) = e(x)/modulus

    c(x) ← e(x)/modulus;  σ̃(x) ← φ_p(s(x) · c(x));  τ̃(x) ← φ_p(t(x) · c(x))
    q(x), r(x) ← polynomials in Z_p[x] such that
        σ̃(x) = w^(1)(x) q(x) + r(x) ∈ Z_p[x]
    σ(x) ← r(x);  τ(x) ← φ_p(τ̃(x) + q(x) · u^(1)(x))

    # 4.2. Update the factors and compute the error
    u(x) ← u(x) + τ(x) · modulus;  w(x) ← w(x) + σ(x) · modulus
    e(x) ← a(x) − u(x) · w(x); modulus ← modulus · p }

# 5. Check termination status
if e(x) = 0 then {
    # Factorization obtained – remove contents
    δ ← cont(u(x)); u(x) ← u(x) / δ;  w(x) ← w(x) / (γ / δ)
    # Note: a(x) ← a(x)/γ would restore a(x) to its input value }
    return( u(x), w(x) )
else return( no such factorization exists )
end
```

The unique unit normal (i.e. monic) factorization in $\mathbf{Z}_5[x]$ of this polynomial is

$$\phi_5(a(x)) = x \cdot (x^2 - 2) \in \mathbf{Z}_5[x].$$

We therefore define

$$u^{(1)}(x) = x, \quad w^{(1)}(x) = x^2 - 2$$

and since $u^{(1)}(x)$ and $w^{(1)}(x)$ are relatively prime in $\mathbf{Z}_5[x]$, the Hensel construction may be applied.

Applying Algorithm 6.1 in the form noted above for the monic case, we first apply in step 2 the extended Euclidean algorithm which yields

$$s(x) = -2x, \quad t(x) = 2.$$

The initializations in step 3 yield

$$u(x) = x, \quad w(x) = x^2 - 2, \quad e(x) = 10x^2 - 430x + 5040,$$

and

$$modulus = 5.$$

Step 4 then applies the Hensel construction precisely as outlined in the proof of Hensel's lemma. (For now we are ignoring the second termination condition of the **while**-loop.) The sequence of values computed for $\sigma(x), \tau(x), u(x), w(x)$, and $e(x)$ in step 4 is as follows.

End of iter. no.	$\sigma(x)$	$\tau(x)$	$u(x)$	$w(x)$	$e(x)$
0	–	–	x	$x^2 - 2$	$10x^2 - 430x + 5040$
1	$x - 1$	1	$x + 5$	$x^2 + 5x - 7$	$-450x + 5075$
2	$-x + 2$	1	$x + 30$	$x^2 - 20x + 43$	$125x + 3750$
3	1	0	$x + 30$	$x^2 - 20x + 168$	0

Note that at the end of each iteration step k, $e(x)$ is exactly divisible by $modulus = 5^{k+1}$ as required at the beginning of the next iteration. The iteration terminates with $u(x) = x + 30$ and $w(x) = x^2 - 20x + 168$. We therefore have the factorization over the integers

$$x^3 + 10x^2 - 432x + 5040 = (x + 30)(x^2 - 20x + 168).$$

•

Example 6.6. In this example we shall see that the Hensel construction may apply even when the given polynomial cannot be factored over the integers. Consider the monic polynomial

$$a(x) = x^4 + 1 \in \mathbf{Z}[x]$$

which is irreducible over the integers. Choosing $p = 5$ and applying the modular homomorphism ϕ_5 to $a(x)$ yields $\phi_5(a(x)) = x^4 + 1$. The unique unit normal factorization in $\mathbf{Z}_5[x]$ of this polynomial is

$$x^4 + 1 = (x^2 + 2)(x^2 - 2) \in \mathbf{Z}_5[x].$$

Since the polynomials $u^{(1)}(x) = x^2 + 2$ and $w^{(1)}(x) = x^2 - 2$ are relatively prime in $\mathbf{Z}_5[x]$, the Hensel construction may be applied. In this case we get an infinite sequence of factors

$$a(x) \equiv u^{(k)}(x) w^{(k)}(x) \pmod{p^k}$$

for $k = 1, 2, 3, \ldots$.

If we apply Algorithm 6.1 to this monic case as in Example 6.5, the result of step 2 is

$$s(x) = -1; \;\; t(x) = 1$$

and the initializations in step 3 yield

$$u(x) = x^2 + 2; \;\; w(x) = x^2 - 2;$$

$$e(x) = 5;$$

$$\mathit{modulus} = 5.$$

If we allow the **while**-loop in step 4 to proceed through four iterations (again we are ignoring the second termination condition of the **while**-loop), the sequence of values computed for $\sigma(x), \tau(x), u(x), w(x)$, and $e(x)$ is as follows.

End of iteration no.	$\sigma(x)$	$\tau(x)$	$u(x)$	$w(x)$	$e(x)$
0	–	–	$x^2 + 2$	$x^2 - 2$	5
1	-1	1	$x^2 + 7$	$x^2 - 7$	50
2	-2	2	$x^2 + 57$	$x^2 - 57$	3250
3	-1	1	$x^2 + 182$	$x^2 - 182$	33125
4	2	-2	$x^2 - 1068$	$x^2 + 1068$	1140625

These iterations could be continued indefinitely yielding an infinite sequence of factors satisfying Hensel's lemma. Note that at the end of iteration step k we always have

$$u(x) w(x) \equiv x^4 + 1 \pmod{5^{k+1}}$$

as claimed in Hensel's lemma. However we will never obtain quadratic factors $u(x), w(x) \in \mathbf{Z}[x]$ such that

$$u(x) w(x) = x^4 + 1 \in \mathbf{Z}[x].$$

We remark that our polynomial $a(x)$ in this case factors into two quadratic factors in $\mathbf{Z}_p[x]$ for every prime p (see Exercise 8.12 in Chapter 8). Thus choosing a prime different than 5 will not change this example.

●

The Leading Coefficient Problem

The Hensel construction provides a method for lifting a factorization modulo p up to a factorization modulo p^l for any integer $l \geq 1$. Example 6.6 shows that this construction does not necessarily lead to a factorization over the integers. However if the monic polynomial $a(x) \in \mathbf{Z}[x]$ has the modulo p factorization

$$a(x) \equiv u^{(1)}(x)w^{(1)}(x) \pmod{p}$$

where $u^{(1)}(x), w^{(1)}(x) \in \mathbf{Z}_p[x]$ are relatively prime monic polynomials and if there exists a factorization over the integers

$$a(x) = u(x)w(x) \in \mathbf{Z}[x] \tag{6.57}$$

such that

$$u(x) \equiv u^{(1)}(x) \pmod{p}, \;\; w(x) \equiv w^{(1)}(x) \pmod{p} \tag{6.58}$$

then the Hensel construction must obtain this factorization. Specifically, let l be large enough so that $p^l > 2B$ where B is an integer which bounds the magnitudes of all integer coefficients appearing in $a(x)$ and in any of its possible factors with the particular degrees $\deg(u^{(1)}(x))$ and $\deg(w^{(1)}(x))$. (For a discussion of techniques for computing such a bound B see Mignotte [4].) Then the Hensel construction may be applied to compute monic polynomials $u^{(l)}(x), w^{(l)}(x) \in \mathbf{Z}_{p^l}[x]$ satisfying

$$a(x) \equiv u^{(l)}(x)w^{(l)}(x) \pmod{p^l} \tag{6.59}$$

and

$$u^{(l)}(x) \equiv u^{(1)}(x) \pmod{p}, \;\; w^{(l)}(x) \equiv w^{(1)}(x) \pmod{p} \tag{6.60}$$

and by the corollary to Theorem 6.2 the factors $u^{(l)}(x), w^{(l)}(x) \in \mathbf{Z}_{p^l}[x]$ are uniquely determined by conditions (6.59)-(6.60). Now if there exists a factorization (6.57) satisfying (6.58) then another such monic factorization in $\mathbf{Z}_{p^l}[x]$ is provided by $\phi_{p^l}(u(x))$ and $\phi_{p^l}(w(x))$ and hence by uniqueness

$$u^{(l)}(x) = \phi_{p^l}(u(x)), \;\; w^{(l)}(x) = \phi_{p^l}(w(x)).$$

But, since $p^l > 2B$, we have $\phi_{p^l}(u(x)) = u(x)$ and $\phi_{p^l}(w(x)) = w(x)$, which proves that $u^{(l)}(x)$ and $w^{(l)}(x)$ are the desired factors over the integers.

The above discussion shows that in the monic case, the Hensel construction may be halted when $p^l > 2B$ at which point either $u^{(l)}(x)w^{(l)}(x) = a(x)$ over the integers or else there exists no factorization satisfying (6.57)-(6.58). The second termination condition of the **while**-loop in step 4 of Algorithm 6.1 is, in the monic case, precisely this condition. Note that since the bound B given to Algorithm 6.1 will invariably be very pessimistic, the first termination condition of the **while**-loop is required to avoid extra costly iterations after a factorization has been discovered.

In the non-monic case the situation is not quite so simple. The Hensel construction requires (in step 4.1 of Algorithm 6.1) the solution $\sigma(x), \tau(x) \in \mathbf{Z}_p[x]$ of the polynomial diophantine equation

$$\sigma(x)u^{(1)}(x) + \tau(x)w^{(1)}(x) \equiv c(x) \pmod{p}. \tag{6.61}$$

The solution of this equation is not uniquely determined but uniqueness is (somewhat artificially) imposed by requiring that the solution satisfy

$$\deg(\sigma(x)) < \deg(w^{(1)}(x)) \tag{6.62}$$

(see Theorem 2.6). Noting that the update formulas (in step 4.2 of Algorithm 6.1) are then

$$u(x) \leftarrow u(x) + \tau(x)p; \ \ w(x) \leftarrow w(x) + \sigma(x)p,$$

it is clear that the degree constraint (6.62) implies that the leading coefficient of $w(x)$ is never updated. In the monic case, this is exactly what we want. Moreover, since

$$c(x) = \frac{a(x) - u(x)w(x)}{modulus} \tag{6.63}$$

it follows in the monic case that

$$\deg(c(x)) < \deg(a(x)) = \deg(u^{(1)}(x)) + \deg(w^{(1)}(x))$$

and therefore the solution of (6.61) also satisfies, by Theorem 2.6,

$$\deg(\tau(x)) < \deg(u^{(1)}(x)).$$

It follows that the leading coefficient of $u(x)$ also is never updated in the monic case. Turning to the non-monic case, we must first assume that the chosen prime p does not divide the leading coefficient of $a(x)$. With this assumption we are assured that $u^{(1)}(x)$ and $w^{(1)}(x)$ have "correct" degrees in the sense that $\deg(u^{(1)}(x)) + \deg(w^{(1)}(x)) = \deg(a(x))$. Then we have from (6.63) that

$$\deg(c(x)) \le \deg(a(x)) = \deg(u^{(1)}(x)) + \deg(w^{(1)}(x))$$

from which it follows exactly as in Theorem 2.6 that

$$\deg(\tau(x)) \le \deg(u^{(1)}(x)). \tag{6.64}$$

The degree constraint (6.64) allows the leading coefficient of $u(x)$ to be updated since, unlike (6.62), the inequality here is not strict inequality. Now at the end of each iteration step k, we have the relationship

$$a(x) \equiv u(x)w(x) \pmod{p^{k+1}}$$

and therefore since (6.62) forces the leading coefficient of $w(x)$ to remain unchanged, all of the updating required by the leading coefficient of $a(x)$ is forced onto the leading coefficient of $u(x)$. (Note that any unit in the ring $\mathbf{Z}_{p^k}$ can be multiplied into one factor and its inverse multiplied into the other factor without changing the given relationship.) This is referred to as the *leading coefficient problem* and it can cause the factors in the Hensel construction to never yield a factorization over the integers even when such a factorization over the integers exists. The following example will clarify this leading coefficient problem.

Example 6.7. Consider the problem of factoring the following polynomial over the integers:

$$a(x) = 12x^3 + 10x^2 - 36x + 35 \in \mathbf{Z}[x].$$

In order to understand the leading coefficient problem which arises we start by presenting the correct answer; namely, the complete unit normal factorization of $a(x)$ over the integers is

$$a(x) = u(x)w(x) = (2x+5)(6x^2 - 10x + 7) \in \mathbf{Z}[x].$$

Let us attempt to solve this factorization problem by the method used in Example 6.5. Choosing $p = 5$ and applying the modular homomorphism ϕ_5 to $a(x)$ yields

$$\phi_5(a(x)) = 2x^3 - x \in \mathbf{Z}_5[x].$$

The unique unit normal factorization in $\mathbf{Z}_5[x]$ of this polynomial is

$$\phi_5(a(x)) = 2(x)(x^2+2) \in \mathbf{Z}_5[x]$$

where 2 is a unit in $\mathbf{Z}_5[x]$. Now in order to choose the initial factors $u^{(1)}(x), w^{(1)}(x) \in \mathbf{Z}_5[x]$ to be lifted, we must attach the unit 2 either to the factor x or else to the factor x^2+2. This is precisely the problem of non-uniqueness which exists at each stage of the Hensel construction. At this initial stage we have

$$\phi_5(a(x)) = (2x)(x^2+2) = (x)(2x^2-1) \in \mathbf{Z}_5[x].$$

Since in this problem we are given the answer, we can see that the "correct" images under ϕ_5 of $u(x), w(x) \in \mathbf{Z}[x]$ are

$$u^{(1)}(x) = 2x;\ \ w^{(1)}(x) = x^2 + 2.$$

However it is important to note that this "correct" attachment of units to factors is irrelevant. The other choice for $u^{(1)}(x)$ and $w^{(1)}(x)$ in this example would be equally valid and would lead to the same "leading coefficient problem" which will arise from the above choice.

The polynomials $u^{(1)}(x)$ and $w^{(1)}(x)$ defined above are relatively prime in $\mathbf{Z}_5[x]$ so the Hensel construction may be applied. Let us apply Algorithm 6.1 in the form used for the monic case of Example 6.5 (i.e. the unmodified Hensel construction as presented in the proof of Hensel's lemma). In step 2 of Algorithm 6.1, the extended Euclidean algorithm applied to $u^{(1)}(x)$ and $w^{(1)}(x)$ yields

$$s(x) = x,\ \ t(x) = -2.$$

The initializations in step 3 yield

$$u(x) = 2x,\ \ w(x) = x^2 + 2;$$

$$e(x) = 10x^3 + 10x^2 - 40x + 35;\quad \textit{modulus} = 5.$$

If we allow the **while**-loop in step 4 to proceed through four iterations (again we are ignoring the second termination condition of the **while**-loop), the sequence of values computed for $\sigma(x), \tau(x), u(x), w(x)$, and $e(x)$ is as follows.

$\sigma(x)$	$\tau(x)$	$u(x)$	$w(x)$	$e(x)$
–	–	$2x$	x^2+2	$10x^3+10x^2-40x+35$
$-2x-1$	$2x+1$	$12x+5$	$x^2-10x-3$	$125x^2+50x+50$
$2x+1$	1	$12x+30$	$x^2+40x+22$	$-500x^2-1500x-625$
$-2x-1$	0	$12x+30$	$x^2-210x-103$	$2500x^2+7500x+3125$
$2x+1$	0	$12x+30$	$x^2+1040x+522$	$-12500x^2-37500x-15625$

These iterations could be continued indefinitely yielding an infinite sequence of factors satisfying Hensel's lemma – i.e. at the end of iteration step k we always have

$$u(x)w(x) \equiv a(x) \pmod{5^{k+1}}.$$

However these factors will never satisfy the desired relationship

$$u(x)w(x) = a(x) \in \mathbf{Z}[x]$$

because $w(x)$ is always monic and there does not exist a monic quadratic factor of $a(x)$ over the integers.

●

It is clear in the above example that the leading coefficient of $a(x)$ is completely forced onto the factor $u(x)$ since $w^{(1)}(x)$, and hence each updated $w(x)$, is monic. Noting the correct factorization of $a(x)$ over the integers, we see that the leading coefficient of $a(x)$ needs to be split in the form $12 = 2 \times 6$ with the factor 2 appearing as the leading coefficient of $u(x)$ and the factor 6 appearing as the leading coefficient of $w(x)$. Algorithm 6.1 contains additional statements which will force the leading coefficients to be correct and we now turn to an explanation of these additional operations.

6.6. SPECIAL TECHNIQUES FOR THE NON-MONIC CASE

Relationship of Computed Factors to True Factors

The first step towards solving the leading coefficient problem is the realization that the factors computed by the Hensel construction are "almost" the correct factors over the integers, in the following sense. Let l be large enough so that $p^l > 2B$ where B bounds the magnitudes of all integer coefficients appearing in $a(x)$ and in its factors. Then Theorem 6.4 below proves that the factors $u^{(l)}(x)$ and $w^{(l)}(x)$ computed by the Hensel construction such that

$$u^{(l)}(x)w^{(l)}(x) \equiv a(x) \pmod{p^l}$$

differ from the true factors over the integers only by a unit in the ring $\mathbf{Z}_{p^l}[x]$ (if an appropriate factorization over the integers exists). In Example 6.7 of the preceding section we see by inspection of $a(x)$ and its known factors that $B = 36$ and therefore $l = 3$ is sufficient, so the factors

$$u^{(3)}(x) = 12x + 30; \;\; w^{(3)}(x) = x^2 + 40x + 22$$

computed in iteration step $k = 2$ must be the correct factors apart from units in the ring $\mathbf{Z}_{125}[x]$. Note that

$$u^{(3)}(x)w^{(3)}(x) = 12x^3 + 510x^2 + 1464x + 660 \in \mathbf{Z}[x]$$

so that $u^{(3)}(x)w^{(3)}(x) \neq a(x)$ but

$$u^{(3)}(x)w^{(3)}(x) \equiv a(x) \pmod{5^3}.$$

Now in this example it is known that the correct leading coefficient of $w(x)$ is 6, so we multiply $w^{(3)}(x)$ by 6 in the domain $\mathbf{Z}_{125}[x]$ and correspondingly multiply $u^{(3)}(x)$ by $6^{-1} \in \mathbf{Z}_{125}[x]$ so as to maintain the relationship

$$[6^{-1}u^{(3)}(x)][6w^{(3)}(x)] \equiv a(x) \pmod{5^3}.$$

Since $6^{-1} = 21 \in \mathbf{Z}_{125}[x]$ we obtain the factors

$$u(x) = 21u^{(3)}(x) = 2x + 5 \in \mathbf{Z}_{125}[x];$$

$$w(x) = 6w^{(3)}(x) = 6x^2 - 10x + 7 \in \mathbf{Z}_{125}[x].$$

Then $u(x)w(x) = a(x)$ in the domain $\mathbf{Z}[x]$ and the desired factors have been obtained.

The above example makes use of the knowledge that 6 is the correct leading coefficient of $w(x)$ and it would seem that such knowledge would not be available in the general case. However we will shortly describe a general method which, by slightly altering the original problem, leads to a situation in which the correct leading coefficients of both factors will always be known. For the moment we must prove the result that $u^{(l)}(x)$ and $w^{(l)}(x)$ are associates in the ring $\mathbf{Z}_{p^l}[x]$ of the true factors over the integers. To this end, recall that the units in a polynomial ring are precisely the units in its coefficient ring and therefore we must understand which elements in a ring of the form $\mathbf{Z}_{p^k}$ are units. Unlike the field $\mathbf{Z}_p$ in which every nonzero element is a unit, the ring $\mathbf{Z}_{p^k}$ (for $k > 1$) has some nonzero elements which fail to have multiplicative inverses (e.g. the element $p \in \mathbf{Z}_{p^k}$ is not a unit). The following theorem proves that most of the elements in the ring $\mathbf{Z}_{p^k}$ are units and identifies those elements which are not units.

Theorem 6.3. Let p be a prime integer and let k be any positive integer. An element $a \in \mathbf{Z}_{p^k}$ is a unit in $\mathbf{Z}_{p^k}$ if and only if p does not divide a (in the integral domain $\mathbf{Z}$).

Proof: We first claim that the integer p is not a unit in $\mathbf{Z}_{p^k}$. For $k = 1$, p is the zero element in $\mathbf{Z}_p$ so the claim is obvious. For $k > 1$, if p is a unit in $\mathbf{Z}_{p^k}$ then there exist integers p^{-1} and c such that in the domain $\mathbf{Z}$

$$p \cdot p^{-1} = c \cdot p^k + 1$$

whence

$$p(p^{-1} - c \cdot p^{k-1}) = 1$$

so $p \mid 1$. The latter is impossible so the claim is proved.

In the one direction, suppose $p \mid a$ so that $a = pq$ for some integer q. If a is a unit in $\mathbf{Z}_{p^k}$ then there exists an integer a^{-1} such that

$$a \cdot a^{-1} \equiv 1 \pmod{p^k}.$$

But since $a = pq$ it follows that

$$pqa^{-1} \equiv 1 \pmod{p^k}$$

which implies that p has an inverse modulo p^k. This contradicts the claim proved above.

In the other direction suppose p does not divide a. Then $\mathrm{GCD}(a, p^k) = 1$ since the only nontrivial divisors of p^k are p^i $(1 \le i \le k)$. Therefore the extended Euclidean algorithm can be applied to compute $a^{-1} \pmod{p^k}$.

●

Theorem 6.4. Let $a(x) \in \mathbf{Z}[x]$ be a given polynomial over the integers, let p be a prime integer which does not divide lcoeff$(a(x))$, and let $u^{(1)}(x), w^{(1)}(x) \in \mathbf{Z}_p[x]$ be two relatively prime polynomials over the field $\mathbf{Z}_p$ such that

$$a(x) \equiv u^{(1)}(x) w^{(1)}(x) \pmod{p}. \tag{6.65}$$

Let l be an integer such that $p^l > 2B$ where B bounds the magnitudes of all integer coefficients appearing in $a(x)$ and in any of its possible factors with degrees not exceeding $\max\{\deg(u^{(1)}(x)), \deg(w^{(1)}(x))\}$. Let $u^{(k)}(x)$ and $w^{(k)}(x)$ be factors computed by the Hensel construction such that

$$a(x) \equiv u^{(k)}(x) w^{(k)}(x) \pmod{p^k} \tag{6.66}$$

and

$$u^{(k)}(x) \equiv u^{(1)}(x) \pmod{p}, \quad w^{(k)}(x) \equiv w^{(1)}(x) \pmod{p} \tag{6.67}$$

for $k = 1, 2, \ldots, l$. If there exist polynomials $u(x), w(x) \in \mathbf{Z}[x]$ such that

$$a(x) = u(x) w(x) \in \mathbf{Z}[x] \tag{6.68}$$

and

$$\mathbf{n}(u(x)) \equiv \mathbf{n}(u^{(1)}(x)) \pmod{p}, \quad \mathbf{n}(w(x)) \equiv \mathbf{n}(w^{(1)}(x)) \pmod{p} \tag{6.69}$$

where $\mathbf{n}$ denotes the normalization "make the polynomial monic as an element of the domain $\mathbf{Z}_p[x]$" then the polynomials $u(x)$ and $u^{(l)}(x)$, as well as $w(x)$ and $w^{(l)}(x)$, are associates in the ring $\mathbf{Z}_{p^l}[x]$. More generally, for each $k \ge 1$ the polynomials $\phi_{p^k}(u(x))$ and $u^{(k)}(x)$, as well as $\phi_{p^k}(w(x))$ and $w^{(k)}(x)$, are associates in the ring $\mathbf{Z}_{p^k}[x]$.

Proof: Let $k \ge 1$ be any fixed positive integer. The assumption that p does not divide lcoeff$(a(x))$ implies, by Theorem 6.3, that lcoeff$(a(x))$ is a unit in $\mathbf{Z}_{p^k}[x]$. We may therefore define the monic polynomial

$$\tilde{a}(x) = \mathrm{lcoeff}(a(x))^{-1} a(x) \in \mathbf{Z}_{p^k}[x].$$

Now (6.66) implies that

$$\mathrm{lcoeff}(a(x)) \equiv \mathrm{lcoeff}(u^{(k)}(x)) \cdot \mathrm{lcoeff}(w^{(k)}(x)) \pmod{p^k}$$

so clearly p does not divide $\mathrm{lcoeff}(u^{(k)}(x))$ and p does not divide $\mathrm{lcoeff}(w^{(k)}(x))$ (for otherwise $p \mid \mathrm{lcoeff}(a(x))$), hence we may also define the monic polynomials

$$\tilde{u}^{(k)}(x) = \mathrm{lcoeff}(u^{(k)}(x))^{-1}\, u^{(k)}(x) \in \mathbf{Z}_{p^k}[x],$$

$$\tilde{w}^{(k)}(x) = \mathrm{lcoeff}(w^{(k)}(x))^{-1}\, w^{(k)}(x) \in \mathbf{Z}_{p^k}[x].$$

Obviously we may normalize the polynomials $u^{(1)}(x), w^{(1)}(x) \in \mathbf{Z}_p[x]$ yielding the monic polynomials

$$\tilde{u}^{(1)}(x) = \mathbf{n}(u^{(1)}(x)),\ \ \tilde{w}^{(1)}(x) = \mathbf{n}(w^{(1)}(x)).$$

It is easy to verify that conditions (6.65), (6.66), and (6.67) remain valid when $a(x), u^{(1)}(x), w^{(1)}(x), u^{(k)}(x), w^{(k)}(x)$ are replaced by $\tilde{a}(x)$, $\tilde{u}^{(1)}(x)$, $\tilde{w}^{(1)}(x)$, $\tilde{u}^{(k)}(x), \tilde{w}^{(k)}(x)$, respectively. Then by the corollary to Theorem 6.2, conditions (6.66) and (6.67) in the monic case uniquely determine the monic polynomial factors $\tilde{u}^{(k)}(x), \tilde{w}^{(k)}(x) \in \mathbf{Z}_{p^k}[x]$. Now suppose there exist polynomial factors $u(x), w(x) \in \mathbf{Z}[x]$ satisfying (6.68) and (6.69) and consider the polynomials $\phi_{p^k}(u(x)), \phi_{p^k}(w(x)) \in \mathbf{Z}_{p^k}[x]$. By reasoning as above, we may normalize these two polynomials in the ring $\mathbf{Z}_{p^k}[x]$ yielding monic polynomials $\tilde{u}(x), \tilde{w}(x) \in \mathbf{Z}_{p^k}[x]$ and these monic polynomials provide another factorization in $\mathbf{Z}_{p^k}[x]$ satisfying the monic versions of (6.66) and (6.67). Hence by uniqueness,

$$\tilde{u}^{(k)}(x) = \tilde{u}(x),\ \ \tilde{w}^{(k)}(x) = \tilde{w}(x).$$

It follows that $u^{(k)}(x)$ and $\phi_{p^k}(u(x))$ are associates in the ring $\mathbf{Z}_{p^k}[x]$ and similarly $w^{(k)}(x)$ and $\phi_{p^k}(w(x))$ are associates in the ring $\mathbf{Z}_{p^k}[x]$.

The above proof holds for any fixed $k \ge 1$. In particular when $k = l$, note that since $p^l > 2B$ we have $\phi_{p^l}(u(x)) = u(x)$ and $\phi_{p^l}(w(x)) = w(x)$.

●

The Modified Hensel Construction

The result of Theorem 6.4 can be used to "fix" the Hensel construction so that it will correctly generate the factors over the integers in the non-monic case. To this end we wish to create a situation in which the correct leading coefficients of the factors are known a priori and this can be achieved as follows. Let us assume that the polynomial $a(x) \in \mathbf{Z}[x]$ to be factored is a primitive polynomial. This assumption simply means that to factor an arbitrary polynomial over the integers we will first remove the unit part and the content so that the problem reduces to factoring the primitive part. Let $a(x)$ have the modulo p factorization (6.65) and suppose that there exist factors $u(x), w(x) \in \mathbf{Z}[x]$ satisfying (6.68) and (6.69). The leading coefficients

$$\alpha = \text{lcoeff}(a(x));\;\; \mu = \text{lcoeff}(u(x));\;\; \nu = \text{lcoeff}(w(x))$$

clearly must satisfy

$$\alpha = \mu\nu$$

but at this point we do not know the correct splitting of α into μ and ν. However, if we define the new polynomial

$$\tilde{a}(x) = \alpha a(x)$$

and seek a factorization of $\tilde{a}(x)$ then we have the relationship

$$\tilde{a}(x) = \mu\nu u(x) w(x) = [\nu u(x)][\mu w(x)].$$

In other words, by defining $\tilde{u}(x) = \nu u(x)$ and $\tilde{w}(x) = \mu w(x)$ we see that there exists a factorization

$$\tilde{a}(x) = \tilde{u}(x)\tilde{w}(x) \in \mathbf{Z}[x]$$

in which the leading coefficient of each factor is known to be α.

The Hensel construction can now be modified for the polynomial $\tilde{a}(x)$ so that for any $k \ge 1$, it computes factors $\tilde{u}^{(k)}(x), \tilde{w}^{(k)}(x) \in \mathbf{Z}_{p^k}[x]$ which satisfy not only the conditions of Hensel's lemma but, in addition, satisfy the relationships

$$\tilde{u}^{(k)}(x) = \phi_{p^k}(\tilde{u}(x));\;\; \tilde{w}^{(k)}(x) = \phi_{p^k}(\tilde{w}(x)) \tag{6.70}$$

where $\tilde{u}(x), \tilde{w}(x) \in \mathbf{Z}[x]$ are the (unknown) factors of $\tilde{a}(x)$ over the integers. (Note that the relationships (6.70) do not hold in Example 6.7.) The modification which can be made to the Hensel construction is a simple adjustment of units in each iteration step. For if $u^{(k)}(x)$ and $w^{(k)}(x)$ denote modulo p^k factors of $\tilde{a}(x)$ satisfying the conditions of Hensel's lemma then the modulo p^k factors $\tilde{u}^{(k)}(x)$ and $\tilde{w}^{(k)}(x)$ which maintain the conditions of Hensel's lemma and, in addition, satisfy (6.70) can be defined by

$$\begin{aligned}\tilde{u}^{(k)}(x) &= \phi_{p^k}(\alpha \cdot \text{lcoeff}(u^{(k)}(x))^{-1} \cdot u^{(k)}(x))\,;\\ \tilde{w}^{(k)}(x) &= \phi_{p^k}(\alpha \cdot \text{lcoeff}(w^{(k)}(x))^{-1} \cdot w^{(k)}(x))\,.\end{aligned} \tag{6.71}$$

Note that the modulo p^k inverses appearing here are guaranteed to exist by assuming the condition that p does not divide lcoeff($a(x)$). The associativity relationships stated in Theorem 6.4 have thus been strengthened to the equality relationships (6.70) by employing the knowledge that the correct leading coefficient of each factor is α. Finally when $k = l$, where l is large enough so that $\frac{1}{2}p^l$ bounds the magnitudes of all integer coefficients appearing in $\tilde{a}(x)$ and its factors, the relationships (6.70) become

$$\tilde{u}^{(l)}(x) = \tilde{u}(x),\;\; \tilde{w}^{(l)}(x) = \tilde{w}(x)$$

so the factors of $\tilde{a}(x)$ (which were assumed to exist) have been obtained. Note that if the bound B is defined for the original polynomial $a(x)$ as in Theorem 6.4 then since the modified Hensel construction is being applied to the larger polynomial $\tilde{a}(x)$, we must now require l to be large enough so that

$$p^l > 2B \cdot \mathrm{lcoeff}(a(x)).$$

The final step of this modified Hensel construction is to deduce the factorization of $a(x)$ from the computed factorization

$$\tilde{a}(x) = \tilde{u}(x)\tilde{w}(x) \in \mathbf{Z}[x].$$

Since $a(x)$ was assumed to be primitive, we have the relationship $a(x) = \mathrm{pp}(\tilde{a}(x))$ from which it follows that the desired factors of $a(x)$ are defined by

$$u(x) = \mathrm{pp}(\tilde{u}(x)), \;\; w(x) = \mathrm{pp}(\tilde{w}(x)).$$

The modification of the Hensel construction which is actually used in Algorithm 6.1 is a more efficient adaptation of the above ideas. Before discussing the improved version we consider an example.

Example 6.8. Let us return to the problem of Example 6.7 where the Hensel construction failed to produce the factors over the integers. We have

$$a(x) = 12x^3 + 10x^2 - 36x + 35 \in \mathbf{Z}[x]$$

and

$$a(x) \equiv u^{(1)}(x)w^{(1)}(x) \pmod 5$$

where $u^{(1)}(x) = 2x$, $w^{(1)}(x) = x^2 + 2$. Note that $a(x)$ is a primitive polynomial, and that the prime 5 does not divide the leading coefficient 12. In the new scheme, we define the new polynomial

$$\tilde{a}(x) = 12a(x) = 144x^3 + 120x^2 - 432x + 420.$$

We know that if there exists a factorization of $a(x)$ satisfying conditions (6.68) and (6.69) then there also exists a corresponding factorization such that 12 is the leading coefficient of each factor. The initial factorization

$$\tilde{a}(x) \equiv \tilde{u}^{(1)}(x)\tilde{w}^{(1)}(x) \pmod 5$$

such that the case $k = 1$ of (6.70) is satisfied can be obtained by applying the adjustment (6.71) to the given polynomials $u^{(1)}(x)$ and $w^{(1)}(x)$. We get

$$\tilde{u}^{(1)}(x) = \phi_5(12 \cdot 2^{-1} \cdot (2x)) = 2x,$$

$$\tilde{w}^{(1)}(x) = \phi_5(12 \cdot 1^{-1} \cdot (x^2 + 2)) = 2x^2 - 1.$$

Applying iteration step $k = 1$ of the usual Hensel construction to the polynomial $\tilde{a}(x)$, we get

$$u^{(2)}(x) = \tilde{u}^{(1)}(x) + (-x + 1)5 = -3x + 5,$$

$$w^{(2)}(x) = \tilde{w}^{(1)}(x) + (x - 1)5 = 2x^2 + 5x - 6.$$

Applying the adjustment (6.71) yields

$$\tilde{u}^{(2)}(x) = \phi_{25}(12 \cdot (-3)^{-1} \cdot (-3x + 5)) = 12x + 5,$$

$$\tilde{w}^{(2)}(x) = \phi_{25}(12 \cdot 2^{-1} \cdot (2x^2 + 5x - 6)) = 12x^2 + 5x - 11.$$

In iteration step $k = 2$ we get

$$u^{(3)}(x) = \tilde{u}^{(2)}(x) + (1)5^2 = 12x + 30,$$

$$w^{(3)}(x) = \tilde{w}^{(2)}(x) + (-x + 1)5^2 = 12x^2 - 20x + 14$$

and the adjustment (6.71) leaves the factors unchanged – i.e. $\tilde{u}^{(3)}(x) = 12x + 30$, $\tilde{w}^{(3)}(x) = 12x^2 - 20x + 14$. At this point,

$$\tilde{a}(x) - \tilde{u}^{(3)}(x)\tilde{w}^{(3)}(x) = 0$$

so the iteration halts and the factorization of $\tilde{a}(x)$ has been obtained. Thus the desired factors of the original polynomial $a(x)$ are

$$u(x) = \mathrm{pp}(\tilde{u}^{(3)}(x)) = 2x + 5,$$

$$w(x) = \mathrm{pp}(\tilde{w}^{(3)}(x)) = 6x^2 - 10x + 7.$$

●

Applying a Smaller Multiplier

In the scheme described above and applied in Example 6.8, note that the polynomial $\tilde{a}(x)$ which is actually factored may contain integer coefficients which are significantly larger than the integer coefficients in the original polynomial $a(x)$. This may lead to a decrease in efficiency compared to a scheme which works directly with the original polynomial $a(x)$. One could consider a scheme in which the Hensel construction is applied to the original polynomial $a(x)$ (exactly as in Example 6.7) until l is large enough so that

$$p^l > 2B \cdot \mathrm{lcoeff}(a(x)), \tag{6.72}$$

and then at the end a "restore leading coefficient" operation could be performed. One disadvantage of such a scheme is that the iteration then loses its "automatic" stopping criterion – i.e. it is not generally possible in such a scheme to recognize that enough iterations have been performed prior to satisfying the bound (6.72). This disadvantage is aggravated by two additional facts: (i) in practice the bound B almost always will be a very pessimistic bound; and (ii) each successive iteration step is usually more costly than the previous step (e.g. in Example 6.7 note the growth in the size of the coefficients of $e(x)$ and the factors with successive iteration steps). Therefore the potential saving of costly iterations offered by an iterative scheme which can recognize termination independently of the bound (6.72) can be very significant. An even more serious disadvantage of a scheme using a final "restore leading coefficient" operation arises in the multivariate case since the computation of polynomial inverses is a very nontrivial problem in a non-Euclidean domain.

The problem of coefficient size in the scheme which factors $\tilde{a}(x)$ rather than directly factoring $a(x)$ can be partially alleviated in certain circumstances as follows. Suppose we choose a multiplier γ which is smaller than $\mathrm{lcoeff}(a(x))$ in defining the new polynomial

$$\tilde{a}(x) = \gamma a(x). \tag{6.73}$$

Suppose it is known that γ is a multiple of the leading coefficient of one of the factors to be computed, let us say $u(x)$ – i.e. suppose it is known that

$$\text{lcoeff}(u(x)) \mid \gamma. \tag{6.74}$$

Then the polynomial $\tilde{a}(x)$ defined by (6.73) has a factorization in which the leading coefficients of the factors are known, where as usual we are assuming the existence of an appropriate factorization of the original polynomial $a(x)$. (Note that the choice $\gamma = \text{lcoeff}(a(x))$ used previously is a particular case of a multiplier which satisfies (6.74) .) In order to see this fact, let the assumed factorization of $a(x)$ be

$$a(x) = u(x)w(x) \in \mathbf{Z}[x]$$

and as before let us define the following leading coefficients:

$$\alpha = \text{lcoeff}(a(x)); \;\; \mu = \text{lcoeff}(u(x)); \;\; \nu = \text{lcoeff}(w(x)).$$

In addition, by (6.74) we may define the integer

$$\beta = \gamma / \mu.$$

Then the polynomial $\tilde{a}(x)$ defined by (6.73) satisfies the following relationship:

$$\tilde{a}(x) = \beta\mu u(x)w(x) = [\beta u(x)][\mu w(x)].$$

Hence by defining $\tilde{u}(x) = \beta u(x)$ and $\tilde{w}(x) = \mu w(x)$ we see that there exists a factorization

$$\tilde{a}(x) = \tilde{u}(x)\tilde{w}(x) \in \mathbf{Z}[x]$$

in which

$$\text{lcoeff}(\tilde{u}(x)) = \beta\mu = \gamma; \;\; \text{lcoeff}(\tilde{w}(x)) = \mu\nu = \alpha$$

where α is the known integer $\text{lcoeff}(a(x))$ and where γ has been specified. It is this generalization of the previously discussed scheme which is implemented in Algorithm 6.1, where γ is an optional input. If γ is unspecified on input then step 1 of the algorithm sets $\gamma = \text{lcoeff}(a(x))$ by default. It might seem that the specification of a γ smaller than $\text{lcoeff}(a(x))$ satisfying (6.74) would be impossible for most practical problems. However it turns out that in the application of the Hensel lifting algorithm to the important problem of polynomial GCD computation, the specification of γ is always possible (see Chapter 7). Finally, note that by (6.73) the termination condition (6.72) (for the case when the factorization of $a(x)$ does not exist) can be changed to the condition

$$p^l > 2B \cdot \gamma.$$

The Replace_lc Operation

The design of Algorithm 6.1 has now been fully explained except for one very significant modification. The scheme we have described (and applied in Example 6.8) requires that formulas (6.71) be applied to adjust units in each iteration step. However it can be seen that step 4 of Algorithm 6.1 contains no such adjustment of units in each iteration of the **while**-loop. Indeed step 4 of Algorithm 6.1 is simply an implementation of the pure unmodified

Hensel construction. The reason that Algorithm 6.1 is able to avoid the extra cost of adjusting units in each iteration stems from the yet-to-be-explained "replace_lc" operation appearing in step 3. This new operation is an ingenious modification described by Yun[8] and attributed to a suggestion by J. Moses. Consider the polynomial $\tilde{a}(x)$ defined by (6.73) and consider its modulo p factors $\tilde{u}^{(1)}(x), \tilde{w}^{(1)}(x) \in \mathbf{Z}_p[x]$ adjusted (as in step 1 of Algorithm 6.1) so that

$$\tilde{u}^{(1)}(x) = \phi_p(\tilde{u}(x)),\ \ \tilde{w}^{(1)}(x) = \phi_p(\tilde{w}(x)) \tag{6.75}$$

where $\tilde{u}(x)$ and $\tilde{w}(x)$ are the factors of $\tilde{a}(x)$ over the integers as discussed above such that

$$\mathrm{lcoeff}(\tilde{u}(x)) = \gamma,\ \ \mathrm{lcoeff}(\tilde{w}(x)) = \mathrm{lcoeff}(a(x)). \tag{6.76}$$

Writing the modulo p factors in the form

$$\tilde{u}^{(1)}(x) = \mu_m x^m + \mu_{m-1}x^{m-1} + \cdots + \mu_0;$$

$$\tilde{w}^{(1)}(x) = \nu_n x^n + \nu_{n-1}x^{n-1} + \cdots + \nu_0$$

where $\mu_m \neq 0$ and $\nu_n \neq 0$, it follows from (6.75) and (6.76) that $\mu_m = \phi_p(\gamma)$ and $\nu_n = \phi_p(\mathrm{lcoeff}(a(x)))$. Now suppose that the factors $\tilde{u}^{(1)}(x)$ and $\tilde{w}^{(1)}(x)$ are changed by simply replacing the leading coefficients μ_m and ν_n by γ and $\alpha = \mathrm{lcoeff}(a(x))$, respectively. To this end we define the algorithmic operation *replace_lc* as follows:

> Given a polynomial $a(x) \in \mathrm{R}[x]$ over a coefficient ring R and given an element $r \in \mathrm{R}$, the result of the operation replace_lc$(a(x), r)$ is the polynomial obtained from $a(x)$ by replacing the leading coefficient of $a(x)$ by r.

In this algorithmic notation, the polynomials $\tilde{u}^{(1)}(x)$ and $\tilde{w}^{(1)}(x)$ are replaced by the polynomials replace_lc$(\tilde{u}^{(1)}(x), \gamma)$ and replace_lc$(\tilde{w}^{(1)}(x), \alpha)$. Let $\tilde{u}^{(1)}(x)$ and $\tilde{w}^{(1)}(x)$ now denote the modified factors – i.e.

$$\begin{aligned} \tilde{u}^{(1)}(x) &= \gamma x^m + \mu_{m-1}x^{m-1} + \cdots + \mu_0; \\ \tilde{w}^{(1)}(x) &= \alpha x^n + \nu_{n-1}x^{n-1} + \cdots + \nu_0. \end{aligned} \tag{6.77}$$

Then the leading coefficients of $\tilde{u}^{(1)}(x)$ and $\tilde{w}^{(1)}(x)$ are no longer represented as elements of the field $\mathbf{Z}_p$ in the usual representation, but nonetheless we still have the property

$$\tilde{a}(x) \equiv \tilde{u}^{(1)}(x)\tilde{w}^{(1)}(x) \pmod{p}.$$

The Hensel construction can therefore be applied using (6.77) as the initial factors.

Let us consider the form of the successive factors which will be computed by the Hensel construction based on (6.77). Using the notation of step 4 of Algorithm 6.1, we first compute

$$c(x) = \frac{e(x)}{p} = \frac{\tilde{a}(x) - \tilde{u}^{(1)}(x)\tilde{w}^{(1)}(x)}{p}.$$

Note that the domain of this computation is $\mathbf{Z}[x]$. Since $\mathrm{lcoeff}(\tilde{a}(x)) = \gamma\alpha$, it is clear from

(6.77) that we will have

$$\deg(c(x)) < \deg(\tilde{a}(x)) = \deg(\tilde{u}^{(1)}(x)) + \deg(\tilde{w}^{(1)}(x)).$$

This strict inequality implies that the Hensel construction will then perform exactly as in the monic case in the following sense. The solution $\sigma(x), \tau(x) \in \mathbf{Z}_p[x]$ of the polynomial diophantine equation solved in step 4.1 of the algorithm will satisfy (as usual) the condition

$$\deg(\sigma(x)) < \deg(\tilde{w}^{(1)}(x))$$

and, in addition, we will have the following condition

$$\deg(\tau(x)) < \deg(\tilde{u}^{(1)}(x))$$

(see Theorem 2.6). Therefore when the factors are updated in step 4.2 of the algorithm the leading coefficients of both factors will remain unchanged. This is a desirable property since the leading coefficients are already known to be the correct integer coefficients. By the same reasoning, each successive iteration of the Hensel construction will also leave the leading coefficients unchanged. Finally since the successive factors computed by this scheme must satisfy Hensel's lemma, Theorem 6.4 guarantees that after a sufficient number of iterations the computed factors will be associates of the true factors of $\tilde{a}(x)$ over the integers (if such factors exist). However, since the computed factors have the same leading coefficients as the true factors over the integers, they can be associates only if they are identically equal. Therefore the desired factors of $\tilde{a}(x)$ will be computed (and will be recognized) by the iteration in step 4 of Algorithm 6.1 and no further adjustment of units is required.

Example 6.9. Consider once again the problem of factoring the non-monic polynomial of Example 6.7:

$$a(x) = 12x^3 + 10x^2 - 36x + 35 \in \mathbf{Z}[x].$$

This time we will apply Algorithm 6.1 in its full generality. The input to the algorithm is the primitive polynomial $a(x)$, the prime $p = 5$ (note that p does not divide lcoeff($a(x)$)), and the two relatively prime modulo 5 factors of $a(x)$ given by $u^{(1)}(x) = 2x$ and $w^{(1)}(x) = x^2 + 2$. The value of the bound B required by the algorithm is not needed in this example because the iterations will terminate by finding a factorization. The integer γ is undefined on input.

In step 1 of Algorithm 6.1 the following values are assigned:

$$\alpha = 12, \quad \gamma = 12,$$

$$a(x) = 144x^3 + 120x^2 - 432x + 420,$$

$$u^{(1)}(x) = 2x, \text{ and } w^{(1)}(x) = 2x^2 - 1.$$

In step 2 the extended Euclidean algorithm is applied to $u^{(1)}(x)$ and $w^{(1)}(x)$ in the domain $\mathbf{Z}_5[x]$ yielding

$$s(x) = x, \quad t(x) = -1.$$

In step 3 the leading coefficients of $u^{(1)}(x)$ and $w^{(1)}(x)$ are replaced by the correct integer coefficients. This gives

$$u(x) = 12x, \; w(x) = 12x^2 - 1,$$
$$e(x) = 120x^2 - 420x + 420,$$

and

$$modulus = 5.$$

In step 4 the sequence of values computed for $\sigma(x), \tau(x), u(x), w(x)$, and $e(x)$ is as follows.

End of iter. no.	$\sigma(x)$	$\tau(x)$	$u(x)$	$w(x)$	$e(x)$
0	–	–	$12x$	$12x^2 - 1$	$120x^2 - 420x + 420$
1	$x - 2$	1	$12x + 5$	$12x^2 + 5x - 11$	$-325x + 475$
2	$-x + 1$	1	$12x + 30$	$12x^2 - 20x + 14$	0

Finally in step 5 we obtain

$$\delta = \text{cont}(12x + 30) = 6,$$

$$u(x) = \frac{12x + 30}{6} = 2x + 5,$$

$$w(x) = \frac{12x^2 - 20x + 14}{12/6} = 6x^2 - 10x + 7.$$

Note that this computation was essentially equivalent to the computation in Example 6.8 except that we have avoided the cost of adjusting units in each iteration step.

●

6.7. THE MULTIVARIATE GENERALIZATION OF HENSEL'S LEMMA

We return now to the general multivariate lifting problem which was discussed at the beginning of Section 6.4. Specifically, we wish to find multivariate polynomials $u(x_1, \ldots, x_v), w(x_1, \ldots, x_v) \in \mathbf{Z}[x_1, \ldots, x_v]$ which satisfy equations (6.39)-(6.40) where $u_0(x_1)$ and $w_0(x_1)$, the images mod $< \mathrm{I}, p >$, are given. Here $\mathrm{I} = < x_2 - \alpha_2, \ldots, x_v - \alpha_2 >$ is the kernel of a multivariate evaluation homomorphism and p is a prime integer.

A Homomorphism Diagram

We consider the lifting process in two separate stages. Firstly, the solution in $\mathbf{Z}_p[x_1]$ is lifted to the solution in $\mathbf{Z}_{p^l}[x_1]$ for some sufficiently large l such that the ring $\mathbf{Z}_{p^l}$ can be identified with $\mathbf{Z}$ for the particular problem being solved. This first stage of lifting is accomplished by the univariate Hensel lifting algorithm (Algorithm 6.1). Secondly, the solution in $\mathbf{Z}_{p^l}[x_1]$ is lifted to the desired solution in $\mathbf{Z}_{p^l}[x_1, \ldots, x_v]$ (which is identified with the original domain $\mathbf{Z}[x_1, \ldots, x_v]$) by the multivariate Hensel lifting algorithm to be described. The latter algorithm is given the solution mod $< \mathrm{I}, p^l >$ and, using an iteration analogous to the univariate case, it lifts to the solution mod $< \mathrm{I}^{k+1}, p^l >$ for $k = 1, 2, \ldots, d$ where d is the maximum total degree in the indeterminates $x_2, \ldots, x_v$ of any term in the solution polynomials.

Figure 6.2 shows a homomorphism diagram for solving a problem using the univariate and multivariate Hensel lifting algorithms. It should be noted that the order of the univariate and multivariate operations has been reversed compared with the homomorphism diagram of Figure 6.1 presented at the end of Section 6.3.

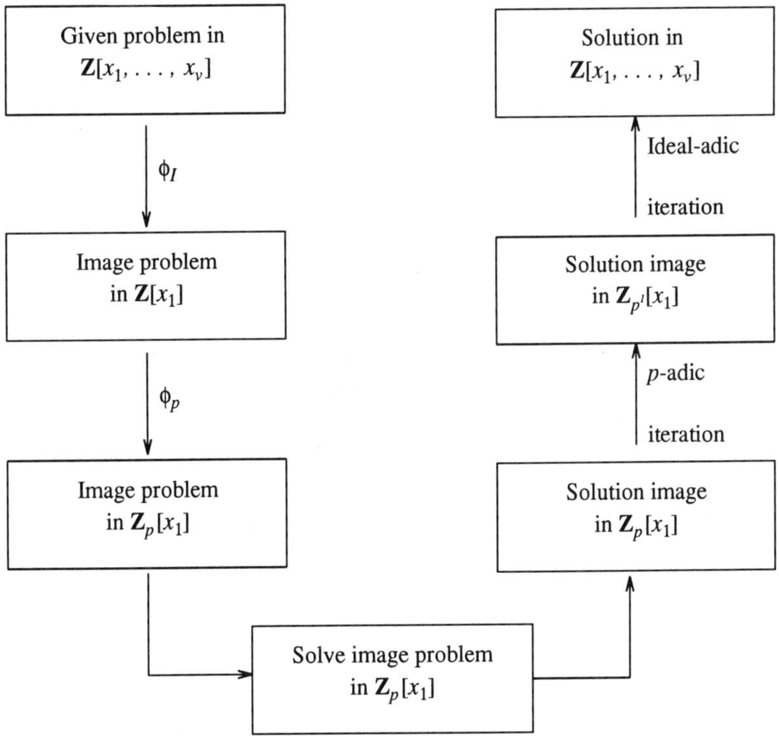

Figure 6.2. Homomorphism diagram for univariate and multivariate Hensel constructions.

In the setting of Section 6.3 (solving a polynomial equation $F(u) = 0$ via Newton's iteration), the computation could be organized in either order. In the current setting (solving a bivariate polynomial equation $F(u, w) = 0$ via Hensel constructions), there is a fundamental reason for organizing the computation in the order specified by Figure 6.2. Before pursuing this point some further remarks about the diagram in Figure 6.2 should be noted. In the box at the end of the arrow labeled "p-adic iteration", the domain is specified as $\mathbf{Z}_{p^l}[x_1]$. As we have already noted, l will be chosen to be sufficiently large for the particular problem so that the ring $\mathbf{Z}_{p^l}$ can be identified with $\mathbf{Z}$ (an identification we have made in the box following the multivariate Hensel construction). The specification of the domain in the form $\mathbf{Z}_{p^l}[x_1]$ is deliberate, in order to emphasize the fact (to be seen shortly) that the domain of the operations required by the multivariate Hensel construction is $\mathbf{Z}_{p^l}[x_1]$ and not $\mathbf{Z}[x_1]$. Another point to be noted about the organization of this diagram is that the multivariate problem has been conceptually separated from the univariate problem, such that a diagram for the multivariate problem could read: "Apply the homomorphism ϕ_I, solve the univariate problem in $\mathbf{Z}[x_1]$ (by any method of your choosing), and finally apply the multivariate Hensel construction." However the operations for the multivariate Hensel construction require that the univariate domain $\mathbf{Z}[x_1]$ must be replaced by a domain $\mathbf{Z}_{p^l}[x_1]$ for some prime p and integer l, even if the univariate problem was solved by a non-Hensel method.

Recall from Section 6.4 that the basic computation to be performed in applying a step of a Hensel iteration, i.e. Newton's iteration applied to the equation

$$F(u, w) = a(x_1, \ldots, x_v) - uw = 0 ,$$

is to solve a polynomial diophantine equation of the form

$$A^{(k)}\Delta u^{(k)} + B^{(k)}\Delta w^{(k)} = C^{(k)} \tag{6.78}$$

for the correction terms $\Delta u^{(k)}, \Delta w^{(k)}$ (where $A^{(k)}, B^{(k)}, C^{(k)}$ are given polynomials). Now if the order of the univariate and multivariate lifting steps is to be that of Figure 6.1 (i.e. I-adic lifting preceding p-adic lifting) then during I-adic lifting equation (6.78) will have to be solved in the domain $\mathbf{Z}_p[x_1]$, and during p-adic lifting equation (6.78) will have to be solved in the domain $\mathbf{Z}_p[x_1, \ldots, x_v]$. As we have already seen in the development of Algorithm 6.1, Theorem 2.6 shows how to solve equation (6.78) in the Euclidean domain $\mathbf{Z}_p[x_1]$. However the necessity to solve equation (6.78) in the multivariate polynomial domain $\mathbf{Z}_p[x_1, \ldots, x_v]$ poses serious difficulties. Theorem 2.6 does not apply because this multivariate domain is certainly not a Euclidean domain. It is possible to develop methods to solve equation (6.78) in multivariate domains but the computational expense of these methods makes the Hensel construction impractical when organized in this way. On the other hand, if the computation is organized as specified in the diagram of Figure 6.2 then during p-adic lifting equation (6.78) will be solved in the Euclidean domain $\mathbf{Z}_p[x_1]$, and during I-adic lifting equation (6.78) will be solved in the ring $\mathbf{Z}_{p^l}[x_1]$. Again we have the apparent difficulty that the latter ring is not a Euclidean domain. However this univariate polynomial ring is "nearly a Euclidean domain" in the sense that the ring $\mathbf{Z}_{p^l}$ is "nearly a field" (see Theorem 6.3 in the preceding section). The constructive proof of Theorem 2.6 (which is based on applying the extended Euclidean algorithm) will often remain valid for solving equation (6.78) in the univariate polynomial ring $\mathbf{Z}_{p^l}[x_1]$, with a little luck in the choice of the prime p for the particular problem being solved. In general though, we cannot guarantee the existence of the inverses required by the extended Euclidean algorithm when it is applied in the ring $\mathbf{Z}_{p^l}[x_1]$.

Polynomial Diophantine Equations in $\mathbf{Z}_{p^l}[x_1]$

A general solution to the problem of solving polynomial diophantine equations in the ring $\mathbf{Z}_{p^l}[x_1]$ is obtained by applying Newton's iteration to lift the solution in $\mathbf{Z}_p[x_1]$ up to a solution in $\mathbf{Z}_{p^l}[x_1]$. The "extended Euclidean" problem to be solved is to find polynomials $s^{(l)}(x_1), t^{(l)}(x_1) \in \mathbf{Z}_{p^l}[x_1]$ which satisfy the equation

$$s^{(l)}(x_1)\cdot u(x_1) + t^{(l)}(x_1)\cdot w(x_1) \equiv 1 \pmod{p^l} \tag{6.79}$$

where $u(x_1), w(x_1) \in \mathbf{Z}_{p^l}[x_1]$ are given polynomials such that $\phi_p(u(x_1)), \phi_p(w(x_1))$ are relatively prime polynomials in the Euclidean domain $\mathbf{Z}_p[x_1]$. The equation to which we will apply Newton's iteration is

$$G(s,t) = s\cdot u(x_1) + t\cdot w(x_1) - 1 = 0.$$

Proceeding as in previous sections, if we have the order-k p-adic approximations $s^{(k)}, t^{(k)}$ to the solution pair $\tilde{s}, \tilde{t}$ and if we obtain correction terms $\Delta s^{(k)}, \Delta t^{(k)}$ which satisfy the equation

$$G_s(s^{(k)}, t^{(k)})\cdot\Delta s^{(k)} + G_t(s^{(k)}, t^{(k)})\cdot\Delta t^{(k)} \equiv -G(s^{(k)}, t^{(k)}) \pmod{p^{k+1}} \tag{6.80}$$

then the order-$(k+1)$ p-adic approximations are given by

$$s^{(k+1)} = s^{(k)} + \Delta s^{(k)}, \quad t^{(k+1)} = t^{(k)} + \Delta t^{(k)}.$$

Writing the correction terms in the form

$$\Delta s^{(k)} = s_k(x_1)p^k, \quad \Delta t^{(k)} = t_k(x_1)p^k$$

where $s_k(x_1), t_k(x_1) \in \mathbf{Z}_p[x_1]$, substituting for the partial derivatives, and dividing through by p^k, equation (6.80) becomes

$$u(x_1)\cdot s_k(x_1) + w(x_1)\cdot t_k(x_1) \equiv \frac{1 - s^{(k)}\cdot u(x_1) - t^{(k)}\cdot w(x_1)}{p^k} \pmod{p}. \tag{6.81}$$

The order-1 p-adic approximations $s^{(1)}, t^{(1)} \in \mathbf{Z}_p[x_1]$ for the solution of equation (6.79) are obtained by the extended Euclidean algorithm (or, in the context of Figure 6.2, they have already been computed in Algorithm 6.1 for the univariate Hensel construction). For $k = 1, 2, \ldots, l-1$, equation (6.81) can be solved for the correction terms $s_k(x_1), t_k(x_1) \in \mathbf{Z}_p[x_1]$ by Theorem 2.6, thus generating the desired solution of equation (6.79).

The following theorem shows that we can solve, in the ring $\mathbf{Z}_{p^l}[x_1]$, the polynomial diophantine equations which arise in the multivariate Hensel construction.

Theorem 6.5. For a prime integer p and a positive integer l, let $u(x_1), w(x_1) \in \mathbf{Z}_{p^l}[x_1]$ be univariate polynomials satisfying

(i) $p \nmid \text{lcoeff}(u(x_1))$ and $p \nmid \text{lcoeff}(w(x_1))$,

(ii) $\phi_p(u(x_1))$ and $\phi_p(w(x_1))$ are relatively prime polynomials in $\mathbf{Z}_p[x_1]$.

Then for any polynomial $c(x_1) \in \mathbf{Z}_{p^l}[x_1]$ there exist unique polynomials $\sigma(x_1), \tau(x_1) \in \mathbf{Z}_{p^l}[x_1]$ such that

$$\sigma(x_1) \cdot u(x_1) + \tau(x_1) \cdot w(x_1) \equiv c(x_1) \pmod{p^l} \tag{6.82}$$

and

$$\deg(\sigma(x_1)) < \deg(w(x_1)). \tag{6.83}$$

Moreover, if $\deg(c(x_1)) < \deg(u(x_1)) + \deg(w(x_1))$ then $\tau(x_1)$ satisfies

$$\deg(\tau(x_1)) < \deg(u(x_1)). \tag{6.84}$$

Proof: We first show existence. The extended Euclidean algorithm can be applied to compute polynomials $s^{(1)}(x_1), t^{(1)}(x_1) \in \mathbf{Z}_p[x_1]$ satisfying the equations

$$s^{(1)}(x_1) \cdot u(x_1) + t^{(1)}(x_1) \cdot w(x_1) \equiv 1 \pmod{p}. \tag{6.85}$$

By Theorem 2.6, equation (6.81) can be solved for polynomials $s_k(x_1), t_k(x_1) \in \mathbf{Z}_p[x_1]$ for successive integers $k \geq 1$, where we define

$$s^{(k)}(x_1) = s^{(1)}(x_1) + s_1(x_1)p + \cdots + s_{k-1}(x_1)p^{k-1};$$

$$t^{(k)}(x_1) = t^{(1)}(x_1) + t_1(x_1)p + \cdots + t_{k-1}(x_1)p^{k-1};$$

and we must prove that

$$s^{(k)}(x_1) \cdot u(x_1) + t^{(k)}(x_1) \cdot w(x_1) \equiv 1 \pmod{p^k}. \tag{6.86}$$

We will prove (6.86) by induction. The case $k = 1$ is given by equation (6.85). Suppose (6.86) holds for some $k \geq 1$. Then noting that

$$s^{(k+1)}(x_1) = s^{(k)}(x_1) + s_k(x_1)p^k$$

and

$$t^{(k+1)}(x_1) = t^{(k)}(x_1) + t_k(x_1)p^k$$

where $s_k(x_1)$ and $t_k(x_1)$ are the solutions of equation (6.81), we have

$$s^{(k+1)}(x_1)u(x_1) + t^{(k+1)}(x_1)w(x_1) =$$

$$s^{(k)}(x_1)u(x_1) + t^{(k)}(x_1)w(x_1) + p^k[s_k(x_1)u(x_1) + t_k(x_1)w(x_1)] \equiv 1 \pmod{p^{k+1}}$$

where we have applied equation (6.81) after multiplying it through by p^k. Thus (6.86) is proved for all $k \geq 1$, and in particular for $k = l$.

Now the desired polynomials $\sigma(x_1), \tau(x_1) \in \mathbf{Z}_{p^l}[x_1]$ satisfying equation (6.82) can be calculated exactly as in the proof of Theorem 2.6. Specifically, the polynomials

$$\tilde{\sigma}(x_1) = s^{(l)}(x_1)c(x_1) \text{ and } \tilde{\tau}(x_1) = t^{(l)}(x_1)c(x_1)$$

satisfy equation (6.82) and then to reduce the degree we apply Euclidean division of $\tilde{\sigma}(x_1)$ by $w(x_1)$ yielding $q(x_1), r(x_1) \in \mathbf{Z}_{p^l}[x_1]$ such that

$$\tilde{\sigma}(x_1) \equiv w(x_1)q(x_1) + r(x_1) \pmod{p^l}$$

where $\deg(r(x_1)) < \deg(w(x_1))$. This division step will be valid in the ring $\mathbf{Z}_{p^l}[x_1]$ because condition (i) guarantees that lcoeff$(w(x_1))$ is a unit in the ring $\mathbf{Z}_{p^l}$ (see Theorem 6.3). Finally, defining

$$\sigma(x_1) = r(x_1) \text{ and } \tau(x_1) = \tilde{\tau}(x_1) + q(x_1)u(x_1) \in \mathbf{Z}_{p^l}[x_1]$$

equation (6.82) and the degree constraint (6.83) are readily verified.

To show uniqueness, let $\sigma_1(x_1), \tau_1(x_1) \in \mathbf{Z}_{p^l}[x_1]$ and $\sigma_2(x_1), \tau_2(x_1) \in \mathbf{Z}_{p^l}[x_1]$ be two pairs of polynomials satisfying (6.82) and (6.83). Subtracting the two different equations of the form (6.82) yields

$$(\sigma_1(x_1) - \sigma_2(x_1))u(x_1) \equiv -(\tau_1(x_1) - \tau_2(x_1))w(x_1) \pmod{p^l}. \tag{6.87}$$

Also, the degree constraint (6.83) satisfied by $\sigma_1(x_1)$ and $\sigma_2(x_1)$ yields

$$\deg(\sigma_1(x_1) - \sigma_2(x_1)) < \deg(w(x_1)). \tag{6.88}$$

Taking the congruence (6.87) modulo p we have a relationship in the domain $\mathbf{Z}_p[x_1]$ which, together with condition (ii), implies that $\phi_p(w(x_1))$ divides $\phi_p(\sigma_1(x_1) - \sigma_2(x_1))$ in the domain $\mathbf{Z}_p[x_1]$. Noting from condition (i) that $\phi_p(w(x_1))$ has the same degree as $w(x_1)$, (6.88) implies that

$$\sigma_1(x_1) - \sigma_2(x_1) \equiv 0 \pmod{p}$$

and then it follows from (6.87) that

$$\tau_1(x_1) - \tau_2(x_1) \equiv 0 \pmod{p}.$$

We now claim that the polynomials $\sigma_1(x_1) - \sigma_2(x_1)$ and $\tau_1(x_1) - \tau_2(x_1)$ satisfying (6.87) are divisible by p^k for all positive integers $k \le l$. The proof is by induction. The case $k = 1$ has just been proved. Suppose that they are divisible by p^k for some $k < l$. Then we may define the polynomials

$$\alpha(x_1) = (\sigma_1(x_1) - \sigma_2(x_1)) / p^k \text{ and } \beta(x_1) = (\tau_1(x_1) - \tau_2(x_1)) / p^k$$

and, dividing through by p^k in congruence (6.87) we have

$$\alpha(x_1)u(x_1) \equiv -\beta(x_1)w(x_1) \pmod{p^{l-k}}.$$

By repeating the argument used above, we conclude that

$$\alpha(x_1) \equiv 0 \pmod{p} \ \text{ and } \ \beta(x_1) \equiv 0 \pmod{p},$$

i.e. $\sigma_1(x_1) - \sigma_2(x_1)$ and $\tau_1(x_1) - \tau_2(x_1)$ are divisible by p^{k+1}, which proves the claim.

Finally, we have proved that

$$\sigma_1(x_1) \equiv \sigma_2(x_1) \pmod{p^l} \ \text{ and } \ \tau_1(x_1) \equiv \tau_2(x_1) \pmod{p^l}$$

which proves uniqueness in the ring $\mathbf{Z}_{p^l}[x_1]$.

Finally we need to show that the degree constraint (6.84) holds. From (6.82) we can write

$$\tau(x_1) \equiv (c(x_1) - \sigma(x_1)u(x_1)) \,/\, w(x_1) \pmod{p^l}$$

and the division here is valid in the ring $\mathbf{Z}_{p^l}[x_1]$ because lcoeff$(w(x_1))$ is a unit in $\mathbf{Z}_{p^l}$, by condition (i). By this same condition, we have

$$\deg(\tau(x_1)) = \deg(c(x_1) - \sigma(x_1)u(x_1)) - \deg(w(x_1)). \tag{6.89}$$

Now if $\deg(c(x_1)) \ge \deg(\sigma(x_1)u(x_1))$ then from (6.89)

$$\deg(\tau(x_1)) \le \deg(c(x_1)) - \deg(w(x_1)) < \deg(u(x_1))$$

as long as $\deg(c(x_1)) < \deg(u(x_1)) + \deg(w(x_1))$ as stated. Otherwise if $\deg(c(x_1)) < \deg(\sigma(x_1)u(x_1))$ (in which case the stated degree bound for $c(x_1)$ also holds because of (6.83)) then from (6.89)

$$\deg(\tau(x_1)) = \deg(\sigma(x_1)u(x_1)) - \deg(w(x_1)) < \deg(u(x_1))$$

where the last inequality follows from (6.83). Thus (6.84) is proved.

●

Multivariate Hensel Construction

We are now ready to develop the multivariate generalization of Hensel's lemma. We pose the problem of finding multivariate polynomials $u(x_1, \ldots, x_v), w(x_1, \ldots, x_v) \in \mathbf{Z}_{p^l}[x_1, \ldots, x_v]$ which satisfy the congruence

$$a(x_1, \ldots, x_v) - u \cdot w \equiv 0 \pmod{p^l} \tag{6.90}$$

such that

$$\begin{aligned} u(x_1, \ldots, x_v) &\equiv u^{(1)}(x_1) \pmod{<\mathrm{I}, p^l>}; \\ w(x_1, \ldots, x_v) &\equiv w^{(1)}(x_1) \pmod{<\mathrm{I}, p^l>}; \end{aligned} \tag{6.91}$$

where $u^{(1)}(x_1), w^{(1)}(x_1) \in \mathbf{Z}_{p^l}[x_1]$ are given univariate polynomials which satisfy (6.90) modulo I. Here, p is a prime integer, l is a positive integer, $a(x_1, \ldots, x_v) \in \mathbf{Z}_{p^l}[x_1, \ldots, x_v]$ is a given multivariate polynomial, and $\mathrm{I} = <x_2 - \alpha_2, \ldots, x_v - \alpha_v>$ is the kernel of a multivariate evaluation homomorphism. Denoting the desired solution polynomials by $\bar{u}$ and $\bar{w}$, we will develop these solutions in their I-adic forms:

$$\begin{aligned} \bar{u} &= u^{(1)} + \Delta u^{(1)} + \Delta u^{(2)} + \cdots + \Delta u^{(d)}; \\ \bar{w} &= w^{(1)} + \Delta w^{(1)} + \Delta w^{(2)} + \cdots + \Delta w^{(d)} \end{aligned} \tag{6.92}$$

where d is the maximum total degree of any term in $\bar{u}$ or $\bar{w}$, $u^{(1)} = \phi_I(\bar{u})$, $w^{(1)} = \phi_I(\bar{w})$, and $\Delta u^{(k)}, \Delta w^{(k)} \in I^k$, for $k = 1, 2, \ldots, d$. From Section 6.2 we know that the I-adic representation of the polynomial $\bar{u}$ is precisely the multivariate Taylor series representation (6.14). The k-th correction term $\Delta u^{(k)} \in I^k$ is the term in (6.14) of total degree k with respect to I and it is represented by k nested summations in the form

$$\Delta u^{(k)} = \sum_{i_1=2}^{v} \sum_{i_2=i_1}^{v} \cdots \sum_{i_k=i_{k-1}}^{v} u_{\mathbf{i}}(x_1)(x_{i_1} - \alpha_{i_1})(x_{i_2} - \alpha_{i_2}) \cdots (x_{i_k} - \alpha_{i_k}) \tag{6.93}$$

where $\mathbf{i} = (i_1, \ldots, i_k)$ is a vector subscript and $u_{\mathbf{i}}(x_1) \in \mathbf{Z}_{p^l}[x_1]$. Similarly, in the I-adic representation of $\bar{w}$ the k-th correction term takes the form

$$\Delta w^{(k)} = \sum_{i_1=2}^{v} \sum_{i_2=i_1}^{v} \cdots \sum_{i_k=i_{k-1}}^{v} w_{\mathbf{i}}(x_1)(x_{i_1} - \alpha_{i_1})(x_{i_2} - \alpha_{i_2}) \cdots (x_{i_k} - \alpha_{i_k}) \tag{6.94}$$

where $w_{\mathbf{i}}(x_1) \in \mathbf{Z}_{p^l}[x_1]$.

Our problem now is to compute, for each $k = 1, 2, \ldots, d$, the k-th correction terms $\Delta u^{(k)}, \Delta w^{(k)}$ in (6.92). Let $u^{(k)}, w^{(k)}$ denote the order-k I-adic approximations to $\bar{u}, \bar{w}$ given by the first k terms in (6.92). Letting $F(u, w)$ denote the left-hand-side of (6.90), Newton's iteration for solving $F(u, w) = 0$ in $\mathbf{Z}_{p^l}[x_1, \ldots, x_v]$ takes the form of the congruence equation

$$w^{(k)} \Delta u^{(k)} + u^{(k)} \Delta w^{(k)} \equiv a(x_1, \ldots, x_v) - u^{(k)} \cdot w^{(k)} \pmod{< I^{k+1}, p^l >} \tag{6.95}$$

which must be solved for the correction terms $\Delta u^{(k)}, \Delta w^{(k)} \in I^k$ and then

$$u^{(k+1)} = u^{(k)} + \Delta u^{(k)}, \quad w^{(k+1)} = w^{(k)} + \Delta w^{(k)}$$

will be order-(k+1) I-adic approximations to $\bar{u}, \bar{w}$. Now since $u^{(k)}, w^{(k)}$ are order-k I-adic approximations we have

$$a(x_1, \ldots, x_v) - u^{(k)} w^{(k)} \in I^k$$

and therefore the right-hand-side of (6.95) may be expressed in the form

$$\sum_{i_1=2}^{v} \sum_{i_2=i_1}^{v} \cdots \sum_{i_k=i_{k-1}}^{v} c_{\mathbf{i}}(x_1)(x_{i_1} - \alpha_{i_1})(x_{i_2} - \alpha_{i_2}) \cdots (x_{i_k} - \alpha_{i_k})$$

for some coefficients $c_{\mathbf{i}}(x_1) \in \mathbf{Z}_{p^l}[x_1]$. Substituting into (6.95) this nested-summation representation for the right-hand-side and also the nested-summation representations (6.93) and (6.94), the congruence (6.95) may be solved by separately solving the following congruence for each term in the I-adic representation

$$w^{(k)} u_{\mathbf{i}}(x_1) + u^{(k)} w_{\mathbf{i}}(x_1) \equiv c_{\mathbf{i}}(x_1) \pmod{< I, p^l >}$$

yielding the desired I-adic coefficients $u_{\mathbf{i}}(x_1), w_{\mathbf{i}}(x_1) \in \mathbf{Z}_{p^l}[x_1]$ which define the correction terms (6.93) and (6.94). Note that since this is a congruence modulo I we may apply the evaluation homomorphism ϕ_I to the left-hand-side, yielding the following polynomial diophantine equation to solve in the ring $\mathbf{Z}_{p^l}[x_1]$ for each term in the I-adic representation

$$w^{(1)}(x_1)u_{\mathbf{i}}(x_1) + u^{(1)}(x_1)w_{\mathbf{i}}(x_1) \equiv c_{\mathbf{i}}(x_1) \pmod{p^l} \tag{6.96}$$

where $u^{(1)}(x_1)$, $w^{(1)}(x_1) \in \mathbf{Z}_{p^l}[x_1]$ are the given polynomials in the problem (6.90) and (6.91) being solved. Theorem 6.5 states the conditions under which the congruence (6.96) has a unique solution $u_{\mathbf{i}}(x_1)$, $w_{\mathbf{i}}(x_1) \in \mathbf{Z}_{p^l}[x_1]$.

The following theorem formally proves the validity of the above method which is known as the *multivariate Hensel construction.*

Theorem 6.6 (Multivariate Hensel Construction). Let p be a prime integer, let l be a positive integer, and let $a(x_1, \ldots, x_v) \in \mathbf{Z}_{p^l}[x_1, \ldots, x_v]$ be a given multivariate polynomial. Let $\mathrm{I} = < x_2-\alpha_2, \ldots, x_v-\alpha_v >$ be the kernel of a multivariate evaluation homomorphism such that $p \nmid \mathrm{lcoeff}(\phi_{\mathrm{I}}(a(x_1, \ldots, x_v)))$. Let $u^{(1)}(x_1)$, $w^{(1)}(x_1) \in \mathbf{Z}_{p^l}[x_1]$ be two univariate polynomials which satisfy

(i) $a(x_1, \ldots, x_v) \equiv u^{(1)}(x_1)w^{(1)}(x_1) \pmod{< \mathrm{I}, p^l >}$,

(ii) $\phi_p(u^{(1)}(x_1))$ and $\phi_p(w^{(1)}(x_1))$ are relatively prime in $\mathbf{Z}_p[x_1]$.

Then for any integer $k \ge 1$ there exist multivariate polynomials $u^{(k)}, w^{(k)} \in \mathbf{Z}_{p^l}[x_1, \ldots, x_v] / \mathrm{I}^k$ such that

$$a(x_1, \ldots, x_v) \equiv u^{(k)}w^{(k)} \pmod{< \mathrm{I}^k, p^l >} \tag{6.97}$$

and

$$u^{(k)} \equiv u^{(1)}(x_1) \pmod{< \mathrm{I}, p^l >}, \quad w^{(k)} \equiv w^{(1)}(x_1) \pmod{< \mathrm{I}, p^l >}. \tag{6.98}$$

Proof: The proof is by induction on k. The case $k = 1$ is given by condition (i). Assume for $k \ge 1$ that we have $u^{(k)}, w^{(k)} \in \mathbf{Z}_{p^l}[x_1, \ldots, x_v] / \mathrm{I}^k$ satisfying (6.97) and (6.98). Define

$$e^{(k)} = a(x_1, \ldots, x_v) - u^{(k)}w^{(k)} \in \mathbf{Z}_{p^l}[x_1, \ldots, x_v] / \mathrm{I}^{k+1} \tag{6.99}$$

and from (6.97) it follows that $e^{(k)} \in \mathrm{I}^k$. Define the polynomial coefficients $c_{\mathbf{i}}(x_1) \in \mathbf{Z}_{p^l}[x_1]$ by expressing $e^{(k)}$ in I-adic form:

$$e^{(k)} = \sum_{i_1=2}^{v} \sum_{i_2=i_1}^{v} \cdots \sum_{i_k=i_{k-1}}^{v} c_{\mathbf{i}}(x_1)(x_{i_1}-\alpha_{i_1})(x_{i_2}-\alpha_{i_2}) \cdots (x_{i_k}-\alpha_{i_k}). \tag{6.100}$$

By Theorem 6.5 (noting that since $p \nmid \mathrm{lcoeff}(\phi_{\mathrm{I}}(a(x_1, \ldots, x_v)))$, condition (i) implies the first condition of Theorem 6.5 and condition (ii) is the second required condition), we can find unique polynomials $\sigma_{\mathbf{i}}(x_1)$, $\tau_{\mathbf{i}}(x_1) \in \mathbf{Z}_{p^l}[x_1]$ such that

$$\sigma_{\mathbf{i}}(x_1)u^{(1)}(x_1) + \tau_{\mathbf{i}}(x_1)w^{(1)}(x_1) \equiv c_{\mathbf{i}}(x_1) \pmod{p^l} \tag{6.101}$$

and

$$\deg(\sigma_{\mathbf{i}}(x_1)) < \deg(w^{(1)}(x_1)), \tag{6.102}$$

for each index $\mathbf{i}$ which appears in the I-adic representation of $e^{(k)}$. Then by defining

$$u^{(k+1)} = u^{(k)} + \sum_{i_1=2}^{v} \sum_{i_2=i_1}^{v} \cdots \sum_{i_k=i_{k-1}}^{v} \tau_{\mathbf{i}}(x_1)(x_{i_1}-\alpha_{i_1})(x_{i_2}-\alpha_{i_2}) \cdots (x_{i_k}-\alpha_{i_k});$$
$$w^{(k+1)} = w^{(k)} + \sum_{i_1=2}^{v} \sum_{i_2=i_1}^{v} \cdots \sum_{i_k=i_{k-1}}^{v} \sigma_{\mathbf{i}}(x_1)(x_{i_1}-\alpha_{i_1})(x_{i_2}-\alpha_{i_2}) \cdots (x_{i_k}-\alpha_{i_k}) \tag{6.103}$$

we have by performing multiplication modulo I^{k+1} and using equations (6.99) - (6.101)

$$u^{(k+1)}w^{(k+1)} \equiv u^{(k)}w^{(k)} + \sum_{i_1=2}^{v} \cdots \sum_{i_k=i_{k-1}}^{v} (\sigma_{\mathbf{i}}(x_1)\, u^{(1)}(x_1) + \tau_{\mathbf{i}}(x_1)\, w^{(1)}(x_1))\, (x_{i_1}-\alpha_{i_1}) \cdots (x_{i_k}-\alpha_{i_k})$$
$$\bmod < \mathrm{I}^{k+1}, p^l >)$$
$$\equiv u^{(k)}w^{(k)} + e^{(k)} \pmod{< \mathrm{I}^{k+1}, p^l >}$$
$$\equiv a(x_1, \ldots, x_v) \pmod{< \mathrm{I}^{k+1}, p^l >}.$$

Thus (6.97) holds for $k + 1$. Also, from (6.103) it is clear that

$$u^{(k+1)} \equiv u^{(k)} \pmod{< \mathrm{I}, p^l >}, \quad w^{(k+1)} \equiv w^{(k)} \pmod{< \mathrm{I}, p^l >}$$

and therefore since (6.98) holds for k it also holds for $k + 1$.

•

The multivariate Hensel construction of Theorem 6.6 generates unique factors $u^{(k)}$, $w^{(k)}$ in the case where $a(x_1, \ldots, x_v)$ is "monic with respect to x_1"; i.e. in the case where the coefficient in $a(x_1, \ldots, x_v)$ of $x_1^{d_1}$ is 1, where d_1 denotes the degree in x_1. For in such a case, we may choose $u^{(k)}$ and $w^{(k)}$ each to be "monic with respect to x_1" and uniqueness follows just as in the univariate case. This result is stated as the following corollary, whose proof is a straightforward generalization of the proof of the corollary to Theorem 6.2 and is omitted.

Corollary (Uniqueness of the Multivariate Hensel construction). In Theorem 6.6, if the given polynomial $a(x_1, \ldots, x_v) \in \mathbf{Z}_{p^l}[x_1, \ldots, x_v]$ has leading coefficient 1 with respect to the indeterminant x_1 and correspondingly if the univariate factors $u^{(1)}(x_1), w^{(1)}(x_1) \in \mathbf{Z}_{p^l}[x_1]$ are chosen to be monic, then for any integer $k \geq 1$ conditions (6.97) and (6.98) uniquely determine factors $u^{(k)}, w^{(k)} \in \mathbf{Z}_{p^l}[x_1, \ldots, x_v] / \mathrm{I}^k$ which each have leading coefficient 1 with respect to the indeterminate x_1.

•

6.8. THE MULTIVARIATE HENSEL LIFTING ALGORITHM

The algorithm which follows directly from Theorem 6.6 has some deficiencies which must be corrected before we can present an efficient algorithm for the multivariate Hensel construction. One such deficiency is the *leading coefficient problem*. For this problem, we will adopt a solution which is directly analogous to the solution developed in Section 6.6 and implemented in Algorithm 6.1 for the univariate case. Less obvious are the efficiency problems associated with the construction presented in the proof of Theorem 6.6. This construction exhibits poor performance in cases where some of the evaluation points α_j are nonzero and this problem is sometimes called the *bad-zero problem*. We will examine this problem now.

The Bad-Zero Problem

The source of the performance problems is the requirement in the proof of Theorem 6.6 that the error $e^{(k)}$ must be expressed in the I-adic form (6.100). This step can lead to very large intermediate expression swell resulting in an exponential cost function for the algorithm. The following example will serve to illustrate.

Example 6.10. Let $p = 5$, $l = 1$,

$$a(x,y,z) = x^2y^4z - xy^9z^2 + xyz^3 + 2x - y^6z^4 - 2y^5z,$$

and $I = <y-1, z-1>$. Noting that

$$a(x,y,z) \equiv x^2 + 2x + 2 \pmod{<I, 5>}$$

we have

$$a(x,y,z) \equiv (x-2)(x-1) \pmod{<I, 5>}.$$

Choosing $u^{(1)}(x) = x - 2$ and $w^{(1)}(x) = x - 1$, the conditions of Theorem 6.6 are satisfied. Since $a(x,y,z)$ is not monic we might expect the Hensel construction to fail to produce true factors in $\mathbf{Z}[x,y,z]$, but in this example the factor $w(x,y,z)$ is monic so the Hensel construction will succeed even though we are ignoring the leading coefficient problem.

The effect of representing the error at each step of the iteration in I-adic form can be seen by considering the I-adic form of $a(x,y,z)$

$$\begin{aligned} a(x,y,z) \equiv\ & (x^2+2x+2) - (x^2+1)(y-1) + (x^2+x-1)(z-1) + (x^2-x)(y-1)^2 \\ & - (x^2-1)(y-1)(z-1) + (2x-1)(z-1)^2 - (x^2-x)(y-1)^3 + (x^2-2x)(y-1)^2(z-1) \\ & - (x+1)(y-1)(z-1)^2 + (x+1)(z-1)^3 + (x^2-x)(y-1)^4 + (-x^2+2x)(y-1)^3(z-1) \\ & - x(y-1)^2(z-1)^2 + (x+1)(y-1)(z-1)^3 - (z-1)^4 - (x-2)(y-1)^5 \\ & + (x^2-2x)(y-1)^4(z-1) + x(y-1)^3(z-1)^2 - (y-1)(z-1)^4 + (x-1)(y-1)^6 \\ & - (2x+1)(y-1)^5(z-1) - x(y-1)^4(z-1)^2 - x(y-1)^7 + (2x+1)(y-1)^6(z-1) \end{aligned}$$

$$- (x+1)(y-1)^5(z-1)^2 + x(y-1)^8 + (-2x)(y-1)^7(z-1) + (x-1)(y-1)^6(z-1)^2$$
$$+ (y-1)^5(z-1)^3 - x(y-1)^9 + 2x(y-1)^8(z-1) - x(y-1)^7(z-1)^2$$
$$+ (y-1)^6(z-1)^3 - (y-1)^5(z-1)^4 - 2x(y-1)^9(z-1) + x(y-1)^8(z-1)^2$$
$$- (y-1)^6(z-1)^4 - x(y-1)^9(z-1)^2 \pmod 5.$$

We see that the I-adic representation contains 38 terms compared with 6 terms in the original expanded representation (which is an I-adic representation with respect to the ideal $I = \langle y, z \rangle$). The number of polynomial diophantine equations of the form (6.101) which must be solved is proportional to the number of terms in the I-adic form of $a(x, y, z)$.

Carrying out the Hensel construction for this example, the factors are developed in I-adic form as follows:

$$u^{(7)} = (x-2) + (-x+1)(y-1) + (x-2)(z-1) + (x)(y-1)^2 + (-x-2)(y-1)(z-1)$$
$$+ (-2)(z-1)^2 + (-x)(y-1)^3 + (x)(y-1)^2(z-1) + (-2)(y-1)(z-1)^2$$
$$+ (z-1)^3 + (x)(y-1)^4 + (-x)(y-1)^3(z-1) + (1)(y-1)(z-1)^3$$
$$+ (x)(y-1)^4(z-1);$$

$$w^{(7)} = (x-1) + (-1)(z-1) + (-1)(y-1)^5 + (-1)(y-1)^5(x-1).$$

Expressing these factors in expanded form (and noting that the coefficient arithmetic is being done modulo 5), we have

$$u^{(7)} \equiv xy^4z + yz^3 + 2 \pmod 5;$$
$$w^{(7)} \equiv x - y^5z \pmod 5.$$

At this point the iteration can be halted because

$$e^{(7)} = a(x, y, z) - u^{(7)} \cdot w^{(7)} = 0.$$

Again note that there are many more terms in the I-adic representation of the factors than in the expanded representation.

●

It is clear from Example 6.10 that the use of nonzero evaluation points can cause a severe case of intermediate expression swell. However it is not always possible to choose the evaluation points to be zero because in the applications of the Hensel construction (see Chapter 7), a necessary condition is that the leading coefficient must not vanish under the evaluation homomorphism. The original implementation of the multivariate Hensel construction (the EZ-GCD algorithm) degraded significantly on problems requiring nonzero evaluation points.

One method of dealing with the I-adic representation in an implementation of the multivariate Hensel construction is to initially perform the change of variables

$$x_j \leftarrow x_j + \alpha_j,\ 2 \le j \le v,$$

if the ideal is $I = < x_2 - \alpha_2, \ldots, x_v - \alpha_v >$. The required I-adic representation is then a straightforward expanded representation based on the new ideal $< x_2, \ldots, x_v >$. However it is important to note that this method suffers from the problem of intermediate expression swell exactly as exhibited in Example 6.10. For in the original polynomial $a(x,y,z)$ in Example 6.10, the result of performing the change of variables:

$$y \leftarrow y + 1;\ \ z \leftarrow z + 1$$

and then expanding, is precisely the 38-term form of $a(x,y,z)$ displayed in the example, with $y - 1$ replaced by y and $z - 1$ replaced by z.

An improvement to the algorithm can be obtained by avoiding the change of variables (or any other explicit representation of the I-adic form) as follows. At iteration step k let $e^{(k)}$ be represented as a multivariate polynomial in expanded form. It is desired to compute the coefficients $c_{\mathbf{i}}(x_1)$ appearing in (6.100), the I-adic representation of $e^{(k)}$, for all order-k vector indices

$$\mathbf{i} = (i_1, i_2, \ldots, i_k).$$

Noting that some of the indices in the vector $\mathbf{i}$ may be repeated, let the term in $e^{(k)}$ corresponding to a particular vector index $\mathbf{i}$ be of the form

$$c_{\mathbf{i}}(x_1)\,(x_{j_1} - \alpha_{j_1})^{n_1}\,(x_{j_2} - \alpha_{j_2})^{n_2} \cdots (x_{j_m} - \alpha_{j_m})^{n_m}$$

where all factors appearing here are distinct. Then the coefficient $c_{\mathbf{i}}(x_1)$ can be computed directly from the expanded representation of $e^{(k)}$ by using the differentiation formula

$$c_{\mathbf{i}}(x_1) = \frac{1}{n_1! \cdots n_m!}\,\phi_I\left(\left(\frac{\partial}{\partial x_{j_1}}\right)^{n_1} \cdots \left(\frac{\partial}{\partial x_{j_m}}\right)^{n_m} e^{(k)} \right). \tag{6.104}$$

This leads to an organization of the main iteration loop of the Hensel construction which can be expressed as follows (where d is the maximum total degree with respect to the indeterminates $x_2, \ldots, x_v$ over all terms in the input polynomial $a(x_1, \ldots, x_v)$).

> **for** k **from** 1 **to** d **while** $e^{(k)} \ne 0$ **do** {
>   **for** each order-k vector index $(i_1, \ldots, i_k)$ with $2 \le i_1 \le \cdots \le i_k \le v$ **do** {
>   Calculate $c_{\mathbf{i}}(x_1)$ using (6.104)
>   Solve equation (6.101) for $\sigma_{\mathbf{i}}(x_1)$ and $\tau_{\mathbf{i}}(x_1)$
>   Update $u^{(k)}$ and $w^{(k)}$ according to (6.103) }
>   Update $e^{(k)}$ }

This organization of the iteration loop is in contrast to the organization which follows more directly from the proof of Theorem 6.6, using the "change of variables" concept, as follows.

```
Substitute x_j ← x_j + α_j (2 ≤ j ≤ v) in a(x_1, . . . , x_v)
for k from 1 to d while e^(k) ≠ 0 do {
   for each term of total degree k appearing in the expanded form of e^k do {
      Pick off the coefficient c_i(x_1)
      Solve equation (6.101) for σ_i(x_1) and τ_i(x_1)
      Update u^(k) and w^(k) according to (6.103) }
   Update e^(k) }
Substitute x_j ← x_j − α_j (2 ≤ j ≤ v) in u^(k) and w^(k).
```

(In both of the above program segments, it is understood that

$$e^{(k)} = a(x_1, \ldots, x_v) - u^{(k)}w^{(k)}$$

computed in $\mathbf{Z}[x_1, \ldots, x_v]$ in expanded form.)

A careful examination of these two organizations of the iteration loop shows that neither one is fully satisfactory for dealing with sparse multivariate polynomials. Recall our observation at the beginning of this chapter that, in practice, multivariate polynomials are generally sparse and the advantage of the Hensel construction over Chinese remainder (interpolation) algorithms is the ability to take advantage of sparseness. In the approach which applies the change of variables, there is potentially a serious loss of sparsity because the representation of the polynomial $a(x_1, \ldots, x_v)$ after substituting the change of variables can have many more terms than the original representation (see Example 6.10). Note, however, that after this substitution step, the iteration then goes on to perform calculations only for terms that *actually appear* in the expanded form of $e^{(k)}$. In contrast, in the approach which avoids the change of variables but uses instead the differentiation formula (6.104), the inner **for**-loop iterates over *all possible* order-k vector indices $\mathbf{i} = (i_1, \ldots, i_k)$ and, in practice, a large proportion of the coefficients $c_{\mathbf{i}}(x_1)$ will be found to be zero. Since the differentiations and substitutions required by formula (6.104) can be performed relatively efficiently for polynomials (particularly if it is programmed to "remember" computed derivatives since higher-order derivatives rely on lower-order derivatives) and since we would program the inner loop to check if $c_{\mathbf{i}}(x_1) = 0$ and avoid any additional work in that case, the method using formula (6.104) is generally preferable. However, the overhead of calculating $c_{\mathbf{i}}(x_1)$ for all possible choices of the vector index $\mathbf{i}$ is significant and the cost of this overhead grows exponentially in the number of variables, independently of the sparsity of the polynomials. In particular, note that in the (relatively common) case where all of the evaluation points are zero the method using (6.104) will be much more costly than the direct approach.

Polynomial Diophantine Equations in $\mathbf{Z}_{p^l}[x_1, \ldots, x_j]$

A significantly more efficient organization of the multivariate Hensel construction was developed by Wang [6] and he called it the EEZ-GCD (Enhanced EZ) algorithm. The main feature of the new algorithm is that it uses a variable-by-variable approach to avoid the "exponential overhead" discussed above. In the context of Figure 6.2, the multivariate Hensel construction lifting the solution from $\mathbf{Z}_{p^l}[x_1]$ to $\mathbf{Z}[x_1, \ldots, x_v]$ is replaced by a sequence of $v-1$ single-variable Hensel constructions to lift the solution

from $\mathbf{Z}_{p^l}[x_1]$ to $\mathbf{Z}_{p^l}[x_1, x_2]$,

from $\mathbf{Z}_{p^l}[x_1, x_2]$ to $\mathbf{Z}_{p^l}[x_1, x_2, x_3]$,

.

.

.

from $\mathbf{Z}_{p^l}[x_1, \ldots, x_{v-1}]$ to $\mathbf{Z}_{p^l}[x_1, \ldots, x_{v-1}, x_v]$.

(As usual, p^l is chosen large enough so that the final solution over the ring $\mathbf{Z}_{p^l}$ is equated with the desired solution over $\mathbf{Z}$.)

Recall that the basic computation to be performed in applying a step of a Hensel iteration is to solve a polynomial diophantine equation in the "base domain". For the univariate Hensel construction in Figure 6.2, the "base domain" is $\mathbf{Z}_p[x_1]$ and Theorem 2.6 gives a method for solving the polynomial diophantine equations. For the "base domain" $\mathbf{Z}_{p^l}[x_1]$, we developed a method in Theorem 6.5 for solving the polynomial diophantine equations. In order to carry out the variable-by-variable Hensel construction, we need a method for solving polynomial diophantine equations in multivariate "base domains" $\mathbf{Z}_{p^l}[x_1, \ldots, x_j]$ and we turn now to the development of such a method. Just as in the proof of Theorem 6.5, we will apply Newton's iteration to the problem and indeed we will employ a variable-by-variable technique for solving this sub-problem.

The polynomial diophantine equation to be solved is to find multivariate polynomials $\sigma_j(x_1, \ldots, x_j), \tau_j(x_1, \ldots, x_j) \in \mathbf{Z}_{p^l}[x_1, \ldots, x_j]$ such that

$$\sigma_j(x_1, \ldots, x_j)\, u(x_1, \ldots, x_j) + \tau_j(x_1, \ldots, x_j)\, w(x_1, \ldots, x_j)$$
$$\equiv c(x_1, \ldots, x_j) \pmod{< \mathrm{I}_j^{d+1}, p^l >} \qquad (6.105)$$

where $\mathrm{I}_j = < x_2 - \alpha_2, \ldots, x_j - \alpha_j >$, d is the maximum total degree of the solution polynomials with respect to the indeterminates $x_2, \ldots, x_j$, and $u(x_1, \ldots, x_j), w(x_1, \ldots, x_j), c(x_1, \ldots, x_j) \in \mathbf{Z}_{p^l}[x_1, \ldots, x_j]$ are given polynomials with $\phi_{<\mathrm{I},p>}(u(x_1, \ldots, x_j))$ and $\phi_{<\mathrm{I},p>}(w(x_1, \ldots, x_j))$ relatively prime polynomials in the Euclidean domain $\mathbf{Z}_p[x_1]$. The equation to which we will apply Newton's iteration is

$$G(\sigma_j, \tau_j) = \sigma_j\, u(x_1, \ldots, x_j) + \tau_j\, w(x_1, \ldots, x_j) - c(x_1, \ldots, x_j) = 0.$$

Choosing the particular variable x_j for lifting and proceeding as in previous sections, if we have the order-k approximations $\sigma_j^{(k)}, \tau_j^{(k)}$ satisfying

$$G(\sigma_j^{(k)}, \tau_j^{(k)}) \equiv 0 \pmod{< (x_j - \alpha_j)^k, I_{j-1}^{d+1}, p^l >}$$

and if we obtain correction terms $\Delta\sigma_j^{(k)}, \Delta\tau_j^{(k)}$ which satisfy the equation

$$G_{\sigma_j}(\sigma_j^{(k)}, \tau_j^{(k)})\, \Delta_j^{(k)} + G_{\tau_j}(\sigma_j^{(k)}, \tau_j^{(k)}) \Delta\tau_j^{(k)} \equiv -G(\sigma_j^{(k)}, \tau_j^{(k)}) \tag{6.106}$$
$$\pmod{< (x_j - \alpha_j)^{k+1}, I_{j-1}^{d+1}, p^l >}$$

then

$$\sigma_j^{(k+1)} = \sigma_j^{(k)} + \Delta\sigma_j^{(k)}, \quad \tau_j^{(k+1)} = \tau_j^{(k)} + \Delta\tau_j^{(k)}$$

will be order-$(k+1)$ approximations satisfying

$$G(\sigma_j^{(k+1)}, \tau_j^{(k+1)}) \equiv 0 \pmod{< (x_j - \alpha_j)^{k+1}, I_{j-1}^{d+1}, p^l >}.$$

Writing the correction terms in the form

$$\Delta\sigma_j^{(k)} = s_{j,k}(x_1, \ldots, x_{j-1})\, (x_j - \alpha_j)^k, \quad \Delta\tau_j^{(k)} = t_{j,k}(x_1, \ldots, x_{j-1})\, (x_j - \alpha_j)^k$$

where $s_{j,k}(x_1, \ldots, x_{j-1}), t_{j,k}(x_1, \ldots, x_{j-1}) \in \mathbf{Z}_{p^l}[x_1, \ldots, x_{j-1}]$, substituting for the partial derivatives, and dividing through by $(x_j - \alpha_j)^k$, equation (6.106) becomes

$$u(x_1, \ldots, x_j)\, s_{j,k}(x_1, \ldots, x_{j-1}) + w(x_1, \ldots, x_j)\, t_{j,k}(x_1, \ldots, x_{j-1})$$
$$\equiv \frac{c(x_1, \ldots, x_j) - \sigma_j^{(k)} u(x_1, \ldots, x_j) - \tau_j^{(k)} w(x_1, \ldots, x_j)}{(x_j - \alpha_j)^k}$$
$$\pmod{< (x_j - \alpha_j), I_{j-1}^{d+1}, p^l >}. \tag{6.107}$$

Note that $I_{j-1} = < x_2 - \alpha_2, \ldots, x_{j-1} - \alpha_{j-1} >$, the interpretation of I_1 is as the empty ideal, and note that the above development has assumed $j > 1$ since if $j = 1$ then the solution of the polynomial diophantine equation (6.105) is given by Theorem 6.5.

We thus have a recursive algorithm for solving the polynomial diophantine equation (6.105). The order-1 approximations $\sigma_j^{(1)}, \tau_j^{(1)}$ with respect to the ideal $< x_j - \alpha_j >$ are obtained by solving equation (6.105) modulo $< x_j - \alpha_j >$, i.e. by solving the $(j - 1)$-variable problem

$$\sigma_{j-1}(x_1, \ldots, x_{j-1})\, u(x_1, \ldots, x_{j-1}, \alpha_j) + \tau_{j-1}(x_1, \ldots, x_{j-1})\, w(x_1, \ldots, x_{j-1}, \alpha_j)$$
$$\equiv c(x_1, \ldots, x_{j-1}, \alpha_j) \pmod{< I_{j-1}^{d+1}, p^l >}$$

and then setting

$$\sigma_j^{(1)} = \sigma_{j-1};\ \ \tau_j^{(1)} = \tau_{j-1}.$$

Then for $k = 1, 2, \ldots, d$, we solve equation (6.107) which, noting that it is to be solved modulo $< x_j - \alpha_j >$, takes the form of the $(j-1)$-variable problem

$$u(x_1, \ldots, x_{j-1}, \alpha_j)\, s_{j,k}(x_1, \ldots, x_{j-1}) + w(x_1, \ldots, x_{j-1}, \alpha_j)\, t_{j,k}(x_1, \ldots, x_{j-1})$$
$$\equiv e_k(x_1, \ldots, x_{j-1}) \pmod{< \mathrm{I}_{j-1}^{d+1}, p^l >}$$

where $e_k(x_1, \ldots, x_{j-1})$ denotes the coefficient of $(x_j - \alpha_j)^k$ in the $< x_j - \alpha_j >$-adic representation of the polynomial

$$e(x_1, \ldots, x_j) = c(x_1, \ldots, x_j) - \sigma_j^{(k)}\, u(x_1, \ldots, x_j) - \tau_j^{(k)}\, w(x_1, \ldots, x_j).$$

The base of the recursion is the univariate polynomial diophantine equation in $\mathbf{Z}_{p^l}[x_1]$ which can be solved by the method of Theorem 6.5.

In the algorithm based on the above development, the solution of equation (6.105) will satisfy the degree constraint

$$\deg_1(\sigma_j(x_1, \ldots, x_j)) < \deg_1(w(x_1, \ldots, x_j))$$

(where $\deg_1$ is the "degree in x_1" function) since the solution of the univariate case of (6.105) satisfies such a constraint (by Theorem 6.5), as does the solution of the univariate case of equation (6.107) which defines the correction terms, leading by induction to the general result.

The recursive algorithm for solving multivariate polynomial diophantine equations is presented as Algorithm 6.2. The conditions which must be satisfied by the input polynomials are the conditions required by Theorem 6.5 for the univariate case at the base of the recursion. The algorithm presented here is a generalization of the algorithm discussed above, to allow for a multi-term polynomial diophantine equation rather than being restricted to a two-term equation. This will allow us to present the multivariate Hensel lifting algorithm in a form which lifts multiple factors at once, rather than being restricted to just two factors.

At the base of the recursion, Algorithm 6.2 invokes procedure UnivariateDiophant which is presented as Algorithm 6.3. Procedure UnivariateDiophant solves a multi-term generalization of the polynomial diophantine equation considered in Theorem 6.5. The algorithm is organized such that it invokes two sub-procedures, MultiTermEEAlift and EEAlift, which are also presented as part of Algorithm 6.3. Procedure EEAlift implements a generalization of the extended Euclidean algorithm such that the solution in $\mathbf{Z}_p[x]$ is lifted up to a solution in $\mathbf{Z}_{p^k}[x]$. Procedure MultiTermEEAlift implements a multi-term generalization of the extended Euclidean algorithm over $\mathbf{Z}_{p^k}[x]$. Note that procedure MultiTermEEAlift invokes procedure MultivariateDiophant, which might appear to lead to an endless recursion; however, the invocation from MultiTermEEAlift is specifically for a two-term polynomial diphantine equation and therefore it will not cause a re-invocation of MultiTermEEAlift.

The key to the generalization of the two-term case discussed throughout this chapter, to the multi-term case presented in the following algorithms, is contained in the MultiTermEEAlift procedure. This procedure implements a multi-term extended Euclidean algorithm over $\mathbf{Z}_{p^k}[x]$ by reducing to the two-term case, as follows. Suppose that we are given $r>2$ polynomials $a_1(x), \ldots, a_r(x) \in \mathbf{Z}_{p^k}[x]$. The multi-term version of the extended Euclidean equation is specified in terms of the r polynomials $b_i(x)$, $i = 1, \ldots, r$ defined by

$$b_i(x) = a_1(x) \times \cdots \times a_{i-1}(x) \times a_{i+1}(x) \times \cdots \times a_r(x) .$$

The task is to compute polynomials $s_j(x)$, $j = 1, \ldots, r$ such that

$$s_1(x)\, b_1(x) + \cdots + s_r(x)\, b_r(x) \equiv 1 \pmod{p^k} \tag{6.108}$$

with $\deg(s_j(x)) < \deg(a_j(x))$. The algorithm proceeds as follows. Define

$$\beta_0(x) = 1$$

and then for each j from 1 to $r-1$ solve the two-term equation

$$\beta_j(x) \times a_j(x) + s_j(x) \times \prod_{i=j+1}^{r} a_i(x) \equiv \beta_{j-1}(x) \pmod{p^k} \tag{6.109}$$

for $\beta_j(x)$ and $s_j(x)$. Finally, define

$$s_r(x) = \beta_{r-1}(x) .$$

It is straightforward to verify that the polynomials $s_j(x)$, $j = 1, \ldots, r$ defined by this process satisfy equation (6.108). For example, consider the final three terms on the left hand side of equation (6.108), which can be written in the form

$$s_{r-2}(x) \times (\prod_{i=1}^{r-3} a_i(x))\, a_{r-1}(x)\, a_r(x) + s_{r-1}(x) \times (\prod_{i=1}^{r-2} a_i(x))\, a_r(x) + s_r(x) \times \prod_{i=1}^{r-1} a_i(x) .$$

Replacing $s_r(x)$ by $\beta_{r-1}(x)$ in the final term, and using case $j = r-1$ of (6.109), we see that the final two terms collapse together into the term

$$\beta_{r-2}(x) \times \prod_{i=1}^{r-2} a_i(x) .$$

Then using case $j = r-2$ of (6.109), this term combines with the preceding term to yield

$$\beta_{r-3}(x) \times \prod_{i=1}^{r-3} a_i(x) .$$

It is clear that the terms on the left hand side of equation (6.108) continue collapsing in this manner until we are left with the term $\beta_0(x)$ which was defined to be 1, showing that equation (6.108) is satisfied. It can also be seen that the desired degree constraint $\deg(s_j(x)) < \deg(a_j(x))$ is the natural degree constraint satisfied by the solution of the two-term equation (6.109), for each j.

Algorithm 6.2. Multivariate Polynomial Diophantine Equations.

```
procedure MultivariateDiophant(a, c, I, d, p, k)
  # Solve in the domain Z_{p^k}[x_1, . . . , x_v] the multivariate polynomial
  # diophantine equation
  #   σ_1 × b_1 + · · · + σ_r × b_r ≡ c  (mod <I^{d+1}, p^k>)
  # where, in terms of the given list of polynomials a_1, . . . , a_r,
  # the polynomials b_i, i = 1, . . . , r, are defined by:
  #   b_i = a_1 × · · · × a_{i-1} × a_{i+1} × · · · × a_r .
  # The unique solution σ_i, i = 1, . . . , r, will be computed such that
  #   degree(σ_i, x_1) < degree(a_i, x_1) .
  #
  # Conditions: p must not divide lcoeff(a_i mod I), i = 1, . . . , r;
  # a_i mod <I, p>, i = 1, . . . , r, must be pairwise relatively prime
  # in Z_p[x_1];  and degree(c, x_1) <  sum(degree(a_i, x_1), i = 1,...,r).
  #
  # INPUT:
  #   (1) A list a of r > 1 polynomials in the domain Z_{p^k}[x_1, . . . , x_v].
  #   (2) A polynomial c ∈ Z_{p^k}[x_1, . . . , x_v].
  #   (3) I, a list of equations [x_2 = α_2, x_3 = α_3, . . . , x_v = α_v]
  #       (possibly null, in which case it is a univariate problem)
  #       representing an evaluation homomorphism;
  #       mathematically, we view it as the ideal
  #         I = <x_2 − α_2, x_3 − α_3, . . . , x_v − α_v> .
  #   (4) A nonnegative integer d specifying the maximum total degree
  #       with respect to x_2, . . . , x_v of the desired result.
  #   (5) A prime integer p.
  #   (6) A positive integer k specifying that the coefficient arithmetic
  #       is to be performed modulo p^k.
  #
  # OUTPUT:
  #   The value returned is the list σ = [σ_1, . . . , σ_r].
  #
  # Remark:  The mod operation must use the symmetric representation.
```

Algorithm 6.2 (continued). Multivariate Polynomial Diophantine Equations.

```
# 1. Initialization.
r ← number of polynomials in a
ν ← 1 + number of equations in I
x_ν ← lhs(I_{ν−1});  α_ν ← rhs(I_{ν−1})

if ν > 1 then {
   # 2.1. Multivariate case.
   A ← product(a_i, i = 1,...,r)
   for j from 1 to r do { b_j ← A / a_j }
   anew ← substitute(x_ν=α_ν, a);  cnew ← substitute(x_ν=α_ν, c)
   Inew ← updated list I with x_ν = α_ν deleted
   σ ← MultivariateDiophant(anew, cnew, Inew, d, p, k)
   e ← (c − sum(σ_i b_i, i = 1,...,r)) mod p^k
   monomial ← 1
   for m from 1 to d while e ≠ 0 do {
      monomial ← monomial × (x_ν − α_ν)
      cm ← coeff of (x_ν − α_ν)^m in the Taylor expansion of e about x_ν = α_ν
      if cm ≠ 0 then {
         Δs ← MultivariateDiophant(anew, cm, Inew, d, p, k)
         Δs ← Δs × monomial    # element-by-element operations
         σ ← σ + Δs            # element-by-element operations
         e ← (e − sum(Δs_i b_i, i = 1,...,r)) mod p^k  } } }
else {
   # 2.2. Univariate case.
   x_1 ← the variable appearing in a
   # Method: For each power of x_1, call UnivariateDiophant.
   σ ← zero list of length r
   for each term z in c do {
      m ← degree(z, x_1);  cm ← lcoeff(z)
      Δs ← UnivariateDiophant(a, x_1, m, p, k)
      Δs ← Δs × cm         # element-by-element operations
      σ ← σ + Δs           # element-by-element operations  } }

return(σ mod p^k)
end
```

Algorithm 6.3. Univariate Polynomial Diophantine Equations.

```
procedure UnivariateDiophant(a, x, m, p, k)
  # Solve in Z_{p^k}[x] the univariate polynomial diophantine equation
  #    σ_1 × b_1 + ... + σ_r × b_r ≡ x^m (mod p^k)
  # where, in terms of the given list of polynomials a_1, ..., a_r,
  # the polynomials b_i, i = 1, ..., r, are defined by:
  #    b_i = a_1 × ... × a_{i-1} × a_{i+1} × ... × a_r.
  # The unique solution σ_1, ..., σ_r, will be computed such that
  #    deg(σ_i) < deg(a_i) .
  #
  # Conditions: p must not divide lcoeff(a_i), i = 1, ..., r;
  #   a_i mod p, i = 1, ..., r, must be pairwise relatively prime in Z_p[x] .
  #
  # OUTPUT:
  #   The value returned is the list σ = [σ_1, ..., σ_r].

  r ← number of polynomials in a
  if r > 2 then {
     s ← MultiTermEEAlift(a, p, k);  result ← [ ]
     for j from 1 to r do {
          result ← append(result, rem(x^m s_j, a_j) mod p^k) } }
  else {
     s ← EEAlift(a_2, a_1, p, k); q ← quo(x^m s_1, a_1) mod p^k
     result ← [ rem(x^m s_1, a_1) mod p^k, (x^m s_2 + q a_2) mod p^k ] }
  return(result)
end

# MultiTermEEAlift computes s_1, ..., s_r such that
#    s_1 × b_1 + ... + s_r × b_r ≡ 1 (mod p^k)
# with deg(s_j) < deg(a_j) where, in terms of the given list of
# polynomials a_1, ..., a_r, the polynomials b_i are defined by:
#    b_i = a_1 × ... × a_{i-1} × a_{i+1} × ... × a_r, i = 1, ..., r .
#
# Conditions: p must not divide lcoeff(a_i), i = 1, ..., r;
#   a_i mod p, i = 1, ..., r, must be pairwise relatively prime in Z_p[x] .
```

Algorithm 6.3 (continued). Univariate Polynomial Diophantine Equations.

procedure MultiTermEEAlift(a, p, k)

$r \leftarrow$ number of polynomials in a

$q_{r-1} \leftarrow a_r$
for j **from** $r-2$ **by** -1 **to** 1 **do** {
  $q_j \leftarrow a_{j+1} \times q_{j+1}$ }
$\beta_0 \leftarrow 1$
for j **from** 1 **to** $r-1$ **do** {
  $\sigma \leftarrow$ MultivariateDiophant($[\, q_j, a_j \,], \beta_{j-1}, [\,], 0, p, k$)
  $\beta_j \leftarrow \sigma_1$; $s_j \leftarrow \sigma_2$ }
$s_r \leftarrow \beta_{r-1}$

return($[\, s_1, \ldots, s_r \,]$)
end

\# EEAlift computes s, t such that $s\, a + t\, b \equiv 1 \pmod{p^k}$
\# with $\deg(s) < \deg(b)$ and $\deg(t) < \deg(a)$.
\# Assumption: GCD($a \bmod p$, $b \bmod p$) = 1 in $\mathbf{Z}_p[x]$.

procedure EEAlift(a, b, p, k)

$x \leftarrow$ the variable appearing in a and b
$amodp \leftarrow a \bmod p$; $bmodp \leftarrow b \bmod p$
$s, t \leftarrow$ polynomials in $\mathbf{Z}_p[x]$ computed by Algorithm 2.2 such that

$$s\, amodp + t\, bmodp \equiv 1 \pmod{p}$$

$smodp \leftarrow s$; $tmodp \leftarrow t$; $modulus \leftarrow p$
for j **from** 1 **to** $k-1$ **do** {
  $e \leftarrow 1 - s \times a - t \times b$; $c \leftarrow \dfrac{e}{modulus} \bmod p$
  $\bar{\sigma} \leftarrow smodp \times c$; $\bar{\tau} \leftarrow tmodp \times c$
  $q \leftarrow \mathrm{quo}(\bar{\sigma}, bmodp) \bmod p$
  $\sigma \leftarrow \mathrm{rem}(\bar{\sigma}, bmodp) \bmod p$
  $\tau \leftarrow (\bar{\tau} + q \times amodp) \bmod p$
  $s \leftarrow s + \sigma \times modulus$; $t \leftarrow t + \tau \times modulus$
  $modulus \leftarrow modulus \times p$ }

return($[\, s, t \,]$)
end

We now present the multivariate Hensel lifting algorithm as Algorithm 6.4. We present the multi-factor lifting algorithm rather than restricting to the case where only two factors are to be lifted. The main work is performed by invoking procedure MultivariateDiophant which was presented as Algorithm 6.2, and since it has been designed to solve multi-term polynomial diophantine equations, the multi-factor Hensel lifting algorithm follows easily.

Algorithm 6.4. Multivariate Hensel Lifting Algorithm.

procedure MultivariateHensel(a, I, p, l, u, lcU)

\# INPUT:

\# (1) A multivariate polynomial $a(x_1, \ldots, x_v) \in \mathbf{Z}[x_1, \ldots, x_v]$
\# which is primitive as a polynomial in the special variable x_1.

\# (2) I, a list of equations $[x_2 = \alpha_2, x_3 = \alpha_3, \ldots, x_v = \alpha_v]$
\# representing the evaluation homomorphism used; mathematically,
\# we view it as the ideal $I = \langle x_2 - \alpha_2, x_3 - \alpha_3, \ldots, x_v - \alpha_v \rangle$
\# and the following condition must hold: $\mathrm{lcoeff}(a, x_1) \neq 0 \pmod{I}$.

\# (3) A prime integer p which does not divide lcoeff(a mod I).

\# (4) A positive integer l such that $p^l/2$ bounds the magnitudes of all
\# integers appearing in a and in any of its factors to be computed.

\# (5) A list u of $n>1$ univariate polynomials in $\mathbf{Z}_{p^l}[x_1]$ which are
\# pairwise relatively prime in the Euclidean domain $\mathbf{Z}_p[x_1]$,
\# such that $a \equiv u_1 \times u_2 \times \cdots \times u_n \pmod{\langle I, p^l \rangle}$.

\# (6) A list lcU of the n correct multivariate leading coefficients
\# corresponding to the univariate factors u.
\#

\# OUTPUT:

\# (1) If there exist n polynomials $U_1, U_2, \ldots, U_n \in \mathbf{Z}[x_1, \ldots, x_v]$
\# such that $a = U_1 \times U_2 \times \cdots \times U_n$ and for each $i = 1, 2, \ldots, n$:
\# $U_i / \mathrm{lcoeff}(U_i, x_1) \equiv u_i / \mathrm{lcoeff}(u_i, x_1) \pmod{\langle I, p^l \rangle}$
\# (where the divisions here are in the ring of integers mod p^l),
\# then the list $U = [U_1, U_2, \ldots, U_n]$ will be the value returned.

\# (2) Otherwise, the value returned will signal "no such factorization".
\#

\# Remark: The mod operation must use the symmetric representation.

Algorithm 6.4 (continued). Multivariate Hensel Lifting Algorithm.

```
# 1. Initialization for the multivariate iteration

ν ← 1 + number of equations in I
A_ν ← a

for j from ν by −1 to 2 do {
    x_j ← lhs(I_{j−1});  α_j ← rhs(I_{j−1})
    A_{j−1} ← substitute(x_j = α_j, A_j) mod p^l }

maxdeg ← max(degree(a, x_i), i = 2, ..., ν)
U ← u;  n ← number of polynomials in u

# 2. Variable-by-variable Hensel iteration
for j from 2 to ν do {
    U1 ← U;  monomial ← 1
    for m from 1 to n do {
        if lcU_m ≠ 1 then {
            coef ← substitute({I[j], ..., I[ν−1]}, lcU_m) mod p^l
            U ← updated list U with lcoeff(U_m, x_1) replaced by coef  } }
    e ← A_j − product(U_i, i = 1, ..., n)
    for k from 1 to degree(A_j, x_j) while e ≠ 0 do {
        monomial ← monomial × (x_j−α_j)
        c ← coeff of (x_j − α_j)^k in the Taylor expansion of e about x_j = α_j

        if c ≠ 0 then {
            ΔU ← MultivariateDiophant(U1, c, [I[1], ..., I[j−2]], maxdeg, p, l)
            ΔU ← ΔU × monomial     # element-by-element operations
            U ← (U + ΔU) mod p^l   # element-by-element operations
            e ← (A_j − product(U_i, i = 1, ..., n)) mod p^l  } } }

# 3. Check termination status

if a = product(U_i, i = 1, ..., n) then return(U)
    else return( no such factorization exists )
end
```

Exercises

1. Determine the p-adic representation of the integer $u = -11109234276$, with $p = 43$.

2. Using $p = 43$, determine the p-adic representation of the polynomial $u(x) = 143x^3 - 1253x - 11109234276$.

3. The p-adic integers are defined as the set of all numbers of the form

$$a_0 + a_1 \cdot p + \cdots + a_n \cdot p^n + \cdots$$

with $0 \le a_i < p$. The p-adic integers form a ring which is often viewed as a "completion" of the integers, in much the same way that the reals can be considered as the completion of the rational numbers. In Chapter 6 we have shown that all positive integers have p-adic representations.

(a) Show that for any integer p, the p-adic integer

$$a = (p-1) + (p-1) \cdot p + (p-1) \cdot p^2 + \cdots$$

represents -1, by showing that $a + 1 = 0$. Thus, all negative integers have p-adic representations.

(b) From part (a), we see that the p-adic integers include all integers. They also can include some rational numbers. Give an example of this by finding the 5-adic representation of $-1/4$.

(c) The p-adic integers also include many non-rational numbers. For example, let $\sqrt{11}$ denote a solution of $x^2 - 11 = 0$. Find the first three terms of the 5-adic expansion of $\sqrt{11}$.

4. Let

$$a(x,y,z) = (x^2y^2 + z^2 + 1) \cdot (x^2 - y^2 + z^2 - 1)$$

and let I be the ideal $<y-1, z>$. Express $a(x,y,z)$ in its I-adic representation.

5. Determine the cube root of

$$x^6 - 531x^5 + 94137x^4 - 5598333x^3 + 4706850x^2 - 1327500x + 125000$$

using reduction mod 3 and Newton's iteration.

6. Determine the square root of

$$x^8 + 12768x^6 + 40755070x^4 - 2464224x^2 + 37249$$

using reduction mod 3 and the quadratic Newton's iteration.

7. Solve the equation

$$a(x)^2 - a(x) = x^6 - 2x^4y^2 + x^3z + x^2y^4 + xy^2z + z^2 - 1$$

over $\mathbf{Z}_3[x,y,z]$ using reduction mod I and the ideal-adic Newton's iteration, where I is the ideal $<y-1, z>$.

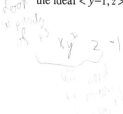

8. Give the details of the quadratic ideal-adic Newton's iteration in the multivariate case. Prove that the convergence is indeed quadratic.

9. An *idempotent* of a ring R is an element $r \in R$ satisfying $r^2 = r$. Show how one can use Newton's iteration to determine an idempotent in the ring $\mathbf{Z}[x_1, \ldots, x_v]$. Show that the quadratic Newton's iteration is superior in this case.

10. Let

$$a(x) = 21x^3 + 616x^2 - 8490x + 5539 \in \mathbf{Z}[x].$$

Apply Algorithm 6.1 (Hensel lifting) to lift the mod 5 factors of a

$$u_0 = 2x^2 - 2x - 1, \; w_0 = -2x + 1$$

up to factors $u, w \in \mathbf{Z}[x]$ with

$$u \equiv u_0 \pmod 5), \; w \equiv w_0 \pmod 5).$$

11. Describe how you could do "quadratic Hensel lifting", that is, describe an algorithm which starts with a relatively prime factorization mod p:

$$a(x) = u_0(x) \cdot w_0(x) \pmod p)$$

and gives, at the k-th step, a factorization of the form

$$a(x) = u_k(x) \cdot w_k(x) \pmod{p^{2^k}})$$

with

$$u_k(x) \equiv u_0(x) \pmod p), \; w_k(x) \equiv w_0(x) \pmod p).$$

12. Let

$$a(x) = x^4 - 394x^3 - 4193x^2 + 126x + 596 .$$

Use the quadratic Hensel lifting algorithm of the previous exercise to lift the mod 3 factors

$$u_0 = x^2 + x + 1, \; w_0 = x^2 + x - 1$$

to factors $u, w \in \mathbf{Z}[x]$ with

$$u \equiv u_0 \pmod 3), \; w \equiv w_0 \pmod 3).$$

13. Let

$$a(x) = 18x^5 - 126x^4 + 174x^3 - 1080x^2 + 1722x - 528 .$$

Factor a by reducing mod 7 and then using Hensel lifting.

14. In this question, we are concerned with generalizing Hensel's lemma to the case where there are more than two relatively prime factors. We proceed as follows.

(a) Show that, if $a(x) = u_1(x) \cdot u_2(x) \cdot u_3(x) \pmod p)$ (a relatively prime factorization) then there exist polynomials $a_1(x), a_2(x), a_3(x)$ satisfying

$$a_1(x)u_2(x)u_3(x) + a_2(x)u_1(x)u_3(x) + a_3(x)u_1(x)u_2(x) = 1 \pmod{p}$$

and $\deg(a_i) < \deg(u_i)$.

(b) Show how a factorization as in part (a) may be lifted to a factorization modulo p^k. The lifting does not change the leading coefficients of $u_1(x)$ and $u_2(x)$.

(c) Generalize part (b) to the case of an arbitrary number of factors.

(d) Repeat the above using *quadratic* lifting (cf. Exercise 11) instead of linear lifting.

15. Let $a(x,y,z)$ be the polynomial from Exercise 4, and let

$$b(x,y,z) = (x^2y^2 + z^2 + 1)\cdot(x^2y^2 + 2xyz + 4xy + z^2 + 4z + 4) .$$

In this question we will determine the GCD of the expanded versions of these two polynomials.

(a) Choosing the ideal $\mathrm{I} = < y-1, z >$, determine

$$\mathrm{GCD}(a,b) \bmod \mathrm{I} \in \mathbf{Z}[x,y,z]/\mathrm{I} .$$

(b) Note that we can equate $\mathbf{Z}$ and $\mathbf{Z}_{13}$ for this problem. Carry out the multivariate Hensel construction given in the proof of Theorem 6.6 to lift the GCD from $\mathbf{Z}[x,y,z]/\mathrm{I}$ up to the true GCD in $\mathbf{Z}[x,y,z]$.

16. Suppose we know in advance that for two polynomials $a, b \in \mathbf{Z}[x_1, \ldots, x_v]$ we can find polynomials q and r such that

$$a = b\cdot q + r \quad \text{with } \deg_v(r) < \deg_v(b).$$

Show how this problem can be solved by Newton's iteration and the Hensel construction. Give an example of your method using

$$a(x,y) = (x-1)y^3 + (3x^2-7x+4)y^2 + (x^2+2x-1)y + (3x^3-x^2-4x+7)$$

and

$$b(x,y) = (x-1)y^2 + (x^2+x-1) .$$

For the reductions use the prime $p = 3$ and the ideal $\mathrm{I} = < y >$.

17. Suppose that when we reduce mod p for some prime p we obtain a factorization in which the two components are *not* relatively prime. Explain what could be done in this case.

18. (M. Monagan)

(a) Let $u^{(k)}$ denote a k-th order p-adic approximation and let u_k denote the k-th term in it, i.e.

$$u^{(k+1)} = u_0 + u_1p + \cdots + u_kp^k.$$

Show that in Algorithm 6.1 the error computation (in $\mathbf{Z}[x]$) $e = a - u^{(k+1)}\cdot w^{(k+1)}$ requires $O(n^2m^2)$ operations where n is the degree and m is the size of the coefficients.

(b) Show that, if

$$u^{(k+1)} = u^{(k)} + u_k p^k, \quad w^{(k+1)} = w^{(k)} + w_k p^k$$

then the error at the k + 1-st step can be computed by

$$e^{(k+1)} = e^{(k)} - p^k (u^{(k)} w_k + w^{(k)} u_k + p^k u_k w_k).$$

Show that computing the error using the above formula results in one full order of magnitude efficiency gain over the existing method.

(c) Computing the error term as in part (b) above is not quite right if replace_lc is being used. Why is this the case? How can the above be altered to ensure that $u(x)$ and $w(x)$ have the correct leading coefficients (mod p^k)?

References

1. M. Lauer, "Computing by Homomorphic Images," pp. 139-168 in *Computer Algebra - Symbolic and Algebraic Computation*, ed. B. Buchberger, G.E. Collins and R. Loos, Springer-Verlag (1982).
2. J.D. Lipson, "Newton's Method: A Great Algebraic Algorithm," pp. 260-270 in *Proc. SYMSAC '76*, ed. R.D. Jenks, ACM Press (1976).
3. R. Loos, "Rational Zeros of Integral Polynomials by p-Adic Expansions," *SIAM J. on Computing*, **12** pp. 286-293 (1983).
4. M. Mignotte, "Some Useful Bounds.," pp. 259-263 in *Computer Algebra - Symbolic and Algebraic Computation*, ed. B. Buchberger, G.E. Collins and R. Loos, Springer-Verlag (1982).
5. A. Miola and D.Y.Y. Yun, "The Computational Aspects of Hensel-Type Univariate Greatest Common Divisor Algorithms," *(Proc. EUROSAM '74) ACM SIGSAM Bull.*, **8**(3) pp. 46-54 (1974).
6. P.S. Wang, "An Improved Multivariate Polynomial Factoring Algorithm," *Math. Comp.*, **32** pp. 1215-1231 (1978).
7. P.S. Wang, "The EEZ-GCD Algorithm," *ACM SIGSAM Bull.*, **14** pp. 50-60 (1980).
8. D.Y.Y. Yun, "The Hensel Lemma in Algebraic Manipulation," Ph.D. Thesis, M.I.T. (1974).
9. H. Zassenhaus, "Hensel Factorization I," *J. Number Theory*, **1** pp. 291-311 (1969).
10. R. Zippel, "Newton's Iteration and the Sparse Hensel Algorithm," pp. 68-72 in *Proc. SYMSAC '81*, ed. P.S. Wang, ACM Press (1981).

CHAPTER 7

POLYNOMIAL GCD COMPUTATION

7.1. INTRODUCTION

In many respects the problem of computing the greatest common divisor of two polynomials is a fundamental concern of algebraic manipulation. Once pioneer computer algebra systems (such as ALPAK or PM) had routines for polynomial operations, the natural progression was to develop routines for the manipulation of rational functions. It soon became apparent that rational manipulation leads to a severe problem of intermediate expression swell. For example, consider the problem of adding two rational functions

$$\frac{a(x)}{b(x)} + \frac{c(x)}{d(x)} = \frac{a(x)\cdot d(x) + b(x)\cdot c(x)}{b(x)\cdot d(x)}.$$

In the absence of any simplification, the degree of the numerator or denominator of the sum is potentially twice that of either of its components. Thus, for example, adding sixteen rational functions, each having numerator and denominator of degree 10 would produce (in the absence of any cancellation) a single rational function having both numerator and denominator of degree 160.

A natural step when implementing a package for rational function manipulation is to try to remove common factors of the numerator and denominator, and thus reduce the size of the expression. The most benefit would be achieved, of course, by removing the largest such factor common to both numerator and denominator, that is, to calculate the greatest common divisor of the two polynomials.

As mentioned in Chapter 2, one can always compute the GCD using a variation of Euclid's algorithm. The basic Euclid's algorithm has been known for centuries, is easily understood, and is easily implemented. However, we show in this chapter that this algorithm has a fundamental flaw for many of the problems which arise in computer algebra. Several improved algorithms have been developed over the past 25 years for computing polynomial GCD's. These new algorithms have come about through careful study of the nature of the GCD problem and application of more sophisticated techniques. In many ways the evolution of GCD algorithms for computer algebra systems mirrors the evolution of symbolic manipulation as a whole.

Polynomial GCD computations also turn up as subproblems in many places besides the simple arithmetic operations on rational functions. For example, they play a prominent role in polynomial factorization and in symbolic integration. They also arise when finding inverses in finite Galois fields and in simple algebraic extension fields.

7.2. POLYNOMIAL REMAINDER SEQUENCES

Texts on modern algebra frequently invoke Euclid's algorithm as a constructive proof of the existence of GCDs and for hand calculation it remains a serviceable tool. We may view Euclid's algorithm for calculating the GCD of two polynomials $A(x)$ and $B(x)$ having coefficients from a field F as the construction of a sequence of remainders. That is, if $\deg(A(x)) \geq \deg(B(x))$, then Euclid's algorithm constructs a sequence of polynomials $R_0(x), R_1(x), \ldots, R_k(x)$ where $R_0(x) = A(x)$, $R_1(x) = B(x)$ and

$$R_0(x) = R_1(x) \cdot Q_1(x) + R_2(x) \quad \text{with } \deg(R_2(x)) < \deg(R_1(x))$$

$$R_1(x) = R_2(x) \cdot Q_2(x) + R_3(x) \quad \text{with } \deg(R_3(x)) < \deg(R_2(x))$$

$$\cdots \qquad \cdots$$

$$R_{k-2}(x) = R_{k-1}(x) \cdot Q_2(x) + R_k(x) \quad \text{with } \deg(R_k(x)) < \deg(R_{k-1}(x))$$

$$R_{k-1}(x) = R_k(x) \cdot Q_k(x).$$

Then $R_k(x) = \mathrm{GCD}(A(x), B(x))$ when it is normalized appropriately to make it unit normal. The extended Euclidean algorithm is the same as Euclid's algorithm, except that it computes the remainder sequence $\{R_i(x)\}$ along with the sequences $\{S_i(x)\}$ and $\{T_i(x)\}$ satisfying

$$R_i(x) = A(x) \cdot S_i(x) + B(x) \cdot T_i(x)$$

for all i. Here

$$S_{i+1}(x) = S_{i-1}(x) - S_i(x) \cdot Q_i(x)$$

$$T_{i+1}(x) = T_{i-1}(x) - T_i(x) \cdot Q_i(x).$$

The quotient $Q_i(x)$ is defined by the division

$$R_{i-1}(x) = R_i(x) \cdot Q_i(x) + R_{i+1}(x)$$

and the initial conditions for these sequences are

$$S_0(x) = T_1(x) = 1, \quad S_1(x) = T_0(x) = 0.$$

Example 7.1. Let $A(x)$ and $B(x)$ be polynomials from $\mathbf{Z}_{23}[x]$ given by

$$A(x) = x^8 + x^6 - 3x^4 - 3x^3 + 8x^2 + 2x - 5,$$

$$B(x) = 3x^6 + 5x^4 - 4x^2 - 9x - 2.$$

Then the sequence of remainders which results from applying Euclid's algorithm to this pair is given by

$$R_2(x) = 2x^4 - 5x^2 - 8,$$

$$R_3(x) = -x^2 - 9x + 2,$$

$$R_4(x) = 10x - 8,$$

$$R_5(x) = -4.$$

Therefore, $A(x)$ and $B(x)$ are relatively prime since their greatest common divisor is a unit in $\mathbf{Z}_{23}[x]$.

●

Example 7.2. Let $A(x), B(x) \in \mathbf{Z}[x]$ be defined by

$$A(x) = x^8 + x^6 - 3x^4 - 3x^3 + 8x^2 + 2x - 5,$$

$$B(x) = 3x^6 + 5x^4 - 4x^2 - 9x + 21.$$

(This is the traditional example to illustrate polynomial GCD algorithms first used by Knuth [8] in 1969.) Since $\mathbf{Z}$ is not a field, we need to work with $\mathbf{Q}[x]$ in order to apply Euclid's algorithm. The resulting sequence of remainders determined by applying Euclid's algorithm to this pair is given by

$$R_2(x) = -\frac{5}{9}x^4 + \frac{1}{9}x^2 - \frac{1}{3},$$

$$R_3(x) = -\frac{117}{25}x^2 - 9x + \frac{411}{25},$$

$$R_4(x) = \frac{233150}{19773}x - \frac{102500}{6591},$$

$$R_5(x) = -\frac{1288744821}{543589225}.$$

Therefore, $A(x)$ and $B(x)$ are relatively prime since their greatest common divisor is a unit in $\mathbf{Q}[x]$.

●

Note the growth in the size of the coefficients when using Euclid's algorithm in Example 7.2. Indeed, this example illustrates the major problem with Euclid's algorithm as a computational tool in a computer algebra setting. If we wished to apply Euclid's algorithm to the multivariate case $\mathbf{Z}[x_1, \ldots, x_n]$ then we would need to work in $\mathbf{Z}(x_1, \ldots, x_{n-1})[x_n]$ (since Euclid's algorithm requires the coefficient domain to be a field). In the latter case our coefficients are rational functions and so to perform coefficient arithmetic we need to recursively apply the GCD algorithm. Thus we lose on two scores: recursive invocation of the algorithm and the inevitable growth of the rational function coefficients.

The above approach also has a second problem. When looking for common factors of two polynomials $A(x)$ and $B(x)$, both having integer coefficients, it is most natural to want the factors to come from the same domain, in this case $\mathbf{Z}[x]$. This is also true when working with polynomials over such domains as $\mathbf{Z}[x_1, \ldots, x_n]$. In either case it does not seem to make sense that one must work with the quotient field of the coefficient domain with its extra GCD overhead. More generally, let $A(x)$ and $B(x)$ be two polynomials from R$[x]$ with R a UFD. We wish to find their greatest common factor in R$[x]$ using only arithmetic in the domain R$[x]$, rather than working with the quotient field of R as our polynomial coefficients.

The simplest method of determining a GCD while calculating only in R[x] is to build a sequence of pseudo-remainders using pseudo-division, rather than quotient field polynomial division. Thus for $A(x)$ and $B(x)$ in R[x] we have

$$\alpha \cdot A(x) = Q(x) \cdot B(x) + R(x)$$

where

$$\alpha = \text{lcoeff}(B)^{\delta+1}, \quad \text{and} \quad \delta = \deg(A) - \deg(B).$$

The scaling factor α allows $Q(x)$ and $R(x)$ to be in R[x]. As noted in Chapter 2, we write

$$Q(x) = \text{pquo}(A(x), B(x)), \quad \text{and} \quad R(x) = \text{prem}(A(x), B(x)),$$

to denote that $Q(x)$ and $R(x)$ are the pseudo-quotient and pseudo-remainder, respectively, of $A(x)$ and $B(x)$. By replacing the remainder with the pseudo-remainder, we calculate the GCD of two polynomials from R[x] with all the coefficient arithmetic taking place in R[x].

Example 7.3. Let $A(x)$ and $B(x)$ be the polynomials from Example 7.2. Consider the sequence of pseudo-remainders formed by direct pseudo-division at each step. In this case we have

$$R_2(x) = -15x^4 + 3x^2 - 9,$$

$$R_3(x) = 15795x^2 + 30375x - 59535,$$

$$R_4(x) = 1254542875143750x - 1654608338437500,$$

$$R_5(x) = 12593338795500743100931141992187500,$$

which again implies that $A(x)$ and $B(x)$ are relatively prime.

●

The preceding method does indeed attain our goal of obtaining a GCD while at the same time working entirely in **Z**[x]. However, it happens that our coefficients grow even more drastically than before. In fact, determining a GCD using the pseudo-remainder method has been (rightfully) described as the WWGCD algorithm (WW=world's worst). The only comparably bad method would be to work over the quotient field but not to apply GCD's to reduce the coefficients.

Since the major problem with using the WWGCD algorithm is caused by the exponential growth in the size of the coefficients, a natural modification to the process would be to remove the content of the coefficients at every step. This is the primitive Euclidean algorithm of Chapter 2.

Example 7.4. Let $A(x)$ and $B(x)$ be the polynomials from Example 7.2. Using pseudo-remainders and content removal at each step gives

$$R_2(x) = 5x^4 - x^2 + 3,$$

$$R_3(x) = 13x^2 + 25x - 49,$$

$$R_4(x) = 4663x - 6150,$$

$$R_5(x) = 1.$$

●

This process is clearly the best in terms of keeping the size of the coefficients of the remainders to a minimum. The problem with this method, however, is that each step requires a significant number of GCD operations in the coefficient domain. While this does not appear costly in the above example, the extra cost is prohibitive when working over coefficient domains of multivariate polynomials.

The two previous methods are examples of

Definition 7.1. Let $A(x), B(x)$ be polynomials from R[x], with $\deg(A) \geq \deg(B)$. A *polynomial remainder sequence (PRS)* for A and B is a sequence of polynomials $R_0(x), R_1(x), \ldots, R_k(x)$ from R[x] satisfying

(a) $R_0(x) = A(x), R_1(x) = B(x)$,

(b) $\alpha_i \cdot R_{i-1}(x) = Q_i(x) \cdot R_i(x) + \beta_i \cdot R_{i+1}(x)$ with $\alpha_i, \beta_i \in$ R, (7.1)

(c) $\text{prem}(R_{k-1}, R_k) = 0$.

●

If the original polynomials $A(x)$ and $B(x)$ are primitive polynomials then conditions (a), (b), and (c) imply that the primitive part of $R_k(x)$ is equal to $\text{GCD}(A(x), B(x))$ (cf. Exercise 7.5). As noted in Chapter 2, we can separate the GCD computation into a computation of the GCD of the contents and the GCD of the primitive parts.

The usual PRS has $\alpha_i = r_i^{\delta_i+1}$ where $r_i = \text{lcoeff}(R_i(x))$ and $\delta_i = \deg(R_{i-1}) - \deg(R_i)$. Example 7.3 determines a PRS with $\beta_i(x) = 1$. This is usually called the *Euclidean PRS*. Example 7.4 gives a second example of a PRS where $\alpha_i = r_i^{\delta_i+1}$ and $\beta_i = \text{cont}(\text{prem}(R_{i-1}(x), R_i(x)))$. It is usually referred to as the *primitive PRS* (since all polynomials in the sequence are primitive, that is, their contents are 1).

In general most PRS's for A and B differ only in the amount of common factor β_i removed in each update step. The Euclidean PRS and the primitive PRS represent opposite extremes of such PRS's. The former removes no common factors of the coefficients while the latter removes all common factors. The general goal of constructing a PRS for a given $A(x)$ and $B(x)$ is to choose the β_i in such a way that keeps the size of the coefficients of the sequence as small as possible. This last condition comes with the caveat that the process of keeping the coefficients as small as possible is to be as inexpensive as possible. Thus we would like to get a sequence of polynomials which has size close to the primitive PRS but without the cost required by this sequence.

In this context we give examples of two PRS's that accomplish this goal.

Example 7.5. The *reduced PRS* satisfies

$$\alpha_i = r_i^{\delta_i+1},\ \beta_1 = 1,\ \beta_i = \alpha_{i-1} \quad \text{for } 2 \le i \le k. \tag{7.2}$$

When the polynomials $A(x)$ and $B(x)$ of Example 7.2 are used, the reduced PRS yields the sequence

$$R_2(x) = -15x^4 + 3x^2 - 9,$$

$$R_3(x) = 585x^2 + 1125x - 2205,$$

$$R_4(x) = -18885150x + 24907500,$$

$$R_5(x) = 527933700.$$

●

In Example 7.5, the coefficient growth is considerably less than that found using the Euclidean PRS. At the same time it does not do any coefficient GCD calculations, only simple division. The reduced PRS algorithm works best in the special case when the remainder sequence is *normal.* By this we mean a sequence of remainders whose degrees differ by exactly 1 at each stage, that is $\delta_i = 1$ for all i. The reduced PRS algorithm, in the case where the sequence is normal, dates back to Sylvester [15] in 1853. In this case, it provides an acceptable method of determining the GCD of two polynomials, with a remainder sequence that has coefficients growing approximately linearly in size (cf. Exercise 7.13).

There is, however, no way of knowing, a priori, that a given pair of polynomials will result in a normal PRS. In addition, in the *abnormal* case, where two successive polynomials in the sequence have degrees differing by more than one (Example 7.5 is one example of an abnormal PRS), the resulting coefficient growth of the reduced PRS can be exponential in n, the degree of the input polynomials. To overcome the problem of exponential coefficient growth in an abnormal PRS, Collins [4] and Brown [1] independently developed the subresultant PRS algorithm. This is an example of a PRS in which the coefficients grow approximately linearly in size even in the abnormal case.

Example 7.6. The *subresultant PRS* satisfies

$$\alpha_i = r_i^{\delta_i+1},\ \beta_1 = (-1)^{\delta_1+1},\ \beta_i = -r_{i-1}\cdot\psi_i^{\delta_i} \quad \text{for } 2 \le i \le k, \tag{7.3}$$

where the ψ_i are defined by

$$\psi_1 = -1, \ \ \psi_i = (-r_{i-1})^{\delta_{i-1}} \cdot \psi_{i-1}^{\ 1-\delta_{i-1}} \ \text{ for } 2 \le i \le k. \tag{7.4}$$

When the polynomials $A(x)$ and $B(x)$ of Example 7.2 are used, the resulting subresultant PRS is

$$\begin{aligned} R_2(x) &= 15x^4 - 3x^2 + 9, \\ R_3(x) &= 65x^2 + 125x - 245, \\ R_4(x) &= -9326x + 12300, \\ R_5(x) &= 260708. \end{aligned}$$

●

As was the case in Example 7.5, the cost of keeping the coefficient growth down in this example does not include any coefficient GCD calculations. Also, in both examples all the coefficient arithmetic takes place in R (rather than having coefficients from Q(R), the quotient field of R).

7.3. THE SYLVESTER MATRIX AND SUBRESULTANTS

Mathematician #1: Okay, so there are three steps to your algorithm. Step one is the input and step three is the output. What is step two?

Mathematician #2: Step two is when a miracle occurs.

Mathematician #1: Oh, I see. Uh, perhaps you could explain that second step a bit more?

The preceding section described several PRS constructions for calculating the GCD of two polynomials from R[x], where R is an arbitrary UFD. In particular, the last two algorithms described, the reduced PRS and the subresultant PRS, both meet the criterion of performing all of the arithmetic inside the domain R[x] while at the same time keeping the cost of controlling the coefficient growth to a minimum. If our only interest was in the implementation of these particular algorithms we would essentially be finished. Either algorithm is easily implemented by simple modifications to Algorithms 2.1 or 2.3. However, in terms of explaining why these methods work, or indeed, convincing a reader that they do indeed work, the previous section does not even provide simple hints.

In order to proceed further in the understanding and development of GCD algorithms, we must look more closely at the structure of the problem itself. Many of the results studied in this section have their origins in the late nineteenth century when mathematicians such as Sylvester and Trudi were developing the theory of equations. This subject was later called the theory of algebraic curves and is the foundation of modern Algebraic Geometry.

Definition 7.2. Let $A(x), B(x) \in R[x]$ be nonzero polynomials with $A(x) = \sum_{i=0}^{m} a_i x^i$ and $B(x) = \sum_{i=0}^{n} b_i x^i$. The *Sylvester matrix* of A and B is the $m+n$ by $m+n$ matrix

$$M = \begin{bmatrix} a_m & a_{m-1} & \cdots\cdots & a_1 & a_0 & & & \\ & a_m & a_{m-1} & \cdots\cdots & a_1 & a_0 & & \\ & & \cdots\cdots & \cdots\cdots & \cdots\cdots & \cdots\cdots & & \\ & & & a_m & \cdots\cdots & \cdots\cdots & a_0 \\ b_n & b_{n-1} & \cdots\cdots & b_1 & b_0 & & \\ & b_n & b_{n-1} & \cdots\cdots & b_1 & b_0 & \\ & & \cdots\cdots & \cdots\cdots & \cdots\cdots & \cdots\cdots & \\ & & & b_n & \cdots\cdots & \cdots\cdots & b_0 \end{bmatrix} \qquad (7.5)$$

where the upper part of the matrix consists of n rows of coefficients of $A(x)$, the lower part consists of m rows of coefficients of $B(x)$, and the entries not shown are zero.

●

Definition 7.3. The *resultant* of $A(x)$ and $B(x) \in R[x]$ (written res(A,B)) is the determinant of the Sylvester matrix of A, B. We also define res$(0,B) = 0$ for nonzero $B \in R[x]$, and res$(A,B) = 1$ for nonzero coefficients $A, B \in R$. We write $\text{res}_x(A,B)$ if we wish to include the polynomial variable (this is important when the coefficient domain is another polynomial domain such as $\mathbf{Z}[y]$).

●

Example 7.7. For the polynomials

$$A(x) = 3x^4 + 3x^3 + x^2 - x - 2, \qquad B(x) = x^3 - 3x^2 + x + 5$$

from $\mathbf{Z}[x]$, we have

$$\text{res}(A, B) = \det(M) = 0$$

where the Sylvester matrix is

$$M = \begin{bmatrix} 3 & 3 & 1 & -1 & -2 & 0 & 0 \\ 0 & 3 & 3 & 1 & -1 & -2 & 0 \\ 0 & 0 & 3 & 3 & 1 & -1 & -2 \\ 1 & -3 & 1 & 5 & 0 & 0 & 0 \\ 0 & 1 & -3 & 1 & 5 & 0 & 0 \\ 0 & 0 & 1 & -3 & 1 & 5 & 0 \\ 0 & 0 & 0 & 1 & -3 & 1 & 5 \end{bmatrix}.$$

●

The origin of the resultant lies with Sylvester's criterion for determining when two polynomials have a non-trivial common factor. This criterion simply states that two polynomials, $A(x)$ and $B(x)$, have a non-trivial common factor if and only if $\text{res}(A,B) = 0$. Its validity follows as a corollary to

Theorem 7.1. Let $A(x), B(x) \in R[x]$ be polynomials of degree m, $n > 0$, respectively. Then there exist polynomials $S(x), T(x) \in R[x]$ with $\deg(S) < n$, $\deg(T) < m$ such that

$$A(x)S(x) + B(x)T(x) = \text{res}(A,B) \ . \tag{7.6}$$

Proof: For $A(x)$ and $B(x)$ (with coefficients a_i, b_i, respectively), we form the $m+n$ equations

$$\begin{aligned}
a_m x^{m+n-1} + a_{m-1}x^{m+n-2} + \cdots + a_0 x^{n-1} &= x^{n-1}A(x), \\
a_m x^{m+n-2} + \cdots + a_1 x^{n-1} + a_0 x^{n-2} &= x^{n-2}A(x), \\
\cdots \qquad & \quad \cdots \\
a_m x^m + a_{m-1}x^{m-1} + \cdots + a_0 &= A(x), \\
\\
b_n x^{m+n-1} + b_{n-1}x^{m+n-2} + \cdots + b_0 x^{m-1} &= x^{m-1}B(x), \\
\cdots \qquad & \quad \cdots \\
b_n x^n + b_{n-1}x^{n-1} + \cdots + b_0 &= B(x).
\end{aligned} \tag{7.7}$$

In matrix form, equation (7.7) can be written as

$$M \cdot \begin{bmatrix} x^{m+n-1} \\ \cdot \\ x \\ 1 \end{bmatrix} = \begin{bmatrix} x^{n-1}A(x) \\ \cdot \\ \cdot \\ B(x) \end{bmatrix}.$$

Using Cramer's rule to solve for the last component, 1, gives

$$\det \begin{bmatrix} a_m & a_{m-1} & \cdots\cdots & a_1 & a_0 & & & x^{n-1}A(x) \\ & a_m & a_{m-1} & \cdots\cdots & a_1 & a_0 & & \\ & & \cdots\cdots & \cdots\cdots & \cdots\cdots & \cdots\cdots & & \\ & & & a_m & \cdots\cdots & \cdots\cdots & a_1 & A(x) \\ b_n & b_{n-1} & \cdots\cdots & b_1 & b_0 & & & x^{m-1}B(x) \\ & b_n & b_{n-1} & \cdots\cdots & b_1 & b_0 & & \\ & & \cdots\cdots & \cdots\cdots & \cdots\cdots & \cdots\cdots & & \\ & & & b_n & \cdots\cdots & \cdots\cdots & b_1 & B(x) \end{bmatrix} = \det(M). \tag{7.8}$$

Expanding the determinant on the left hand side of equation (7.8) by minors along the last column then gives the result.

●

Corollary (Sylvester's Criterion). Let $A(x)$, $B(x) \in$ R$[x]$, R a UFD. Then $A(x)$ and $B(x)$ have a non-trivial common factor if and only if $\text{res}(A,B) = 0$.

Proof: If $\text{res}(A,B) \neq 0$, then Theorem 7.1 implies that any divisor of both $A(x)$ and $B(x)$ must divide the resultant. Since the resultant is a constant, the only divisors of both polynomials must have degree 0, and hence there are no non-trivial divisors.

Conversely, suppose $\text{res}(A,B) = 0$ so that (7.6) becomes

$$A(x)\cdot S(x) = -B(x)\cdot T(x).$$

If there are no non-trivial common factors of $A(x)$ and $B(x)$, then it is easy to see that $B(x) \mid S(x)$. But this is impossible since Theorem 7.1 implies that $\deg(S) < n = \deg(B)$.

●

The proof of Theorem 7.1 shows clearly that the Sylvester matrix itself plays a major role in the theory of GCD algorithms. Its importance to polynomial remainder sequences (and hence to polynomial GCD calculations) lies in the ability to represent the polynomial equation

$$A(x)\cdot S(x) + B(x)\cdot T(x) = R(x) \tag{7.9}$$

in terms of a linear system of equations having as a coefficient matrix the Sylvester matrix of $A(x)$ and $B(x)$. As such, one has all the power and tools of linear algebra (e.g. determinants, row and/or column operations) at one's disposal. As an example, we state without proof the following.

Theorem 7.2 (Laidacker [9]). If the Sylvester matrix is triangularized to row echelon form, using only row operations, then the last nonzero row gives the coefficients of the polynomial GCD (over the quotient field of R).

●

Polynomial remainder sequences provide a number of solutions to equations of the form (7.8), with right hand sides of varying degrees. In particular, one may obtain linear systems of equations having submatrices of M as the coefficient matrices. As a result, certain submatrices of the Sylvester matrix also play a prominent role in GCD calculations. For example, let M_j be the submatrix of M formed by deleting the last j rows of A terms, the last j rows of B terms and the last $2j$ columns. Clearly $M_0 = M$. Then we have the following generalization of Sylvester's criterion.

Theorem 7.3. The degree of a GCD of two polynomials $A(x)$, $B(x)$ from R[x] is equal to the first j such that

$$\det(M_j) \neq 0.$$

Proof: The proof of Theorem 7.3 closely parallels the proof of Theorem 7.1. We leave the proof as a (difficult) exercise for the reader (cf. Exercise 7.6).

●

The quantity $\det(M_j)$ is usually denoted by $\mathrm{res}^{(j)}(A,B)$ and is referred to as the j-th *principal resultant* of A and B.

Generalizing further, let M_{ij} be the $(m+n-2j)\times(m+n-2j)$ matrix determined from M by deleting:

(a) rows $n-j+1$ to n (each having coefficients of $A(x)$);

(b) rows $m+n-j+1$ to $m+n$ (each having coefficients of $B(x)$);

(c) columns $m+n-2j$ to $m+n$, except for column $m+n-i-j$.

This gives

$$M_{ij} = \begin{bmatrix} a_m & a_{m-1} & \cdots\cdots & a_1 & a_{2j-n} & a_{i+j-n+1} \\ & a_m & a_{m-1} & \cdots\cdots & & \\ & & \cdots\cdots & \cdots\cdots\ \cdot\cdot & & \cdot\cdot \\ & & & a_m & a_{j+1} & a_i \\ b_n & b_{n-1} & \cdots\cdots & b_1 & b_{2j-m} & b_{i+j-m+1} \\ & b_n & b_{n-1} & \cdots\cdots & & \\ & & \cdots\cdots & \cdots\cdots\ \cdot\cdot & & \cdot\cdot \\ & & & b_n & b_{j+1} & b_i \end{bmatrix} \qquad (7.10)$$

where the coefficients with negative subscripts are zero. Clearly M_{jj} is the same as what we previously called M_j.

Definition 7.3. The j-th *subresultant* of $A(x)$ and $B(x)$ is the polynomial of degree j defined by

$$S(j,A,B) = \det(M_{0j}) + \det(M_{1j})\cdot x + \cdots + \det(M_{jj})\cdot x^j , \tag{7.11}$$

for $0 \le j \le n$.

●

Notice that we may also write the j-th subresultant as a determinant

$$S(j,A,B) = \det \begin{bmatrix} a_m & a_{m-1} & \cdots\cdots & a_1 & a_{2j-n} & x^{n-j-1}\cdot A(x) \\ & a_m & a_{m-1} & \cdots\cdots & & \\ & & \cdots\cdots & \cdots\cdots \; .. & & .. \\ & & & a_m & a_{j+1} & A(x) \\ b_n & b_{n-1} & \cdots\cdots & b_1 & b_{2j-m} & x^{m-j-1}\cdot B(x) \\ & b_n & b_{n-1} & \cdots\cdots & & \\ & & \cdots\cdots & \cdots\cdots \; .. & & .. \\ & & & b_n & b_{j+1} & B(x) \end{bmatrix} . \tag{7.12}$$

Since the determinant on the left of equation (7.12) may be expanded as

$$\sum_{i=0}^{n} \det \begin{bmatrix} a_m & a_{m-1} & \cdots\cdots & a_1 & a_{2j-n} & a_{i+j-n+1}\cdot x^i \\ & a_m & a_{m-1} & \cdots\cdots & & \\ & & \cdots\cdots & \cdots\cdots \; .. & & .. \\ & & & a_m & a_{j+1} & a_i\cdot x^i \\ b_n & b_{n-1} & \cdots\cdots & b_1 & b_{2j-m} & b_{i+j-m+1}\cdot x^i \\ & b_n & b_{n-1} & \cdots\cdots & & \\ & & \cdots\cdots & \cdots\cdots \; .. & & .. \\ & & & b_n & b_{j+1} & b_i\cdot x^i \end{bmatrix} = \sum_{i=0}^{n} \det(M_{ij})\, x^j$$

the equivalence of the forms (7.11) and (7.12) follows from the fact that $\det(M_{ij}) = 0$ for $i > j$ (since in these cases M_{ij} has two repeated columns).

Example 7.8. Let $A(x)$ and $B(x)$ be as in Example 7.7. Then we have already determined that

$$S(0,A,B) = \mathrm{res}(A,B) = 0.$$

The other subresultants are calculated as follows.

$$S(1,A,B) = \det \begin{bmatrix} 3 & 3 & 1 & -1 & x \cdot A(x) \\ 0 & 3 & 3 & 1 & A(x) \\ 1 & -3 & 1 & 5 & x^2 B(x) \\ 0 & 1 & -3 & 1 & xB(x) \\ 0 & 0 & 1 & -3 & B(x) \end{bmatrix} = 1192x + 1192.$$

$$S(2,A,B) = \det \begin{bmatrix} 3 & 3 & A(x) \\ 1 & -3 & xB(x) \\ 0 & 1 & B(x) \end{bmatrix} = 34x^2 - 28x - 62.$$

$$S(3,A,B) = \det\,[B(x)] = x^3 - 3x^2 + x + 5.$$

Note that by Sylvester's criterion, $A(x)$ and $B(x)$ have a non-trivial common factor.

●

Example 7.9. Let $A(x)$ and $B(x)$ be the polynomials from Example 7.2. Using straightforward determinant calculations we calculate the subresultants as

$$S(0,A,B) = 260708 = \mathrm{res}(A,B),$$

$$S(1,A,B) = 9326x - 12300,$$

$$S(2,A,B) = 169x^2 + 325 - 637,$$

$$S(3,A,B) = 65x^2 + 125x - 245,$$

$$S(4,A,B) = 25x^4 - 5x^2 + 15,$$

$$S(5,A,B) = 15x^4 - 3x^2 + 9,$$

$$S(6,A,B) = 3 \cdot B(x) = 9x^6 + 15x^4 - 12x^2 - 27x + 63.$$

●

The importance of subresultants becomes clear when one expands the determinant (7.12) by minors along the last column. We obtain

$$S(j,A,B) = A(x) \cdot S_j(x) + B(x) \cdot T_j(x) \tag{7.13}$$

where $S_j(x)$ and $T_j(x)$ are polynomials of degree at most $n-j-1$ and $m-j-1$, respectively. Note the similarity of the correspondence of equations (7.6), the coefficient matrix M and the resultant in the proof of Theorem 7.1 with the correspondence of equation (7.9), the j-th principal submatrix M_j and the j-th subresultant (7.12). Indeed, this correspondence provides the motivation for Definition 7.3. For any PRS $\{R_0(x), \ldots, R_k(x)\}$ with $\deg(R_i(x)) = n_i$, we obtain solutions to the equations

$$A(x) \cdot U_i(x) + B(x) \cdot V_i(x) = \gamma_i \cdot R_i(x) \tag{7.14}$$

with $\gamma_i \in R$ and where $U_i(x)$ and $V_i(x)$ are of degree $n-n_{i-1}$ and $m-n_{i-1}$, respectively. Theorem 2.6 implies that for each i we have

$$R_i(x) = s_i \cdot S(n_{i-1}-1, A, B) \tag{7.15}$$

where s_i comes from the quotient field of R.

When s_i is from R, rather than Q(R), we obtain a divisor for $R_i(x)$ without any need for coefficient GCD calculations. Clearly, the constant s_i depends on the choices of α_i and β_i defining the update condition (7.1). Determining s_i in terms of the α's and the β's will then give candidates for these divisors. Thus, we need to determine the s_i in terms of the α's and β's. We consider first one division step.

Lemma 7.1. Suppose

$$A(x) = Q(x) \cdot B(x) + R(x)$$

with $\deg(A) = m$, $\deg(B) = n$, $\deg(Q) = n-m$, $\deg(R) = k$ and $m \ge n > k$. Let b and r denote the leading coefficients of $B(x)$ and $R(x)$, respectively. Then

$$S(j,A,B) = (-1)^{(m-j)(n-j)} \begin{cases} b^{m-k} \cdot S(j,B,R) & 0 \le j < k \\ b^{m-k} \cdot r^{n-k-1} \cdot R(x) & j = k \\ 0 & k < j < n-1 \\ b^{m-n+1} \cdot R(x) & j = n-1. \end{cases} \tag{7.16}$$

Proof (following Brown and Traub [2]): If $Q(x) = \sum_{i=0}^{n-m} q_i x^i$, then equating powers in our division equation gives

$$(1, -q_{m-n}, \ldots, -q_0) \cdot \begin{bmatrix} a_m & & \cdots & a_0 \\ b_n & \cdots & b_0 & \\ & & & \\ & b_m & \cdots & b_0 \end{bmatrix} = (r_m, \ldots, r_0) \tag{7.17}$$

where $r_{k+1} = \cdots = r_m = 0$ and coefficients with subscripts out of range are zero. Selecting the first p columns of (7.17) and rearranging gives

$$(1, -q_{m-n}, \ldots, -q_0) \cdot \begin{bmatrix} a_m & \cdots & a_{p-n+2j} & x^{n-j-p}A(x) \\ b_n & \cdots & b_{p-m+2j} & x^{m-j-p}B(x) \\ & & & \\ b_m & \cdots & b_{p-n+2j} & B(x) \end{bmatrix}$$

$$= (0, \ldots, 0, r_m, \ldots, r_{m-p+1}, x^{n-j-p}R(x)) \tag{7.18}$$

for all j and p with $0 \le j < n$ and $1 \le p \le n-j$. Since the left hand side represents the p-th row of the A portion of (7.12) and a linear combination of the rows p through $p+m-n$ of the B portion, we can replace the p-th row of the A portion by the right hand side of (7.18) without affecting the value of the determinant. Doing this for all the A rows from 1 to $n-j$, and rearranging the order of the rows gives

$$S(j,A,B) = (-1)^{(n-j)(m-j)} \det \begin{bmatrix} b_n & \cdots & & & b_{2j-m} & x^{m-j-1}B(x) \\ & & & & & \cdot \\ & & & & & \cdot \\ & b_n & & \cdots & b_{j+1} & B(x) \\ r_m & & \cdots & & r_{2j-n} & x^{n-j-1}R(x) \\ & & & & & \cdot \\ & & & & & \cdot \\ & & r_m & \cdots & r_{j+1} & R(x) \end{bmatrix}. \tag{7.19}$$

When $j \ge k$, all the elements below the diagonal are zero. Since the determinant of an upper triangular matrix is the product of the diagonal entries, the last three identities in equation (7.16) hold. On the other hand, when $j < k$, the determinant on the right of equation (7.19) is the determinant of the matrix

$$\left[\begin{array}{ccc|c} b_n & & \cdot & \cdot\ \cdot \\ & \cdot & & \\ & & b_n & \cdot\ \cdot \\ \hline & & & \\ & 0 & & S^* \\ & & & \end{array}\right] \tag{7.20}$$

where S^* is a $k+n$ by $k+n$ matrix whose determinant is $S(j,B,R)$. Using standard properties of determinants gives the first identity and completes the proof.

●

Example 7.10. For the polynomials $A(x)$ and $B(x)$ from Example 7.7, we have

$$A(x) = (3x+12)\cdot B(x) + (34x^2 - 28x - 62).$$

Hence, for example,

$$S(0,A,B) = S(0,B,R) = 0,$$

$$S(1,A,B) = S(1,B,R) = 1192x - 1192,$$

$$S(2,A,B) = R(x) = 34x^2 - 28x - 62.$$

●

Translating Lemma 7.1 into our PRS situation gives

Lemma 7.2. Let $\{R_0(x), \ldots, R_k(x)\}$ be a PRS in R$[x]$ defined by the division relations

$$\alpha_i \cdot R_{i-1}(x) = Q_i(x) \cdot R_i(x) + \beta_i \cdot R_{i+1}(x), \quad 0 < i \le k.$$

Let $n_i = \deg(R_i(x))$, $\delta_i = n_{i-1} - n_i$, $\gamma_i = \delta_i + \delta_{i+1}$ and $r_i = \mathrm{lcoeff}(R_i(x))$. Then

$$S(j, R_{i-1}, R_i) \alpha_i^{n_i - j} (-1)^{(n_i - j)(n_{i-1} - j)} = \begin{cases} r_i^{\gamma_i} \cdot \beta_i^{n_i - j} \cdot S(j, R_i, R_{i+1}) & 0 \le j < n_{i+1} \\ r_i^{\gamma_i} \cdot r_{i+1}^{\delta_{i+1} - 1} \cdot \beta_i^{\delta_{i+1}} \cdot R_{i+1}(x) & j = n_{i+1} \\ 0 & n_{i+1} < j < n_i - 1 \\ r_i^{\delta_i + 1} \cdot \beta_i \cdot R_{i+1}(x) & j = n_i - 1. \end{cases} \tag{7.21}$$

Proof: Using the identity

$$S(j, aA, bB) = a^{n-j} \cdot b^{m-j} \cdot S(j, A, B)$$

along with the results of Lemma 7.2, we get

$$\alpha_i^{(n_i - j)} \cdot S(j, R_{i-1}, R_i) = S(j, \alpha_i R_{i-1}, R_i)$$

$$= (-1)^{(n_{i-1} - j)(n_i - j)} \cdot \begin{cases} r_i^{n_{i-1} - n_{i+1}} \cdot S(j, R_i, \beta_i \cdot R_{i+1}) & 0 \le j < n_{i+1} \\ r_i^{n_{i-1} - n_{i+1}} \cdot \beta_i \cdot r_{i+1}^{n_i - n_{i+1} - 1} \cdot \beta_i R_{i+1}(x) & j = n_{i+1} \\ 0 & n_{i+1} < j < n_i - 1 \\ r_i^{n_{i-1} - n_i + 1} \cdot \beta_i R_{i+1}(x) & j = n_i - 1. \end{cases}$$

$$= (-1)^{(n_{i-1} - j)(n_i - j)} \cdot \begin{cases} r_i^{\gamma_i} \cdot \beta_i^{n_i - j} \cdot S(j, R_i, R_{i+1}) & 0 \le j < n_{i+1} \\ r_i^{\gamma_i} \cdot \beta_i^{\delta_{i+1}} \cdot r_{i+1}^{\delta_{i+1} - 1} \cdot R_{i+1}(x) & j = n_{i+1} \\ 0 & n_{i+1} < j < n_i - 1 \\ r_i^{\delta_i + 1} \cdot \beta_i \cdot R_{i+1}(x) & j = n_i - 1. \end{cases}$$

which gives Lemma 7.2. ●

Example 7.11. Suppose $\{R_0(x), \cdots, R_k(x)\}$ is a *normal* Euclidean PRS. Then $\delta_i = 1$, $\gamma_i = 2$, $(-1)^{(n_i - j)(n_{i-1} - j)} = 1$, $\alpha_i = r_i^2$, and $\beta_i = 1$. In this case the subresultants are given by

$$S(j,R_{i-1},R_i)\cdot(r_i^2)^{n_i-j} = \begin{cases} r_i^2\cdot S(j,R_i,R_{i+1}) & 0 \le j < n_{i+1} \\ r_i^2\cdot R_{i+1}(x) & j = n_{i+1}\ . \end{cases} \tag{7.22}$$

Hence $R_2(x) = S(n_2,R_0,R_1)$, $R_3(x) = S(n_3,R_1,R_2)$, and so on. Notice that we can use equation (7.22) to further simplify the pseudo-remainders. For example, we have

$$\begin{aligned} R_3(x) &= S(n_3,R_1,R_2) = \{\ S(n_3,R_1,R_2)\cdot r_1^2\ \}\ /\ r_1^2 \\ &= S(n_3,R_0,R_1)\cdot r_1^4/r_1^2 = S(n_3,R_0,R_1)\cdot r_1^2. \end{aligned}$$

We can simplify $R_3(x)$ by dividing all the coefficients by r_1^2. It is not hard to check that at the i-th stage a similar reduction gives

$$R_{i+1}(x) = S(n_{i+1},R_0,R_1)\cdot r_1^2 \cdots r_{i-1}^2.$$

Thus at this stage the pseudo-remainder is carrying coefficients which can be simplified by division by $r_1^2 \cdots r_{i-1}^2$.

●

Example 7.11 illustrates the use of subresultants to determine known common factors from a given PRS. Of course, the real interest of such information would be to apply this knowledge to obtain a reduced remainder at every step.

Example 7.12. We will use Lemma 7.2 to build an entirely new PRS for the previous example. We do this by first applying pseudo-division and then reducing the result by dividing out known factors. We would start by setting $R_3(x) = \overline{R}_3(x)/r_1^2$, where $\overline{R}_3(x)$ is the pseudo-remainder of $R_1(x)$ and $R_2(x)$. The PRS would then have β values $\beta_1 = 1$, $\beta_2 = r_1^2$. Proceeding one more step, let us pseudo-divide $R_3(x)$ into $R_2(x)$ to get $\overline{R}_4(x)$. Using Lemma 7.2 with $\beta_3 = 1$ gives

$$\overline{R}_4(x) = S(n_4,R_2,R_3) = S(n_4,R_1,R_2)\cdot r_2^2.$$

Again, this implies that there is a common factor of the coefficients of the pseudo-remainder $\overline{R}_4(x)$, hence the natural candidate for the next member of this reduced sequence would be

$$R_4(x) = \overline{R}_4(x)\ /\ r_2^2$$

that is, we get a new member of the PRS with $\beta_3 = r_2^2$. Continuing in this manner we obtain a simplified PRS with $\beta_i = r_{i-1}^2$ for all $i \ge 2$. In the case of a normal PRS, this construction is equivalent to both the reduced PRS and the subresultant PRS described in Examples 7.5 and 7.6 of the previous section.

●

Lemma 7.2 relates each member of a PRS to a constant times the subresultant of the previous two members of the PRS. When the PRS is normal, Example 7.12 shows how to use this lemma to obtain known divisors that can be used to simplify the PRS at every step.

When the PRS is abnormal, a similar simplification is also possible. However, to obtain the reduction we need to go further by relating each member of the PRS to the subresultants of the first two elements of the PRS, $R_0(x)$ and $R_1(x)$. Indeed, from Lemma 7.2 we have

$$\begin{aligned}
R_2(x) &= d_{22}\cdot S(n_1-1, R_0, R_1) \\
R_3(x) &= d_{32}\cdot S(n_2-1, R_1, R_2) = d_{33}\cdot S(n_2-1, R_0, R_1) \\
R_4(x) &= d_{42}\cdot S(n_3-1, R_2, R_3) = d_{43}\cdot S(n_3-1, R_1, R_2) = d_{44}\cdot S(n_3-1, R_0, R_1) \\
&\quad \cdots \qquad\qquad\qquad\qquad\qquad\qquad \cdots
\end{aligned}$$

and our goal is to determine α_j and β_k so that $d_{ii} \in$ R. Theorem 7.4 uses Lemmas 7.1 and 7.2 to account for all the subresultants $S(j,R_0,R_1)$ for $0 \le j < n_j$. This then gives explicit representations for the d_{ii} above.

Theorem 7.4 (Fundamental Theorem of PRS).

$$S(j,R_0,R_1) = \begin{cases} \eta_i\cdot R_i(x) & j = n_{i-1}-1 \\ \tau_i\cdot R_i(x) & j = n_i \\ 0 & \text{otherwise} \end{cases} \tag{7.23}$$

where η_i and τ_i are defined by

$$\eta_i = (-1)^{\phi_i}\cdot r_{i-1}^{\,1-\delta_{i-1}} \prod_{p=1}^{i-1} \{ (\frac{\beta_p}{\alpha_p})^{n_p-n_{i-1}+1} \cdot r_p^{\,\gamma_p} \}, \tag{7.24}$$

$$\tau_i = (-1)^{\sigma_i}\cdot r_i^{\,\delta_i-1} \prod_{p=1}^{i-1} \{ (\frac{\beta_p}{\alpha_p})^{n_p-n_i} \cdot r_p^{\,\gamma_p} \} \tag{7.25}$$

with

$$\phi_i = \sum_{p=1}^{i-1} (n_p-n_{i-1}+1)(n_{p-1}-n_{i-1}+1), \quad \sigma_i = \sum_{p=1}^{i-1} (n_{p-1}-n_i)(n_p-n_i).$$

Proof: When $0 \le j < n_{i+1}$ and $1 \le i < k$, Lemma 7.2 gives

$$S(j,R_{i-1},R_i)\cdot\alpha_i^{(n_i-j)} = (-1)^{(n_i-j)(n_{i-1}-j)} r_i^{\gamma_i}\cdot\beta_i^{(n_i-j)}\cdot S(j,R_i,R_{i+1}).$$

Iterating this identity, we get

$$S(j,R_0,R_1)\cdot\prod_{p=1}^{i}\alpha_p^{(n_p-j)} = S(j,R_i,R_{i+1})\prod_{p=1}^{i}[r_p^{\gamma_p}\beta_p^{(n_p-j)}(-1)^{(n_p-j)(n_{p-1}-j)}] \tag{7.26}$$

for $0 \le j < n_{i+1}$ and $1 \le i \le k-1$.

When $i = k-1$, equation (7.26) implies

$$c_j\cdot S(j,R_0,R_1) = d_j\cdot S(j,R_{k-1},R_k), \quad 0 < j < n_k \tag{7.27}$$

for some constants c_j and d_j. But by the definition of a PRS, $\text{prem}(R_{k-1},R_k) = 0$; hence

$$S(j,R_0,R_1) = 0, \quad 0 < j < n_k$$

and so we are left with the case when $1 \le i < k-1$.

When $j = n_{i+1}$, equation (7.21) and the iteration procedure gives

$$S(j,R_0,R_1)\cdot\prod_{p=1}^{i}\alpha_p^{(n_p-j)}$$

$$= \prod_{p=1}^{i-1}[r_p^{\gamma_p}\beta_p^{(n_p-j)}(-1)^{(n_p-j)(n_{p-1}-j)}]S(j,R_{i-1},R_i)\alpha_i^{(n_i-j)}$$

$$= \prod_{p=1}^{i-1}[r_p^{\gamma_p}\beta_p^{(n_p-j)}(-1)^{(n_p-j)(n_{p-1}-j)}](-1)^{(n_i-j)(n_{i-1}-j)}r_i^{\gamma_i}r_{i+1}^{\delta_{i+1}-1}\beta_i^{i+1}R_{i+1}(x)$$

$$= \prod_{p=1}^{i}[r_p^{\gamma_p}\beta_p^{(n_p-j)}(-1)^{(n_p-j)(n_{p-1}-j)}]r_{i+1}^{\delta_{i+1}-1}R_{i+1}(x). \tag{7.28}$$

Thus

$$S(n_{i+1},R_0,R_1) = \tau_{i+1}\cdot R_{i+1}(x)$$

where τ is given by equation (7.25).

When $n_{i+1} < j < n_i-1$, Lemma 7.2 combined with equation (7.21) and iteration yields

$$S(j,R_0,R_1) = 0.$$

Finally, if $j = n_i-1$, then Lemma 7.1 and a similar iteration process gives

$$S(j,R_0,R_1)\prod_{p=1}^{i}\alpha_p^{(n_p-n_i+1)}$$

$$= r_i^{\delta_i+1}\cdot\beta_i\cdot R_{i+1}(x)\cdot\prod_{p=1}^{i-1}[r_p^{\gamma_p}\beta_p^{(n_p-j)}(-1)^{(n_p-j)(n_{p-1}-j)}]$$

$$= r_i^{1-\delta_{i+1}}\cdot\prod_{p=1}^{i}[r_p^{\gamma_p}\beta_p^{(n_p-n_i+1)}(-1)^{(n_p-n_i+1)(n_{p-1}-n_i+1)}\cdot R_{i+1}(x)].$$

Translating $i+1$ to i and rearranging terms gives

$$S(n_{i-1}-1,R_0,R_1) = \eta_i\cdot R_i(x) \tag{7.29}$$

with η_i given by (7.24).

●

As a corollary to Theorem 7.4, we can show that the reduced PRS is indeed a valid PRS (Example 7.12 is a special case). To see this, recall that the update coefficients for the reduced PRS are defined by

$$\alpha_i = r_i^{\delta_i+1}, \text{ for } i \ge 1,$$

and

$$\beta_1 = 1, \ \ \beta_i = \alpha_{i-1}, \ \text{for} \ \ i \ge 2.$$

To show that the PRS is valid we need only show that $R_i(x) \in \mathbf{R}[x]$ for all i.

By the fundamental theorem, we have

$$R_i(x) = s_i \cdot S(n_{i-1}-1, A, B)$$

where

$$s_i = 1 / \eta_i = (-1)^{\phi_i} \cdot r_{i-1}^{\delta_i - 1} \cdot \prod_{p=1}^{i-1} [(\alpha_p / \beta_p)^{n_p - n_{i-1}+1} \cdot r_p^{-\gamma_p}]$$

Since

$$s_i = (-1)^{\phi_i} \cdot r_{i-1}^{\delta_i - 1} \cdot \alpha_1^{n_2 - n_{i-1}+1} \cdot \prod_{p=2}^{i-1} [(\alpha_p / \alpha_{p-1})^{n_p - n_{i-1}+1} \cdot r_p^{-\gamma_p}]$$

$$= (-1)^{\phi_i} \cdot r_{i-1}^{\delta_i - 1} \cdot \prod_{p=1}^{i-2} [\alpha_p^{\delta_{p+1}} \cdot r_p^{-\gamma_p}] \cdot \alpha_{i-1} \cdot r_{i-1}^{-\gamma_{i-1}}$$

$$= (-1)^{\phi_i} \cdot \prod_{p=1}^{i-2} r_p^{\delta^p (\delta_{p+1} - 1)} \ \in \ \mathbf{R}.$$

Since also $S(n_i, A, B) \in \mathbf{R}[x]$, this implies $R_i(x) \in \mathbf{R}[x]$ for all i.

Similarly, we may use the fundamental theorem to demonstrate the validity of the subresultant PRS defined in Example 7.6. We use the same argument as above. Indeed, the basic idea of the subresultant PRS is to choose β_i so that

$$R_i(x) = S(n_i, A, B),$$

that is, so that $\tau_i = 1$ for all i. We leave the proof of validity as an exercise for the reader (cf. Exercise 7.12).

As a third corollary to the fundamental theorem, we can establish bounds on the growth of the coefficients of the subresultants. Let $S_i(x)$ be the i-th subresultant in the PRS for $R_0(x)$ and $R_1(x)$, that is,

$$S_i(x) = S(n_i - 1, R_0, R_1), \ \text{for} \ \ 1 < i \le k.$$

Following Brown [1] set

$$m_i = 1/2 \cdot (n_0 + n_1 + 2) - n_i$$

$$= 1/2 \cdot (n_0 + n_1 - 2(n_i - 1)) \ \text{for} \ \ 1 < i < k.$$

Then m_i is an approximate measure of the degree loss at each step through the PRS. It increases monotonically from $m_2 \ge 1$ to $m_k \le (n_0 + n_1)/2$. Since each coefficient of $S_i(x)$ is a determinant of order $2 \cdot m_i$, we can derive bounds on their size in terms of m_i.

Suppose that the coefficients of $R_0(x)$ and $R_1(x)$ are integers bounded in magnitude by c. Using Hadamard's inequality

$$|\det(a_{ij})| \le \prod_i (\sum_j a_{ij}^2)^{1/2}$$

we see that the coefficients of $S_i(x)$ are bounded in magnitude by

$$(2 \cdot m_i \cdot c^2)^{m_i}.$$

Taking logarithms, the length of the coefficients is

$$m_i \cdot [\ \log(c) + \log(2 \cdot m_i)\ ].$$

Although the growth permitted by these bounds is faster than linear, the first term is usually larger than the second in most common cases. Therefore the nonlinear coefficient growth is usually not observed. In any case, $n \cdot \log n$ growth is considerably slower than the exponential growth observed in the Euclidean PRS.

If the coefficients of $R_0(x)$ and $R_1(x)$ are polynomials in $x_1, \ldots, x_v$ over the integers with degree at most e_j in x_j, then the coefficients of $T_i(x)$ have degree at most $2m_i \cdot e_j$ in x_j for $0 < j < v$. If the polynomial coefficients have at most t terms each and have integer coefficients of size c then the integer coefficients of $T_i(x)$ are bounded in size by

$$(2m_i \cdot c^2 \cdot t^2)^{m_i}. \qquad (7.30)$$

Taking logarithms, the integer coefficients are then seen to be bounded in length by

$$w_i \cdot [\ 2c + \log(2 \cdot w_i) + 2 \log t\ ].$$

This bound is a generalization of the previous one. The term in $\log t$ reflects the fact that the coefficients of a polynomial product are sums of products of coefficients of the given polynomials.

The Primitive Part Revisited

Once we have constructed a PRS, $\{R_0(x), R_1(x), \ldots, R_k(x)\}$ the desired GCD is $G(x)$, the primitive part of $R_k(x)$. Since $G(x)$ divides both $R_0(x)$ and $R_1(x)$, its coefficients are likely to be smaller. On the other hand, coefficient growth in the PRS implies that the coefficients of $R_k(x)$ are larger than those of $R_0(x)$ and $R_1(x)$, and therefore also of $G(x)$. In other words, $R_k(x)$ is likely to have a very large content. There is a method for removing some of this content without computing any GCD's of the coefficients.

Let $g = \text{lcoeff}(G(x))$ and $r = \text{GCD}(r_0, r_1)$ (recall that we use the notation $r_i = \text{lcoeff}(R_i(x))$ for a PRS). Now, since $G(x) \mid R_0(x)$ and $G(x) \mid R_1(x)$ this implies $g \mid r_0$ and $g \mid r_1$, hence also $g \mid r$. Let $R(x) = (r/g) \cdot G(x)$. Clearly $R(x)$ has r as its leading coefficient and $G(x)$ as its primitive part. Furthermore, $R(x) = (r \cdot R_k(x))/r_k$. Since r divides the first column of M_{ij}, it also divides $S(j, R_0, R_1)$. By the fundamental theorem of PRS this implies that r also divides r_k. Therefore, $R(x) = R_k(x)/(r_k/r)$.

This implies that the large factor r_k/r can be removed from $R_k(x)$ for the price of computing r and performing two divisions. It remains to remove the content of $R(x)$ which divides r.

If for some reason r does not divide r_k (e.g. applying the primitive PRS algorithm) we may compute $R(x)$ directly by applying the formula $R(x) = (r \cdot R_k(x))/r_k$. Alternatively $f = \mathrm{GCD}(r, r_k)$ can be removed from r and r_k before this formula is used, to get $H(x) = R_k(x)/(r_k/h)$ and then $G(x)$ is the primitive part of $H(x)$.

7.4. THE MODULAR GCD ALGORITHM

The problem of coefficient growth that was exhibited with the Euclidean PRS is similar to the growth that one encounters when using naive methods to solve linear equations over an integral domain. As we have seen in Chapter 5, mapping the entire problem to a simpler domain via homomorphisms can lead to better algorithms which avoid problems with coefficient growth. Of course, a single homomorphism does not usually retain all the information necessary to solve the problem in the original domain. However, in certain cases a single reduction is enough when solving the GCD problem.

Lemma 7.3. Let R and R′ be UFD's with $\phi : \mathrm{R} \to \mathrm{R}'$ a homomorphism of rings. This induces a natural homomorphism, also denoted by ϕ, from R[x] to R′[x]. Suppose $A(x)$, $B(x) \in \mathrm{R}[x]$ and $C(x) = \mathrm{GCD}(A(x), B(x))$ with $\phi(\mathrm{lcoeff}(C(x))) \neq 0$. Then

$$\deg(\,\mathrm{GCD}(\phi(A(x)), \phi(B(x)))\,) \geq \deg(\,\mathrm{GCD}(A(x), B(x))).$$

Proof: Let the cofactors of $C(x)$ be $P(x)$, $Q(x) \in \mathrm{R}[x]$ with

$$A(x) = P(x) \cdot C(x), \quad B(x) = Q(x) \cdot C(x).$$

Since ϕ is a ring homomorphism, we have

$$\phi(A(x)) = \phi(P(x)) \cdot \phi(C(x)), \quad \phi(B(x)) = \phi(Q(x)) \cdot \phi(C(x))$$

so that $\phi(C(x))$ is a common factor of both $\phi(A(x))$ and $\phi(B(x))$. Therefore $\phi(C(x))$ divides into their GCD. Since $\phi(\mathrm{lcoeff}(C(x))) \neq 0$, we have

$$\deg(\mathrm{GCD}(\phi(A(x)), \phi(B(x)))) \geq \deg(\phi(C(x))) = \deg(\mathrm{GCD}(A(x), B(x))).$$

●

The condition that the homomorphic image of the leading coefficient of the GCD not be zero is usually determined by checking that the homomorphic images of of A and B do not decrease in degree. Notice that the concepts of leading coefficient, degree (in this case a degree vector denoted by ∂) and GCD are all well-defined in multivariate domains (cf. Section 2.6). Using these corresponding notions, we may easily generalize Lemma 7.3 to multivariate domains.

As a corollary to Lemma 7.3, suppose ϕ is a homomorphism in which the images of both $A(x)$ and $B(x)$ are relatively prime. Then, as long as the GCD of the two polynomials does not have its degree reduced under the homomorphism, Lemma 7.3 states that

$$0 = \deg(\mathrm{GCD}(\phi(A(x)),\phi(B(x)))) \geq \deg(\mathrm{GCD}(A(x),B(x)))$$

and hence $A(x)$ and $B(x)$ will be relatively prime.

Example 7.13. The polynomials $A(x)$ and $B(x)$ from Example 7.2 map under the modular homomorphism $\phi_{23} : \mathbf{Z}[x] \to \mathbf{Z}_{23}[x]$ to the polynomials in Example 7.1. Since the polynomials in Example 7.1 are relatively prime in $\mathbf{Z}_{23}[x]$ the polynomials must be relatively prime considered over $\mathbf{Z}[x]$.

●

The advantages of this method are clear. The particular problem is mapped to a domain which has more algebraic structure (e.g. a Euclidean domain) allowing for a wider range of algorithms (e.g. Euclid's algorithm). In addition, the arithmetic is simpler because in some sense all the arithmetic is done in "single precision", rather than in domains with arbitrarily large coefficients. The problem, of course, is that the price for simpler arithmetic is information loss, resulting in either incomplete solutions or in solutions that are of a different form than the desired solution.

Example 7.14. The polynomials $A(x)$ and $B(x)$ from Example 7.2 map under the modular homomorphism $\phi_2 : \mathbf{Z}[x] \to \mathbf{Z}_2[x]$ to the polynomials

$$\phi_2(A(x)) = x^8 + x^6 + x^4 + x^3 + x^2 + 1$$

and

$$\phi_2(B(x)) = x^6 + x^4 + x + 1.$$

Calculating the GCD in $\mathbf{Z}_2[x]$ gives

$$\mathrm{GCD}(\phi_2(A(x)),\phi_2(B(x))) = x + 1$$

so we do not get complete information from this homomorphism.

●

Homomorphisms such as the one in Example 7.14 are called *unlucky* homomorphisms. Fortunately, in the cases of interest to us, unlucky homomorphisms do not occur too often. We will return to this point later in the section.

The original idea of using homomorphisms in GCD calculation was suggested by Joel Moses as a fast heuristic method for checking if two polynomials are relatively prime (as in our previous example). Subsequently Collins [4] (in the univariate case) and Brown [1] (in the multivariate case) developed an algorithm that determines a GCD using homomorphic reductions even in the non-relatively prime case. The central observation lies in the fact that, if ϕ is not an unlucky homomorphism, and we set $C(x) = \mathrm{GCD}(A(x), B(x))$, then $\phi(C(x))$ and $\mathrm{GCD}(\phi(A(x)), \phi(B(x)))$ are associates, that is

$$\phi(\mathrm{GCD}(A(x),B(x))) = c \cdot \mathrm{GCD}(\phi(A(x)),\phi(B(x)))$$

for some constant c. If c were known in advance, then the image of $C(x)$ under the homomorphism would be known. Repeating such a process for a number of homomorphisms yields a set of homomorphic images of $C(x)$ which could then be inverted by the Chinese remainder algorithm.

The problem is that the constant c is rarely known in advance, unless of course either $A(x)$ or $B(x)$ is monic (in which case we know that $C(x)$ is also monic). Otherwise, we may proceed as follows. By removing the content of $A(x)$ and $B(x)$, we may assume that these two polynomials are primitive (the correct GCD will then be the product of the GCD of the contents and the primitive parts). The leading coefficient of $C(x)$ certainly divides the coefficient GCD of the leading coefficients of $A(x)$ and $B(x)$. If g denotes this GCD, then $g = u \cdot \mathrm{lcoeff}(C(x))$ for some constant u. If we normalize $\phi(A(x))$ and $\phi(B(x))$ to have their leading coefficients be $\mathrm{lcoeff}(\phi(C(x)))$, then we would have $c = u$. The Chinese remainder algorithm would then produce $c \cdot C(x)$, rather than $C(x)$, but $C(x)$ could then be determined by taking the primitive part of $c \cdot C(x)$. Note that the above argument is also valid in the case of multivariate domains, with the corresponding notions of leading coefficient and degree corresponding to lexicographical ordering of terms.

Once we have thrown away all unlucky homomorphisms and normalized correctly, the only problem remaining is that of inverting a number of homomorphisms for a particular value. This is done by the Chinese remainder algorithm.

Example 7.15. Let $A(x), B(x) \in \mathbf{Z}[x]$ be given by

$$A(x) = x^4 + 25x^3 + 145x^2 - 171x - 360,$$

$$B(x) = x^5 + 14x^4 + 15x^3 - x^2 - 14x - 15.$$

Both $A(x)$ and $B(x)$ are monic so there is no problem normalizing our image results.

Reducing mod 5 and using the obvious notation gives

$$A_5(x) = x^4 - x, \quad B_5(x) = x^5 - x^4 - x^2 + x.$$

Calculating the GCD in $\mathbf{Z}_5[x]$ gives

$$\mathrm{GCD}(A_5(x), B_5(x)) = x^4 - x.$$

Reducing mod 7 gives

$$A_7(x) = x^4 - 3x^3 - 2x^2 - 3x - 3, \quad B_7(x) = x^5 + x^3 - x^2 - 1$$

and calculating the GCD in $\mathbf{Z}_7[x]$ gives

$$\mathrm{GCD}(A_7(x), B_7(x)) = x^2 + 1.$$

Since this degree is less than the mod 5 reduction, we know that the mod 5 reduction is an example of a bad reduction. Therefore we throw away the mod 5 calculation.

Continuing, we reduce mod 11 to obtain

$$A_{11}(x) = x^4 + 3x^3 + 2x^2 + 5x + 3,$$

$$B_{11}(x) = x^5 + 3x^4 + 4x^3 - x^2 - 3x - 4$$

with the GCD in $\mathbf{Z}_{11}[x]$ given by

$$\mathrm{GCD}(A_{11}(x), B_{11}(x)) = x^2 + 3x + 4.$$

Since both reductions have the same degree we make the assumption that both are good reductions and so

$$\mathrm{GCD}(A(x), B(x)) = x^2 + ax + b$$

for integers a and b. Since

$$a \equiv 0 \bmod 7, \;\; a \equiv 3 \bmod 11$$

and

$$b \equiv 1 \bmod 7, \;\; b \equiv 4 \bmod 11$$

the integer CRA of Chapter 5 gives the unique values of a and b in the range $-38 \le a, b \le 38$ as

$$a = 14, \;\; b = 15.$$

Checking our candidate GCD by division shows that we indeed have determined the desired GCD. The reason that we check our answer at this stage, instead of choosing another prime, will be explained later in the section. Similarly, why a simple division test determines the validity of our answer will be explained later.

●

We can summarize the standard form of a multiple-homomorphism algorithm as follows:

(a) bound the number of homomorphic images needed;

(b) extract the next image;

(c) perform the image algorithm;

(d) if the image result is well-formed incorporate it into the inverse image result; otherwise return to step (b);

(e) if sufficient image results have been constructed then halt, otherwise return to step (b).

Step (d) states that when we run into an unlucky homomorphism we throw the homomorphism away, that is, we simply ignore the results from this "bad reduction" (as we did in Example 7.15). This is plausible as long as we believe the as yet unproven claim that such bad reductions are rare.

In the case of polynomial GCD's the coefficient homomorphisms we use are:

(i) $\phi_m : \mathbf{Z} \rightarrow \mathbf{Z}_m$ – the modular homomorphism, which maps integers into their remainder modulo m. Normally m is chosen to be a prime so that $\mathbf{Z}_m$ is a finite field. This restricts our integer coefficients to be of finite size and therefore we do not need to worry about the growth of integer coefficients.

(ii) $\phi_{w-b} : \mathrm{R}[w] \to \mathrm{R}$ – the evaluation homomorphism, which maps polynomials in a variable w to their value at $w = b$. This restricts our polynomial coefficients to be of degree 0 and therefore stops the growth in the degree of the coefficients.

As described in Chapter 5, these homomorphisms can be inverted by applying the Chinese remainder algorithm to a collection of homomorphic images.

Example 7.16. Let $A(x,y,z)$ and $B(x,y,z) \in \mathbf{Z}[x,y,z]$ be given by

$$A(x,y,z) = 9x^5 + 2x^4yz - 189x^3y^3z + 117x^3yz^2 + 3x^3 - 42x^2y^4z^2 + 26x^2y^2z^3$$
$$+ 18x^2 - 63xy^3z + 39xyz^2 + 4xyz + 6,$$
$$B(x,y,z) = 6x^6 - 126x^4y^3z + 78x^4yz^2 + x^4y + x^4z + 13x^3 - 21x^2y^4z - 21x^2y^3z^2$$
$$+ 13x^2y^2z^2 + 13x^2yz^3 - 21xy^3z + 13xyz^2 + 2xy + 2xz + 2.$$

Both A and B are listed in lexicographically decreasing order of their exponent vectors (cf. Definition 2.14).

Let $C(x,y,z)$ be the GCD of A and B. Since C divides B and $\deg_y(B) = 4$ we know that $\deg_y(C) \le 4$. Similarly, $\deg_z(C) \le 3$. We use the three moduli $p_1 = 11$, $p_2 = 13$ and $p_3 = 17$ which together using the integer CRA will cover all coefficients in the range [-1215,1215]. Because the leading coefficients of A and B have a nontrivial common factor of 3, we normalize our images to ensure that the resulting mod p multivariate GCD's have leading coefficient 3.

To calculate the GCD in $\mathbf{Z}_{11}[x,y,z]$, we consider the polynomials

$$A_{11}(x,y,z) = -2x^5 + 2x^4yz - 2x^3y^3z - 4x^3yz^2 + 3x^3 + 2x^2y^4z^2$$
$$+ 4x^2y^2z^3 - 4x^2 + 3xy^3z - 5xyz^2 + 4xyz - 5,$$
$$B_{11}(x,y,z) = -5x^6 - 5x^4y^3z + x^4yz^2 + x^4y + x^4z + 2x^3 + x^2y^4z + x^2y^3z^2$$
$$+ 2x^2y^2z^2 + 2x^2yz^3 + xy^3z + 2xyz^2 + 2xy + 2xz + 2$$

as polynomials in x and y having coefficients from the domain $\mathbf{Z}_{11}[z]$. In this context the GCD of the leading coefficients of the two polynomials is 1, hence no normalization need be done at this stage. To compute the GCD in this domain we evaluate the polynomials at four arbitrary points from $\mathbf{Z}_{11}$ and compute the GCD's recursively. Let $z = 2$ be one such random choice. We calculate the GCD of

$$A_{11}(x,y,2) = -2x^5 + 4x^4y - 4x^3y^3 + 5x^3y + 3x^3 - 3x^2y^4 - x^2y^2$$
$$- 4x^2 - 5xy^3 - 2xy - 3xy - 5,$$
$$B_{11}(x,y,2) = -5x^6 + x^4y^3 + 4x^4y + x^4y + 2x^4 + 2x^3 + 2x^2y^4 + 4x^2y^3$$
$$- 3x^2y^2 + 5x^2y + 2xy^3 - 3xy + 2xy + 4x + 2$$

obtained by applying the homomorphism $\phi_{<z-2>}$, viewing these as polynomials in x with coefficients from $\mathbf{Z}_{11}[y]$. In this case, the leading coefficients of the two polynomials are relatively prime, hence no normalization need take place. We compute the GCD of these bivariate polynomials by evaluating the GCD in five homomorphic images, that is, evaluating the polynomials at five distinct y points and computing the GCD's (this time in univariate domains). For example, taking $y = 3$ as a random value of y we calculate the GCD of the two modular polynomials in $\mathbf{Z}_{11}[x]$ using Euclid's algorithm to obtain

$$GCD(A_{11}(x,3,2), B_{11}(x,3,2)) = x^3 + x + 2 \pmod{11}.$$

Similarly, evaluating our polynomials from $\mathbf{Z}_{11}[x,y]$ at the random y points $y = 5$, $y = -4$, , $y = -2$, and $y = 2$ we obtain

$$GCD(A_{11}(x,5,2), B_{11}(x,5,2)) = x^3 - 4x + 2 \pmod{11},$$

$$GCD(A_{11}(x,-4,2), B_{11}(x,-4,2)) = x^3 + 5x + 2 \pmod{11},$$

$$GCD(A_{11}(x,-2,2), B_{11}(x,-2,2)) = x^3 + x + 2 \pmod{11},$$

$$GCD(A_{11}(x,2,2), B_{11}(x,2,2)) = x^3 - 5x + 2 \pmod{11}.$$

Interpolating the individual coefficients at the five y values gives the image of our GCD in $\mathbf{Z}_{11}[x,y]$ under the evaluation homomorphism $\phi_{<z-2>}$ as

$$x^3 + 2xy^3 - 3xy + 2 \pmod{11}.$$

Repeating the above with the three additional random z values $z = -5$, $z = -3$, and $z = 5$ gives the images in $\mathbf{Z}_{11}[x,y,z]$ as

$$x^3 - 5xy^3 - 5xy + 2 \pmod{11},$$

$$x^3 - 3xy^3 - 4xy + 2 \pmod{11},$$

and

$$x^3 + 5xy^3 - 5xy + 2 \pmod{11},$$

respectively. Interpolating the individual coefficients at the four z values gives

$$x^3 + xy^3z + 2xyz^2 + 2 \pmod{11}.$$

After normalizing so that our leading coefficient is 3, the candidate for the image of our GCD in $\mathbf{Z}_{11}[x,y,z]$ is

$$3x^3 + 3xy^3z - 5xyz^2 - 5 \pmod{11}.$$

Repeating the same process, we obtain the candidate for the image of our GCD in $\mathbf{Z}_{13}[x,y,z]$ as

$$3x^3 + 2xy^3z + 6 \pmod{13}$$

and in $\mathbf{Z}_{17}[x,y,z]$ as

$$3x^3 + 5xy^3z + 5xyz^2 + 6 \pmod{17}.$$

Using the CRA with each coefficient gives our candidate for the GCD in $\mathbf{Z}[x,y,z]$ as

$$3x^3 - 63xy^3z + 39xyz^2 + 6.$$

Removing the integer content gives

$$x^3 - 21xy^3z + 13xyz^2 + 2. \tag{7.31}$$

Dividing (7.31) into both A and B verifies that it is indeed the GCD.

●

Overview of the Algorithms

Basically we have two very similar algorithms for our homomorphism techniques and one basis algorithm.

MGCD - This algorithm reduces the multivariate integer GCD problem to a series of multivariate finite field GCD problems by applying modular homomorphisms.

PGCD - This algorithm reduces the k-variate finite field GCD problem to a series of $(k-1)$-variate finite field GCD problems by applying evaluation homomorphisms. This algorithm is used recursively.

UGCD - Everything is reduced to the univariate finite field case. Since $\mathbf{Z}_p$ is a field, $\mathbf{Z}_p[x]$ is a Euclidean domain and therefore Euclid's algorithm (Algorithm 2.1) can be applied. Since $\mathbf{Z}_p$ is a *finite* field we no longer have any problem with coefficient growth.

In terms of the outline of homomorphism algorithms presented in Chapter 5, there are still some problems to be solved.

(a) How many homomorphic images are needed?

If $C(x) = \text{GCD}(A(x), B(x))$ then we know that $C(x) \mid A(x)$ and $C(x) \mid B(x)$. This can be used as an easy bound on the number of images to use. Namely, we continue constructing new image GCD's $C^*(x)$ and incorporating them into the problem domain GCD $C(x)$. Once $C(x) \mid A(x)$ and $C(x) \mid B(x)$ we stop. This involves a divide-if-divisible test.

A variation on this idea is to notice that

$$\text{lcoeff}(C) \mid \text{GCD}(\text{lcoeff}(A), \text{lcoeff}(B)).$$

Testing this equality is even cheaper. Another simple test is to divide $C(x)$ into $A(x)$ and $B(x)$ only when the Chinese remainder algorithm leaves our $C(x)$ unchanged for one iteration. If we continued Example 7.15 with another prime, say 13, then the resulting invocation of the CRA would again give $x^2 + 14x + 15$, and hence a division check would only be done at this stage.

For a-priori bounds we can argue that

$$\deg_i(C) \le \min(\deg_i(A), \deg_i(B)), \text{ for } 1 \le i \le k\ ,$$

because $C(x)$ must divide both polynomials regardless which x_i is the main variable. This is used as a bound in PGCD. In MGCD the analogous bound is that, in all likelihood,

Algorithm 7.1. Modular GCD Algorithm.

procedure MGCD(A, B)
  # Given $A, B \in \mathbf{Z}[x_1, \ldots, x_k]$, nonzero, we determine the GCD of the
  # two polynomials via modular reduction.
  # Remove integer content
  $a \leftarrow \mathrm{icont}(A)$; $b \leftarrow \mathrm{icont}(B)$; $A \leftarrow A/a$; $B \leftarrow B/b$
  # Compute coefficient bound for GCD(A, B)
  $c \leftarrow \mathrm{igcd}(a,b)$; $g \leftarrow \mathrm{igcd}(\mathrm{lcoeff}(A), \mathrm{lcoeff}(B))$
  $(q, H) \leftarrow (0,0)$; $n \leftarrow \min(\deg_k(A), \deg_k(B))$
  limit $\leftarrow 2^n \cdot |g| \cdot \min(\|A\|_\infty, \|B\|_\infty)$

  while true **do** { $p \leftarrow \mathrm{New}(\mathit{LargePrime})$
    while $p \mid g$ **do** $p \leftarrow \mathrm{New}(\mathit{LargePrime})$
    $A_p \leftarrow A \bmod p$; $B_p \leftarrow B \bmod p$
    $g_p \leftarrow g \bmod p$; $C_p \leftarrow \mathrm{PGCD}(A_p, B_p, p)$; $m \leftarrow \deg_k(C_p)$;
    # Normalize so that $g_p = \mathrm{lcoeff}(C_p)$
    $C_p \leftarrow g_p \cdot \mathrm{lcoeff}(C_p)^{-1} \cdot C_p$
    # Test for unlucky homomorphisms
    if $m < n$ **then** {
      $(q, H) \leftarrow (p, C_p)$; $n \leftarrow m$ }
    elseif $m = n$ **then** {
      # Test for completion. Update coefficients of
      # GCD candidate H and modulus q via integer CRA.
      for all coefficients h_i in H **do** {
        $h_i \leftarrow \mathrm{IntegerCRA}([q, p], [h_i, (C_p)_i])$
        $q \leftarrow q \cdot p$ }
    if $q > \mathit{limit}$ **then** {
      # Remove integer content of result and do division check
      $C \leftarrow \mathrm{pp}(H)$
      if $C \mid A$ **and** $C \mid B$ **then**
        return($c \cdot C$) }
    elseif $m = 0$ **then**
      return(c) }
end

$$\|C\|_\infty \le 2^{\min(\deg A, \deg B)} \min(\|A\|_\infty, \|B\|_\infty) \,|\, \gcd(\, \mathrm{lcoeff}(A), \mathrm{lcoeff}(B)) \,| \,.$$

Both of these bounds are used in addition to the a-postiori test in the completion step of the algorithms.

(b) How do we ensure that the image result is well formed?

Here the problem is to detect and exclude unlucky homomorphisms. As is shown in the next subsection, unlucky homomorphisms occur when some subresultants fall in the kernel of the homomorphism ϕ. Thus a simple a-priori necessary but not sufficient condition is to ensure that

$$\mathrm{lcoeff}(A) \text{ and } \mathrm{lcoeff}(B) \notin \ker \phi \,.$$

As explained earlier, a simple a-postiori test is the degree anomaly check. A bad homomorphism will *always* give a GCD of too high a degree.

Unlucky Homomorphisms

It is possible to derive a bound on the number of unlucky homomorphisms and show that they are very unlikely to occur. First we look at how they arise.

Theorem 7.5. Let A, B be nonzero primitive polynomials in $\mathbf{Z}[x_1, \ldots, x_k]$ and $C = \mathrm{GCD}(A,B)$. Let $n_i = \deg_i(C)$ and $w_i = \mathrm{icont}(\, S_i(n_i,A,B)\,)$, the integer content of the n_i-th subresultant with x_i as the main variable. Then every unlucky prime divides $w = \prod_{i=1}^{k} w_i$.

Proof: Let $S_i'(j,A,B)$ be the j-th subresultant over $\mathbf{Z}_p$ with x_i as the main variable. Then

$$S_i'(j,A,B) = S_i(j,A,B) \bmod p$$

for all $0 < j \le \min(\deg_i(A), \deg_i(B))$. The fact that p is an unlucky prime will be disclosed by the degree anomaly and the GCD will have degree $m > n_i$ in x_i for some i. Hence $S_i'(n_i,A,B) = 0$, that is, $p \mid S_i(n_i,A,B)$. Conversely, if $p \mid S_i(n_i,A,B)$ then $S_i'(n_i,A,B) = 0$ and the PRS terminates for some $m > n$. Therefore p is an unlucky prime if and only if $p \mid S_i(n_i,A,B)$ for some $1 \le i \le k$. Let w_i be the integer content of $S_i(n_i,A,B)$. Then p is unlucky if and only if $p \mid \prod_{i=0}^{k} w_i$ which proves Theorem 7.5.

●

Theorem 7.5 says that there are only a finite number of primes p which induce unlucky homomorphisms. Indeed, it also allows us to derive an upper bound on the number of unlucky primes.

Algorithm 7.2. Multivariate GCD Reduction Algorithm.

```
procedure PGCD(A, B, p)

  # Given A, B ∈ Z_p[x_1, . . . , x_k],
  # PGCD(A, B, p) calculates the GCD of A and B.

  if k = 1 then { # Call univariate GCD algorithm
     C ← UGCD(A, B, p)
     if deg(C) = 0 then C ← 1
     return(C) }

  # Determine content of A and B considered as multivariate
  # polynomials in Z_p[x_1, . . . , x_{k-1}] with coefficients from Z_p[x_k]

  a ← cont(A);  b ← cont(B);   A ← A/a;  B ← B/b
  c ← UGCD(a, b, p);  g ← UGCD(lcoeff(A), lcoeff(B), p)

  # Notice that both c and g are in Z_p[x_k]

  # Main loop:
  (q, H) ← (1, 1);  n ← min(deg_k(A), deg_k(B));  limit ← n + deg_k(g)
  while true do {
     b ← New(Member Z_p) with g(b) ≠ 0
     A_b ← A mod (x_k − b);   B_b ← B mod (x_k − b)
     C_b ← PGCD(A_b, B_b, p);   m ← deg_{k-1}(C_b); g_b ← g(b)
     # Normalize C_b so that lcoeff(C_b) = g_b
     C_b ← g_b·lcoeff(C_b)^{-1} C_b

     # Test for unlucky homomorphism
     if m < n then { (q, H) ← (1, 1);  n ← m }
     elseif m = n then
        # Use previous result to continue building H via
        # polynomial interpolation (i.e. via polynomial CRA)
        (q, H) ← PolyInterp(q, H, b, C_b, p)

     # Test for completion
     if lcoeff(H) = g then {
        C ← pp(H)
        if C | A and C | B then return(c·C)
        elseif m = 0 then return(c) } }

end
```

Corollary. Let u be the number of unlucky primes $p > a > 2$. Set

$$c = \max(\|A\|_\infty, \|B\|_\infty),\ \ m = 1/2 \max_{1 \le i \le k} (\deg_i(A), \deg_i(B)),\ \ t = \max_{1 \le i \le k} t_i$$

where t_i is the number of coefficient terms of A, B viewed as polynomials in x_i. Then $u < m \cdot k \log_a (2mc^2t^2)$.

Proof: By equation (7.30), $w_i \le (2m \cdot c^2t^2)^m$ for each i. Thus, $w \le (2m \cdot c^2t^2)^{mk}$. Let P be the product of all unlucky primes $p > a$. Then from Theorem 7.5, we have

$$a^u < P \le w, \quad \text{i.e. } u < \log_a w \le m \cdot k \log_a (2mc^2t^2) \ .$$

●

The corollary to Theorem 7.5 implies that it is in our best interests to choose p as large as possible. On the other hand if $p \ge B$, the word size of our computer, then we must return to multi-precision integer arithmetic and we lose the benefits of a modular algorithm. A good compromise is to choose

$$a = (B + 1)/2 < p < B$$

If B=2^{32} (say) then the prime density theorem indicates that there are about

$$n = 2^{32}/32 - 2^{31}/31 \approx 2^{27}$$

or about 100 million such primes. Thus there is no shortage.

We can also argue that we would be very unlikely to encounter a bad prime. If our prime p is fixed and w is chosen at random then the probability that p divides w is $p^{-1} < a^{-1}$. Therefore the probability that p is unlucky is less than $k/a = k/2^{32}$ (say) for k variables.

Just as there are unlucky primes for the integer portion of the algorithm, there can be unlucky points of evaluation (b-values) for the polynomial portion. We can state the analogous theorem.

Theorem 7.6. Let A and B be nonzero polynomials in $\mathbf{Z}_p[x_1, \ldots, x_k]$ with $C = \mathrm{GCD}(A,B)$. Let $n_i = \deg_i(C)$ and $w_i = \mathrm{pcont}(\ S_i(n_i,A,B))$ be the polynomial content in $\mathbf{Z}_p[x_1, \ldots, x_k]$ of the subresultant with respect to x_i. Then an unlucky b-value is a root of $\prod_{i=1}^{k} w_i$ in $\mathbf{Z}_p$.

●

This gives us a bound on the number of unlucky b-values.

Corollary. Let u be the total number of unlucky b-values. If

$$m = 1/2 \max_{1 \le i \le k-1} (\deg_i(A), \deg_i(B)) \ \text{ and } \ e = \max(\deg_k(A), \deg_k(B))$$

then $u \le 2m \cdot e(k-1)$.

●

Again we can argue that if p is very large then encountering a bad b-value is very unlikely. If b is chosen at random among the elements of **Z** then the probability of it being bad is $u/p \le 2m \cdot e(k-1)/p$. If $p \gg 1$ then $u/p \ll 1$.

7.5. THE SPARSE MODULAR GCD ALGORITHM

The modular GCD algorithm of the previous section solves a single problem by the usual method of solving a number of easier problems in much simpler domains. The problem with this method, of course, is the large number of domains that may need to be used, indeed it is exponential in the number of variables. For many problems, many more domains are used than necessary, especially when the input has a sparse, rather than a dense, structure. This is especially true in the case of multivariate polynomials were the nonzero terms are generally small in comparison to the number of terms possible. In this section we discuss an algorithm, SparseMod, for calculating the GCD of two multivariate polynomials over the integers. This algorithm first appeared in the 1979 Ph.D. thesis of R. Zippel [19]. The process is actually a general technique for approaching sparse problems using probabilistic techniques.

The sparse modular methods are based on the simple observation that evaluating a polynomial at a random point will almost never yield zero if the point is chosen from a large enough set. The sparse modular GCD algorithm determines the GCD by constructing an alternating sequence of dense and sparse interpolations. A dense interpolation assumes that the resulting polynomial of degree n has all its coefficients as unknown and hence requires $n + 1$ evaluation points to determine it (e.g. via Newton interpolation). Sparse interpolation assumes that, although the resulting polynomial may be of degree n, there are only t unknown coefficients with $t \ll n$. The resulting unknown coefficients are then determined by a linear system of equations once $t + 1$ evaluation points are given. The problem with sparse interpolation, of course, is that one must decide in advance which coefficients will be assumed to be zero.

For example, consider the computation of a GCD, $C(x,y,z)$ of two multivariate polynomials $A(x,y,z)$ and $B(x,y,z)$. Let y_0, z_0 be random values and compute, as in the modular method, the univariate polynomial

$$C(x,y_0,z_0) = \mathrm{GCD}\,(A(x,y_0,z_0), B(x,y_0,z_0))$$
$$= c_n(y_0, z_0)\,x^n + \cdots + c_1(y_0, z_0)\,x + c_0(y_0, z_0).$$

The problem now is to determine $c_i(y,z)$ for $i = 0, \ldots, n$. Let d be a bound for the degree of the y value in $C(x,y,z)$. Using dense polynomial interpolation at the $d + 1$ values $c_i(y_0,z_0)$, $c_i(y_1,z_0), \ldots, c_i(y_d,z_0)$ we obtain $c_i(y,z_0)$.

The result is a polynomial with a certain number of terms, t. Those terms which are zero for (x,y,z_0) are assumed to be identically zero. Let d now denote a bound for the degree of the z value in $C(x,y,z)$. The algorithm now proceeds by performing d sparse interpolations to produce the images $c_i(y,z_1), \ldots, c_i(y,z_d)$. Dense interpolation is then used to construct $c_i(y,z)$. This is done for every i with the final result pieced together to obtain $C(x,y,z)$.

This is our candidate for the GCD. The correctness of the result is checked by division into both $A(x,y,z)$ and $B(x,y,z)$.

Example 7.17. Let $A(x,y,z)$ and $B(x,y,z) \in \mathbf{Z}[x,y,z]$ be given by

$$A(x,y,z) = x^5 + 2yzx^4 + (13yz^2 - 21y^3z + 3)x^3 + (26y^2z^3 - 42y^4z^2 + 2)x^2 + (39yz^2 - 63y^3z + 4yz)x + 6,$$

$$B(x,y,z) = x^6 + (13yz^2 - 21y^3z + z + y)x^4 + 3x^3 + (13yz^3 + 13y^2z^2 - 21y^3z^2 - 21y^4z)x^2 + (13yz^2 - 21y^3z + 2z + 2y)x + 2\,.$$

Because A and B have leading coefficient 1, we need not normalize our GCD's since they will also be monic. We will show how to obtain the coefficient of the x term in the GCD. Choose two primes, $p = 11$ and $q = 17$, and determine the images of the GCD in both $\mathbf{Z}_{11}[x,y,z]$ and $\mathbf{Z}_{17}[x,y,z]$. As in Example 7.16 we can deduce that the GCD has at most degree 4 in y and 3 in z. For $z = 2$ we perform our GCD calculations in the Euclidean domain $\mathbf{Z}_{11}[x]$ at the five random y values 1, 3, 5, -4, 4 obtaining

$$\mathrm{GCD}(A(x,1,2), B(x,1,2)) = x^3 - x + 2 \pmod{11},$$

$$\mathrm{GCD}(A(x,3,2), B(x,3,2)) = x^3 + x + 2 \pmod{11},$$

$$\mathrm{GCD}(A(x,5,2), B(x,5,2)) = x^3 + 4x + 2 \pmod{11},$$

$$\mathrm{GCD}(A(x,-4,2), B(x,-4,2)) = x^3 + 5x + 2 \pmod{11},$$

$$\mathrm{GCD}(A(x,4,2), B(x,4,2)) = x^3 - 5x + 2 \pmod{11}.$$

Using dense interpolation gives the image of our GCD in $\mathbf{Z}_{11}[x,y]$ under the evaluation homomorphism $\phi_{<z-2>}$ as

$$x^3 + (-3y + 2y^3)x + 2 \pmod{11}. \tag{7.32}$$

Rather than repeating the above calculations for an independent random choice of z, we use (7.32) as a model and decide that our GCD is of the form

$$x^3 + (ay + by^3)x + 2,$$

that is, that there are no y^0x and y^2x terms. (This is of course much more plausible when the prime p is chosen to be much larger than 11; however we are only presenting an example of the steps of the algorithm rather than attempting a full scale justification for the assumptions made.) Thus, let $z = -5$ be a random z value different than 2 and suppose $y = -3$ and $y = 2$ are two random y values. Calculating as before we obtain

$$\mathrm{GCD}\,(A(x,-3,-5), B(x,-3,-5)) = x^3 - 4x + 2 \pmod{11},$$

$$\mathrm{GCD}\,(A(x,2,-5), B(x,2,-5)) = x^3 + 5x + 2 \pmod{11}.$$

To determine the linear coefficient we have the two equations

$$-3a - 5b = -4 \qquad (\text{mod } 11),$$

$$2a - 3b = \ 4 \qquad (\text{mod } 11)$$

which has a solution $a = -5$, $b = -5$. Thus under the evaluation homomorphism $\phi_{<z+5>}$, our candidate for the image of our GCD in $\mathbf{Z}_{11}[x,y]$ is given by

$$x^3 + (-5y - 5y^3)x + 2 \qquad (\text{mod } 11).$$

A similar pair of sparse interpolations have images under $\phi_{<z+3>}$, and $\phi_{<z-5>}$ given by

$$x^3 + (-4y - 3y^3)x + 2 \qquad (\text{mod } 11),$$

$$x^3 + (-5y + 5y^3)x + 2 \qquad (\text{mod } 11),$$

respectively. Dense interpolation gives the candidate of the image of the GCD in $\mathbf{Z}_{11}[x,y,z]$ as

$$x^3 + (2yz^2 + y^3z)x + 2 \qquad (\text{mod } 11).$$

In a similar fashion, we can also obtain the candidate for the image of the GCD in $\mathbf{Z}_{17}[x,y,z]$ by sparse interpolation. Indeed, we assume that the GCD of A and B in $Z_{17}[x,y,z]$ is probably of the form

$$x^3 + (cyz^2 + dy^3z)x + 2 \qquad (\text{mod } 17). \tag{7.33}$$

Evaluating at the random points (in $\mathbf{Z}_{17}$) $y = 7$, $z = -4$ and $y = -2$, $z = 4$, respectively, we obtain

$$\text{GCD}\,(A(x,7,-4), (B(x,7,-4)) = x^3 + 8x + 2 \qquad (\text{mod } 17), \tag{7.34}$$

$$\text{GCD}\,(A(x,-2,4), B(x,-2,4)) = x^3 + x + 2 \qquad (\text{mod } 17). \tag{7.35}$$

Using sparse interpolation combining (7.33), (7.34) and (7.35) gives $c = -4$, $d = -4$, so our candidate for the image of our GCD in $Z_{17}[x,y,z]$ is

$$x^3 + (-4yz^2 - 4y^3z)x + 2 \qquad (\text{mod } 17).$$

Using the integer CRA applied to each coefficient gives our prospective GCD in $\mathbf{Z}[x,y,z]$ as

$$x^3 + (13yz^2 - 21y^3z)x + 2.$$

Note that this method required only 13 univariate GCD calculations. Using the modular method on this example would require approximately 40 univariate GCD calculations.

●

The algorithm is probabilistic, with each sparse interpolation based on the assumption that coefficients which are zero in one dense interpolation are probably zero in all cases. The probability that the computed GCD is incorrect can be made arbitrarily small by using a large enough range for the evaluation points. Note that since at each stage of the sparse modular algorithm a series of independent sparse interpolations are performed, the algorithm lends itself well to parallel implementations (cf. Watt [17]).

7.6. GCD'S USING HENSEL LIFTING: THE EZ-GCD ALGORITHM

It is intuitively clear that an algorithm such as MGCD does far too much work in the case of sparse problems. This is particularly true in sparse problems with many variables. While it can be argued that sparse problems form a set of measure zero in the space of all problems, the sparse ones tend to be the most common in practice. Therefore we need a method which copes well with them.

A good method is one developed by Moses and Yun [13] which uses Hensel's lemma. It was Zassenhaus [18] who originally proposed the use of Hensel's lemma to construct polynomial factors over the integers. It was (partly) in his honor that Moses and Yun named their method the Extended Zassenhaus GCD (EZ-GCD) Algorithm.

The fact that one could use Hensel's lemma in GCD calculations should be clear, since the main application of Hensel's construction in the preceding chapter was to lift a mod p univariate factorization in $\mathbf{Z}_p[x]$ to a factorization in the multivariate domain $\mathbf{Z}[x,y_1, \ldots, y_k]$. Thus, the basic idea is to reduce two polynomials A and B to a modular univariate representation and to determine their GCD in the simpler domain. This gives a factorization of both reduced polynomials. With a little luck, a GCD and one of its cofactors will be relatively prime. If $\mathrm{GCD}(A,B) = C$ where $A = C \cdot P$ and $B = C \cdot Q$ then hopefully C and P, or C and Q are relatively prime. The algorithm proceeds on this assumption and applies *one* evaluation homomorphism for each variable and one modular homomorphism for the integer coefficients of A and B:

$$A_I = A \bmod I\,, \quad B_I = B \bmod I. \tag{7.36}$$

Now the GCD, C_I and the cofactors P_I and Q_I of A_I and B_I, respectively, are constructed. If $\mathrm{GCD}(C_I,P_I) = 1$ (say) then a finite number of applications of Hensel's lemma will disclose C and P.

Using this Hensel-type technique has a number of advantages:

(a) It behaves like the modular algorithm for relatively prime A and B. Thus, relatively prime polynomials A, B will be disclosed by $\deg(C_I) = 0$ and therefore are discovered very quickly.

(b) If I is chosen so that as many b-values as possible are zero, then the algorithm preserves sparsity and can cope with sparse polynomials quite well.

(c) If A and B are dense, then the algorithm has the evaluation-interpolation style of the modular algorithm, which usually proves effective in solving dense problems.

Overview of the Algorithm

As was the case with the modular algorithm, the method is divided into several subalgorithms:

EZ-GCD: The main routine which computes the GCD over the integers of the multivariate polynomials $A(x,\mathbf{y})$, $B(x,\mathbf{y})$ by applying modular and evaluation homomorphisms to map these polynomials into $A_I(x)$ and $B_I(x)$ in $\mathbf{Z}_p[x]$.

UGCD: The same basis routine as in the modular algorithm. It computes the GCD of $A(x)$ and $B(x)$ using Euclid's algorithm (Algorithm 2.1).

SGCD: The special GCD routine which is invoked if the GCD of $A(x)$ and $B(x)$ is not relatively prime to its cofactors.

EZ-LIFT: The generalized Hensel routine which lifts the GCD in $\mathbf{Z}_p[x]$ up to C, the multivariate GCD over the integers. This lifting is carried out in two stages:

(i) First the univariate stage where images in $\mathbf{Z}_p[x]$ are lifted to $\mathbf{Z}_q[x]$ where $q = p^k$.

(ii) The multivariate stage, where images in $\mathbf{Z}_q[x]$ are lifted to $\mathbf{Z}_q[x,\mathbf{y}]$. (See Chapter 6 for details.)

Example 7.18. Let

$$A(x,y) = y^2 + 2xy - 3y + x^2 - 3x - 4,$$

$$B(x,y) = y^2 + 2xy + 5y + x^2 - 5x + 4.$$

Note first that both $A(x,y)$ and $B(x,y)$ are primitive. Since 0 is the best evaluation point, we try evaluating at $y = 0$, obtaining $I = \langle y \rangle$ and

$$\phi_I(A) = x^2 - 3x - 4, \;\; \phi_I(B) = x^2 - 5x + 4.$$

Calculating the univariate GCD gives

$$\text{GCD}(\phi_I(A),\phi_I(B)) = x + 1.$$

Notice that evaluation at $y = 0$ does not reduce the degrees of either A or B, so that this is not an invalid evaluation.

When we repeat the above with the evaluation $y = 1$, we again obtain a univariate GCD of degree one, hence we decide that we will try lifting one of the factorizations. In particular, we decide to lift the factorization

$$\phi_I(A) = (x + 1)\cdot[\, \phi_I(A) \,/\, (x + 1) \,] = (x + 1)\cdot(x - 4).$$

that is, lift

$$A(x,y) \equiv (x + 1)\cdot(x - 4) \mod I.$$

Notice that these two factors are indeed relatively prime, hence Hensel's construction is valid. Lifting these factors gives the factorization

$$A(x,y) = (x + 1 + y)\cdot(x - 4 + y).$$

We check that $x + 1 + y$ is also a factor of $B(x,y)$. Since it is, we conclude that

$$\text{GCD}(A,B) = x + y + 1.$$

•

Algorithm 7.3. The Extended Zassenhaus GCD Algorithm.

```
procedure EZ-GCD(A,B)
  # Given two polynomials A, B ∈ Z[x,y_1, . . . , y_k]
  # with deg_x(A) ≥ deg_x(B), we compute the triple
  # <A/C ,B/C ,C> where C = GCD(A,B), using Hensel lifting.

  # Compute the content, primitive part, lcoeff, GCD etc. all viewing A
  # and B as polynomials over the coefficient domain Z[y_1, . . . , y_k].

  a ← cont(A); b ← cont(B); A ← A/a; B ← B/a
  g ← GCD(a,b); a ← a/g; b ← b/g

  # Find a valid evaluation prime
  p ← New(Prime) with lcoeff(A) mod p ≠ 0 and lcoeff(B) mod p ≠ 0

  # Find a valid evaluation point b = (b_1 , . . . , b_k)
  # with 0 ≤ b_i < p and as many b_i's = 0 as possible.

  b ← New(EvaluationPoint) with lcoeff(A)(b) ≠ 0 and lcoeff(B)(b) ≠ 0
  A_I ← A(b) mod p;  B_I ← B(b) mod p
  C_I ← UGCD(A_I,B_I);  d ← deg_x(C_I)
  if d = 0 then return(<Aa,Bb,g>)

  # Double check the answer: Choose a new prime and evaluation point
  p' ← New(Prime) with lcoeff(A) mod p' ≠ 0 and lcoeff(B) mod p' ≠ 0
  c ← New(EvaluationPoint) with lcoeff(A)(c) ≠ 0 and lcoeff(B)(c) ≠ 0
  A_I' ← A(c) mod p'; B_I' ← B(c) mod p'
  C_I' ← UGCD(A_I',B_I');  d_I' ← deg_x(C_I')

  if d_I' < d then {
     # Previous evaluation was bad, try again
     A_I ← A_I'; B_I ← B_I'; C_I ← C_I'; d ← d_I'; b ← c

     goto double check step }
  elseif d_I' > d then {
     # This evaluation was bad; repeat double check step
     goto double check step }
```

Algorithm 7.3 (continued). The EZ-GCD Algorithm.

\# Test for special cases
if $d = 0$ **then return**($< A \cdot a, B \cdot b, g >$)
if $d = \deg_x(B)$ **then** {
  if $B \mid A$ **then return**($< a \cdot A/B, b, B \cdot g >$)
  else {
   \# Bad evaluation, repeat the double check
   $d \leftarrow d - 1$;
   goto double check step } }

\# Check for relatively prime cofactors
if UGCD(B_I,C_I) = 1 **then** {
  $U_I \leftarrow B$; $H_I \leftarrow B_I/C_I$; $c \leftarrow b$ }
elseif UCGD(A_I,C_I) = 1 **then** {
  $U_I \leftarrow A$; $H_I \leftarrow A_I/C_I$; $c \leftarrow a$ }
else return(SGCD($A, B, \mathbf{b}, p$))

\# Lifting step
$U_I \leftarrow c \cdot U_I$; $c_I \leftarrow c(\mathbf{b}) \bmod p$; $C_I \leftarrow c_I \cdot C_I$
$(C, E) \leftarrow$ EZ_LIFT($U_I, C_I, H_I, \mathbf{b}, p, c$)
if $U_I = C \cdot E$ **then goto** double check step

\# Final check
$C \leftarrow \mathrm{pp}(C)$
if $C \mid B$ **and** $C \mid A$ **then return**($< a \cdot A/C, b \cdot B/C, g \cdot C >$)
else goto double check step

end

The Common Divisor Problem

There are a number of problems which must be overcome by the algorithm. The common divisor problem occurs when the image GCD is not relatively prime to its cofactors; i.e.

$$\mathrm{GCD}(A_I/C_I,\ C_I) \neq 1 \text{ and } \mathrm{GCD}(B_I/C_I,\ C_I) \neq 1. \tag{7.37}$$

There are a number of suggestions for resolving the situation:

(a) D. Yun originally suggested performing a square-free decomposition of either A or B (see Chapter 8). This involves computing the GCD of the polynomial and its derivative. Hence it may be invoked recursively and so a special routine SGCD was set up for the process.

(b) Paul Wang [16] suggested dividing out the offending common factor. Let:

$$H_I = A_I / C_I \text{ and } F = \mathrm{GCD}(H_I,\, C_I) \tag{7.38}$$

$$P = H_I / F \;\;,\;\; Q = C_I / F \text{ and so } A_I = P \cdot Q \cdot F^2 .$$

P and Q are mutually relatively prime, and hopefully relatively prime with F_I. Then $A \equiv P \cdot Q \cdot F^2 \bmod (I,p)$ can be lifted using a parallel version of Hensel's lemma. This lifting produces:

$$A = P \cdot Q \cdot F^2 \;\text{ and so } C = \mathrm{GCD}(A,B) = F \cdot Q. \tag{7.39}$$

F can be computed from F^2 using an efficient polynomial n-th root routine. If $\mathrm{GCD}(P,F) \neq 1$ or $\mathrm{GCD}(Q,F) \neq 1$ then the same trick can be applied again until a set of pair-wise relatively prime factors is found.

(c) The simplest strategy is due to David Spear. Notice that if both equations (7.37) are true then there is an infinite set of integers (a,b) such that:

$$P_I = aA_I + bB_I \text{ and } \mathrm{GCD}(P_I/C_I,\, C_I) = 1.$$

If $P = aA + bB$ then the congruence $P = C_I(P_I/C_I) \bmod (p,I)$ can be readily solved by Hensel's lemma. The pair (a,b) can be determined by trial and error.

Unlucky Homomorphisms

As with the modular method, unlucky homomorphisms cause problems for the algorithm. Again we argue probabalistically that if p is very large then finding one is very unlikely. In any event EZ-GCD takes a number of precautions:

(i) A final division check is made in the last step. If this check fails then we know that we have been deceived up to now, and we try again. If the check succeeds then we know that we have found our GCD.

(ii) The algorithm requires that at least two evaluations yield GCD's of the same degree d. We know that a bad homomorphism will only give a GCD of too large a degree and so our sequence of trial degrees must converge to the correct one.

(iii) Two special cases are checked:

- (a) If d=0 then we know immediately that the primitive parts of the inputs A and B are relatively prime, and no further work need be done.
- (b) If d=deg(B) $\leq$ deg(A) then either B is our GCD (since we made both A and B primitive in the initial step and we can test this by division, or we have a bad homomorphism and the GCD must have degree less than deg(B).

Example 7.19. Let

$$A(x,y) = (x+1) \cdot y^2 - 1, \;\; B(x,y) = y + (x+1).$$

Choosing the evaluation point $x = 0$, that is, I = <x> and calculating the GCD of the resulting univariate problem gives

$$\mathrm{GCD}(\phi_I(A),\phi_I(B)) = \mathrm{GCD}(y^2 - 1, y + 1) = y + 1.$$

Choosing a second value, say $x = 1$ so $I = \langle x-1 \rangle$ gives

$$\mathrm{GCD}(\phi_I(A),\phi_I(B)) = \mathrm{GCD}(2y^2 - 1, y + 2) = 1.$$

Thus we can deduce that A and B are relatively prime. Notice that if we applied Hensel lifting after the first factorization, then we know that Hensel lifting will only change the variables in the ideal. In particular it will not change y. Hence we would never obtain the correct answer. We note that the probability of two evaluations giving (incorrect) GCD's of degree 1 is very rare, hence the double check will usually find these bad evaluations.

●

Bounding the Number of Lifting Steps

This can be dealt with as before.

(a) For the integer coefficients a bound on the size of the lifted coefficient is:

$$q \geq 2\|F\|_\infty \cdot \|\mathrm{lcoeff}(F)\|_\infty$$

where the lifting is performed using:

$$F \equiv CH \pmod{I}$$

(b) For the multivariate lifting a bound on the number of lifting steps is:

$$k \geq \deg(F) + \deg(\mathrm{lcoeff}(F)) + 1$$

where $\deg(A)$ is the maximum total degree of any monomial in the polynomial A.

The Leading Coefficient Problem

Another problem which is peculiar to the Hensel construction is that it tends not to update the leading coefficients of the factors correctly. The normalization technique used in the modular algorithm can be used to force the issue, but this can produce very dense coefficients from a sparse polynomial. See Chapter 6 for a discussion of techniques for dealing with this problem in the univariate case.

Yun observed that if the leading coefficient of some multiple of the factor is known, then the leading coefficients can be updated correctly. In particular, for the problem $C = \mathrm{GCD}(A,B)$ we know that $\mathrm{lcoeff}(C)$ must divide $\gamma = \mathrm{GCD}(\mathrm{lcoeff}(A), \mathrm{lcoeff}(B))$. This is the value γ which can be passed into Algorithm 6.1 of the preceding chapter.

The Nonzero b-value Problem

As pointed out in the initial description of the algorithm, an effort is made to choose as many values $b_i = 0$ as possible. This retains the sparsity of a problem and leads to considerable computational efficiency. However, if such b-values lead to a bad homomorphism, then they must be discarded and nonzero b-values used. When this occurs, the EZ-GCD algorithm has the same exponential performance seen in the modular algorithm for sparse situations. The class of problems:

$$\mathrm{GCD}\left((x+1)^2 \cdot \prod_{i=1}^{k} y_i \,,\, (x^2-1) \cdot \prod_{i=1}^{k} y_i \right) = (x+1) \cdot \prod_{i=1}^{k} y_i$$

will cause this sort of problem.

7.7. A HEURISTIC POLYNOMIAL GCD ALGORITHM

The original motivation for modular reduction in GCD problems was to obtain a fast method for deciding if two polynomials are relatively prime. Thus, we reduce the polynomials modulo some prime p, and calculate the GCD of the reduced polynomials in the simpler domain to see if they are relatively prime. If the answer is positive, then we are able to deduce from Lemma 7.3 that the original polynomials were themselves relatively prime. However, before the advent of algorithms such as MGCD or EZ-GCD, nothing could be deduced in the case where the reduction did not end up with relatively prime polynomials. However, because the reduction modulo a prime was so inexpensive when compared to the existing GCD methods of the day, this approach did provide a good heuristic. Thus, there were enough times where the method did (very quickly) deduce that two polynomials were relatively prime, that it compensated for the extra overhead of performing the modular reduction in the case where the two polynomials were not relatively prime. This was true even when a number of primes were used.

In this section, we present a heuristic approach by which nontrivial GCD's can be computed. For example, if $A(x)$ and $B(x)$ are two polynomials in $\mathbf{Z}[x]$, and ξ is any integer, then the mapping

$$\phi : \mathbf{Z}[x] \to \mathbf{Z}, \quad \phi(A(x)) = A(\xi)$$

is a ring homomorphism. We would like to investigate the possibility that there is some sort of equivalent version of Lemma 7.3 for such a homomorphism even though $\mathbf{Z}$ is not a polynomial domain. In particular, if the images of two polynomials are relatively prime, then we would like to claim (with a large degree of certainty) that the original polynomials are relatively prime. Thus, for example, let $A(x)$ and $B(x)$ be the two polynomials from Example 7.2 and ϕ the homomorphism that evaluates at $\xi = 3$, that is, $\phi(P(x)) = P(3)$ for any polynomial $P(x) \in \mathbf{Z}[x]$. Then the images of $A(x)$ and $B(x)$ under this homomorphism become

$$\alpha = \phi(A(x)) = A(3) = 7039, \qquad \beta = \phi(B(x)) = B(3) = 2550.$$

Calculating the integer GCD gives $\mathrm{igcd}(\alpha,\beta) = 1$. Thus we would like to deduce that $A(x)$ and $B(x)$ are themselves relatively prime.

Of course, as it is given at present, the above description does not work. For example, if we let

$$\hat{A}(x) = (x-2)\cdot A(x), \quad \hat{B}(x) = (x-2)\cdot B(x)$$

where $A(x)$ and $B(x)$ are from from Example 7.2, then again the images are relatively prime. In this case the common factor has been lost under this homomorphism. However, it turns out that it is not the method that is at fault here, but rather the choice of the evaluation point.

Putting aside the technical difficulties for now, we note that the method has many advantages. The main advantage of course is that rather than doing any GCD calculations in $\mathbf{Z}[x]$ (expensive), we calculate our GCD over the integers (inexpensive). The overhead of this "heuristic" is the cost of two polynomial evaluations, along with an integer GCD. It is inexpensive enough that one can try this method four or five times in the hopes that a suitable point can determine the relative primeness (if the two are indeed relatively prime). When it fails to deduce a firm answer, we could continue with more standard methods.

As was the case with modular reduction, there are a number of problems with this approach, not the least of which is determining a priori, a set of good evaluation points. Furthermore, as was the case with modular reduction or Hensel lifting, when the images are not relatively prime we would like to avoid totally wasting the results. In particular we would like to construct, if possible, the desired GCD from the image GCD.

For the remainder of this section we describe a new heuristic algorithm, GCDHEU, for the computation of polynomial GCD's based on the above ideas. This approach is found to be very efficient for problems in a small number of variables. As demonstrated above, the algorithm can be viewed as a modular-type algorithm in that it uses evaluation and interpolation, but only a single evaluation per variable is used. The heuristic algorithm can be incorporated into a reorganized form of the EZ-GCD algorithm such that the base of the EZ-GCD algorithm, rather than a univariate GCD algorithm, is GCDHEU which is often successful for problems in up to four variables. The heuristic approach originally appeared in a paper by Char, Geddes and Gonnet [3]. Based on an earlier announcement of the results, improvements to this approach were given by Davenport and Padget [5]. Using a variation of the original heuristic approach, Schonhage [14] gave a probabilistic algorithm for GCD's of polynomials having integer coefficients. This section is primarily based on the original paper of Char, Geddes, and Gonnet.

Single-Point Evaluation and Interpolation

Consider the problem of computing $C(x) = \mathrm{GCD}(A(x),B(x))$ where $A(x), B(x) \in \mathbf{Z}[x]$ are univariate polynomials. Let $\xi \in \mathbf{Z}$ be a positive integer which bounds twice the magnitudes of all coefficients appearing in $A(x)$ and $B(x)$ and in any of their factors. Let $\phi_{x-\xi} : \mathbf{Z}[x] \to \mathbf{Z}$ denote the substitution $x = \xi$ (i.e. the evaluation homomorphism whose kernel is the ideal $I = <x-\xi>$) and let

$$\alpha = \phi_{x-\xi}(A(x)), \quad \beta = \phi_{x-\xi}(B(x)).$$

Define $\gamma = \mathrm{igcd}(\alpha,\beta)$ and suppose for the moment that the following relationship holds (this development will be made mathematically rigorous below):

$$\gamma = \phi_{x-\xi}(C(x)).$$

Our problem now is to reconstruct the polynomial $C(x)$ from its image γ under the evaluation $x = \xi$.

The reconstruction of $C(x)$ from γ will be accomplished by a special kind of interpolation which exploits the fact that ξ is assumed to be larger than twice the magnitudes of the coefficients appearing in $C(x)$. The required interpolation scheme is equivalent to the process of converting the integer γ into its ξ-adic representation:

$$\gamma = c_0 + c_1\xi + c_2\xi^2 + \cdots + c_d\xi^d, \tag{7.40}$$

where d is the smallest integer such that $\xi^{d+1} > 2|\gamma|$, and $-\xi/2 < c_i \le \xi/2$ for $0 \le i \le d$. This can be accomplished by the simple loop:

$e \leftarrow \gamma$
for i **from** 0 **while** $e \ne 0$ **do** {
  $c_i \leftarrow \phi_\xi(e)$
  $e \leftarrow (e - c_i)/\xi$ }

where $\phi_\xi : \mathbf{Z} \to \mathbf{Z}_\xi$ is the standard "mod ξ" function using the "symmetric representation" for the elements of $\mathbf{Z}_\xi$. Our claim is that, under appropriate conditions yet to be specified, the coefficients c_i are precisely the coefficients of the desired GCD

$$C(x) = c_0 + c_1x + c_2x^2 + \cdots + c_dx^d.$$

Example 7.20. Let

$$A(x) = 6x^4 + 21x^3 + 35x^2 + 27x + 7,$$
$$B(x) = 12x^4 - 3x^3 - 17x^2 - 45x + 21$$

and $\xi = 100$ (why this is a good choice will be explained later). Then

$$A(100) = 621352707 \quad \text{and} \quad B(100) = 1196825521$$

with

$$\text{igcd}(621352707, 1196825521) = 30607.$$

Using the above loop, we get $c_0 = 7$, $c_1 = 6$ and $c_2 = 3$, hence the candidate for the GCD is

$$C(x) = 3x^2 + 6x + 7.$$

One may verify that this is indeed the correct GCD of $A(x)$ and $B(x)$.

●

The method outlined above generalizes immediately to multivariate GCD's through recursive application of evaluation/interpolation.

This construction will be made precise by Theorem 7.7. First we need the following lemma. In the univariate case, this lemma follows immediately from

Cauchy's Inequality [11].

Let $P = a_0 + a_1 x + \cdots + a_d x^d$, $a_d \neq 0$, $d \geq 1$ be a univariate polynomial over the complex field **C**. Then any root α of P satisfies

$$|\alpha| < 1 + \frac{\max(|a_0|, |a_1|, \ldots, |a_{d-1}|)}{|a_d|}.$$

●

We also use the following

Lemma 7.4. Let $P \in \mathbf{Z}[x_1, x_2, \ldots, x_k]$ be a nonzero polynomial in one or more variables. Let x denote one of the variables x_k and let $\alpha \in \mathbf{Z}$ be any integer. If $(x - \alpha) \mid P$ then $|\alpha| \leq \|P\|_\infty$.

Proof: Let $P = (x - \alpha)\, Q$ for some $Q \in \mathbf{Z}[x_1, x_2, \ldots, x_k]$. Write Q in the form

$$Q = q_0 + q_1 x + \cdots + q_d x^d$$

where $x = x_j$ is the particular variable appearing in the divisor $x - \alpha$ and $q_i \in \mathbf{Z}[x_1, \ldots, x_{j-1}, x_{j+1}, \ldots, x_k]$ for $0 \leq i \leq d$. Then

$$P = -\alpha\, q_0 + (q_0 - \alpha\, q_1)\, x + \cdots + (q_{d-1} - \alpha\, q_d)\, x^d + q_d\, x^{d+1}.$$

By definition of the norm function $\|\ \|_\infty$, we have the following inequalities:

$$|-\alpha\, q_0| \leq \|P\|_\infty;$$
$$|q_{i-1} - \alpha\, q_i| \leq \|P\|_\infty, \text{ for } 1 \leq i \leq d;$$
$$|q_d| \leq \|P\|_\infty.$$

Now if $|\alpha| > \|P\|_\infty$, the first inequality above implies $q_0 = 0$, and then the second set of inequalities imply $q_i = 0$, for $1 \leq i \leq d$. But this implies that $Q = 0$ which is impossible since P is nonzero, yielding the desired contradiction.

●

In the following theorem, we note that it is possible for γ computed by the method described earlier in this section to be larger than $\phi_{x_k - \xi}(C)$, so we denote the polynomial reconstructed from γ by G which may differ from C. The theorem proves that a simple division check will determine whether or not G is a greatest common divisor of A and B.

Theorem 7.7. Let $A, B \in \mathbf{Z}[x_1, x_2, \ldots, x_k]$ be nonzero polynomials. Let $\xi > 2$ be a positive integer which bounds twice the norm of A, B, and any of their factors in the domain $\mathbf{Z}[x_1, x_2, \ldots, x_k]$. Let $\gamma = \mathrm{GCD}(\alpha, \beta)$ where $\alpha = \phi_{x_k - \xi}(A)$ and $\beta = \phi_{x_k - \xi}(B)$, and let G denote the polynomial such that

$$\phi_{x_k-\xi}(G) = \gamma \tag{7.41}$$

and whose coefficients $g_i \in \mathbf{Z}[x_1,x_2, \ldots, x_{k-1}]$ are the ξ-adic coefficients of γ defined by (7.40). Then G is a greatest common divisor of A and B if and only if

$$G \mid A \text{ and } G \mid B. \tag{7.42}$$

Proof: The "only if" proposition is immediate. To prove the "if" proposition, suppose that (7.42) holds. Let $C = \mathrm{GCD}(A,B) \in \mathbf{Z}[x_1,x_2, \ldots, x_k]$. We have $G \mid C$ by (7.42). Let $C = G\,H$ for some $H \in \mathbf{Z}[x_1,x_2, \ldots, x_k]$. Then $\phi_{x_k-\xi}(C) = \phi_{x_k-\xi}(G)\,\phi_{x_k-\xi}(H)$ $= \gamma\,\phi_{x_k-\xi}(H)$, by (7.41) and the fact that $\phi_{x_k-\xi}$ is a homomorphism. Since $C \mid A$ and $C \mid B$ it follows that $\phi_{x_k-\xi}(C) \mid \alpha$ and $\phi_{x_k-\xi}(C) \mid \beta$ whence $\phi_{x_k-\xi}(C) \mid \gamma$. We therefore conclude that

$$\gamma\,\phi_{x_k-\xi}(H) \mid \gamma,$$

implying that $\phi_{x_k-\xi}(H) = \pm 1$. It follows that the polynomial $H - 1$ (or $H + 1$) either is zero or else has a linear factor $x_k - \xi$. In the latter case, by Lemma 7.4, we must have

$$\|H - 1\|_\infty \ge \xi \text{ or } \|H + 1\|_\infty \ge \xi.$$

This is impossible since, H being a factor of A and of B, by the definition of ξ we have

$$\|H\|_\infty \le \xi/2$$

from which it follows that

$$\|H \pm 1\|_\infty \le \xi/2 + 1 < \xi$$

(since $\xi > 2$). Therefore $H = \pm 1$ and so G is an associate of C. ●

Example 7.20 gives two polynomials A and B along with an evaluation point $\xi = 100$ that satisfies the conditions of Theorem 7.7. Thus to verify that the C generated in the example is indeed the correct GCD one simply divides C into both A and B. Since C does indeed divide into both it is the GCD.

Choosing a Small Evaluation Point

In order to develop an efficient heuristic algorithm for GCD computation, let us note some properties of the algorithm implied by Theorem 7.7. For multivariate polynomials A and B, suppose that the first substitution is $x_j = \eta$ for $\eta \in \mathbf{Z}$. After the evaluation of A and B at $x_j = \eta$, the algorithm will be applied recursively until an integer GCD computation can be performed. The size of the evaluation point used at each level of recursion depends on the size of the polynomial coefficients at that level. This can grow rapidly with each recursive step. Specifically, if the original polynomials contain k variables and if the degree in each variable is d, then it is easy to verify that the size of the integers in the integer GCD computation at the base of the recursion is $O(\eta^{d^k})$. Thus, it is clear that this algorithm will be unacceptable for problems in many variables with nontrivial degrees. An important aspect of the heuristic is that it must check and give up quickly if proceeding would generate unacceptably large integers.

It must be noted, however, that integers which contain hundreds of digits can be manipulated relatively efficiently in most symbolic computation systems. Thus, we find in practice that the algorithm described here becomes non-competitive only when the size of the integers grows to a few thousand digits. For example, if $k = 2$, $d = 5$, and $\eta = 100$ in the above notation then the integers will grow to approximately 50 digits in length. Solving such a problem by the method of this section would be relatively trivial on conventional systems.

In view of the above remarks on the growth in the size of the integers, it is important to choose the evaluation point reasonably small at each stage of the recursion. In particular, we will not choose a standard upper bound [11] on the size of the coefficients that can appear in the factors as might be implied by Theorem 7.7. Our heuristic algorithm is allowed to have some probability of failure, and the only essential condition is that the division checks of Theorem 7.7 must be guaranteed to detect incorrect results. So we pose the question: "How small can we choose the evaluation point ξ and yet be guaranteed that division checks are sufficient to detect incorrect results?"

Example 7.21. Suppose we have

$$A(x) = x^3 - 9x^2 - x + 9 = (x+1)(x-9)(x-1),$$

$$B(x) = \quad x^2 - 8x - 9 = (x+1)(x-9).$$

If we choose the evaluation point $\xi = 10$, then $\phi_{x-10}(a) = 99$, $\phi_{x-10}(b) = 11$, and $\gamma = \text{igcd}(99,11) = 11$. The 10-adic representation of 11 is simply $1 \times 10 + 1$ and therefore $G = x + 1$ is computed as the proposed GCD(A,B). Doing division checks, we find that $G \mid A$ and $G \mid B$. However, G is not the correct answer but is only a factor of the true GCD. What has happened is that the factor $H = x - 9$ has disappeared under the mapping since $\phi_{x-10}(H) = 1$.

●

Theorem 7.8 proves that division checks are guaranteed to detect incorrect results as long as the evaluation point is chosen to be strictly greater than $1 + \min(\|A\|_\infty, \|B\|_\infty)$. (Note that in Example 7.21, $\xi = 1 + \min(\|A\|_\infty, \|B\|_\infty)$, which is not a *strict* bound.) This is a little weaker requirement than Theorem 7.7 since we no longer need to deal with the norm of each possible factor of A and B. First, we need the following lemma dealing with the size of the factors of a polynomial when evaluated at an integer value.

Lemma 7.5. Let $P \in \mathbf{Z}[x_1, x_2, \ldots, x_k]$ be a non-constant polynomial in one or more variables and let $\delta \in \mathbf{Z}$ be a given positive integer. Let x denote one of the variables x_j and let $\alpha \in \mathbf{Z}$ be an integer satisfying

$$|\alpha| \geq \|P\|_\infty + \delta + 1. \qquad (7.43)$$

If $Q \in \mathbf{Z}[x_1, x_2, \ldots, x_k]$ is a non-constant polynomial such that

$$Q \mid P \quad \text{and} \quad \phi_{x-\alpha}(Q) \in \mathbf{Z} \tag{7.44}$$

then

$$\| \phi_{x-\alpha}(Q) \|_\infty > \delta. \tag{7.45}$$

Proof: First consider the univariate case $k = 1$. Let the complete factorization of P over the complex field **C** be

$$P = c \prod_{i=1}^{d} (x - r_j) \tag{7.46}$$

for $r_i \in \mathbf{C}$, where $d \geq 1$ is the degree of P and $c \neq 0$ is the leading coefficient of P. Then by Cauchy's inequality,

$$|r_i| < 1 + \frac{\|P\|_\infty}{|c|}.$$

For any $\alpha \in \mathbf{Z}$ we have

$$\| \phi_{x-\alpha}(P) \|_\infty \geq |c| \prod_{i=1}^{d} \big| |\alpha| - |r_i| \big| .$$

Now if α satisfies (7.43) then for each i,

$$\big| |\alpha| - |r_i| \big| > (1 + \|P\|_\infty + \delta) - (1 + \frac{|P|}{|c|}) \geq \delta \tag{7.47}$$

(where we have used the fact that $|c| \geq 1$) yielding

$$\| \phi_{x-\alpha}(P) \|_\infty > |c| \, \delta^d \geq \delta.$$

Similarly, any non-constant polynomial $Q \in \mathbf{Z}[x]$ satisfying (7.44) has a factorization over **C** consisting of one or more of the linear factors in (7.46) and therefore (7.47) implies

$$\| \phi_{x-\alpha}(Q) \|_\infty > \delta$$

as claimed.

Turning to the multivariate case $k > 1$, we can choose values $\alpha_i \in \mathbf{Z}$ $(i \neq j)$ for all variables x_i except $x = x_j$, such that if

$$\mathrm{I} = \langle x_1 - \alpha_1, \ldots, x_{j-1} - \alpha_{j-1}, x_{j+1} - \alpha_{j+1}, \ldots, x_k - \alpha_k \rangle$$

denotes the kernel of the corresponding evaluation homomorphism then $P^{(1)} = \phi_{\mathrm{I}}(P)$ is a univariate polynomial in x satisfying

$$\| P^{(1)} \|_\infty \geq \| P \|_\infty .$$

Note that this can always be achieved by choosing the values $\alpha_i \in \mathbf{Z}$ arbitrarily large and arbitrarily distant from each other. Now if $Q \in \mathbf{Z}[x_1, x_2, \ldots, x_k]$ is a non-constant polynomial satisfying (7.44) then $\phi_{\mathrm{I}}(Q) \mid P^{(1)}$ and therefore, by the univariate case already proved,

$$\| \phi_{x-\alpha} (\phi_{\mathrm{I}} (Q)) \|_\infty > \| P^{(1)} \|_\infty \geq \| P \|_\infty .$$

But

$$\phi_{x-\alpha} (\phi_{\mathrm{I}} (Q)) = \phi_{\mathrm{I}} (\phi_{x-\alpha} (Q)) = \phi_{x-\alpha} (Q)$$

since the order is irrelevant in the application of evaluation homomorphisms, and by (7.45), so the proof is complete.

●

Theorem 7.8. Let $A, B \in \mathbf{Z}[x_1, x_2, \ldots, x_k]$ be nonzero polynomials and let $\xi \in \mathbf{Z}$ be a positive integer satisfying

$$\xi > 1 + \min(\| A \|_\infty , \| B \|_\infty).$$

Let $\gamma = \mathrm{GCD}(\alpha, \beta)$ where $\alpha = \phi_{x_v - \xi} (a)$ and $\beta = \phi_{x_v - \xi} (b)$, and let G denote the polynomial formed from the ξ-adic expansion of γ such that

$$\phi_{x_k - \xi} (G) = \gamma.$$

G is a greatest common divisor of A and B if and only if

$$G \mid A \text{ and } G \mid B.$$

Proof: The first half of the proof of Theorem 7.8 remains valid for Theorem 7.7, yielding the conclusion that $C = G\, H$ where $\phi_{x_v - \xi} (H) = \pm 1$. Noting that H is a factor of A and of B, let P denote the polynomial A or B with minimum norm. The conditions of Lemma 7.5 are satisfied for P with $x = x_v$, $\alpha = \xi$, and $\delta = 1$, and therefore if H is a non-constant polynomial we have

$$| \phi_{x_v - \xi} (H) | > 1$$

contradicting the above conclusion. Thus H must be a constant, whence $H = \pm 1$, proving that G is an associate of g.

●

The Heuristic GCD Algorithm

Theorem 7.8 places a lower bound on the size of the evaluation point which guarantees recognition of incorrect results by division checks. However, there exist problems where this approach fails to find the true GCD via the ξ-adic expansion method, no matter what evaluation point is used. We now wish to correct the algorithm so that there will always be a reasonable probability of success for any problem.

Example 7.22. Suppose that

$$A(x) = x^3 - 3x^2 + 2x \; = (x - 2)\,(x - 1)\,x$$

$$B(x) = x^3 + 6x^2 + 11x + 6 = (x + 1)\,(x + 2)\,(x + 3)$$

Note that the norms are given by $\| A \|_\infty = 3$ and $\| B \|_\infty = 11$. By Theorem 7.8 we can choose

$\xi = 5$. We get $\phi_{x-5}(A) = 60$, $\phi_{x-5}(B) = 336$, and $\gamma = \text{igcd}(60,336) = 12$. The ξ-adic representation of 12 is $2\times5 + 2$ and therefore $G(x) = 2x+2$ is computed as the proposed GCD(A,B). Of course the true GCD is 1 and division checks will detect that this result is incorrect.

Trying a second time, with a larger evaluation point, say $\xi = 20$, we get $\phi_{x-20}(A) = 6840$, $\phi_{x-20}(B) = 10626$, and $\gamma = \text{igcd}(6840,10626) = 6$. The ξ-adic representation of 6 is 6 and therefore $G = 6$ is computed as the proposed GCD(A,B). Again the division checks will detect that 6 is not a factor of the original polynomials. Thus if we stop after two evaluations we will return the answer fail for this example.

●

In the case of Example 7.22, the polynomials A and B will always have a common factor, when evaluated, of at least 6. When a small evaluation point is chosen, as above, then this extraneous integer factor will be interpolated to an extraneous polynomial factor, yielding an incorrect result. Even if the evaluation point is sufficiently large then any extraneous integer factor will remain as an integer content in the interpolated polynomial.

We are therefore led to the concept of removing the integer content from the polynomials. We will therefore impose the condition that the input polynomials A and B are *primitive* with respect to **Z** that is, the integer content has been removed from A and from B. Correspondingly, we will remove the integer content from the computed polynomial G before test dividing because the divisors of a primitive polynomial must be primitive (cf. Exercise 7.4). Now it becomes crucial for us to ensure that when we remove the integer content from G we are not removing any factors that correspond to factors of the true GCD C. For if a factor of C evaluates to an integer that is small relative to ξ (specifically, less than $\xi/2$) then such an integer may remain as part of the integer content in the interpolated polynomial and will be discarded. We are then back to the situation where the division checks may succeed even though the computed G is not a greatest common divisor. Theorem 7.9 shows how large we must now choose ξ so that, even when the integer content is removed, the division check will give a true answer.

Theorem 7.9. Let $A, B \in \mathbf{Z}[x_1,x_2, \ldots ,x_v]$ be nonzero polynomials which are primitive with respect to **Z**. Let $\xi \in \mathbf{Z}$ be a positive integer satisfying

$$\xi > 1 + 2\min(\|A\|_\infty, \|B\|_\infty). \tag{7.48}$$

Let $\gamma = \text{GCD}(\alpha,\beta)$ where $\alpha = \phi_{x_v-\xi}(A)$ and $\beta = \phi_{x_v-\xi}(B)$, and let G denote the polynomial formed from the ξ-adic expansion of γ and satisfying

$$\phi_{x_k-\xi}(G) = \gamma.$$

With pp(G) denoting the result of dividing G by its integer content, pp(G) = GCD(A,B) if and only if

$$\text{pp}(G) \mid A \text{ and } \text{pp}(G) \mid B.$$

Proof: Proceeding as in the first half of the proof of Theorem 7.7, we let $C = \text{GCD}(A,B)$ and we can conclude that $C = \text{pp}(G)\,H$ where

$$\phi_{x_k-\xi}\,(\text{pp}(G))\cdot\phi_{x_k-\xi}\,(H) \mid \gamma.$$

Now if we denote the integer content of G by κ then $\text{pp}(G) = G/\kappa$ so we have

$$\frac{\gamma}{\kappa}\,\phi_{x_k-\xi}\,(H) \mid \gamma.$$

Multiplying through by κ yields $\phi_{x_k-\xi}\,(H) \mid \kappa$ from which we conclude that $\phi_{x_k-\xi}\,(H) \in \mathbf{Z}$ and furthermore

$$|\,\phi_{x_k-\xi}\,(H)\,| \le \kappa \le \xi/2\,, \tag{7.49}$$

the latter inequality coming from the fact that, by construction, the coefficients of G are bounded in magnitude by $\xi/2$. Noting that H is a factor of A and of B, let P denote the polynomial A or B with minimum norm. The conditions of Lemma 7.5 are satisfied for P with $x = x_k$, $\alpha = \xi$, and $\delta = \xi/2$, as can be seen by writing $\xi = \xi/2 + \xi/2$ and noting that (7.48) implies

$$\xi\,/2 \ge 1 + \|P\|_\infty\,.$$

Therefore, if H is a non-constant polynomial we have

$$|\,\phi_{x_k-\xi}\,(H)\,| > \xi/2$$

contradicting (7.49). Thus H must be a constant, whence $H = \pm 1$ because C is primitive, proving that $\text{pp}(G)$ is an associate of C.

●

We are now ready to present algorithm GCDHEU. The algorithm assumes that the input polynomials are primitive and that the integer content will be removed from the output returned. It uses the result of Theorem 7.9 in choosing the evaluation points so that the division checks constitute a valid checking mechanism. The division checks being used in the algorithm are based on division of polynomials over the field of *rational numbers*, which is equivalent to removing the integer content from the divisor and then doing test division over the integers (noting that the dividend is already primitive).

To ensure that the calculation does not become too expensive (it is only a heuristic after all) we check on the size of the integers that would be generated if the computation were allowed to proceed, and the return mechanism in the algorithm is, in this case, indicated by Return_To_Top_Level. This needs to be a more ''drastic'' return mechanism than the ordinary returns appearing otherwise in the algorithm because of the recursive nature of the algorithm and the fact that there is no point in continuing computation on a problem that has lead to such large integers. Upon a return of failure, we increase the size of the evaluation point.

An exact doubling algorithm to compute the next evaluation point will produce a sequence of values whose suitability would be highly correlated. This would not be good, as failure of the first point would tend to imply that later choices would also fail. We thus wish the prime decomposition of successive evaluation points to have no obvious pattern (i.e.

Algorithm 7.4. GCD Heuristic Algorithm.

```
procedure GCDHEU(A,B)
  # Given polynomials A, B ∈ Z[x_1, . . . , x_k] we use
  # a heuristic method for trying to determine G = GCD(A,B)

  vars ← Indeterminates(A) ∪ Indeterminates(B)
  if SizeOf( vars) = 0 then return(igcd(A, B))
  else x ← vars[1]

  ξ ← 2·min(|A|_∞ ,|B|_∞ ) + 2

  to 6 do {
    if length(ξ) · max(deg_x(A), deg_x(B)) > 5000 then
      Return_To_Top_Level(fail_flag)

    γ ← GCDHEU(φ_{x−ξ} (A),φ_{x−ξ} (B))
    if γ ≠ fail_flag then
      # Generate polynomial G from ξ-adic expansion of γ
      G ← 0
      for i from 0 while γ ≠ 0 do {
        g_i ← φ_ξ (γ)
        G ← G + g_i·x^i
        γ ← (γ − g_i) / ξ }
      if G | A and G | B then return(G)

    # Create a new evaluation point using square of golden ratio
    ξ ← iquo ( ξ × 73794, 27011 ) }

  return(fail_flag)

end
```

some "randomness"). To achieve this, we would like to avoid having the result of the product be an integer, so that truncation will happen. To ensure that truncation will happen most of the time, we would like to pick a multiplier α such that α, α^2, α^3, . . . , are never "close" to a "small" rational. A good "small rational" approximation means that one of the first convergents in the continued fraction decomposition for α^i is large. By these criteria "poor" candidates for α would be $\frac{2000001}{1000000}$ or $\frac{1414213}{1000000}$. The value we select for α is one such that the first convergents for α, α^2, . . . , α^6 are very small. This selection was done from a random set of candidates.

We remark that the heuristic technique presented in this last section generalizes to many other types of calculations, including such calculations as determinants of polynomial matrices and fast methods for algebraic number computations (cf. Geddes, Gonnet, and Smedley [6]). Indeed, there is now an entire theory of the use of these heuristics to reduce the problem of intermediate expression swell in computer algebra. We refer the reader to Monagan [12].

Exercises

1. (a) Use the Euclidean PRS to determine the GCD of

$$A(x) = 2x^6 + 5x^5 + 7x^4 + 3x^3 + 6x^2 - 2x + 1,$$

$$B(x) = 3x^5 + 3x^4 + 6x^3 - x^2 + 3x - 4.$$

   (b) Repeat the calculation using the primitive PRS.

2. Repeat the calculation in Exercise 1 using

$$A(x) = x^4 + x^3 - w,$$

$$B(x) = x^3 + 2x^2 + 3wx - w + 1,$$

   with w an unknown.

3. Show that the coefficients of a Euclidean PRS can grow exponentially in n, the degree of the input polynomials.

4. Prove Gauss' lemma : The product of two primitive polynomials is itself primitive.

5. Show that a PRS does indeed calculate the GCD of two polynomials in R[x], R a UFD.

6. Prove Theorem 7.3 by:
   (a) Using the approach in Theorem 7.1.
   (b) Using subresultants.

7. Calculate, using determinants, the subresultants for the polynomials from Exercise 1.

8. Calculate the reduced PRS for the polynomials from Exercise 1.

9. Calculate the subresultant PRS for the polynomials from Exercise 1. Compare this sequence with the quantities calculated in Exercise 8.

10. Let $\{R_i(x)\}$ be the reduced PRS from Example 7.5. For each $i \geq j$ determine d_{ij} where

$$R_i(x) = d_{ij} \cdot S(n_{i-1}, R_{i-j}, R_{i-j+1}).$$

Set up a table of such values.

11. Repeat Exercise 10, using $\{R_i(x)\}$ from Example 7.6 (the subresultant PRS).

12. Verify that the subresultant PRS algorithm is valid, that is, produces a PRS with coefficients in R for two polynomials from R[x].

13. What is the rate of growth of the coefficients of a *normal* remainder sequence if we are using the reduced PRS algorithm? Notice in this case we have $\beta_2 = 1$, $\beta_i = r_{i-2}^2$ for $i \geq 3$.

14. Program both the reduced PRS and subresultant PRS in your favorite computer algebra system. Use these algorithms to calculate the entire PRS for each method, when applied to the following pairs of polynomials:

(a)
$$A(x) = 1206x^{10} + 1413x^9 - 1201x^8 + 2506x^7 + 4339x^6$$
$$+ 12x^5 + 1405x^4 + 415x^3 + 1907x^2 - 588x - 1818,$$
$$B(x) = 402x^9 + 203x^8 + 402x^7 + 103x^6 - 704x^5$$
$$+ 1706x^4 - 1196x^3 - 313x^2 + 710x - 383.$$

(b)
$$A(x) = 3x^9 + 5x^8 + 7x^7 - 3x^6 - 5x^5 - 7x^4 + 3x^3 + 5x^2 + 7x - 2,$$
$$B(x) = -x^8 + x^5 + x^2 + x + 1.$$

15. Program the MGCD algorithm in your favorite computer algebra system. Test this on the polynomials in Exercise 14. *univariate in //[x]*

16. Program the EZ-GCD algorithm in your favorite computer algebra system. Test this on the polynomials

$$A(x,y) = x^3y + 3yz - 4,$$
$$B(x,y) = -3x^2z + 2y^2 + 1.$$

Compare this with the MGCD algorithm of Exercise 15 along with the PRS algorithms determined previously.

17. Using your EZ-GCD algorithm from the previous exercise, test the results using the polynomials from Exercises 2.18 and 2.19 from Chapter 2. Also test the algorithm using the polynomials

$$A(x,y,z) = (z^4 + (y^4 + x^4)\cdot z^3 + a^4)\cdot(z^4 - 2z + y^4 + x^4 + a^4),$$
$$B(x,y,z) = (z^4 + (y^4 + x^4)\cdot z^3 + a^4)\cdot(z^4 - y^4 + 2y + x^4 + a^4),$$

and the polynomials

$$A(x,y,z) = (z^4 + (y^4 + x^4)\cdot z^3 + a^4)\cdot(z^4 - 2z + y^4 + x^4 + a^4),$$
$$B(x,y,z) = (z^4 + (y^4 + x^4)\cdot z^3 + a^4 + 1)\cdot(z^4 - y^4 + 2y + x^4 + a^4).$$

18. Program the GCDHEU heuristic in your favorite computer algebra system. Test this on the polynomials

$$A(x) = P(x)^3\cdot Q(x)^2\cdot R(x)^4, \text{ and } B(x) = P(x)^6\cdot Q(x)^3\cdot R(x)$$

where $P(x)$, $Q(x)$ and $R(x)$ are given by

$$P(x) = 704984x^4 - 995521x^3 - 918115x^2 + 903293x + 342709,$$
$$Q(x) = 8685x^5 + 7604x^4 - 2020x^3 - 5255x^2 + 2517x + 3120,$$
$$R(x) = 544x^6 - 566x^5 + 892x^4 - 58x^3 - 335x^2 - 175x + 443.$$

Compare this with your implementations of your PRS methods and your EZ-GCD algorithm.

19. Prove the following: Let b be a positive integer, and let S be the set of polynomials $a(x) = \sum_{i=0}^{k} a_i x^i \in \mathbf{Z}[x]$ such that $b \geq \| a \|_\infty$. Show that the evaluation homomorphism ϕ_b is one to one from S to its image.

20. Suppose that $A(x)$ and $B(x)$ are from the ring of algebraic numbers $\mathbf{Z}[x]/\langle P(x)\rangle$, $P(x)$ an irreducible polynomial having only integer coefficients. Modify the GCDHEU algorithm to create a heuristic algorithm for the GCD($A(x)$,$B(x)$). *Hint:* Choose an integer m and let $n = P(m)$. Consider the homomorphism

$$\phi : \mathbf{Z}[x]/\langle P(x)\rangle \to \mathbf{Z}_n, \quad \phi(C(x)) = C(m) \bmod n.$$

21. How can the GCD algorithms be modified to find the cofactors D_1, D_2 of the inputs A, B where $A = C\cdot D_1$, $B = C\cdot D_2$, GCD(A,B) = C?

22. The *Sturm* sequence of a polynomial $A(x) \in \mathbf{Q}[x]$ is a sequence of polynomials $\{A_0(x), \ldots, A_k(x)\}$ defined by

$$A_0(x) = A(x), \quad A_1(x) = A'(x)$$
$$A_i(x) = -\operatorname{rem}(A_{i-2}(x), A_{i-1}(x)).$$

A Sturm sequence is usually of interest in the special case when GCD($A(x)$,$A'(x)$) = 1 in which case A(x) is *square-free* (cf. Chapter 8). One example of the importance of a

Sturm sequence is that the number of sign variations of such a sequence at the end-points of an interval tells how many real roots of A(x) lie inside that interval. Describe algorithms to calculate the Sturm sequence of a polynomial.

References

1. W.S. Brown, ''On Euclid's Algorithm and the Computation of Polynomial Greatest Divisors,'' *J. ACM*, **18** pp. 476-504 (1971).

2. W.S. Brown and J.F. Traub, ''On Euclid's Algorithm and the Theory of Subresultants,'' *J. ACM*, **18** pp. 505-514 (1971).

3. B.W. Char, K.O. Geddes, and G.H. Gonnet, ''GCDHEU: Heuristic Polynomial GCD Algorithm Based on Integer GCD Computation,'' *J. Symbolic Comp.*, **9** pp. 31-48 (1989).

4. G.E. Collins, ''Subresultants and Reduced Polynomial Remainder Sequences,'' *J. ACM*, **14**(1) pp. 128-142 (1967).

5. J.H. Davenport and J.A. Padget, ''HEUGCD: How Elementary Upperbounds Generate Cheaper Data,'' pp. 11-28 in *Proc. EUROCAL '85, Vol. 2, Lecture Notes in Computer Science* **204**, ed. B.F. Caviness, Springer-Verlag (1985).

6. K.O. Geddes, G.H. Gonnet, and T.J. Smedley, ''Heuristic Methods for Operations with Algebraic Numbers,'' pp. 475-480 in *Proc. ISSAC '88, Lecture Notes in Computer Science* **358**, ed. P. Gianni, Springer-Verlag (1988).

7. W. Habicht, ''Eine Verallgemeinerung des Sturmschen Wurzelzaehlverfahlens,'' *Commentarii Mathematici Helvetici*, **21** pp. 99-116 (1948).

8. D.E. Knuth, *The Art of Computer Programming, Volume 2: Seminumerical Algorithms (second edition)*, Addison-Wesley (1981).

9. M.A. Laidacker, ''Another Theorem Relating Sylvester's Matrix and the Greatest Common Divisor,'' *Mathematics Magazine*, **42** pp. 126-128 (1969).

10. R. Loos, ''Generalized Polynomial Remainder Sequences,'' pp. 115-137 in *Computer Algebra - Symbolic and Algebraic Computation*, ed. B. Buchberger, G.E. Collins and R. Loos, Springer-Verlag (1982).

11. M. Mignotte, ''Some Useful Bounds.,'' pp. 259-263 in *Computer Algebra - Symbolic and Algebraic Computation*, ed. B. Buchberger, G.E. Collins and R. Loos, Springer-Verlag (1982).

12. M. Monagan, ''Signatures + Abstract Data Types = Computer Algebra - Intermediate Expression Swell,'' Ph.D. Thesis, Dept. of CS, Univ. of Waterloo (1989).

13. J. Moses and D.Y.Y.Yun, ''The EZGCD Algorithm,'' pp. 159-166 in *Proc. ACM Annual Conference*, (1973).

14. A. Schonhage, ''Probabilistic Computation of Integer Polynomial GCDs,'' *J. of Algorithms*, **9** pp. 265-271 (1988).

15. J.J. Sylvester, ''On a Theory of the Syzygetic Relations of Two Rational Integral Functions Comprising an Application to the Theory of Sturm Functions and that of the Greatest Algebraic Common Measure,'' *Philisophical Transactions*, **143** pp. 407-548 (1853).

16. P.S. Wang, ''The EEZ-GCD Algorithm,'' *ACM SIGSAM Bull.*, **14** pp. 50-60 (1980).

17. S.M. Watt, ''Bounded Parallelism in Computer Algebra,'' Ph.D. Thesis, Dept. of CS, Univ. of Waterloo (1986).

18. H. Zassenhaus, ''Hensel Factorization I,'' *J. Number Theory*, **1** pp. 291-311 (1969).

19. R.E. Zippel, ''Probabilistic Algorithms for Sparse Polynomials,'' Ph.D. Thesis, M.I.T. (1979).

CHAPTER 8

POLYNOMIAL FACTORIZATION

8.1. INTRODUCTION

The problem of factoring polynomials arises in numerous areas in symbolic computation. Indeed, it plays a critical role as a subproblem to many other problems including simplification, symbolic integration and the solution of polynomial equations. Polynomial factorization also plays a significant role in such diverse fields as algebraic coding theory, cryptography and number theory.

In Chapter 6 it was shown how homomorphism techniques reduce factoring problems involving multivariate polynomials over the integers to univariate factoring problems modulo a prime. Hensel lifting is then used to obtain a factorization in the larger domain. However, the problem of factoring over the integers modulo a prime p has yet to be resolved. This chapter will complete the process by describing two well-known algorithms, Berlekamp's algorithm and distinct-degree factorization, for the factorization of two polynomials having coefficients from a Galois field GF(q), where $q = p^m$, p a prime. Factorization modulo a prime is just the special case $m = 1$. For additional information on the subject of factorization over finite fields, we refer the reader to the texts by Berlekamp [2], Lidl and Niederreiter [13], Knuth [7] or McEliece [14].

Once we are able to factor multivariate polynomials over the integers, it is a simple matter to factor multivariate polynomials having coefficients from the rational number field. However, for applications such as symbolic integration we need more, namely the ability to factor polynomials having coefficients from an algebraic extension of the rationals, that is, where the coefficients come from such domains as $\mathbf{Q}(\sqrt{2})$ or $\mathbf{Q}(\sqrt{2},\sqrt{3})$. Therefore in this chapter we describe an algorithm due to Trager (which in turn is a variation of an algorithm that dates back to Kronecker) to factor polynomials over algebraic number fields.

8.2. SQUARE-FREE FACTORIZATION

In this section we develop an algorithm for determining the square-free factorization of a polynomial defined over a unique factorization domain. From this decomposition, we obtain all the repeated factors of a polynomial. This effectively reduces the factorization problem to one of factoring those polynomials known to have no repeated factors.

Definition 8.1. Let $a(x) \in R[x]$ be a primitive polynomial over a unique factorization domain R. Then $a(x)$ is *square-free* if it has no repeated factors, that is, if there exists no $b(x)$ with $\deg(b(x)) \geq 1$ such that

$$b(x)^2 \mid a(x).$$

The *square-free factorization* of $a(x)$ is

$$a(x) = \prod_{i=1}^{k} a_i(x)^i \tag{8.1}$$

where each $a_i(x)$ is a square-free polynomial and

$$\text{GCD}(a_i(x), a_j(x)) = 1 \text{ for } i \neq j.$$

●

Note that some of the $a_i(x)$ in the square-free factorization may be 1. Thus, for example,

$$a(x) = (x^2+1)\cdot(x^2-1)^4(x^3+3x)^5$$

is a square-free factorization with $a_2(x) = a_3(x) = 1$. Note that while the components are pairwise relatively prime, the components themselves are not necessarily completely factored. Of course, once an algorithm has been given to determine the square-free factorization of a polynomial, then a complete factorization of any polynomial can be determined. Algorithms for factoring square-free polynomials are presented later in the chapter. Besides its usefulness in polynomial factoring, square-free factorization is also of central importance in symbolic integration (cf. Chapters 11 and 12). The primary goal of this section is to show that such a factorization is easy to accomplish.

To determine whether a polynomial is square-free or not, the concept of the derivative of a polynomial becomes useful. Recall from Chapter 2 that for any polynomial $a(x) = a_0 + \cdots + a_n x^n$ we may formally define the derivative by

$$a'(x) = a_1 + 2a_2 x + \cdots + n\, a_n x^{n-1}. \tag{8.2}$$

Of course, this definition of a derivative has useful properties which are familiar from calculus. In particular as pointed out in Chapter 2, it satisfies

(a) Linearity: $(a(x)+b(x))' = a'(x)+b'(x)$, $(c\cdot a(x))' = c\cdot a'(x)$;

(b) The product rule: $(a(x)\cdot b(x))' = a'(x)\cdot b(x) + a(x)\cdot b'(x)$;

(c) The power rule: $(a(x)^n)' = n a(x)^{n-1}\cdot a'(x)$.

Theorem 8.1. Let $a(x)$ be a primitive polynomial in $R[x]$, R a unique factorization domain of characteristic 0. Let $c(x) = \text{GCD}(a(x), a'(x))$. Then $a(x)$ has repeated factors if and only if $c(x)$ is not 1.

Proof: Suppose that $a(x)$ is a polynomial which has repeated factors. Then we can write

$$a(x) = b(x)^2 \cdot w(x)$$

so

$$a'(x) = 2 \cdot b(x) \cdot b'(x) w(x) + b(x)^2 \cdot w'(x) = b(x) \cdot \hat{w}(x)$$

where $\hat{w}(x) \in R[x]$. Therefore $a(x)$ and $a'(x)$ have a nontrivial factor, and hence the GCD is not 1.

Conversely, suppose $c(x)$ is nontrivial but that $a(x)$ is square-free. Let the factorization of $a(x)$ be

$$a(x) = p_1(x) \cdot p_2(x) \cdots p_k(x)$$

where each $p_i(x)$ is irreducible, $\deg(p_i(x)) \geq 1$, and

$$\mathrm{GCD}(p_i(x), p_j(x)) = 1 \quad \text{for} \quad i \neq j. \tag{8.3}$$

Then the derivative is given by

$$a'(x) = p_1'(x) \cdot p_2(x) \cdots p_k(x) + \cdots + p_1(x) \cdot p_2(x) \cdots p_k'(x). \tag{8.4}$$

Suppose $p_i(x) \mid c(x)$. There must be at least one such i, say it is $i = 1$. Then $p_1(x) \mid a'(x)$, which from (8.4) implies

$$p_1(x) \mid p_1'(x) \cdot p_2(x) \cdots p_k(x).$$

From (8.3), we have

$$p_1(x) \mid p_1'(x).$$

Since the degree of $p_1(x)$ is greater than the degree of $p_1'(x)$, this can only happen if

$$p_1'(x) = 0. \tag{8.5}$$

But for domains of characteristic 0, equation (8.5) holds if and only if $p_1(x)$ is a constant, a contradiction.

●

Theorem 8.1 gives a simple method for determining when a polynomial has repeated factors. Consider now the problem of obtaining the square-free factors. Let $a(x)$ have the square-free factorization given by (8.1). Taking the derivative gives

$$a'(x) = \sum_{i=1}^{k} a_1(x) \cdots i \cdot a_i(x)^{i-1} a_i'(x) \cdots a_k(x)^k \tag{8.6}$$

and hence

$$c(x) = \mathrm{GCD}(a(x), a'(x)) = \prod_{i=2}^{k} a_i(x)^{i-1}. \tag{8.7}$$

If

$$w(x) = a(x)/c(x) = a_1(x)\cdot a_2(x) \cdots a_k(x) \tag{8.8}$$

then $w(x)$ is the product of the square-free factors without their multiplicities. Calculating

$$y(x) = \mathrm{GCD}(c(x), w(x)) \tag{8.9}$$

and then noticing that

$$a_1(x) = w(x)/y(x) \tag{8.10}$$

gives the first square-free factor. Finding the second square-free factor of $a(x)$ is the same as determining the first square-free factor of $c(x)$. The common factors of this polynomial and its derivative are determined by a simple division

$$\mathrm{GCD}(c(x), c'(x)) = \prod_{i=3}^{k} a_i(x)^{i-2} = c(x)/y(x).$$

The product of the remaining square-free factors is of course just $y(x)$. These observations lead to an efficient iterative scheme for determining the square-free factorization of $a(x)$.

Algorithm 8.1. Square-Free Factorization.

```
procedure SquareFree(a(x))

    # Given a primitive polynomial a(x) ∈ R[x], R a UFD
    # with characteristic zero, we calculate the
    # square-free factorization of a(x).

    i ← 1;  Output ← 1;  b(x) ← a'(x)
    c(x) ← GCD(a(x),b(x));   w(x) ← a(x)/c(x)
    while c(x) ≠ 1 do {
        y(x) ← GCD(w(x),c(x));     z(x) ← w(x)/y(x)
        Output ← Output ·z(x)^i;   i ← i + 1
        w(x) ← y(x);       c(x) ← c(x)/y(x) }
    Output ← Output ·w(x)^i
    return(Output)
end
```

Example 8.1. Let $a(x)$ be a polynomial in $\mathbf{Z}[x]$ defined by

$$a(x) = x^8 - 2x^6 + 2x^2 - 1.$$

Entering the procedure, we first calculate

$$b(x) = a'(x) = 8x^7 - 12x^5 + 4x, \;\; c(x) = x^4 - 2x^2 + 1 \;\; \text{and} \;\; w(x) = x^4 - 1.$$

Since $c(x) \neq 1$ we go through the main loop. After the first loop the variables are

$$y(x) = x^2 - 1, \quad z(x) = \text{Output} = x^2 + 1, \quad i = 2, \;\; w(x) = c(x) = x^2 - 1.$$

Entering the main loop the second time gives

$$y(x) = x^2 - 1, \quad z(x) = 1, \;\; i = 3, \;\; w(x) = x^2 - 1, \;\; \text{and} \;\; c(x) = 1$$

while Output remains unchanged. Since $c(x) = 1$, we exit the main loop and return

$$\text{Output} = \text{Output}\cdot w(x)^3 = (x^2 + 1)\cdot(x^2 - 1)^3.$$

This is the desired square-free factorization.

●

The algorithm presented above is simple, easy to understand, and computes the correct quantities. However, there are other, more efficient algorithms for square-free factorization in the case of characteristic 0 domains. In particular, we include a square-free algorithm due to D. Yun[22] which provides a second method for obtaining the square-free factorization. The complexity of determining the square-free factorization of a polynomial $a(x)$ can be shown to be equivalent to the complexity of twice the cost of GCD($a(x)$,$a'(x)$) (cf. Exercise 8.5).

Starting with the square-free factorization of $a(x)$

$$a(x) = a_1(x)\cdot a_2(x)^2 \cdots a_k(x)^k$$

we obtain

$$a'(x) = a_1'(x)\cdot a_2(x)^2 \cdots a_k(x)^k + \cdots + k\cdot a_1(x)\cdot a_2(x)^2 \cdots a_k(x)^{k-1} a_k'(x)$$

and

$$c(x) = \text{GCD}(a(x),a'(x)) = a_2(x)\cdot a_3(x)^2 \cdots a_k(x)^{k-1}.$$

Let $w(x) = a(x)/c(x) = a_1(x) \cdots a_k(x)$ be the product of the square-free factors and

$$\begin{aligned} y(x) &= a'(x)/c(x) \\ &= a_1'(x)\cdot a_2(x) \cdots a_k(x) + \cdots + k\cdot a_1(x) \cdots a_{k-1}(x) a_k'(x). \end{aligned}$$

Notice that both $w(x)$ and $y(x)$ add little overhead relative to the GCD operation. Letting

$$\begin{aligned} z(x) &= y(x) - w'(x) \\ &= a_1(x)\cdot a_2'(x) \cdots a_k(x) + \cdots + (k-1) a_1(x) \cdots a_{k-1}(x) a_k'(x) \\ &= a_1(x)\cdot[\, a_2'(x) \cdots a_k(x) + \cdots + (k-1) a_2(x) \cdots a_{k-1}(x) a_k'(x)\,] \end{aligned}$$

we get the first square-free term by calculating

$$a_1(x) = \text{GCD}(w(x),z(x)).$$

So far the process has followed in a manner similar to our previous method, except for the added derivative calculation. As before, the natural next step is to determine the square-free decomposition of $c(x)/w(x)$. In this second step, the corresponding $w(x)$, $y(x)$ and $z(x)$ are found by

$$w(x) = w(x)/a_1(x) = a_2(x) \cdots a_k(x),$$

$$y(x) = \frac{z(x)}{a_1(x)} = a_2{}'(x) \cdots a_k(x) + \cdots + (k-1)a_2(x) \cdots a_{k-1}(x)a_k{}'(x)$$

and

$$z(x) = y(x) - w'(x)$$

with the second square-free factor of $a(x)$ determined by GCD($w(x)$,$z(x)$). The advantage of Yun's method is that for the price of one differentiation, the GCD calculation inside the main loop of the process is considerably simpler. This is especially noticeable in the case of fields such as **Q**, where coefficient growth becomes a real concern during GCD calculations.

Algorithm 8.2. Yun's Square-Free Factorization.

```
procedure SquareFree2(a(x))
    # Given a primitive polynomial a(x) ∈ R[x], R a
    # UFD of characteristic zero, we calculate the square-free
    # factorization of a(x) using Yun's algorithm.

    i ← 1;  Output ← 1
    b(x) ← a'(x), c(x) ← GCD(a(x),b(x))

    if c(x) = 1 then w(x) ← a(x)
    else {
        w(x) ← a(x)/c(x)
        y(x) ← b(x)/c(x)
        z(x) ← y(x) − w'(x)
        while z(x) ≠ 0 do {
            g(x) ← GCD(w(x),z(x));    Output ← Output·g(x)^i
            i ← i + 1; w(x) ← w(x)/g(x)
            y(x) ← z(x)/g(x);    z(x) ← y(x) − w'(x) } }

    Output ← Output ·w(x)^i
    return(Output)
end
```

Example 8.2. Let $a(x)$ be the same polynomial in $\mathbf{Q}[x]$ from Example 8.1. We will use

$$a(x) = (x^2 + 1)\cdot(x^2 - 1)^3 \tag{8.11}$$

and demonstrate how the second algorithm "picks off" the individual square-free factors.

At the initialization step, we determine that the greatest common factor $c(x)$ is given by $c(x) = (x^2 - 1)^2$. Since $c(x) \neq 1$, we calculate

$$w(x) = (x^2 + 1)(x^2 - 1), \quad y(x) = 2x(x^2 - 1) + 3(x^2 + 1)(2x),$$

and $z(x) = 2(x^2 + 1)(2x)$. Entering the **while**-loop, we get Output $= g(x) = x^2 + 1$. The updating part of the loop results in

$$i = 2, \quad w(x) = (x^2 - 1), \quad y(x) = 4x, \quad z(x) = 2x.$$

Going through the **while**-loop a second time we get $g(x) = 1$, Output $= x^2 + 1$, with the updates given by

$$i = 3, \quad w(x) = (x^2 - 1), \quad y(x) = 2x, \quad z(x) = 0.$$

Since $z(x) = 0$ we exit the loop. Since $w(x)$ holds the last square-free factor, the output is given by (8.11).

●

8.3. SQUARE-FREE FACTORIZATION OVER FINITE FIELDS

Suppose that we now wish to determine a square-free factorization over a domain having a nonzero characteristic. In particular, we want to determine a square-free factorization for polynomials whose coefficients come from the field GF(q), a Galois field of order $q = p^m$, where p is a prime.

A procedure to determine a square-free factorization in GF(q)[x] follows the same process as in the preceding section. As before, one determines the derivative and then computes the GCD of the polynomial and its derivative. If this GCD is one, then the polynomial is square-free. If it is not one then the GCD is again divided into the original polynomial, provided that the GCD is not zero. In domains of characteristic zero, $\mathrm{GCD}(a(x), a'(x)) \neq 0$ for any nontrivial polynomial $a(x)$. Such is not the case for polynomials defined over finite fields.

Example 8.3. In GF(13)[x] we have

$$a(x) = x^{13} + 1 \quad \Rightarrow \quad a'(x) = 13\cdot x^{12} = 0$$

since the field has characteristic 13. However, in this example we may obtain a square-free factorization for $a(x)$ by noticing that

$$(x + 1)^{13} = x^{13} + \binom{13}{1}x^{12} + \cdots + \binom{13}{12}x + 1 = x^{13} + 1 = a(x)$$

where the last equality holds because 13 divides $\binom{13}{i}$ for all $1 \leq i \leq 12$.

●

The process for obtaining a square-free factorization in Example 8.3 generalizes to all Galois fields. We first require some basic facts from finite field theory.

Lemma 8.1. Let GF(q) be a Galois field of order $q = p^m$, p a prime. For any $r, s \in$ GF(q) we have

$$r^q = r, \tag{8.12}$$

$$r^{1/p} = r^{q/p} = r^{p^{m-1}}. \tag{8.13}$$

$$(r + s)^{p^j} = r^{p^j} + s^{p^j} \;,\; j = 0, 1, \ldots, m. \tag{8.14}$$

Proof: For any $r \in$ GF(q) the set $\{1, r, r^2, \ldots\}$ is a finite subgroup of the multiplicative group of GF(q). Since the latter group has order $q - 1$, Lagrange's theorem (cf. Herstein [5]) implies that the order of the subgroup (which is the same as the order of r) must divide $q - 1$. Therefore,

$$r^{q-1} = 1,$$

from which (8.12) is a direct consequence. Equation (8.13) follows from the observation

$$(r^{p^{m-1}})^{p} = r^{p^m} = r.$$

Expanding the left hand side of (8.14) gives

$$r^{p^j} + \binom{p^j}{1} r^{p^j-1} \cdot s + \cdots + \binom{p^j}{p^j-1} r \cdot s^{p^j-1} + s^{p^j} = r^{p^j} + s^{p^j}$$

where the last equality follows from the fact that the characteristic, p, divides $\binom{p^j}{k}$ for all $1 \le k \le p^j - 1$.

●

Using Lemma 8.1, we obtain

Theorem 8.3. Let $a(x) = a_0 + a_1 x + \cdots + a_n x^n$ be a polynomial of degree n in GF(q)[x] satisfying $a'(x) = 0$. Then $a(x) = b(x)^p$ for some polynomial $b(x)$.

Proof: Since $a'(x) = 0$ the only nonzero powers of x in $a(x)$ must be divisible by p. Therefore

$$a(x) = a_0 + a_p x^p + a_{2p} x^{2p} + \cdots + a_{kp} x^{kp}$$

for some integer k. Let

$$b(x) = b_0 + b_1 x + \cdots + b_k x^k$$

where

$$b_i = {a_{ip}}^{1/p} = {a_{ip}}^{p^{m-1}}$$

(with the last equality holding from (8.13)). Repeated use of Lemma 8.1 gives

$$b(x)^p = b_0^p + b_1^p x^p + \cdots + b_k^p x^{kp}$$
$$= a_0 + a_p x^p + \cdots + a_{kp} x^{kp} = a(x) .$$

●

Note that the proof of Theorem 8.3 not only shows the existence, but in fact also shows how to construct $b(x) = a(x)^{1/p}$ when $a'(x) = 0$. Theorem 8.3, along with the previous discussion on square-free decompositions, leads to an algorithm for square-free factorization.

Algorithm 8.3. Finite Field Square-Free Factorization.

```
procedure SquareFreeFF(a(x),q)

   # Given a monic polynomial a(x) ∈ GF(q)[x], with GF(q) a
   # Galois field  of order q = p^m, we calculate the
   # square-free factorization of a(x).

   i ← 1;  Output ← 1; b(x) ← a'(x)
   if b(x) ≠ 0 then {
      c(x) ← GCD(a(x),b(x))
      w(x) ← a(x)/c(x)
      while w(x) ≠ 1 do {
         y(x) ← GCD(w(x),c(x));  z(x) ← w(x)/y(x)
         Output ← Output ·z(x)^i; i ← i + 1
         w(x) ← y(x); c(x) ← c(x)/y(x) }
      if c(x) ≠ 1 then {
         c(x) ← c(x)^(1/p)
         Output ← Output · ( SquareFreeFF(c(x)) )^p }}
   else {
      a(x) ← a(x)^(1/p)
      Output ← ( SquareFreeFF(a(x)) )^p }
   return(Output)

end
```

Example 8.4. Let $a(x)$ be a polynomial in $GF(3)[x] = \mathbf{Z}_3[x]$ defined by

$$a(x) = x^{11} + 2x^9 + 2x^8 + x^6 + x^5 + 2x^3 + 2x^2 + 1.$$

Then

$$a'(x) = 2x^{10} + x^7 + 2x^4 + x$$

and

$$c(x) = \mathrm{GCD}(a(x), a'(x)) = x^9 + 2x^6 + x^3 + 2.$$

Since $c(x) \neq 0$ we have $w(x) = x^2+2$ and we enter the **while**-loop. After one loop we have

$$y(x) = x+2, \quad z(x) = x+1, \quad \text{Output}= x+1,$$

with updates

$$i = 2, \quad w(x) = x+2, \quad \text{and} \quad c(x) = x^8+x^7+x^6+x^2+x+1.$$

The second time through the loop gives

$$y(x) = x+2, \quad z(x) = 1, \quad \text{Output}= x+1,$$

with updates

$$i = 3, \ w(x) = x+2, \quad \text{and} \quad c(x) = x^7+2x^6+x+2.$$

The third time through the loop also does not change Output. For the fourth time through the loop we get

$$y(x) = 1, \quad z(x) = x+2, \quad \text{Output} = (x+1)\cdot(x+2)^4,$$

with updates

$$i = 5, \ w(x) = 1, \quad \text{and} \quad c(x) = x^6+1.$$

Since $w(x) = 1$, we exit the **while**-loop. Since $c(x) \neq 1$, it must be a perfect cube. The cube root of $c(x)$ is just $x^2 + 1$, and invoking the square-free procedure recursively determines that it is square-free. Therefore, cubing the resulting square-free factorization and combining it with the output to that point gives the square-free decomposition of $a(x)$ as

$$\text{Output} = (x + 1)\cdot(x^2 + 1)^3\cdot(x + 2)^4$$

which is the square-free factorization of $a(x)$.

●

Example 8.5. Let the symbols A, B, C, and D represent square-free polynomials in GF(3)[x] and set

$$a = A\cdot B^3\cdot C^5\cdot D^9.$$

We will show how SquareFreeFF calculates the square-free factorization by tracing part of the algorithm at the symbolic level.

Initially, the algorithm determines

$$b = a' = A'\cdot B^3\cdot C^5\cdot D^9 + 2\cdot A\cdot B^3\cdot C^4\cdot C'\cdot D^9$$

hence the greatest common divisor is

$$c = B^3\cdot C^4\cdot D^9.$$

The square-free terms (not including those having exponents divisible by the modulus) is

$$w = a/c = A\cdot C.$$

Entering the **while**-loop, we have $y = \mathrm{GCD}(c,w) = C$, $z = w/C = A$ so Output $= A$ and the remaining variables are updated by

$$i = 2,\ \ w = C,\ \ c = B^3 \cdot C^3 \cdot D^9.$$

Entering the **while**-loop for the second time gives $y = C,\ z = 1$, so Output remains unchanged, while the updated variables become

$$i = 3,\ w = C,\ c = B^3 \cdot C^2 \cdot D^9.$$

By the time the loop has been exited for the fourth time, the updated variables are

$$i = 5,\ w = C,\ c = B^3 \cdot D^9$$

with Output still set to A. Inside the **while**-loop for the fifth time, however, we have $y = 1,\ \ z = C$, hence Output $= A \cdot C^5$. The resulting updates for this step are

$$i = 6,\ w = 1,\ c = B^3 \cdot D^9.$$

Since $w = 1$, the algorithm exits the **while**-loop. Since $c \neq 1$, a square-free decomposition is recursively determined for the cube root of c, that is for $B \cdot D^3$. Working through the algorithm gives this as $B \cdot D^3$, and this factorization cubed is included in Output. Finally, the algorithm returns Output $= A \cdot B^3 \cdot C^5 \cdot D^9$, which is the desired final form. ●

A similar modification must also be made in order to carry the square-free algorithm of Yun over to the finite field case (cf. Exercise 8.6).

8.4. BERLEKAMP'S FACTORIZATION ALGORITHM

In this section we describe a well known algorithm due to Berlekamp [1] which factors polynomials in GF(q)[x], where GF(q) is a Galois field of order $q = p^m$, p a prime. This algorithm is a wonderful exhibition of the elegance of computational algebra, combining finite field theory and vector spaces over finite fields to obtain a desired factorization.

Suppose that $a(x) \in$ GF(q)[x] is the polynomial to be factored. By the previous section we may solve the problem in the case where $a(x)$ has already been made square-free. Notice first that the residue ring $V = GF(q)[x]/\langle a(x)\rangle$ is a vector space over the field GF(q) having dimension n, where n is the degree of $a(x)$. Let

$$W = \{\ v(x) \in GF(q)[x]\ :\ v(x)^q = v(x) \bmod a(x)\ \}. \tag{8.15}$$

Then W can be identified with the set

$$\{\ [v(x)]\ \in V\ :\ [v(x)]^q = [v(x)]\ \}. \tag{8.16}$$

With a slight abuse of notation we will also call this set W and identify a residue class $[v(x)]$ with its unique representative of degree less than n. The set W plays a central role in Berlekamp's algorithm.

Theorem 8.4. The subset W is a subspace of the vector space V.

Proof: Suppose $v_1(x)$ and $v_2(x)$ are in W. Using arguments similar to the proof of equation (8.14) we get

$$(v_1(x)+v_2(x))^q = v_1(x)^q + v_2(x)^q = v_1(x)+v_2(x) \tag{8.17}$$

so $v_1(x)+v_2(x)$ is also in W. If $c \in$ GF(q) and $v(x) \in$ W, then using equation (8.12) gives

$$(c\cdot v(x))^q = c^q \cdot v(x)^q = c\cdot v(x) \tag{8.18}$$

hence $c\cdot v(x)$ is in V. Equations (8.17) and (8.18) show that W is a subspace of V.

●

Example 8.6. Suppose $a(x)$ is irreducible in GF(q)[x]. Then, in addition to being a vector space over GF(q), V is also a field. As such, the polynomial

$$p(z) = z^q - z \in \mathrm{V}[z] \tag{8.19}$$

can have at most q roots in V. From Lemma 8.1 every element of GF(q) is a root of (8.19) and, since there are q such elements, these account for all the roots. Thus W consists of all constant polynomials modulo $a(x)$ and so can be identified with GF(q) itself. As a result, W is a subspace of dimension one in V.

●

Example 8.6 gives a criterion for determining when $a(x)$ is irreducible. Calculate a basis for W. If there is only one basis element then $a(x)$ is irreducible. But, what if W has dimension greater than one? In this case the Chinese remainder theorem will provide a method that determines W in a manner similar to Example 8.6.

Suppose that $a(x)$ is square-free with a factorization given by

$$a(x) = a_1(x)\cdot \cdots \cdot a_k(x)$$

where the $a_i(x)$'s are irreducible and pairwise relatively prime. For each i, let

$$\mathrm{V}_i = \mathrm{GF}(q)[x]/\langle a_i(x)\rangle .$$

By the Chinese remainder theorem, the mapping

$$\phi : \mathrm{V} \rightarrow \mathrm{V}_1 \times \cdots \times \mathrm{V}_k$$

defined by

$$\phi(v(x) \bmod a(x)\,) = (v(x) \bmod a_1(x)\,, \ldots, \,v(x) \bmod a_k(x))$$

is a ring isomorphism. Note that

$$v(x)^q \equiv v(x) \bmod a(x) \;\Rightarrow\; v(x)^q \equiv v(x) \mod a_i(x)$$

for all $i \in \{\,1, \ldots, k\,\}$. Thus, ϕ induces a ring homomorphism

$$\phi_{\mathrm{W}} : \mathrm{W} \rightarrow \mathrm{W}_1 \times \cdots \times \mathrm{W}_k \tag{8.20}$$

where, for each i,

$$\mathrm{W}_i = \{\, s \in \mathrm{V}_i \;:\; s^q = s \,\}.$$

Since $a_i(x)$ is irreducible, each V_i is a field. Thus, as in Example 8.6, each W_i can be identified with the ground field, GF(q).

Theorem 8.5. The induced mapping ϕ_W in (8.20) is a ring (and hence a vector space) isomorphism. In particular, the dimension of W is k, the number of irreducible factors of $a(x)$.

Proof: That ϕ_W is one-to-one follows since ϕ has this property. We need to show that ϕ_W is onto. Let

$$(s_1, \ldots, s_k) \in W_1 \times \cdots \times W_k.$$

Since ϕ is onto, there exists a $v(x) \in V$ such that

$$\phi(v(x)) = (s_1, \ldots, s_k).$$

We need only show that $v(x) \in W$. But

$$\phi(v(x)^q) = (s_1^q, \ldots, s_k^q) = (s_1, \ldots, s_k) = \phi(v(x)).$$

Since ϕ is one-to-one, this implies $v(x)^q = v(x)$, hence $v(x) \in W$ as required.

Since the ring isomorphism is the same as a vector space isomorphism in this case, the second part of Theorem 8.5 follows from the observation that each W_i has dimension one as a vector space over GF(q), hence $W_1 \times \cdots \times W_k$ has dimension k.

●

Theorem 8.5 is useful in that it gives the number of factors in $a(x)$. However, it still leaves open the question of how to calculate the factors knowing W, or for that matter, how to calculate W itself. We begin by answering the first question.

Theorem 8.6. Let $a(x)$ be a square-free polynomial in GF(q)[x] and let $v(x)$ be a nonconstant polynomial in W. Then

$$a(x) = \prod_{s \in GF(q)} GCD(v(x) - s, a(x)).$$

Proof: The polynomial $x^q - x$ factors in GF(q)[x] as

$$x^q - x = \prod_{s \in GF(q)} (x - s)$$

so

$$v(x)^q - v(x) = \prod_{s \in GF(q)} (v(x) - s).$$

Since $a_i(x)$ divides into $a(x)$ for all i, we have

$$a_i(x) \mid v(x)^q - v(x) = \prod_{s \in GF(q)} (v(x) - s). \tag{8.21}$$

Note that

$$GCD(v(x) - s, v(x) - t) = 1 \tag{8.22}$$

for $s \neq t$. Equations (8.21) and (8.22) imply that, for a given i, $a_i(x)$ must divide $v(x) - s_i$ for exactly one s_i. Therefore

$$a_i(x) \mid \mathrm{GCD}(a(x), v(x) - s_i) \quad i = 1, \ldots, k$$

and so

$$a(x) \mid \prod_{i=1}^{k} \mathrm{GCD}(v(x) - s_i, a(x)) \mid \prod_{s \in \mathrm{GF}(q)} \mathrm{GCD}(v(x) - s, a(x)). \tag{8.23}$$

Clearly

$$\mathrm{GCD}(v(x) - s, a(x)) \mid a(x)$$

for each $s \in \mathrm{GF}(q)$, which, combined with (8.22), gives

$$\prod_{s \in \mathrm{GF}(q)} \mathrm{GCD}(v(x) - s, a(x)) \mid a(x). \tag{8.24}$$

Equations (8.23) and (8.24) prove Theorem 8.6.

●

Theorem 8.6 reduces the problem of factoring $a(x)$ to the problem of determining all $v(x)$ in W, together with a series of GCD calculations. However, if there are k factors then finding the q^k elements of W is somewhat prohibitive. Since W is a vector space of dimension k, it is enough to calculate k linearly independent basis vectors and then apply Theorem 8.6. This makes the problem more tractable, hence we direct our attention to describing a basis for W.

For any polynomial $v(x) \in \mathrm{GF}(q)[x]$, we have

$$\begin{aligned} v(x)^q &= (v_0 + v_1 x + \cdots + v_{n-1} x^{n-1})^q \\ &= v_0{}^q + v_1{}^q x^q + \cdots + v_{n-1}{}^q x^{q(n-1)} \\ &= v_0 + v_1 x^q + \cdots + v_{n-1} x^{q(n-1)} = v(x^q) \end{aligned}$$

since every $v_i \in \mathrm{GF}(q)$ satisfies $v_i{}^q = v_i$. Therefore we may write

$$\mathrm{W} = \{\ v(x) \in \mathrm{GF}(q)[x] \ :\ v(x^q) - v(x) = 0 \mod a(x)\ \}$$

which gives W as a solution space of a system of n equations in n unknowns, n the degree of $a(x)$. We may determine the coefficient matrix of the system of equations by letting Q be the $n \times n$ matrix whose entries $q_{i,j}$ (for $0 \le i, j \le n-1$) are determined by

$$x^{q \cdot j} = q_{j,0} + q_{j,1} x + \cdots + q_{j,n-1} x^{n-1} \mod a(x)\,; \tag{8.25}$$

that is, the matrix Q has rows $0, 1, \ldots, n-1$ determined from the remainders of $a(x)$ divided into $x^0, x^q, x^{q \cdot 2}, \ldots, x^{q \cdot (n-1)}$, respectively.

Theorem 8.7. Let W be given by equation (8.15). Then with Q determined from (8.25) we have

$$W = \{ \mathbf{v} = (v_0, \ldots, v_{n-1}) \; : \; \mathbf{v} \cdot (Q - I) = \mathbf{0} \}. \tag{8.26}$$

Proof: The equation

$$v(x^q) - v(x) \equiv 0 \mod a(x) \tag{8.27}$$

is equivalent to

$$\begin{aligned} 0 &\equiv \sum_{j=0}^{n-1} v_j x^{q \cdot j} - \sum_{j=0}^{n-1} v_j x^j \mod a(x) \\ &\equiv \sum_{j=0}^{n-1} v_j [\sum_{i=0}^{n-1} q_{j,i} \cdot x^i] - \sum_{j=0}^{n-1} v_j x^j \mod a(x) \\ &\equiv \sum_{i=0}^{n-1} \{ \sum_{j=0}^{n-1} v_j \cdot q_{j,i} - v_i \} \cdot x^i \mod a(x) \end{aligned}$$

hence

$$\sum_{j=0}^{n-1} v_j \cdot q_{j,i} - v_i \equiv 0 , \text{ for all } i = 0, \ldots, n-1. \tag{8.28}$$

Equation (8.28) is equivalent to

$$(v_0, \ldots, v_{n-1}) \cdot Q - (v_0, \ldots, v_{n-1}) = (0, \ldots, 0),$$

that is,

$$\mathbf{v} \cdot (Q - I) = \mathbf{0}. \tag{8.29}$$

The equivalence of equations (8.27) and (8.29) gives (8.26).

●

To complete Berlekamp's algorithm we need to discuss the three main steps in detail. To generate the matrix Q involves calculating

$$x^q \mod a(x), x^{2q} \mod a(x), \ldots, x^{(n-1)q} \mod a(x) .$$

We may build up the Q matrix by a simple iterative procedure that generates $x^{m+1} \mod a(x)$ given that $x^m \mod a(x)$ has been determined. If

$$a(x) = a_0 + a_1 x + \cdots + a_{n-1} x^{n-1} + x^n$$

and

$$x^m \equiv r_{m,0} + r_{m,1} x + \cdots + r_{m,n-1} x^{n-1} \mod a(x)$$

then (working mod $a(x)$) we have

Algorithm 8.4. Berlekamp's Factoring Algorithm.

```
procedure Berlekamp(a(x),q)
    # Given a square-free polynomial a(x) ∈ GF(q)[x]
    # calculate irreducible factors a_1(x), ..., a_k(x) such
    # that a(x) = a_1(x) ··· a_k(x).

    Q ← FormMatrixQ(a(x),q)
    v^[1], v^[2], ..., v^[k] ← NullSpaceBasis(Q − I))

    # Note: we can ensure that v^[1] = (1, 0, ..., 0).
    factors ← { a(x) }
    r ← 2

    while SizeOf(factors) < k do {

      foreach u(x) ∈ factors do {

        foreach s ∈ GF(q) do {
          g(x) ← GCD(v^[r](x) − s, u(x))
          if g(x) ≠ 1 or g(x) ≠ u(x) then {
              Remove(u(x), factors)
              u(x) ← u(x)/g(x)
              Add({u(x), g(x)}, factors) }
          if SizeOf(factors) = k then return(factors) }
      r ← r + 1 } }

  end
```

$$\begin{aligned} x^{m+1} &\equiv r_{m,0}x + r_{m,1}x^2 + \cdots + r_{m,n-1}x^n \\ &\equiv r_{m,0}x + r_{m,1}x^2 + \cdots + r_{m,n-1}(-a_0 - a_1x - \cdots - a_{n-1}x^{n-1}) \\ &\equiv -r_{m,n-1}a_0 + (r_{m,0} - r_{m,n-1}a_1)x + \cdots + (r_{m,n-2} - r_{m,n-1}a_{n-1})x^{n-1} \\ &\equiv r_{m+1,0} + r_{m+1,1}x + \cdots + r_{m+1,n-1}x^{n-1} \end{aligned}$$

where

$$r_{m+1,0} = -r_{m,n-1}a_0, \quad \text{and} \quad r_{m+1,i} = r_{m,i-1} - r_{m,n-1}a_i$$

for $i = 1, \ldots, n-1$. Thus, we can generate the Q matrix by storing a vector of elements from GF(q)

$$\mathbf{r} \leftarrow (r_0, \ldots, r_{n-1})$$

initialized by

$$\mathbf{r} \leftarrow (1, 0, \ldots, 0)$$

and updated by

$$\mathbf{r} \leftarrow (-r_{n-1} \cdot a_0, r_0 - r_{n-1} \cdot a_1, \ldots, r_{n-2} - r_{n-1} \cdot a_{n-1}).$$

In this process, each new row of Q, i.e. each additional x^{iq}, requires $q \cdot n$ multiplications. Thus the cost of generating the Q matrix is $O(q \cdot n^2)$ field operations.

Algorithm 8.5. Form Q Matrix.

procedure FormMatrixQ($a(x), q$)

\# Given a polynomial $a(x)$ of degree n in GF(q)[x], calculate
\# the Q matrix required by Berlekamp's algorithm.

$n \leftarrow \deg(a(x))$; $\mathbf{r} \leftarrow (1, 0, \ldots, 0)$; Row(0,$Q$)) $\leftarrow \mathbf{r}$
for m **from** 1 **to** $(n-1)q$ **do** {
  $\mathbf{r} \leftarrow (-r_{n-1} \cdot a_0, r_0 - r_{n-1} \cdot a_1, \ldots, r_{n-2} - r_{n-1} \cdot a_{n-1})$
  if $q \mid m$ **then**
    Row($m/q, Q$) $\leftarrow \mathbf{r}$ }
return(Q)
end

Example 8.7. Let $a(x)$ be the polynomial in GF(11)[x] = $\mathbf{Z}_{11}[x]$ given by

$$a(x) = x^6 - 3x^5 + x^4 - 3x^3 - x^2 - 3x + 1.$$

We will determine the Q matrix (of size 6×6) for $a(x)$ using the method described above. Row 0 of Q will be given by

$$(1, 0, 0, 0, 0, 0, 0)$$

since $1 \equiv 1 \mod a(x)$. In addition,

$$\begin{aligned}
x &\equiv x && \mod a(x), \\
x^2 &\equiv x^2 && \mod a(x), \\
x^3 &\equiv x^3 && \mod a(x), \\
x^4 &\equiv x^4 && \mod a(x), \\
x^5 &\equiv x^5 && \mod a(x),
\end{aligned}$$

and

$$x^6 \equiv 3x^5 - x^4 + 3x^3 + x^2 + 3x - 1 \mod a(x).$$

Therefore,

$$\begin{aligned} x^7 &\equiv 3x^6 - x^5 + 3x^4 + x^3 + 3x^2 - x \\ &\equiv 3\cdot(3x^5 - x^4 + 3x^3 + x^2 + 3x - 1) - x^5 + 3x^4 + x^3 + 3x^2 - x \\ &\equiv -3x^5 - x^3 - 5x^2 - 3x - 3 \mod a(x). \end{aligned}$$

Continuing in this manner, we obtain $x^8 \mod a(x)$, $x^9 \mod a(x)$, $x^{10} \mod a(x)$ and

$$x^{11} \equiv 5x^5 - 5x^4 - 3x^3 - 3x^2 + 5x + 3 \mod a(x).$$

This gives row 1 of Q as

$$(3, 5, -3, -3, -5, 5).$$

Proceeding as above to calculate $x^{22} \mod a(x), \ldots, x^{55} \mod a(x)$, we get the matrix

$$Q = \begin{bmatrix} 1 & 0 & 0 & 0 & 0 & 0 \\ 3 & 5 & -3 & -3 & -5 & 5 \\ 3 & -5 & -5 & 1 & -1 & 0 \\ -2 & 4 & -1 & 3 & -4 & -2 \\ -4 & -3 & -1 & 0 & 0 & -3 \\ -3 & -1 & -4 & -3 & -1 & -3 \end{bmatrix}.$$

●

Once Q has been determined, there is still the problem of determining a basis for the solution space of $Q - I$. Using elementary column operations, we will column reduce the matrix $Q - I$ to a matrix L which is in *triangular idempotent form.* Such a matrix will be lower-triangular and have only a 0 or a 1 on the main diagonal. Furthermore, if the i-th diagonal element is a 1 then it is the only nonzero entry in that *row.* If the i-th diagonal element is a 0, then the i-th *column* is 0. For example, the lower-triangular matrix

$$L = \begin{bmatrix} 1 & 0 & 0 & 0 & 0 \\ 3 & 0 & 0 & 0 & 0 \\ 0 & 0 & 1 & 0 & 0 \\ 2 & 0 & 4 & 0 & 0 \\ 0 & 0 & 0 & 0 & 1 \end{bmatrix} \qquad (8.30)$$

is in triangular idempotent form. It is not hard to prove that such a matrix is indeed idempotent, that is, satisfies

$$L^2 = L$$

(cf. Exercise 8.7).

When the matrix $Q - I$ is column reduced to a matrix L, the resulting solution spaces are of course the same. The advantage of column reducing a matrix to one that is in triangular idempotent form is given by:

Theorem 8.8. Let L be a matrix in triangular idempotent form. Then the nonzero rows of $I - L$ form a basis for the solution space

$$S = \{ \mathbf{v} \in \mathrm{V} : \mathbf{v} \cdot L = \mathbf{0} \}.$$

Proof: Let $\mathbf{v}^{[1]}, \ldots, \mathbf{v}^{[k]}$ be the set of nonzero rows of $I - L$. Since L is idempotent, we have

$$(I - L) \cdot L = 0$$

so

$$\mathbf{v}^{[i]} \cdot L = \mathbf{0}$$

for all i. Hence $\mathbf{v}^{[1]}, \ldots, \mathbf{v}^{[k]}$ are all in the space S.

These vectors are also linearly independent. For, suppose there are constants $c_1, \ldots, c_k$ such that

$$c_1 \cdot \mathbf{v}^{[1]} + \cdots + c_k \cdot \mathbf{v}^{[k]} = \mathbf{0}.$$

Then there exists a nonzero vector $\mathbf{v}$ such that

$$\mathbf{v} \cdot (I - L) = \mathbf{0} \ , \text{ i.e. } \ \mathbf{v} = \mathbf{v} \cdot L \ .$$

Here the j-th component of $\mathbf{c}$ is zero if the j-th row of $I - L$ is zero, and is c_i if the i-th nonzero row is row j. Suppose that row j is the i-th nonzero row of $I - L$. From the definition of triangular idempotent form, we have that column j of L is the zero column. Therefore the j-th component of $\mathbf{v} \cdot L$, that is, c_i, is zero. Since this occurs for all i in which there is a nonzero row of $I - L$, we must have $c_1 = \cdots = c_k = 0$ proving that the vectors are indeed linearly independent.

Let l_{ii} be the i-th diagonal entry of L. Then from the defining properties of a triangular idempotent matrix, it is clear that l_{ii} is one if and only if the i-th row of $I - L$ is the zero row. Similarly, l_{ii} is zero if and only if the i-th column of $I - L$ is the same as the negative of the i-th column of the identity. Therefore the rank of L is $n - k$, and so S has dimension k. Since the nonzero rows of $I - L$ form a set of k linearly independent vectors in S, they will form a basis.

●

Example 8.8. The matrix L of equation (8.30) is in triangular idempotent form. The rank of L is 3, hence the dimension of the solution space of L is 2. The two nonzero rows of $I - L$ are

$$\mathbf{v}^{[1]} = (-3, 1, 0, 0, 0) \quad \text{and} \quad \mathbf{v}^{[2]} = (-2, 0, -4, 1, 0).$$

●

An algorithm to determine a basis for the null space of the matrix $Q - I$ is given by Algorithm 8.6.

Algorithm 8.6. Null Space Basis Algorithm.

```
procedure NullSpaceBasis(M)

  # Given a square matrix M, we return a basis {v^[1], . . . , v^[k]} for the null
  # space { v : v·M = 0 } of M. The algorithm does this
  # by transforming M to triangular idempotent form.

  n ← rowsize(M)
  for k from 1 to n do {

     # Search for pivot element
     for i from k to n while M_ki = 0 do i ← i + 1

     if i ≤ n then {

        # Normalize column i and interchange this with column k
        Column(i,M) ← Column(i,M) ·M_ki^(-1)
        SwitchColumn(i,k,M)

        # Eliminate rest of row k via column operations
        for i to n with i ≠ k do
           Column(i,M) ← Column(i,M) − Column(k,M) ·M_ki }}

  # Convert M to M − I
  for i from 1 to n do M_ii ← M_ii − 1

  # Read off nonzero rows of M
  i ← 0; j ← 1

  while j ≤ n do {

     while Row(j,M)=0 and j ≤ n do j ← j + 1
     if j ≤ n then {
        i ← i + 1
        v^[i] ← Row(j,M) } }

  return( {v^[1] , . . . , v^[i]} )

end
```

Example 8.9. Let Q be the matrix of Example 8.7. We will determine the solution space for

$$Q - I = \begin{bmatrix} 0 & 0 & 0 & 0 & 0 & 0 \\ 3 & 4 & -3 & -3 & -5 & 5 \\ 3 & -5 & 5 & 1 & -1 & 0 \\ -2 & 4 & -1 & 2 & -4 & -2 \\ -4 & -3 & -1 & 0 & -1 & -3 \\ -3 & -1 & -4 & -3 & -1 & -4 \end{bmatrix}$$

using the above method.

We use column 2 to eliminate the rest of the nonzero terms in the second row. At the end of this loop, the matrix looks like

$$\begin{bmatrix} 0 & 0 & 0 & 0 & 0 & 0 \\ 0 & 1 & 0 & 0 & 0 & 0 \\ 4 & -4 & 4 & 0 & 1 & -2 \\ -5 & 1 & 2 & 5 & 1 & 4 \\ 1 & 2 & 5 & -5 & -2 & -2 \\ -5 & -3 & -2 & -1 & -5 & 0 \end{bmatrix}.$$

Since the element in row 3, column 3 is nonzero, we multiply the third column by $4^{-1} = 3$ and use the resulting column to reduce our matrix to

$$\begin{bmatrix} 0 & 0 & 0 & 0 & 0 & 0 \\ 0 & 1 & 0 & 0 & 0 & 0 \\ 0 & 0 & 1 & 0 & 0 & 0 \\ 4 & 3 & -5 & 5 & -5 & 5 \\ -4 & -4 & 4 & -5 & 5 & -5 \\ -3 & -5 & 5 & -1 & 1 & -1 \end{bmatrix}.$$

Multiplying the fourth column by $5^{-1} = -2$ and reducing, we obtain

$$L = \begin{bmatrix} 0 & 0 & 0 & 0 & 0 & 0 \\ 0 & 1 & 0 & 0 & 0 & 0 \\ 0 & 0 & 1 & 0 & 0 & 0 \\ 0 & 0 & 0 & 1 & 0 & 0 \\ 0 & -1 & -1 & -1 & 0 & 0 \\ 0 & 0 & 4 & 2 & 0 & 0 \end{bmatrix}$$

which is a matrix in triangular idempotent form. Since

$$I - L = \begin{bmatrix} 1 & 0 & 0 & 0 & 0 & 0 \\ 0 & 0 & 0 & 0 & 0 & 0 \\ 0 & 0 & 0 & 0 & 0 & 0 \\ 0 & 0 & 0 & 0 & 0 & 0 \\ 0 & 1 & 1 & 1 & 1 & 0 \\ 0 & 0 & -4 & -2 & 0 & 1 \end{bmatrix}$$

a basis for W will be

$$\mathbf{v}^{[1]} = (1, 0, 0, 0, 0, 0), \quad \mathbf{v}^{[2]} = (0, 1, 1, 1, 1, 0), \quad \mathbf{v}^{[3]} = (0, 0, -4, -2, 0, 1).$$

In terms of polynomial representation, the basis is

$$v^{[1]}(x) = 1, \quad v^{[2]}(x) = x^4 + x^3 + x^2 + x \text{ and } v^{[3]}(x) = x^5 - 2x^3 - 4x^2.$$

●

Example 8.10. In this example we finish factoring the polynomial $a(x)$ of Example 8.7 using Berlekamp's algorithm. Since there are three basis vectors for W, it follows that $a(x)$ factors into three irreducible factors (it is easy to check that $a(x)$ is square-free, hence the algorithms of this section are applicable). To complete the algorithm for $a(x)$ it remains to calculate the GCD's. For this we have

$$\text{GCD}(a(x), v^{[2]}(x)) = x + 1,$$

so we have one of the factors $a_1(x) = x + 1$. Letting

$$a(x) = \frac{a(x)}{x+1} = x^5 - 4x^4 + 5x^3 + 3x^2 - 4x + 1$$

we then calculate

$$\text{GCD}(a(x), v^{[2]}(x) + 1) = 1, \quad \text{GCD}(a(x), v^{[2]}(x) + 2) = 1,$$
$$\text{GCD}(a(x), v^{[2]}(x) + 3) = 1, \quad \text{GCD}(a(x), v^{[2]}(x) + 4) = 1,$$
$$\text{GCD}(a(x), v^{[2]}(x) + 5) = 1, \quad \text{GCD}(a(x), v^{[2]}(x) + 6) = 1,$$
$$\text{GCD}(a(x), v^{[2]}(x) + 7) = x^3 + 2x^2 + 3x + 4.$$

Since

$$\frac{a(x)}{x^3 + 2x^2 + 3x + 4} = x^2 + 5x + 3$$

the factorization is

$$a(x) = (x + 1)\cdot(x^2 + 5x + 3)\cdot(x^3 + 2x^2 + 3x + 4).$$

Note that if we had found $\text{GCD}(a(x), v^{[2]}(x)+s) = 1$ for all $s \in \mathbf{Z}_{11}$ then we would have repeated the above process with $v^{[3]}(x)$.

●

Theorem 8.9. Suppose that $q = p^m$ is small enough to fit into a single computer word. Then the cost of Berlekamp's algorithm for computing the factors of a polynomial $a(x)$ of degree n in the domain GF(q) is $O(k \cdot q \cdot n^2 + n^3)$. Here k is the number of factors of $a(x)$. On average, k is approximately $\log(n)$ (Knuth [7]).

Proof: The cost of ensuring that $a(x)$ is a square-free polynomial is $O(n^2)$, the cost of a GCD operation. The cost of generating the Q matrix is $q \cdot n^2$ field multiplications, while the cost of determining a basis for the solution space is the cost of Gaussian elimination, that is, $O(n^3)$ field multiplications. Using the method presented, each factor requires q GCD calculations each at an approximate cost of n^2 operations. Hence, this last step requires about $k \cdot q \cdot n^2$ field operations, giving the total cost for the Berlekamp algorithm as $O(k \cdot q \cdot n^2 + n^3)$ field multiplications.

●

The cost of addition and multiplication in the field GF(p^m) is given in Chapter 4. When p fits inside one computer word, addition and multiplication are proportional to m and m^2, respectively. When p requires more than one computer word, addition and multiplication are proportional instead to $m \cdot \log(p)$ and $m^2 \cdot \log^2(p)$, respectively.

8.5. THE BIG PRIME BERLEKAMP ALGORITHM

When q is large, the Berlekamp method from the previous section requires modification to be a viable procedure. For example, when q is large the cost of generating the Q matrix along with calculating the GCD's exhaustively is dominated by $O(q \cdot k \cdot n^2)$. This procedure is useful only for small values of q. For example, to factor a polynomial of degree 100 over the field GF(3^{14}) having four factors requires on the order of 191 billion field operations. In this section we present a modification of the algorithm of the previous section which makes the process feasible for large q.

For large q, the Q matrix may be generated more efficiently by binary powering. To implement binary powering, first store the values

$$x^n \bmod a(x),\ x^{n+1} \bmod a(x), \ldots, x^{2n-2} \bmod a(x).$$

If

$$x^h \equiv r_0 + r_1 x + \cdots + r_{n-1} x^{n-1} \bmod a(x)$$

then x^{2h} may be generated by

$$x^{2h} \equiv x^h \cdot x^h \equiv \hat{r}_0 + \hat{r}_1 x + \cdots + \hat{r}_{2n-2} x^{2n-2} \bmod a(x)$$

where

$$\hat{r}_i = r_i \cdot r_0 + \cdots + r_0 \cdot r_i \quad (\text{with } r_i = 0 \text{ for } i \geq n).$$

Replacing $x^n \bmod a(x), \ldots, x^{2n-2} \bmod a(x)$ by the values already calculated and stored, gives the coefficients for x^{2h}. Using this method, we generate the first row of Q, i.e. $x^q \bmod a(x)$ in $\log(q)$ steps. Thus, the first row is determined in $\log(q) \cdot n^2$ operations, rather than the $q \cdot n$ operations of the preceding section. However, the subsequent rows, i.e. the representations for $x^{2q}, \ldots, x^{nq}$, may all be determined by multiplication by x^q followed by replacement of $x^n \bmod a(x), \ldots, x^{2n-2} \bmod a(x)$. Thus, each subsequent step also requires n^2 (rather than $q \cdot n$) operations. The total number of operations to generate the Q matrix in this manner is $O(\log(q) \cdot n^2 + n^3)$ field operations. This compares to $O(q \cdot n^2)$ operations using the method of the preceding chapter. When $q = 4782969 = 3^{14}$, for example, the cost of generating the matrix Q using binary powering is about $O(23 \cdot n^2 + n^3)$ operations versus $O(4782969 \cdot n^2)$ operations by the previous method. Binary powering in this case is better for all polynomials of degree less than approximately 4782943. Even when $q = 32 = 2^5$, binary powering generates Q more efficiently for all polynomials of degree less than 27.

Although binary powering is, in general, a significant improvement in generating Q for large q, the gains will be lost if we require the exhaustive methods of the previous section to determine the factors since this cost will be $O(k \cdot q \cdot n^2)$ operations. Furthermore, a greatest common divisor operation is most expensive when the input polynomials are relatively prime. Thus those instances where we gain the least amount of information are also the most expensive to calculate.

The initial attempt to reduce the cost of the GCD step of Berlekamp's algorithm was initiated by Zassenhaus (cf.[23]). His approach was to determine the $s \in GF(q)$ which would give nontrivial GCD's, that is, to obtain only those values of GF(q) which would be useful to the factorization process. For a given $v(x)$, define

$$S = \{ \, s \in GF(q) \; : \; GCD(v(x) - s, a(x)) \neq 1 \, \}$$

and define

$$m_v(x) = \prod_{s \in S} (x - s). \tag{8.31}$$

Theorem 8.10 (Zassenhaus). The polynomial $m_v(x)$ defined by (8.31) is the minimal polynomial for $v(x)$. That is, $m_v(x)$ is the polynomial of least degree such that

$$m_v(v(x)) \equiv 0 \mod a(x). \tag{8.32}$$

Proof: For an arbitrary $i \in \{1, \ldots, k\}$, $a_i(x)$ divides $GCD(v(x) - s, a(x))$ for some s in S. Therefore $a_i(x)$ divides a factor of $m_v(v(x))$, and hence $a(x)$ divides $m_v(v(x))$. Thus equation (8.32) holds.

Suppose now that $m_v(x)$ is not the polynomial of least degree satisfying equation (8.32). In particular, suppose that $m(x)$ is a polynomial of smaller degree. Since $m(x)$ has a smaller degree there must exist an s in S such that

$$m(x) = q(x) \cdot (x - s) + r \tag{8.33}$$

where r is a nonzero constant in GF(q). Since s is in S, one of the factors of $a(x)$ divides $v(x) - s$; say the factor is $a_1(x)$. Then $a_1(x)$ divides $m(v(x))$ since $m(v(x)) \equiv 0 \bmod a(x)$. Substituting $v(x)$ for x in equation (8.33) implies that $a_1(x)$ must divide the nonzero constant r, a contradiction.

●

There is a standard way to compute the minimal polynomial for a given $v(x)$. For each r starting at 1, we determine $v(x)^2 \bmod a(x)$, $v(x)^3 \bmod a(x)$, . . . , $v(x)^r \bmod a(x)$ and solve

$$m_0 + m_1 \cdot v(x) + \cdots + m_r v(x)^r \equiv 0 \mod a(x).$$

The first nontrivial solution gives $m(x)$. Of course there must be at least one nontrivial solution for $r \le k \le q$ since $v(x)^q - v(x) \equiv 0 \mod a(x)$ for all the $v(x)$ that are of interest to us. The process is then completed by factoring $m(x)$, with the corresponding GCD's taken at the roots. We remark that there is also an alternate method for calculating $m(x)$ which uses the notion of a resultant (cf. Exercise 8.16).

Example 8.11. Consider $a(x)$ from Example 8.7. From the first two steps of the Berlekamp algorithm we know that there are three irreducible factors, along with a basis for W. Let

$$v(x) = v^{[2]}(x) = x^4 + x^3 + x^2 + x,$$

an element in W. We will determine the minimal polynomial for $v(x)$. Note that

$$v(x)^2 \equiv -2x^5 + 2x^4 - 5x^3 - x^2 + 2x + 5 \mod a(x).$$

However, there are no nonzero solutions of

$$a + b \cdot v(x) + v(x)^2 \equiv 0 \mod a(x)$$

hence the minimal polynomial has degree at least three. Since the first two steps of Berlekamp's method imply that there are exactly three factors, the set S has at most three elements. Therefore the minimal polynomial must have degree exactly three. Calculating

$$a(x)^3 \equiv x^5 - 5x^4 + 4x^3 + 2x^2 - 5x + 3 \mod a(x),$$

and setting up an equation of the form

$$a + b \cdot v(x) + c \cdot v(x)^2 + v(x)^3 \equiv 0 \mod a(x)$$

we find that there is a solution given by $a = 0$, $b = 4$, $c = -5$. The minimal polynomial is

$$m_v(x) = x^3 - 5x^2 + 4x$$

which factors as

$$m_v(x) = x \cdot (x - 1) \cdot (x - 4).$$

Therefore, the set S consists of $\{0, 1, 4\}$ and so we know that only these need to be checked when applying the GCD calculations.

●

The above method does indeed reduce the number of GCD calculations to only those that are necessary for the calculation of the factors of $a(x)$. However, the reduction does not come for free. Generating the minimal polynomial and the subsequent root finding requires substantial computation. The method is feasible, however, if the number of factors, k, is small in comparison to q (where the exhaustive search method is at its worst).

The previous method ensured that the greatest common divisor operation was only undertaken when it was ensured that an irreducible factor would result from this calculation. Subsequently, D. Cantor and H. Zassenhaus [4] have given an efficient algorithm which calculates GCD's even in some cases where success is not assured. The GCD operations that are done turn out to be successful about one-half the time, so about twice the number of such operations might need to be done. However, determining which greatest common divisor pairs are to be used is straightforward and considerably cheaper than the above method.

Central to the approach of Cantor and Zassenhaus is the observation that, for odd q, we have the factorization

$$x^q - x = x\cdot(x^{(q-1)/2} - 1)\cdot(x^{(q-1)/2} + 1) \tag{8.34}$$

and hence any $v(x)$ in W satisfies

$$v(x)\cdot(v(x)^{(q-1)/2} - 1)\cdot(v(x)^{(q-1)/2} + 1) = v(x)^q - v(x) \equiv 0 \mod a(x).$$

The nontrivial common factors of $v(x)^q - v(x)$ and $a(x)$ are then spread out amongst $v(x)$, $(v(x)^{(q-1)/2} - 1)$ and $(v(x)^{(q-1)/2} + 1)$. It is reasonable to expect that almost half the nontrivial common factors of $v(x)^q - v(x)$ and $a(x)$ are either factors of $v(x)^{(q-1)/2} - 1$ or $v(x)^{(q-1)/2} + 1$, since both of these are about half the size of $v(x)^q - v(x)$. Indeed, we have

Theorem 8.11. The probability of $\mathrm{GCD}(v(x)^{(q-1)/2} - 1, a(x))$ being nontrivial is

$$1 - (\frac{q-1}{2q})^k - (\frac{q+1}{2q})^k. \tag{8.35}$$

In particular, the probability is at least 4/9.

Proof: Let

$$(s_1, \ldots, s_k),\ s_i \in \mathrm{GF}(q) \tag{8.36}$$

be the modular representation of $v(x)$, and define

$$w(x) = \mathrm{GCD}(v(x)^{(q-1)/2} - 1, a(x)).$$

Then $w(x)$ is nontrivial if either $w(x) \neq 1$ or $w(x) \neq a(x)$.

Suppose $a_i(x)$ is a factor of $w(x)$. This is equivalent to

$$s_i^{(q-1)/2} = 1,$$

that is, the i-th component of $v(x)$ is a quadratic residue of q. In each component subspace W_i there are q elements, exactly $(q - 1)/2$ of which are quadratic residues, a probability of $(q-1)/2q$. When all the components of (8.36) are equally likely, the probability that an element in W has components all of which are quadratic residues is

$$(\frac{q-1}{2q})^k.$$

Every such element satisfies $w(x) = a(x)$, that is, results in a trivial greatest common divisor. Similarly there are $(q + 1)/2$ non-quadratic residues, hence the probability that an element in W has components none of which are quadratic residues is

$$(\frac{q+1}{2q})^k.$$

Every such element satisfies $w(x) = 1$, that is, results in a trivial greatest common divisor. Thus, the probability that a random $v(x)$ in W has a nontrivial $w(x)$ is (8.35). Expanding (8.35) by using the binomial expansion gives the probability as

$$1 - (\frac{1}{2q})^k \{(q-1)^k + (q+1)^k\}$$
$$= 1 - (\frac{1}{2q})^k \{ 2q^k + 2\binom{k}{2} \cdot q^{k-2} + 2\binom{k}{4} \cdot q^{k-4} + \cdots + 2 \}$$
$$= 1 - \frac{1}{2^{k-1}} \{ 1 + \binom{k}{2} \cdot q^{-2} + \binom{k}{4} \cdot q^{-4} + \cdots + q^{-k} \}$$
$$\geq 1 - \frac{1}{2} \{ 1 + \frac{1}{9} \} = \frac{4}{9}$$

where the last inequality holds because $k \geq 2$ and $q \geq 3$.

●

Note that the statement of Theorem 8.11 could easily have replaced $\text{GCD}(v(x)^{(q-1)/2} - 1, a(x))$ by $\text{GCD}(v(x)^{(q-1)/2} + 1, a(x))$ and be equally valid.

The procedure for determining the factors of $a(x)$ given that a basis $v^{[1]}(x), \ldots, v^{[k]}(x)$ has been determined for W is as follows. Let $v(x) = c_1 \cdot v^{[1]}(x) + \cdots + c_k \cdot v^{[k]}(x)$ with each $c_i \in \text{GF}(q)$ a random element of W. If we have already obtained a partial factorization $a_1(x) \cdots a_m(x) \cdot a_{m+1}(x)$ where m is initially 0, then calculate

$$w(x) = \text{GCD}(v(x)^{(q-1)/2} - 1, a_{m+1}(x)).$$

We know that we will find a nontrivial factor about half of the time. If $w(x)$ is trivial, then randomly pick a new $v(x)$. Otherwise decompose $a_{m+1}(x)$ as

$$a_{m+1}(x) = w(x) \cdot (\, v(x)/w(x)\,)$$

and continue the process until we have determined the k irreducible factors.

Example 8.12. Let $a(x)$ be the polynomial from Example 8.7. Let $v^{[1]}(x)$, $v^{[2]}(x)$ and $v^{[3]}(x)$ be the basis for W determined in Example 8.9. Taking a random set of coefficients from $GF(11)$, we might consider the random element in W given by

$$v(x) = 3v^{[1]}(x) - 2v^{[2]}(x) + 5v^{[3]}(x) = 5x^5 - 2x^4 - x^3 - 2x + 3.$$

Then

$$\text{GCD}(a(x), v(x)^5 - 1) = x^5 - 4x^4 + 5x^3 + 3x^2 - 4x + 1$$

with

$$a(x) = (x+1) \cdot (x^5 - 4x^4 + 5x^3 + 3x^2 - 4x + 1)$$

so $a_1(x) = x + 1$, and we know that the second factor must split into two irreducible components.

Taking another random element of W, say

$$v(x) = 2v^{[1]}(x) + 3v^{[2]}(x) + 4v^{[3]}(x)$$
$$= 4x^5 + 3x^4 - 5x^3 - 2x^2 + 3x + 2$$

we obtain

$$\text{GCD}(x^5 - 4x^4 + 5x^3 + 3x^2 - 4x + 1,\ v(x)^5 - 1) = 1$$

so no information is determined from this choice.

Taking a third random element of W, say

$$v(x) = v^{[1]}(x) + 3v^{[2]}(x) - 4v^{[3]}(x) = -4x^5 + 3x^4 - 3x^2 + 3x + 1$$

we obtain

$$\text{GCD}(x^5 - 4x^4 + 5x^3 + 3x^2 - 4x + 1,\ v(x)^5 - 1) = x^2 + 5x + 3.$$

Since $x^5 - 4x^4 + 5x^3 + 3x^2 - 4x + 1$ reduces to

$$(x^2 + 5x + 3)\cdot(x^3 + 2x^2 + 3x + 4)$$

we obtain our factorization

$$a(x) = (x + 1)\cdot(x^2 + 5x + 3)\cdot(x^3 + 2x^2 + 3x + 4).$$

●

Theorem 8.12. The big prime Berlekamp algorithm for factoring a polynomial $a(x)$ of degree n in the domain GF(q) has complexity $O(k\cdot n^2\cdot\log(q)\cdot\log(k) + 2n^3)$ field operations. As before, k represents the number of factors of $a(x)$, which on average is approximately $\log(n)$.

Proof: The cost of determining the Q matrix using binary powering is just $O(n^2\cdot\log(q) + n^3)$ field operations. Determining the basis for W, that is, determining the solution space for the matrix $Q - I$ adds another $O(n^3)$ operations. Every random choice of $v(x)$ requires a random linear combination of the k basis vectors and hence requires $O(k)$ field operations. If the degree of $a_i(x)$ is r, then it takes $O(r^2\cdot\log(q))$ field operations (using binary powering) to calculate $v(x)^{(q-1)/2} - v(x) \bmod a_i(x)$ and a further $O(r^2)$ operations to calculate the corresponding GCD. There are, on average, approximately $O(\log(k))$ random $v(x)$'s and corresponding GCD computations required. The total complexity is therefore bounded by $O(n^2\cdot\log(q)\cdot\log(k))$ field operations for the splitting step. Thus the total cost is given by $O(n^2\cdot\log(q)\cdot\log(k) + 2n^3)$ field operations.

●

We remark that the term "field operations" appearing in Theorem 8.12 depends on the size of the integer q along with the type of representation used to represent the field GF(q). For example, if the integer q is greater than the word-size of the computer then the average arithmetic cost is usually multiplied by $\log^2(q)$.

The previous method does not work in the case $q = 2^m$, since in this case the factorization (8.34) does not hold. However the polynomial $x^{2^m} - x$ can still be factored into two factors, each of about the same degree. For this we need the trace polynomial defined over

GF(2^m) by

$$Tr(x) = x + x^2 + x^4 + \cdots + x^{2^{m-1}}.$$

Lemma 8.2. The trace polynomial $Tr(x)$ defined on GF(2^m) satisfies:

(a) For any v and w in GF(2^m), $Tr(v + w) = Tr(v) + Tr(w)$;

(b) For any v in GF(2^m), $Tr(v) \in GF(2)$;

(c) $x^{2^m} - x = Tr(x) \cdot (Tr(x) + 1)$.

Proof: Statement (a) follows from

$$\begin{aligned} Tr(v + w) &= (v + w) + (v + w)^2 + \cdots + (v + w)^{2^{m-1}} \\ &= (v + w) + (v^2 + w^2) + \cdots + (v^{2^{m-1}} + w^{2^{m-1}}) \quad \text{(cf. Lemma 8.1)} \\ &= (v + v^2 + \cdots + v^{2^{m-1}}) + (w + w^2 + \cdots + w^{2^{m-1}}) \\ &= Tr(v) + Tr(w). \end{aligned}$$

Statement (b) follows from

$$\begin{aligned} Tr(v)^2 &= (v + v^2 + \cdots + v^{2^{m-1}})^2 \\ &= v^2 + v^4 + \cdots + v^{2^{m-1}} + v^{2^m} \\ &= v^2 + v^4 + \cdots + v^{2^{m-1}} + v = Tr(v). \end{aligned}$$

Therefore, for every v in GF(2^m), $Tr(v)$ is a root of the polynomial $x^2 - x$. Since the elements of the field GF(2) give exactly 2 roots of this polynomial in the field GF(2^m), $Tr(v)$ must be one of these elements.

To show statement (c) let α be any element of GF(2^m). From part (b), $Tr(\alpha) = 0$ or 1, and hence α is a root of $Tr(x) \cdot (Tr(x) + 1)$. Therefore

$$x^{2^m} - x = \prod_{\alpha \in GF(2^m)} (x - \alpha) \qquad \text{divides} \quad Tr(x) \cdot (Tr(x) + 1). \tag{8.37}$$

Since both $Tr(x)$ and $Tr(x) + 1$ have degree 2^{m-1}, the degree of their product is 2^m, the same degree as $x^{2^m} - x$. This fact combined with equation (8.37) proves statement (c).

●

Theorem 8.13. The probability of GCD($Tr(v(x))$,$a(x)$), $v(x) \in$ W being nontrivial is

$$1 - (\frac{1}{2})^{k-1}.$$

In particular, the probability is at least 1/2.

Proof: Let $v(x) \in$ W. Proceeding as in the proof of Theorem 8.11, suppose

$$(s_1, \ldots, s_k)$$

is the unique modular representation of $v(x)$. The linearity statement of Lemma 8.2 implies that

$$Tr(v(x) \bmod a_i(x)) = Tr(v(x)) \bmod a_i(x)$$

and so the unique modular representation for $Tr(v(x))$ is

$$(Tr(s_1), \ldots, Tr(s_k)).$$

Also from Lemma 8.2, we know that for each i, $Tr(s_i)$ is in GF(2), that is, it is either 0 or 1. The condition $Tr(s_i) = 0$ is equivalent to

$$a_i(x) \mid Tr(v(x)).$$

Therefore, GCD($Tr(v(x))$,$a(x)$) will be trivial if either all of the components are 1 (in which case the GCD will be 1) or if all the components are 0 (in which case the GCD will be $a(x)$). Since either case occurs with a probability of $(1/2)^k$, the probability that neither occurs is

$$1 - 2\cdot(\frac{1}{2})^k = 1 - (\frac{1}{2})^{k-1}.$$

●

Example 8.13. Let GF(16) be the field defined by $\mathbf{Z}_2[x]/<x^4 + x + 1>$ and let α be a root of $x^4 + x + 1$. Then every element of GF(16) can be written in the form

$$a_0 + a_1\alpha + a_2\alpha^2 + a_3\alpha^3$$

where each a_i is either 0 or 1. Let $a(x)$ be the degree 5 polynomial in GF(16)[x] defined by

$$a(x) = (1 + \alpha + \alpha^3) + \alpha x + (1 + \alpha^2 + \alpha^3)x^3 + (1 + \alpha + \alpha^3)x^4 + x^5.$$

We will factor $a(x)$ using Berlekamp's algorithm and nondeterministic splitting.

Working out the powers of x gives

$$Q = \begin{bmatrix} 1 & 0 & 0 & 0 & 0 \\ 1+\alpha+\alpha^2+\alpha^3 & 0 & 1+\alpha^2 & \alpha^2 & \alpha^2 \\ \alpha+\alpha^2 & \alpha & 1+\alpha^2 & \alpha+\alpha^2+\alpha^3 & 1+\alpha+\alpha^2+\alpha^3 \\ \alpha+\alpha^2+\alpha^3 & 1+\alpha+\alpha^2+\alpha^3 & \alpha^2 & \alpha+\alpha^3 & 1 \\ 0 & \alpha & 1+\alpha+\alpha^3 & \alpha^2 & \alpha+\alpha^2+\alpha^3 \end{bmatrix},$$

while calculation of the solution space of $Q - I$ gives a basis for W as

$$v^{[1]}(x) = 1,\ v^{[2]}(x) = (\alpha+\alpha^2)\, x + (1+\alpha+\alpha^2)\, x^2 + (1+\alpha+\alpha^2+\alpha^3)\, x^3 + x^4.$$

Since the dimension of the solution space is 2, we know that $a(x)$ splits into two irreducible factors. A random linear combination of the basis elements might then look like

Algorithm 8.7. Big Prime Berlekamp Factoring Algorithm.

procedure BigPrimeBerlekamp($a(x)$,p^m)

\# Given a square-free polynomial $a(x) \in \mathrm{GF}(p^m)[x]$
\# calculate irreducible factors $a_1(x), \ldots, a_k(x)$ such
\# that $a(x) = a_1(x) \cdots a_k(x)$ using the big prime
\# variation of Berlekamp's algorithm.

$Q \leftarrow$ BinaryPoweringFormQ($a(x)$)
$\mathbf{v}^{[1]}, \mathbf{v}^{[2]}, \ldots, \mathbf{v}^{[k]} \leftarrow$ NullSpaceBasis($Q - I$)

\# Note: we can ensure that $\mathbf{v}^{[1]} = (1, 0, \ldots, 0)$.
factors $\leftarrow \{\ a(x)\ \}$

while SizeOf(*factors*) $< k$ **do** {

foreach $u(x) \in$ *factors* **do** {
$(c_1, \ldots, c_k) \leftarrow$ RandomCoefficients(GF(p^m))
$v(x) \leftarrow c_1 v^{[1]}(x) + \cdots + c_k v^{[k]}(x)$

if p=2 **then**
$v(x) \leftarrow v(x) + v(x)^2 + \cdots + v(x)^{2^{m-1}}$
else $v(x) \leftarrow v(x)^{(p^m-1)/2} - 1 \mod u(x)$
$g(x) \leftarrow$ GCD($v(x)$,$u(x)$)

if $g(x) \neq 1$ **and** $g(x) \neq u(x)$ **then** {
Remove($u(x)$, *factors*)
$u(x) \leftarrow u(x)/g(x)$
Add($\{u(x), g(x)\}$,*factors*)

if SizeOf(*factors*) $= k$ **then return**(*factors*) } } }

end

$$v(x) = \alpha^3 \cdot v^{[1]}(x) + (\alpha^2 + \alpha^3) \cdot v^{[2]}(x)$$
$$= \alpha^3 + (\alpha + \alpha^2 + \alpha^3)\, x + \alpha\, x^2 + \alpha^3\, x^3 + (\alpha^2 + \alpha^3)\, x^4.$$

Taking the trace gives

$$Tr(v(x)) = v(x) + v(x)^2 + v(x)^4 + v(x)^8$$
$$= (\alpha+\alpha^2+\alpha^3) + x + (\alpha+\alpha^2)\, x^2 + (1+\alpha+\alpha^3)\, x^3 + (1+\alpha+\alpha^2)\, x^4$$

with the greatest common divisor working out to

$$\mathrm{GCD}(a(x), Tr(v(x))) = (1+\alpha^2) + (\alpha^2+\alpha^3)\,x + x^2.$$

This gives one of the proper factors, with division giving the other. The final factorization for $a(x)$ is then

$$(\,(1+\alpha^2) + (\alpha^2+\alpha^3)\,x + x^2)\cdot(\,(1+\alpha^3) + (\alpha+\alpha^3)\,x + (1+\alpha+\alpha^2)\,x^2 + x^3\,).$$

●

We remark that the algebraic manipulation involved in computing Example 8.13 would be extremely difficult to do without the assistance of a computer algebra system. In our case, Example 8.13 was completed with the help of the MAPLE system.

8.6. DISTINCT DEGREE FACTORIZATION

Given a square-free polynomial $a(x) \in \mathrm{GF(q)}$, q a power of a prime, the distinct degree factorization method first obtains a partial factorization

$$a(x) = \prod a_i(x) \tag{8.38}$$

where each $a_i(x)$ is the product of all irreducible factors of $a(x)$ of degree i. The process then uses a second technique to split the individual $a_i(x)$ into their irreducible components of degree i.

To obtain the partial factorization (8.38), we use

Theorem 8.14. The polynomial $p_r(x) = x^{q^r} - x$ is the product of all monic polynomials over $\mathrm{GF}(q)$ whose degree divides r.

Proof: Let $m(x)$ be an irreducible polynomial of degree d in $\mathrm{GF}(q)[x]$ with $d \mid r$, say $r = d \cdot s$. Consider the residue ring

$$\mathrm{F} = \mathrm{GF}(q)[x] \,/\, \langle m(x) \rangle. \tag{8.39}$$

Since $m(x)$ is irreducible, F is a field. Since $m(x)$ has degree d, the field F has q^d elements. From Fermat's little theorem, every $f \in \mathrm{F}$ satisfies

$$f^{q^d} = f$$

and so also

$$f^{q^r} = ((f^{q^d})^{q^d} \cdots)^{q^d} = f. \tag{8.40}$$

In particular consider $f = [x]$, the representative of the polynomial x in F. Equation (8.40) implies

$$[p_r(x)] = [x^{q^r} - x] = [x]^{q^r} - [x] = f^{q^r} - f = 0 \tag{8.41}$$

in F. Equation (8.41) is equivalent to

$$p_r(x) \equiv 0 \mod m(x),$$

that is, $m(x)$ divides $p_r(x)$.

To show the converse, suppose $m(x)$ is irreducible of degree d and divides $p_r(x)$ where $d > r$. Then, with the field F as given by (8.39), we have

$$[x]^{q^r} - [x] = [p_r(x)] = 0 \qquad (8.42)$$

in F. Let

$$f = [f_0 + f_1 \cdot x + \cdots + f_{d-1} \cdot x^{d-1}]$$
$$= f_0 + f_1 \cdot [x] + \cdots + f_{d-1} \cdot [x]^{d-1}, \quad f_i \in \mathrm{GF(q)}$$

be an arbitrary element of F. Using equation (8.42) and Lemma 8.1, we have

$$f^{q^r} = \{ f_0 + f_1 \cdot [x] + \cdots + f_{d-1} \cdot [x]^{d-1} \}^{q^r}$$
$$= f_0^{q^r} + f_1^{q^r} \cdot [x]^{q^r} + \cdots + f_{d-1}^{q^r} \cdot [x]^{q^r \cdot d-1}$$
$$= f_0^{q^r} + f_1^{q^r} \cdot [x] + \cdots + f_{d-1}^{q^r} \cdot [x]^{d-1}$$
$$= f_0 + f_1 \cdot [x] + \cdots + f_{d-1} \cdot [x]^{d-1} = f.$$

But then every element of F is a root of the polynomial $p_r(x)$. Since there are more elements in F than the degree of $p_r(x)$ we obtain a contradiction.

●

Therefore to determine, for example, $a_1(x)$, the product of all the linear factors of $a(x)$, one calculates

$$a_1(x) = \mathrm{GCD}(a(x), x^q - x),$$

and setting

$$a(x) = a(x)/a_1(x)$$

leaves a square-free polynomial having no linear factors. Similarly,

$$a_2(x) = \mathrm{GCD}(a(x), x^{q^2} - x),$$

and we continue the process to obtain $a_3(x)$, and so on. When $a(x)$ is of degree n, we need only search for factors of degree at most $n/2$, at which point if the polynomial remaining is nontrivial it will in fact be irreducible. The partial factorization algorithm is given by Algorithm 8.8.

Example 8.14. Let

$$a(x) = x^{63} + 1 \in \mathrm{GF}(2)[x] \ .$$

Then

$$a_1(x) = \mathrm{GCD}(a(x), x^2 - x) = x + 1,$$

and we set

$$a(x) = \frac{a(x)}{a_1(x)} = \frac{x^{63}+1}{x+1} = x^{62} + x^{61} + \cdots + x^2 + x + 1.$$

Therefore

$$a_2(x) = \mathrm{GCD}(a(x), x^4 - x) = x^2 + x + 1,$$

with

$$a(x) = \frac{a(x)}{a_2(x)} = x^{60} + x^{57} + x^{54} + \cdots + x^6 + x^3 + 1.$$

The product of the irreducible factors of degree 3 is then given by

$$a_3(x) = \mathrm{GCD}(a(x), x^8 - x) = x^6 + x^5 + x^4 + x^3 + x^2 + x + 1,$$

with the updated $a(x)$ given by

$$x^{54} + x^{53} + x^{51} + x^{50} + x^{48} + x^{46} + x^{45} + x^{42} + x^{33} + x^{32} + x^{30} + x^{29} + x^{27}$$
$$+ x^{25} + x^{24} + x^{22} + x^{21} + x^{12} + x^{11} + x^9 + x^8 + x^6 + x^4 + x^3 + x + 1.$$

Calculating further, we obtain

$$\mathrm{GCD}(a(x), x^{16} - x) = 1,\ \ \mathrm{GCD}(a(x), x^{32} - x) = 1,\ \text{and}$$
$$\mathrm{GCD}(a(x), x^{64} - x) = a(x);$$

hence over GF(2)[x] we have the partial factorization

$$x^{63} + 1 = (x+1)\cdot(x^2+x+1)\cdot(x^6+x^5+x^4+x^3+x^2+x+1)\cdot a_6(x)$$

where $a_6(x)$ is the polynomial of degree 54 given above. Therefore $x^{63}+1$ has one irreducible factor of degree 1, $x+1$, one irreducible factor of degree 2, x^2+x+1, two irreducible factors of degree 3 and nine irreducible factors of degree 6.

●

The GCD calculations are done by first reducing $x^{q^j} - x$ modulo $a(x)$ for each j. A natural way to reduce x^{q^j} modulo $a(x)$ is by taking the q-th power of $x^{q^{j-1}}$ mod $a(x)$, that is,

$$(x^{q^j} - x) \bmod a(x) \equiv (\, x^{q^{j-1}} \bmod a(x)\,)^q - x \bmod a(x).$$

For large integers q, a better method pointed out by Lenstra[10] is to calculate the Q matrix from the previous section. For any polynomial $v(x)$ in the ring $\mathrm{V} = \mathrm{GF}(q)/\langle a(x)\rangle$ we then have

$$\mathbf{v}\cdot Q \equiv \mathbf{v}^q \bmod a(x)$$

(cf. Exercise 8.13). Here $\mathbf{v}$ is the vector $(v_0, \ldots, v_{n-1})$ where $v_0 + v_1 x + \cdots + v_{n-1}x^{n-1}$ is the unique representative of $v(x)$ in the residue ring V of smallest degree; $\mathbf{v}^q$ is the corresponding vector of coefficients of the representative of $v(x)^q$ in V.

Part one of the distinct degree algorithm gives the product of all irreducible factors of each degree i. From this partial factorization, we can easily deduce the number of irreducible factors of degree i. If all the irreducible factors of $a(x)$ have distinct degrees, the previous method gives the complete factorization. However, when there is more than one

Algorithm 8.8. Distinct Degree Factorization (Part 1: Partial Factorization).

procedure PartialFactorDD($a(x)$,q)

\# Given a square-free polynomial $a(x)$ in GF(q)[x],
\# we calculate the partial distinct degree factorization
\# $a_1(x) \cdots a_d(x)$ of $a(x)$.

$i \leftarrow 1;\quad w(x) \leftarrow x;\ a_0(x) \leftarrow 1$

while $i \le$ degree($a(x)$)/2 **do** {
  $w(x) \leftarrow w(x)^q \bmod a(x)$
  $a_i(x) \leftarrow$ GCD($a(x)$,$w(x) - x$)
  if $a_i(x) \ne 1$ **then** {
    $a(x) \leftarrow a(x)/a_i(x)$
    $w(x) \leftarrow w(x) \bmod a(x)$ }
  $i \leftarrow i+1$ }
return($a_0(x) \cdots a_{i-1}(x)a(x)$)

end

irreducible factor of degree i say, there remains the problem of splitting the i-th factor, $a_i(x)$, into its irreducible factors.

Thus suppose that $a_i(x)$ has degree greater than i. When q is odd, the method of Cantor and Zassenhaus can again be used. Let $v(x)$ be any polynomial. Since $v(x)^{q^i} - v(x)$ is a multiple of all irreducible polynomials of degree i, it follows that $a_i(x)$ factors as

$$\mathrm{GCD}(a_i(x),v(x))\cdot\mathrm{GCD}(a_i(x),v(x)^{(q^i-1)/2} - 1)\cdot\mathrm{GCD}(a_i(x),v(x)^{(q^i-1)/2} + 1).$$

As was the case with the GCD calculations using the Cantor-Zassenhaus method in the big prime Berlekamp algorithm, $\mathrm{GCD}(a_i(x),v(x)^{(q^i-1)/2} - 1)$ is nontrivial approximately half the time, as long as $v(x)$ is chosen to have degree at most $2\cdot i - 1$.

Example 8.15. Let

$$a(x) = x^{15} - 1 \in \mathrm{GF}(11)[x].$$

Applying the first part of the distinct degree algorithm, we obtain the partial factorization

$$a(x) = a_1(x)\cdot a_2(x) = (x^5 - 1)\cdot(x^{10} + x^5 + 1).$$

Therefore $a(x)$ has five linear factors and five irreducible quadratic factors. To complete the factorizations, we apply the Cantor-Zassenhaus method and obtain for our first random try

$$\mathrm{GCD}(a_1(x),\ (x+4)^5 - 1) = x^2 + 5x + 5.$$

Since

$$a_1(x) = (x^2 + 5x + 5)\cdot(x^3 - 5x^2 - 2x + 2)$$

using $x+8$ as our next random choice gives

$$\mathrm{GCD}(x^2 + 5x + 5,\ (x+8)^5 - 1) = x - 1,$$

with

$$x^2 + 5x + 5 = (x - 1)\cdot(x - 5),$$

and

$$\mathrm{GCD}(x^3 - 5x^2 - 2x + 2,\ (x+8)^5 - 1) = x - 4,$$

with

$$x^3 - 5x^2 - 2x + 2 = (x - 4)\cdot(x^2 - x + 5).$$

Continuing with this method we obtain

$$a_1(x) = (x - 1)\cdot(x - 3)\cdot(x - 4)\cdot(x - 5)\cdot(x + 2).$$

Splitting $a_2(x)$ follows the same random pattern, with

$$\mathrm{GCD}(a_2(x),\ (x+2)^{60} - 1) = x^6 + 3x^5 + 4x^4 - 2x^3 + 5x^2 + 4x - 2$$

and

$$a_2(x) = (x^6 + 3x^5 + 4x^4 - 2x^3 + 5x^2 + 4x - 2)\cdot(x^4 - 3x^3 + 5x^2 - x + 5).$$

To find the factors of degree 2, we try

$$\mathrm{GCD}(x^4 - 3x^3 + 5x^2 - x + 5,\ (x+7)^{60} - 1) = x^2 + 3x - 2,$$

and

$$\mathrm{GCD}(x^6+3x^5+4x^4-2x^3+5x^2+4x-2,(x+7)^{60} - 1) = x^4+2x^3+x^2-5x-2.$$

Since

$$x^4 - 3x^3 + 5x^2 - x + 5 = (x^2 + 3x - 2)\cdot(x^2 + 5x + 3)$$

and

$$x^6+3x^5+4x^4-2x^3+5x^2+4x-2 = (x^4 + 2x^3 + x^2 - 5x - 2)(x^2 + x + 1)$$

we have three three irreducible factors of degree 2 so far. Continuing the factorization using $x^4 + 2x^3 + x^2 - 5x - 2$ we finally end up with

$$(x^2 + 3x - 2)\ (x^2 + 5x + 3)(x^2 + 4x + 5)\cdot(x^2 - 2x + 4)\ (x^2 + x + 1)$$

as our factorization for $a_2(x)$. Thus we obtain a complete factorization for $x^{15} - 1$ in $\mathrm{GF}(11)[x]$.

●

Algorithm 8.9. Distinct Degree Factorization (Part II: Splitting Factors).

procedure SplitDD($a(x), n, p^m$)

\# We assume that $a(x)$ is a polynomial in GF(p^m),
\# made up of factors all of degree n. We split $a(x)$
\# into its complete factorization via Cantor-Zassenhaus method

if $\deg(a,x) \le n$ **then return**($\{a(x)\}$)

\# each factor has degree given by:
$m \leftarrow \deg(a(x),x)/\, n$
factors $\leftarrow \{a(x)\}$

while SizeOf(*factors*) $< m$ **do** {
  $v(x) \leftarrow$ RandomPoly(degree=$2n - 1$)
  if p=2 **then**
    $v(x) \leftarrow v(x) + v(x)^2 + \cdots + v(x)^{2^{n \cdot m - 1}}$
  else
    $v(x) \leftarrow v(x)^{(q^n-1)/2} - 1$
  $g(x) \leftarrow$ GCD($a(x)$,$v(x)$)
  if $g(x) \ne 1$ **and** $g(x) \ne a(x)$ **then**
  factors $\leftarrow$ SplitDD($g(x), n, p^m$) $\cup$ SplitDD($a(x)/g(x), n, p^m$) }
return(*factors*)

end

As was the case in the previous section, the non-deterministic splitting method requires alternate methods in the case where q is even, that is, when $q = 2^m$ for some m. Lemma 8.2 gives a factorization of the form

$$x^{q^i} - x = x^{2^{m \cdot i}} - x = Tr(x) \cdot (Tr(x) + 1)$$

where $Tr(x)$ is the polynomial of degree $2^{m \cdot i - 1}$ defined as in the previous section. Therefore, for an arbitrary polynomial $v(x)$ of degree at most $2i - 1$, we have

$$a_i(x) = \mathrm{GCD}(a_i(x), Tr(v(x))) \cdot \mathrm{GCD}(a_i(x), Tr(v(x)) + 1).$$

By calculating

$$\mathrm{GCD}(a_i(x), Tr(v(x)))$$

for a random $v(x)$ we will obtain an irreducible factor of $a_i(x)$ with a probability of $1 - (1/2)^{r-1}$ where r is the number of irreducible factors of $a_i(x)$.

8.7. FACTORING POLYNOMIALS OVER THE RATIONALS

Consider now the problem of factoring a polynomial in the domain $\mathbf{Q}[x]$. Multiplication by the LCM of the denominators converts this to a factorization problem in the domain $\mathbf{Z}[x]$. Let $a(x) \in \mathbf{Z}[x]$ and let p be a prime which does not divide the leading coefficient of $a(x)$. Previous sections show how to factor the modular polynomial $a(x) \pmod p$ in the domain $\mathbf{Z}_p[x]$. If $a(x)$ is irreducible in $\mathbf{Z}_p[x]$ then $a(x)$ is also irreducible in $\mathbf{Z}[x]$. Otherwise if $a(x) = u_p(x) \cdot v_p(x) \pmod p$ is a factorization in this domain with $\mathrm{GCD}(u_p(x), v_p(x)) = 1$, then this pair of factors can be lifted to a possible factorization of $a(x)$ in $\mathbf{Z}[x]$ using the Hensel method of Chapter 6. If the $u_p(x)$ and $v_p(x)$ are irreducible in $\mathbf{Z}_p[x]$ then $a(x)$ can have at most two factors in $\mathbf{Z}[x]$. However, in general, this is not the case and we must determine both the number of factors of $a(x)$ and how to combine the lifted factors into true factors. This process results in an exponential combinatorial problem that must somehow be resolved.

Example 8.16. Let

$$a(x) = x^{16} + 11x^4 + 121 .$$

Factoring this polynomial in the domain $\mathbf{Z}_{13}[x]$ gives

$$a(x) = u_1(x) \cdot u_2(x) \cdot u_3(x) \cdot u_4(x) \cdot u_5(x) \cdot u_6(x) \pmod{13}$$

where u_1 and u_2 have degree 2, and the other four factors have degree 3. This results in 37 possible distinct factor pairings which could be tried for the initial lifting process.

●

For a given prime, let D_p be the set of the degrees of the factors in the pairings in the mod p reductions. Because of symmetry we need only consider those degrees which are less than or equal to one half the degree of the polynomial. Thus, in the case of Example 8.16, each pairing must have at least one factor whose degree is in the set

$$D_{13} = \{ 2, 3, 4, 5, 6, 7, 8 \} .$$

One method of reducing the number of combinations which need be tried when lifting modular factors and then determining true factorizations is the modular reduction of not one, but a number of primes. The prime with the least number of pairings would then be used in the lifting stage. It is also possible to reduce the number of possible pairings by comparing the degree sets of the various modular reductions.

Example 8.17. Let $a(x)$ be the polynomial from Example 8.16. Reducing mod 23 factors $a(x)$ into eight irreducible factors, each of degree 2. This gives 162 possible distinct pairings to try in the initial lifting stage, with the degree set given by

$$D_{23} = \{ 2, 4, 6, 8 \}$$

This implies that the possible degree set of $a(x)$ must be in the set

$$D_{13,23} = \{\ 2, 4, 6, 8\ \}.$$

Therefore 12 possible pairings are eliminated from the mod 13 reduction of Example 8.16 leaving only 25 to consider. In this case the reductions mod 13 are more useful because of the smaller number of polynomials.

●

Thus we see that reducing mod p a number of times can reduce the number of possible pairings which need to be lifted. This is often a useful method even when there is only a single possible pairing to choose from.

Example 8.18. Again let $a(x)$ be the polynomial from Example 8.16. Then reduction mod 5 results in 2 factors of degree 4 and 12, respectively. Thus we have the degree set

$$D_5 = \{\ 4\ \}.$$

There is only one possible pairing which needs to be lifted. Reduction mod 31 gives two factors each of degree 8 for a degree set

$$D_{31} = \{\ 8\ \}$$

hence a combined degree set is given by

$$D_{5,31} = \{\ \},$$

that is, there are no possible factors of $a(x)$.

●

Determining the number of modular reductions is commonly based on heuristics using a measure of the average number of factors per reduction (determined from the degree of $a(x)$). The modular reductions can continue until the number of reductions exceeds this measure or until the degree analysis stops refining the degree set, whichever comes last. Unfortunately it is also possible that degree analysis will not gain anything. Indeed there are examples of polynomials, known as the Swinnerton-Dyer polynomials, which factor into linear and quadratic factors for every mod p reduction. The polynomial

$$a(x) = x^8 - 40x^6 + 352x^4 - 960x^2 + 576$$

is such an example.

Let p be a particular prime which has been chosen for use with Hensel lifting. Let

$$a(x) \equiv u_1^{(1)}(x) \cdots u_r^{(1)}(x) \pmod{p}$$

be a factorization in $\mathbf{Z}_p[x]$ into irreducible factors. Then for any positive integer k one can simultaneously construct polynomials $u_i^{(k)}(x)$ for $1 \le i \le r$ satisfying

$$a(x) \equiv u_1^{(k)}(x) \cdots u_r^{(k)}(x) \pmod{p^k}$$

and

$$u_i^{(k)}(x) \equiv u_i^{(1)}(x) \pmod{p} \quad i = 1, \ldots, r.$$

Such a process is called parallel Hensel lifting [21] and some of the details were discussed at the end of Chapter 6. Once p^k is larger than a bound on the size of the coefficients which could appear in the factors, we begin the combination step to find true factors. Each such attempt involves multiplying a set of polynomials together to form a single potential factor and then dividing this into $a(x)$. It is often possible to reduce the number of combinations without doing the costly polynomial division and multipications. One such method is to consider the trailing coefficients of the factors. For a given combination of factors, one can multiply the trailing coefficients together and then divide this into the trailing coefficient of $a(x)$. If $a(x)$ is monic and the trailing term division is not exact then the given combination of factors can be eliminated from consideration without the polynomial operations. When $a(x)$ is not monic such an approach requires some modifications (see Exercise 8.21).

Example 8.19. Let

$$a(x) = x^8 + 4x^7 - 2x^6 - 20x^5 + 3x^4 + 44x^3 + 22x^2 - 4x + 34.$$

A degree analysis using the primes 3, 5, 7, 11, 13 , 17 and 19 yields the degree set

$$D_{3,5,7,11,13,17,19} = \{0, 2, 4\}$$

with $p = 19$ the best prime to use for the lifting step. In this case the lifting will continue until the modulus is at least

$$\| a(x) \|_\infty \cdot 2^{\deg(a(x))} = 44 \cdot 2^8 = 11264$$

which implies that a factorization be lifted to $\mathbf{Z}_{19^4}[x] = \mathbf{Z}_{130321}[x]$. Doing this via parallel Hensel lifting gives the factors as

$$u_1^{(4)}(x) = x^2 + 30024x - 30026, \quad u_2^{(4)}(x) = x^2 - 30022x - 60049,$$

$$u_3^{(4)}(x) = x^2 - 30028x + 30026, \quad u_4^{(4)}(x) = x^2 + 30030x + 60055.$$

Combining trailing coefficients, reducing mod 130321 and dividing into the trailing coefficient of $a(x)$ reduces the number of possible factors to

$$u_1^{(4)}(x) \cdot u_3^{(4)}(x) \text{ and } u_2^{(4)}(x) \cdot u_4^{(4)}(x).$$

Expanding these combinations and dividing into $a(x)$ yields the complete factorization

$$a(x) = (x^4 + 4x^3 + 6x^2 + 4x + 2) \cdot (x^4 - 8x^3 + 24x^2 - 32x + 17).$$

●

Multivariate Polynomial Factorization

The problem of factoring a multivariate polynomial in $\mathbf{Q}[x_1, \ldots, x_v]$, or equivalently in $\mathbf{Z}[x_1, \ldots, x_v]$, can now be solved using the various tools which have been developed. An overview of the method is described by the homomorphism diagram in Figure 6.2 of Chapter

6. At the base of the homomorphism diagram is the problem of factoring a univariate polynomial over a finite field, a task which has been described earlier in this chapter. The lifting
process to construct the desired factorization in the original multivariate domain can be
accomplished by the univariate and multivariate Hensel lifting algorithms developed in
Chapter 6.

Note that the polynomial may factor into more factors in an image domain than in the
original multivariate domain, and hence the issue of combining image factors arises. The
discussion above indicates how this problem would be resolved.

In order to invoke the Multivariate Hensel Lifting Algorithm (Algorithm 6.4), it will be
noted that this algorithm requires as input the list of *correct multivariate leading coefficients*.
Specifically, suppose that our task is to factor $a(x_1, \ldots, x_v) \in \mathbf{Z}[x_1, \ldots, x_v]$, x_1 is chosen as
the main variable, and we have determined that, modulo an evaluation homomorphism, the
polynomial factorization is

$$a(x_1, \ldots, x_v) \equiv u_1(x_1) \cdots u_n(x_1) \in \mathbf{Z}[x_1].$$

The leading coefficient of the original polynomial $a(x_1, \ldots, x_v)$, viewed as a polynomial in
the main variable x_1, is a multivariate polynomial in the other $v-1$ variables, in general. The
leading coefficient problem, which was discussed in detail in Chapter 6 for the univariate
case, arises equally in the case of multivariate Hensel lifting. Namely, during Hensel lifting
the leading coefficients of the factors may not be updated properly. However, in the same
manner as was discussed for the univariate case, if we can specify the correct leading coefficients then by forcing the correct leading coefficient onto each factor, the factors will be
lifted properly. This is what Algorithm 6.4 does, hence why the list of correct leading coefficients is required as input to that algorithm.

How can we determine the list of correct leading coefficients for the factors? A method
for this was presented by Wang [20] as follows. Suppose that the original polynomial is of
degree d in x_1 and let

$$a(x_1, \ldots, x_v) = a_d(x_2, \ldots, x_v)x_1^d + \cdots .$$

We first factor the leading coefficient $a_d(x_2, \ldots, x_v)$, which implies a recursive invocation
of our multivariate factorization algorithm. The problem now is to correctly distribute the
factors of the leading coefficient to the polynomial factors $u_1(x_1), \ldots, u_n(x_1)$. Wang's
suggestion is that when choosing the set of evaluation points $\alpha_2, \ldots, \alpha_v \in \mathbf{Z}$ for the variables $x_2, \ldots, x_v$, ensure that the following three conditions hold:

1) $a_d(\alpha_2, \ldots, \alpha_v)$ does not vanish;

2) $a(x_1, \alpha_2, \ldots, \alpha_v)$ has no multiple factors;

3) each factor of the leading coefficient $a_d(x_2, \ldots, x_v)$, when evaluated at $\alpha_2, \ldots, \alpha_v$,
   has a prime number factor which is not contained in the evaluations of the other factors.

The third condition above allows us to identify the correct leading coefficients as follows. Namely, factor the leading coefficients of each image factor

$$u_1(x_1), \ldots, u_n(x_1) \in \mathbf{Z}[x_1].$$

The multivariate leading coefficient $a_d(\alpha_2, \ldots, \alpha_v)$, when evaluated, is an integer whose factors make up the leading coefficients of the $u_i(x_1)$. By using the identifying primes of condition (3) and checking which of the multivariate factors of the leading coefficient $a_d(x_2, \ldots, x_v)$, when evaluated, is divisible by each identifying prime, we can attach the correct multivariate leading coefficient to each factor $u_1(x_1), \ldots, u_n(x_1)$. Algorithm 6.4 can then be invoked to lift the univariate factors up to multivariate factors.

8.8. FACTORING POLYNOMIALS OVER ALGEBRAIC NUMBER FIELDS

Given the information of the previous sections along with the concepts of Chapter 6, we are now in a position to factor any polynomial over $\mathbf{Q}[x]$. While this is the most common domain for factoring, the algorithms for symbolic integration (cf. Chapter 11) require that we be able to factor polynomials having coefficients from an algebraic number field.

There are several algorithms for factoring polynomials over algebraic number fields. We choose to present an algorithm due to Trager[17]. This in turn is a variation of an algorithm originally due to Kronecker (cf. van der Waerden[18]).

Preliminary Definitions

An algebraic number field can be specified by $F[x]/<m(x)>$ where F is a field and $m(x)$ is an irreducible polynomial over $F[x]$. If α is a root of $m(x)$ (in a larger field containing F) and $m(x)$ is of degree n then we have the isomorphism

$$F[x]/<m(x)> = F(\alpha)$$
$$= \{ [f_0 + f_1\alpha + \cdots + f_{n-1}\alpha^{n-1}] : f_i \in F\} .$$

Example 8.20. If $F = \mathbf{Q}$ and $m(x) = x^2 - 2$, then

$$\mathbf{Q}[x]/<x^2 - 2> = \mathbf{Q}(\sqrt{2}) = \{a + b\sqrt{2} : a, b \in \mathbf{Q}\}$$

with addition and multiplication given by

$$(a + b\sqrt{2}) + (c + d\sqrt{2}) = (a + c) + (b + d)\sqrt{2},$$
$$(a + b\sqrt{2}) \cdot (c + d\sqrt{2}) = (ac + 2bd) + (ad + bc)\sqrt{2}.$$

●

Let $m(x)$ be the unique, monic minimal polynomial of α over F. The conjugates of α over F are the remaining distinct roots of $m(x) : \alpha_2, \alpha_3, \ldots, \alpha_n$. Thus, for example, $-\sqrt{2}$ is the conjugate of $\sqrt{2}$ over $\mathbf{Q}$. If $\beta \in F(\alpha)$ is represented by

$$\beta = f_0 + f_1\alpha + \cdots + f_{n-1}\,\alpha^{n-1}$$

then the conjugates of β are $\beta_2, \ldots, \beta_n$ where β_i is represented by

$$\beta_i = f_0 + f_1\alpha_i + \cdots + f_{n-1}\,\alpha_i^{n-1}.$$

Note that conjugation induces a series of isomorphisms

$$\sigma_i : \mathrm{F}(\alpha) \rightarrow \mathrm{F}(\alpha_i) \quad \text{where } \sigma_i(\beta) = \beta_i.$$

Fundamental to computations involving algebraic numbers is

Theorem 8.15. An element $\beta \in \mathrm{F}(\alpha_1, \ldots, \alpha_n)$ is in F if and only if it is invariant under all permutations of the α_i.

Proof: By the fundamental theorem of symmetric functions (cf. Herstein [5]), β can be uniquely expressed in terms of the elementary symmetric functions of $\alpha_1, \ldots, \alpha_n$. These in turn are expressed in terms of the coefficients of the minimal polynomial which lie in F. The result follows directly.

●

For polynomial factorization over an algebraic number field we rely heavily on the Norm function defined as

$$\mathrm{Norm}(\beta) = \beta \cdot \beta_2 \cdots \beta_n$$

that is, the Norm of an element β is the product of all its conjugates. It is not hard to prove that Norm(β) is invariant under conjugation for any β, and hence Norm is a function

$$\mathrm{Norm} : \mathrm{F}(\alpha) \rightarrow \mathrm{F}.$$

The norm also has an alternate definition in terms of resultants. If $q(x)$ is monic then we show in Chapter 9 (cf. Theorem 9.3) that (up to sign differences)

$$\mathrm{res}_x(p,q) = \prod_{x\,:\,q(x)=0} p(x).$$

Therefore, for any β represented by a polynomial $b(\alpha)$ we have

$$\mathrm{Norm}\,(\beta) = \mathrm{res}_x(b(x), m(x))$$

where b, m are considered as polynomials in x. We can extend the definition of Norm to include polynomials over $\mathrm{F}(\alpha)$. Let $p \in \mathrm{F}(\alpha)[z]$. Then we can consider p as a bivariate polynomial in the variables α and z. We define

$$\mathrm{Norm}\,(p) = \mathrm{res}_x(p(x,z), m(x))$$

which results in a polynomial in $\mathrm{F}[z]$. We note that a similar approach extends the Norm to multivariate polynomials over $\mathrm{F}(\alpha)$.

A fundamental property of the Norm function is

$$\text{Norm}(b \cdot c) = \text{Norm}(b) \cdot \text{Norm}(c).$$

Thus any polynomial $a(z)$ which factors in $F(\alpha)[z]$ will result in a factorization of Norm(a) in $F[z]$. Trager's algorithm proceeds by reversing this procedure, that is, by first factoring Norm(a) and then lifting the factors from $F[z]$ to factors of $a(z)$ in $F(\alpha)[z]$.

In order to proceed with this approach, we first need to recognize when a given $a(z) \in F(\alpha)[z]$ is irreducible. To this end we have

Theorem 8.16. Suppose $a(z) \in F(\alpha)[z]$ is irreducible over $F(\alpha)$. Then Norm(a) is a power of an irreducible polynomial over F.

Proof: Suppose $\text{Norm}(a) = b(z) \cdot c(z)$ where $b(z)$, $c(z)$ are relatively prime polynomials from $F[z]$. The polynomial $a(z)$ must divide one of $b(z)$ or $c(z)$ in $F(\alpha)[z]$, say $b(z)$ (since $a(z)$ divides Norm(a) in $F(\alpha)[z]$ and $a(z)$ is irreducible in this domain). Thus

$$b(z) = a(z) \cdot d(z)$$

with $d(z) \in F(\alpha)[z]$ and relatively prime to $a(z)$. Taking conjugates gives

$$b(z) = \sigma_i(a(z)) \cdot \sigma_i(d(z))$$

hence $\sigma_i(a(z))$ is a factor of $b(z)$ for all i. But then

$$\text{Norm}(a) = \prod_i \sigma_i(a) \mid b(z)$$

so $c(z) = 1$. This implies Norm $(a) = b(z)$ and $b(z)$ must either be irreducible or a power of an irreducible element.

●

Theorem 8.16 implies that, if our original polynomial $a(z) \in F(\alpha)[z]$ has the property that Norm(a) is a square-free polynomial in $F[z]$ then $a(z)$ will be irreducible if and only if Norm(a) is irreducible. Similarly, if $a(z)$ factors in $F(\alpha)[z]$ as

$$a(z) = a_1(z) \cdots a_k(z) \tag{8.43}$$

with each $a_i(z)$ irreducible in $F(\alpha)[z]$, then

$$\text{Norm}(a) = \text{Norm}(a_1) \cdot \text{Norm}(a_2) \cdots \text{Norm}(a_k) \tag{8.44}$$

with each Norm(a_i) an irreducible polynomial. When Norm(a) is square-free it can not happen that $\text{Norm}(a_i) = \text{Norm}(a_j)$ for some $i \neq j$ since this would result in a repeated factor in Norm(a). In particular there will be a one to one correspondence between the factors of $a(z)$ over $F(\alpha)$ and the factors of Norm(a) over F. The converse of this is given by

Theorem 8.17. Let $a(z)$ be a polynomial in $F(\alpha)[z]$ with the property that Norm(a) is square-free. Let $p_1(z), \ldots, p_k(z)$ be a complete factorization of Norm(a) over F[z]. Then

$$a(z) = \prod_{i=1}^{k} \text{GCD}(a(z), p_i(z))$$

is a complete factorization of $a(z)$ over F(α)[z].

Proof: Suppose (8.43) is a complete factorization of $a(z)$ in F(α)[z] so that (8.44) is a factorization of Norm(a). Thus, for each i, we have

$$p_i(z) = \text{Norm}(a_j) \tag{8.45}$$

for some j. Since Norm(a) is square-free it cannot happen that Norm(a_j) = Norm(a_h) for some $h \neq j$.

We claim that if $a_j(z)$ and $p_i(z)$ are related by (8.45), then

$$a_j(z) = \text{GCD}(a(z), p_i(z)) \tag{8.46}$$

where the GCD is taken over the domain F(α)[z]. By (8.43), (8.45) and the definition of Norm we see that $a_j(z)$ divides both $a(z)$ and $p_i(z)$ in F(α)[z]. The existence of a larger divisor is equivalent to an $a_h(z)$ dividing both $a(z)$ and $p_i(z)$ in F(α)[z] for some $h \neq j$. Since $a_h(z)$ divides $p_i(z)$, we have that

$$\text{Norm}(a_h) \mid \text{Norm}(p_i)\,. \tag{8.47}$$

Since $p_i(z) \in$ F[z], we have

$$\text{Norm}(p_i) = p_i(z)^n. \tag{8.48}$$

Since Norm(a_h) is irreducible, equations (8.47) and (8.48) imply that Norm(a_h) = $p_i(z)$, a contradiction when $h \neq j$. Thus (8.46) holds and our proof is complete

●

Thus, in the case where Norm(a) is square-free, we can factor $a(z)$ by following the arrows

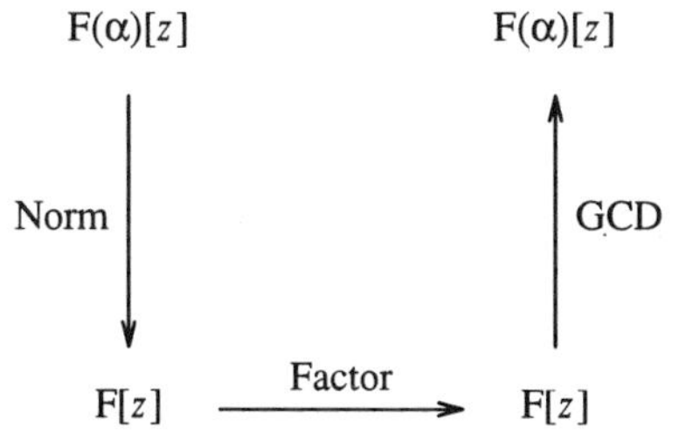

For a given $a(z) \in F(\alpha)[z]$ we can use the previous sections of this chapter to reduce the factoring problem to one where $a(z)$ is already square-free. However the results of this section require, in addition, that Norm(a) be square-free. We accomplish this by finding some s in F such that $b(z) = a(z + s \cdot \alpha)$ is square-free. The polynomial $b(z)$ is then factored using the previous approach into

$$b(z) = b_1(z) \cdots b_k(z).$$

A factorization for $a(z)$ is then given by

$$a(z) = a_1(z) \cdots a_k(z) \quad \text{where} \quad a_i(z) = b_i(z - s \cdot \alpha) .$$

That there always exists an s such that $a(z + s \cdot \alpha)$ is square-free follows from

Theorem 8.18. Let $a(z)$ be a square-free polynomial in $F(\alpha)[z]$. Then Norm($a(z - s \cdot \alpha)$) is square-free except for a finite set of $s \in F$.

Proof: Let

$$\text{Norm}(a) = \prod_{i=1}^{r} p_i(z)^i$$

be the square-free factorization of Norm(a) in $F[z]$. Since $a(z)$ is square-free and divides Norm(a), it will also divide the square-free polynomial

$$p(z) = p_1(z) \quad \cdots \quad p_r(z) .$$

Notice that $p(z) \in F[z]$.

Let the roots of $p(z)$ be $\beta_1, ..., \beta_k$ so that $p(z) = \prod_i (z - \beta_i)$. Since $p(z)$ is square-free, all the β_i are distinct. For any $s \in F$ we have

$$c_s(z) = \text{Norm}(p(z - s \cdot \alpha)) = \prod_j \prod_i (z - (s \cdot \alpha_j + \beta_i)) .$$

The polynomial $c_s(z) \in F[z]$ can have a multiple root in F if and only if

$$s \cdot \alpha_j + \beta_i = s \cdot \alpha_n + \beta_v$$

for some j, i, u, v, that is, if and only if

$$s = \frac{\beta_v - \beta_i}{\alpha_j - \alpha_u} .$$

Thus, for all but a finite number of $s \in F$, we have

$$\text{Norm}(a(z - x \cdot \alpha)) \mid p(z - s \cdot \alpha) \mid c_s(z)$$

with $c_s(z)$ a polynomial in $F[z]$ with no multiple roots. In these cases $a_s(z) = a(z - s \cdot \alpha)$ is a polynomial in $F(\alpha)[z]$ having a square-free norm.

●

The algorithm for polynomial factorization over an algebraic number field is then given by Algorithm 8.10.

Algorithm 8.10. Factorization over Algebraic Number Fields.

```
procedure AlgebraicFactorization(a(z), m(x), α)
  # Given a square-free polynomial a(z) ∈ F(α)[z], α
  # an algebraic number with minimal polynomial m(x)
  # of degree n, we factor a. We consider a as
  # a bivariate polynomial in α and z.

  # Find s such that Norm (a_s(z)) is square-free
  s ← 0; a_s(α,z) ← a(α,z)
  Norm(a_s) ← res_x(m(x),a_s(x,z))

  while deg( GCD(Norm(a_s), Norm(a_s)') ) ≠ 0 do {
      s ← s + 1; a_s(α,z) ← a_s(α,z−α)
      Norm(a_s) ← res_x(m(x),a_s(x,z)) }

  # Factor Norm(a_s) in F[z] and lift results to F(α)[z].
  b ← factors(Norm(a_s))

  if SizeOf(b)=1 then return (a(z))
  else
      foreach a_i(z) ∈ b do {
          a_i(α,z) ← GCD (a_i(z), a_s(α,z))
          a_i(α,z) ← a_i(α,z+s·α)
          substitute(a_i(z) ← a_i(α,z), b) }
  return (b)
end
```

Example 8.21. Let

$$a_\alpha(x) = z^4 + z^3 + (2 + \alpha - \alpha^2)z^2 + (1 + \alpha^2 - 2\alpha^3)z - 2 \in \mathbf{Q}(\alpha)$$

where $\alpha = 3^{1/4}$. Then the minimal polynomial for α is $m(x) = x^4 - 3$ and the norm of $a_\alpha(z)$ is given by

$$\mathrm{Norm}(a_\alpha(z)) = \mathrm{res}_x(f_x(z), m(x))$$

$$= z^{16} + 4z^{15} + 14z^{14} + 32z^{13} + 47z^{12} + 92z^{11} + 66z^{10} + 120z^{9}$$
$$- 50z^8 - 24z^7 - 132z^6 - 40z^5 - 52z^4 - 64z^3 - 64z^2 - 32z + 16.$$

It is easy to check that this norm is squarefree over $\mathbf{Q}[z]$. Factoring the norm over $\mathbf{Q}[z]$, we obtain

$$\mathrm{Norm}(f_\alpha(z)) = g(z) \cdot h(z)$$

where

$$g(z) = z^8 + 4z^7 + 10z^6 + 16z^5 - 2z^4 - 8z^3 - 20z^2 - 8z + 4$$

and

$$h(z) = z^8 + 4z^6 + 9z^4 + 4z^2 + 4 .$$

Taking GCD's in the domain $\mathbf{Q}(\alpha)[z]$, we obtain

$$\mathrm{GCD}(f_\alpha(z), g(z)) = z^2 + (1-\alpha)z + (1-\alpha^2)$$

and

$$\mathrm{GCD}(f_\alpha(z), h(z)) = z^2 + \alpha z + (1+\alpha^2).$$

Thus the factorization of $f_\alpha(z)$ over $\mathbf{Q}(\alpha)[z]$ is

$$f_\alpha(z) = (\, z^2 + (1-\alpha)z + (1-\alpha^2)\,) \cdot (\, z^2 + \alpha z + (1+\alpha^2)\,).$$

●

Algorithm 8.10 has drawbacks in that, although the algorithm maps down to simpler domains to perform its factoring, the resulting polynomials grow quickly both in degree and coefficient size. Obtaining efficient algorithms for factoring polynomials over algebraic number fields is an active area of current research.

Exercises

1. Show that the derivative of a polynomial satisfies linearity, the product rule, the power rule, and the chain rule.

2. Calculate the square-free factorization of
$$x^{10} + 2x^9 + 2x^8 + 2x^7 + x^6 + x^5 + 2x^4 + x^3 + x^2 + 2x + 1 \in \mathbf{Z}[x].$$

3. Calculate the square-free factorization of
$$x^{10} + x^6 + x^5 + x^3 + x^2 + 1 \in \mathbf{Z}_2[x].$$

4. Let $a(x)$ be given symbolically by

$$a = A \cdot B^2 \cdot C^6 \cdot D^{10}$$

where $A, B, C,$ and D are all irreducible polynomials. Apply Algorithm 8.2 to a to obtain its square-free factorization.

5. Show that the running time of Algorithm 8.2 is at most twice the running time required to calculate GCD($a(x)$,$a'(x)$).

6. Generalize Algorithm 8.2 to work for polynomials defined over a Galois field. Apply your algorithm to the symbolic polynomial from Exercise 2 considered as a polynomial over $\mathbf{Z}_2[x]$.

7. Let L be in triangular idempotent form. Show that $L^2 = L$.

8. Determine a basis for the set $\{ \mathbf{v} : \mathbf{v} \cdot M = \mathbf{0} \}$ where

$$M = \begin{bmatrix} 0 & 0 & 0 & 0 & 0 & 0 & 0 & 0 \\ 0 & 1 & 1 & 0 & 0 & 0 & 0 & 0 \\ 0 & 0 & 1 & 0 & 1 & 0 & 0 & 0 \\ 0 & 0 & 0 & 1 & 0 & 0 & 1 & 0 \\ 1 & 0 & 0 & 1 & 0 & 0 & 1 & 0 \\ 1 & 0 & 1 & 1 & 1 & 0 & 0 & 0 \\ 0 & 0 & 1 & 0 & 1 & 1 & 0 & 1 \\ 1 & 1 & 0 & 0 & 1 & 1 & 0 & 1 \end{bmatrix}$$

and where the coefficient arithmetic is in $\mathbf{Z}_2$.

9. Factor the polynomial $a(x) = x^{15} - 1 \in \mathbf{Z}_2[x]$ using Berlekamp's algorithm.

10. Let $a(x) = x^{10} + x^6 = x^5 + x^3 + x^2 + 1 \in \mathbf{Z}_2[x]$. Factor $a(x)$ using Berlekamp's algorithm.

11. Factor the polynomial

$$x^{17} + x^{14} + x^{13} + x^{12} + x^{11} + x^{10} + x^9 + x^8 + x^7 + x^5 + x^4 + x + 1$$

over $\mathbf{Z}_2[x]$.

12. How many factors does $a(x) = x^4 + 1 \in \mathbf{Z}_p[x]$ have when

(a) $p = 2$?

(b) $p \equiv 1 \mod 8$?

(c) $p \equiv 3 \mod 8$?

(d) $p \equiv 5 \mod 8$?

13. Let $a(x)$ be a polynomial of degree n and Q the matrix whose i-th row is the components of $x^{i \cdot q} \mod a(x)$. Let $v(x) = \sum_{j=0}^{n-1} v_j \cdot x^j$ and $\mathbf{v} = (v_0, \ldots, v_{n-1})$. Show that

$$\mathbf{v} \cdot Q = \mathbf{v}^q \mod a(x)$$

where $\mathbf{v}^q = (v_0^q, \ldots, v_{n-1}^q)$.

14. Let $a(x)$ be an *irreducible* polynomial in GF(q)[x] and let Q be its matrix as given in Exercise 13. Show that the characteristic polynomial of Q is given by

$$\det(x\,I - Q) = x^n - 1.$$

Generalize this to the case where $a(x)$ is reducible.

15. Specify an algorithm which implements BinaryPoweringFormQ, that is, the procedure for generating the Q matrix via binary powering.

16. Let $a(x) \in$ GF(q)[x] and $v(x) \in$ W where W is the subspace (8.15). Show that

$$m_v(x) = \text{res}_z(v(z), a(z) - x)$$

where $m_v(x)$ is the minimal polynomial of $v(x)$ (cf. Theorem 8.10) and where the resultant of the two polynomials is with respect to the z variable. Verify this in the case of Example 8.11.

17. Using the distinct degree method, factor the polynomial

$$a(x) = x^8 + x^7 + 2x^6 + 3x^5 + 3x^4 + 3x^3 + 2x^2 + 2x + 1$$

(a) over GF(7),

(b) over GF(19),

(c) over GF(23).

18. What is the complete factorization in **Q**[x] of $a(x) = x^5 + x^4 + x^2 + x + 2$?

19. What is the cost of determining if $a(x) \in$ GF(q)[x] is irreducible if

(a) we use Berlekamp's algorithm?

(b) we use the big prime Berlekamp algorithm?

(c) we use the distinct degree method.

20. Let $q = p^k$ be a power of a prime p. The trace polynomial $Tr(x)$ can be defined over more general Galois fields GF(q^m) by

$$Tr(x) = x + x^q + x^{q^2} + \cdots + x^{q^{m-1}}.$$

(a) For $v, w \in \mathrm{GF}(q^m)$, prove that $Tr(v + w) = Tr(v) + Tr(w)$.

(b) For $c \in \mathrm{GF}(q)$, $v \in \mathrm{GF}(q^m)$, prove that $Tr(c \cdot v) = c \cdot Tr(v)$.

(c) For $v \in \mathrm{GF}(q^m)$, prove that $\mathrm{Tr}(v) \in \mathrm{GF}(q)$.

(d) Prove that

$$x^{q^m} - x = \prod_{s \in \mathrm{GF}(q)} (Tr(x) - s).$$

21. Suppose $a(x) \in \mathbf{Q}[x]$ is not monic. What modifications are required when using trailing coefficients to determine potential factors as suggested for the combination step in Section 8.7?

22. Use Algorithm 8.10 to factor the polynomial

$$x^4 - 2$$

considered as a polynomial over $\mathbf{Q}(\sqrt{2}, \sqrt{-1})$. You should find that you need to factor the polynomial

$$x^{16} - 8x^{14} + 52x^{12} - 264x^{10} + 3078x^8 - 5112x^6 - 6668x^4 + 8968x^2 + 12769$$

over $\mathbf{Q}[x]$.

23. This question determines an easy way to find, for each $\beta \in \mathrm{F}(\alpha)$, α algebraic over F, the minimal polynomial $m_\beta(x)$ for β.

(a) For $\beta \in \mathrm{F}(\alpha)$, let $b(x) = \mathrm{Norm}(x - \beta)$. Show that $b(x) = m_\beta(x)^k$ for some integer k.

(b) Using part (a), determine $m_\beta(x)$.

24. Specify an algorithm which determines the splitting field of a polynomial over $\mathbf{Q}[x]$.

References

1. E.R. Berlekamp, "Factoring Polynomials over Finite Fields," *Bell System Technical Journal*, **46** pp. 1853-1859 (1967).
2. E.R. Berlekamp, *Algebraic Coding Theory*, McGraw-Hill, New York (1968).
3. E.R. Berlekamp, "Factoring Polynomials over Large Finite Fields," *Math. Comp.*, **24** pp. 713-735 (1970).
4. D.G. Cantor and H. Zassenhaus, "A New Algorithm for Factoring Polynomials over a Finite Field," *Math. Comp.*, **36** pp. 587-592 (1981).
5. I.N. Herstein, *Topics in Algebra*, Blaisdell (1964).

6. E. Kaltofen, "Factorization of Polynomials," pp. 95-113 in *Computer Algebra - Symbolic and Algebraic Computation*, ed. B. Buchberger, G.E. Collins and R. Loos, Springer-Verlag (1982).

7. D.E. Knuth, *The Art of Computer Programming, Volume 2: Seminumerical Algorithms (second edition)*, Addison-Wesley (1981).

8. S. Landau, "Factoring Polynomials over Algebraic Number Fields," *SIAM J. of Computing*, **14** pp. 184-195 (1985).

9. D. Lazard, "On Polynomial Factorization," pp. 126-134 in *Proc. EUROCAM '82, Lecture Notes in Computer Science* **144**, Springer-Verlag (1982).

10. A.K. Lenstra, "Factorization of Polynomials," pp. 169-198 in *Computational Methods in Number Theory*, ed. H.W. Lenstra, R. Tijdeman, (1982).

11. A.K. Lenstra, H.W. Lenstra, and L. Lovasz, "Factoring Polynomials with Rational Coefficients," *Math. Ann.*, **261** pp. 515-534 (1982).

12. A.K. Lenstra, "Factoring Polynomials over Algebraic Number Fields," pp. 245-254 in *Proc. EUROCAL '83, Lecture Notes in Computer Science* **162**, ed. H. van Hulzen, Springer-Verlag (1983).

13. R. Lidl and H. Niederreiter, *Introduction to Finite Fields and their Applications*, Cambridge University Press (1986).

14. R.J. McEliece, *Finite Fields for Computer Scientists and Engineers*, Kluwer Academic Publishers (1987).

15. R.T. Moenck, "On the Efficiency of Algorithms for Polynomial Factoring," *Math. Comp.*, **31** pp. 235-250 (1977).

16. M.O. Rabin, "Probabilistic Algorithms in Finite Fields," *SIAM J. Computing*, **9** pp. 273-280 (1980).

17. B. Trager, "Algebraic Factoring and Rational Function Integration," pp. 219-226 in *Proc. SYMSAC '76*, ed. R.D. Jenks, ACM Press (1976).

18. B.L. van der Waerden, *Modern Algebra (Vols. I and II)*, Ungar (1970).

19. P. Wang, "Factoring Multivariate Polynomials over Algebraic Number Fields," *Math. Comp.*, **30** pp. 324-336 (1976).

20. P.S. Wang, "An Improved Multivariate Polynomial Factoring Algorithm," *Math. Comp.*, **32** pp. 1215-1231 (1978).

21. P.S. Wang, "Parallel p-adic Construction in the Univariate Polynomial Factoring Algorithm," *Proc. of MACSYMA Users Conference*, pp. 310-318 (1979).

22. D.Y.Y. Yun, "On Square-Free Decomposition Algorithms," pp. 26-35 in *Proc. SYMSAC '76*, ed. R.D. Jenks, ACM Press (1976).

23. H. Zassenhaus, "Hensel Factorization I," *J. Number Theory*, **1** pp. 291-311 (1969).

CHAPTER 9

SOLVING SYSTEMS OF EQUATIONS

9.1. INTRODUCTION

In this chapter we consider the classical problem of solving (exactly) a system of algebraic equations over a field F. This problem, along with the related problem of solving single univariate equations, was the fundamental concern of algebra until the beginning of the "modern" era (roughly, in the nineteenth century); it remains today an important, widespread concern in mathematics, science and engineering. Although considerable effort has been devoted to developing methods for *numerical* solution of equations, the development of exact methods is also well motivated. Obviously, exact methods avoid the issues of conditioning and stability. Moreover, in the case of nonlinear systems, numerical methods cannot guarantee that all solutions will be found (or prove that none exist). Finally, many systems which arise in practice contain "free" parameters and hence must be solved over non-numerical domains.

We first consider the solution of linear systems. In the present context of exact computation, it is often desirable to imagine the coefficient domain as the quotient field of an integral domain and to organize algorithms so that most of the computation is performed in the integral domain. This aim is motivated by the fact that systems with either rational number or rational function coefficients are arguably most common in practice; and, any system with rational coefficients may be written in terms of integer or integral polynomial coefficients. We therefore develop variants of Gaussian elimination which exploit this view. We then briefly describe a number of other approaches which appear in the literature. Each of the methods discussed may also be applied to the computation of determinants. Familiarity with all of the relevant results of (basic) linear algebra is assumed.

Finally, we discuss the more difficult problem of solving systems of nonlinear equations. This topic is usually covered briefly in basic algebra courses; but, inevitably one encounters only simple systems which may be solved by a combination of substitution, factorization and the known formulae for univariate equations. Although these techniques are extremely useful, the general solution requires a more robust approach. We present a method for nonlinear elimination using the polynomial resultant defined in Chapter 7. It will be useful, for this and later chapters, to develop the theory of resultants and discuss algorithms for their computation. We take the view that the roots of a single polynomial equation are known (although computation with them requires the theory of algebraic extension fields), and therefore that the reduction of a system to a single equation constitutes a solution. It should be pointed out that the technique described here is by no means the only method for solving nonlinear equations. A more recent (but still "classical") approach, which uses a generalized resultant due to Macaulay, is the basis for methods described by Lazard [11] and Canny et al. [5]. (See also the extensive references of the latter.) In Chapter 10, we describe a more modern technique using the theory of Gröbner bases for polynomial ideals.

9.2. LINEAR EQUATIONS AND GAUSSIAN ELIMINATION

Let us consider a linear system of equations over an integral domain D, written

$$A\mathbf{x} = \mathbf{b}, \tag{9.1}$$

where A is an $m \times n$ matrix with entries a_{ij}, $\mathbf{x} = [x_1, \ldots, x_n]^T$, and $\mathbf{b} = [b_1, \ldots, b_m]^T$. As is typical in such discussions, we immediately restrict our attention to the case where A is a square $(n \times n)$ matrix; extensions to other cases will be evident later on. It is well known from linear algebra that the solutions of (9.1) (or a proof that there are none) may be computed by reducing the *augmented matrix*

$$A^{(0)} = [a_{ij}^{(0)}] = [A \mid \mathbf{b}], \quad 1 \le i \le n, \; 1 \le j \le n+1,$$

where $a_{i,n+1}^{(0)} = b_i$, $1 \le i \le n$, to upper echelon form using Gaussian elimination. That is, we apply a series of transformations to $A^{(0)}$:

$$A^{(0)} \to A^{(1)} \to \cdots \to A^{(n-1)}$$

such that (when A is nonsingular) the first n columns of $A^{(n-1)}$ form an upper-triangular matrix, and the solutions of (9.1) are precisely those of

$$A^{(n-1)} \begin{bmatrix} \mathbf{x} \\ -1 \end{bmatrix} = \mathbf{0}.$$

We assume for now that no diagonal entry in the above is zero. (Again, this may be relaxed at a later point by adding a "pivoting" strategy to our algorithms.) The best known formula for the above transformation is the *ordinary Gaussian elimination* formula: compute the entries of $A^{(k)}$ by

$$a_{ij}^{(k)} = a_{ij}^{(k-1)} - \frac{a_{ik}^{(k-1)}}{a_{kk}^{(k-1)}} a_{kj}^{(k-1)}, \quad k+1 \le i \le n, \; k+1 \le j \le n+1, \tag{9.2}$$

which is applied for $k = 1, 2, \ldots, n-1$. We adopt the convention that the elements $a_{lk}^{(k)}$ $(k+1 \le l \le n)$ are assumed to be zero at the end of step k, and all other elements not explicitly updated remain unchanged. Using standard terminology, at step k the element $a_{kk}^{(k-1)}$ is called the *pivot element* and row k is called the *pivot row*.

Example 9.1. Consider the linear system:

$$\begin{aligned} 3x_1 + 4x_2 - 2x_3 + x_4 &= -2\,, \\ x_1 - x_2 + 2x_3 + 2x_4 &= 7\,, \\ 4x_1 - 3x_2 + 4x_3 - 3x_4 &= 2\,, \\ -x_1 + x_2 + 6x_3 - x_4 &= 1\,. \end{aligned}$$

The sequence of transformations of the augmented matrix given by the ordinary Gaussian elimination formula (9.2) is:

$$A^{(0)} = \begin{bmatrix} 3 & 4 & -2 & 1 & -2 \\ 1 & -1 & 2 & 2 & 7 \\ 4 & -3 & 4 & -3 & 2 \\ -1 & 1 & 6 & -1 & 1 \end{bmatrix}; \quad A^{(1)} = \begin{bmatrix} 3 & 4 & -2 & 1 & -2 \\ 0 & -\frac{7}{3} & \frac{8}{3} & \frac{5}{3} & \frac{23}{3} \\ 0 & -\frac{25}{3} & \frac{20}{3} & -\frac{13}{3} & \frac{14}{3} \\ 0 & \frac{7}{3} & \frac{16}{3} & -\frac{2}{3} & \frac{1}{3} \end{bmatrix};$$

$$A^{(2)} = \begin{bmatrix} 3 & 4 & -2 & 1 & -2 \\ 0 & -\frac{7}{3} & \frac{8}{3} & \frac{5}{3} & \frac{23}{3} \\ 0 & 0 & -\frac{20}{7} & -\frac{72}{7} & -\frac{159}{7} \\ 0 & 0 & 8 & 1 & 8 \end{bmatrix}; \quad A^{(3)} = \begin{bmatrix} 3 & 4 & -2 & 1 & -2 \\ 0 & -\frac{7}{3} & \frac{8}{3} & \frac{5}{3} & \frac{23}{3} \\ 0 & 0 & -\frac{20}{7} & -\frac{72}{7} & -\frac{159}{7} \\ 0 & 0 & 0 & -\frac{139}{5} & -\frac{278}{5} \end{bmatrix}.$$

●

Note that if D is not a field, the update formula (9.2) produces fractions as in the above example: when $a_{ij}^{(0)} \in \mathrm{D}$, we have $a_{ij}^{(k)} \in \mathrm{F_D}$ for $k > 0$. In principle, this is inconsequential since the solutions to the given system lie in the quotient field anyway. However, the (hidden) cost of reducing these fractions to lowest terms via GCD computations is a significant one. This is particularly important if the coefficients happen to be rational functions. It is therefore desirable to delay the forming of quotients as far as possible.

If we assume that $d = \det(A) \neq 0$, we know from Cramer's rule that the solutions of (9.1) can be expressed in the form

$$x_i = \frac{x_i^*}{d}\,, \quad 1 \le i \le n\,,$$

where x_i^* is given by

$$x_i^* = \det \begin{bmatrix} a_{11} & a_{12} & \cdots & a_{1,i-1} & b_1 & a_{1,i+1} & \cdots & a_{1n} \\ . & . & . & . & . & . & . & . \\ . & . & . & . & . & . & . & . \\ . & . & . & . & . & . & . & . \\ a_{n1} & a_{n2} & \cdots & a_{n,i-1} & b_n & a_{n,i+1} & \cdots & a_{nn} \end{bmatrix} = (A^{adj}\mathbf{b})_i \, . \tag{9.3}$$

Clearly it is possible to compute the solutions almost entirely within D. In asymptotic terms, however, computing determinants directly (i.e. from cofactor expansions) would seem relatively inefficient since it requires $O(n!)$ operations compared to $O(n^3)$ for Gaussian elimination (cf. Exercise 9.6). Nonetheless, the above concept remains a valid one; we therefore wish to describe variants of the classical scheme (9.2) which compute the quantities x_i^*, d (entirely within D) more efficiently. The basic concepts of so-called "fraction-free" Gaussian elimination were presented by Bareiss [1] in the context of the integral domain **Z**. Later Bareiss [2] extended these ideas to a general integral domain. (Such a generalization was also presented by Lipson [12].)

The simplest means of avoiding fractions during elimination is to simply clear the denominator of (9.2); this yields the *division-free elimination* formula:

$$a_{ij}^{(k)} = a_{kk}^{(k-1)}\, a_{ij}^{(k-1)} - a_{kj}^{(k-1)}\, a_{ik}^{(k-1)} , \tag{9.4}$$

$$k{+}! \le i \le n,\ k{+}1 \le j \le n{+}1 ,$$

for $k = 1, 2, \ldots, n-1$.

Example 9.2. Consider the linear system of Example 9.1. The sequence of transformations of the augmented matrix given by the division-free Gaussian elimination formula (9.4) is:

$$A^{(0)} = \begin{bmatrix} 3 & 4 & -2 & 1 & -2 \\ 1 & -1 & 2 & 2 & 7 \\ 4 & -3 & 4 & -3 & 2 \\ -1 & 1 & 6 & -1 & 1 \end{bmatrix}; \quad A^{(1)} = \begin{bmatrix} 3 & 4 & -2 & 1 & -2 \\ 0 & -7 & 8 & 5 & 23 \\ 0 & -25 & 20 & -13 & 14 \\ 0 & 7 & 16 & -2 & 1 \end{bmatrix};$$

$$A^{(2)} = \begin{bmatrix} 3 & 4 & -2 & 1 & -2 \\ 0 & -7 & 8 & 5 & 23 \\ 0 & 0 & 60 & 216 & 477 \\ 0 & 0 & -168 & -21 & -168 \end{bmatrix}; \quad A^{(3)} = \begin{bmatrix} 3 & 4 & -2 & 1 & -2 \\ 0 & -7 & 8 & 5 & 23 \\ 0 & 0 & 60 & 216 & 477 \\ 0 & 0 & 0 & 35028 & 70056 \end{bmatrix}.$$

●

We see that while the update formula (9.4) avoids producing fractions, the growth in the size of the integers involved is explosive. It is easy to show that if the entries $a_{ij}^{(0)}$ are δ digit integers, the entries $a_{ij}^{(k)}$ may contain $2^k\delta$ digits. Hence, the computing time for the division-free scheme is typically worse than for the ordinary Gaussian elimination scheme.

9.3. FRACTION-FREE GAUSSIAN ELIMINATION

It is useful to consider the formulas (9.2), (9.4) as analogous to the Euclidean algorithm and Euclidean PRS, respectively, of Chapter 7. Clearly, it is possible to reduce the coefficients of (9.4) in the manner of the primitive PRS, by computing

$$\mathrm{GCD}(a_{i,k+1}^{(k)},\ a_{i,k+2}^{(k)},\ \ldots,\ a_{i,n+1}^{(k)})$$

for each $k+1 \le i \le n$, at each stage. This, however, still requires many GCD subcomputations. Fortunately (as with the reduced PRS algorithm) it is possible to divide the results of (9.4) by a previous pivot element *within* the integral domain D. To see how this is possible, suppose that

$$A^{(0)} = \begin{bmatrix} a & b & c & \ldots & d \\ e & f & g & \ldots & h \\ p & q & r & \ldots & s \\ \ldots & \ldots & \ldots & \ldots & \ldots \\ \ldots & \ldots & \ldots & \ldots & \ldots \end{bmatrix}.$$

After two iterations of the division-free formula (9.4) we obtain

$$A^{(1)} = \begin{bmatrix} a & b & c & \ldots & d \\ 0 & (af-be) & (ag-ce) & \ldots & (ah-de) \\ 0 & (aq-bp) & (ar-cp) & \ldots & (as-dp) \\ \ldots & \ldots & \ldots & \ldots & \ldots \\ 0 & \ldots & \ldots & \ldots & \ldots \end{bmatrix};$$

and then

$$A^{(2)} = \begin{bmatrix} a & b & c & \ldots & d \\ 0 & (af-be) & (ag-ce) & \ldots & (ah-de) \\ 0 & 0 & \begin{matrix}(a^2fr-a^2gq-aber \\ +abgp+aceq-acfp)\end{matrix} & \ldots & \begin{matrix}(a^2fs-a^2qh-abes \\ +abhp+adeq-adfp)\end{matrix} \\ \ldots & \ldots & \ldots & \ldots & \ldots \\ 0 & 0 & \ldots & \ldots & \ldots \end{bmatrix}.$$

Notice that the entries of the third row (and all following rows) of $A^{(2)}$ are divisible by the pivot $a = a_{11}^{(0)}$. Therefore the entries of the altered rows of $A^{(2)}$ may be divided by a (within D) before proceeding with the elimination. After another iteration, we would find that the (altered) rows of $A^{(3)}$ are divisible by $a_{22}^{(1)} = af - be$. In fact, such divisions are possible all through the elimination process. Formally, this is expressed as the (*single-step*) *fraction-free elimination* scheme:

$$a_{00}^{(-1)} = 1, \ [a_{ij}^{(0)}] = A^{(0)}, \tag{9.5}$$

$$a_{ij}^{(k)} = (a_{kk}^{(k-1)} a_{ij}^{(k-1)} - a_{kj}^{(k-1)} a_{ik}^{(k-1)}) / a_{k-1,k-1}^{(k-2)}, \tag{9.6}$$

$$k+1 \le i \le n, \ k+1 \le j \le n+1,$$

for $k = 1, 2, \ldots, n-1$ which, according to Bareiss, was known to Jordan.

Example 9.3. For the linear system of Example 9.1, the sequence of transformations of the augmented matrix given by the fraction-free formulas (9.5), (9.6) is:

$$A^{(0)} = \begin{bmatrix} 3 & 4 & -2 & 1 & -2 \\ 1 & -1 & 2 & 2 & 7 \\ 4 & -3 & 4 & -3 & 2 \\ -1 & 1 & 6 & -1 & 1 \end{bmatrix}; \ A^{(1)} = \begin{bmatrix} 3 & 4 & -2 & 1 & -2 \\ 0 & -7 & 8 & 5 & 23 \\ 0 & -25 & 20 & -13 & 14 \\ 0 & 7 & 16 & -2 & 1 \end{bmatrix};$$

$$A^{(2)} = \begin{bmatrix} 3 & 4 & -2 & 1 & -2 \\ 0 & -7 & 8 & 5 & 23 \\ 0 & 0 & 20 & 72 & 159 \\ 0 & 0 & -56 & -7 & -56 \end{bmatrix}; \ A^{(3)} = \begin{bmatrix} 3 & 4 & -2 & 1 & -2 \\ 0 & -7 & 8 & 5 & 23 \\ 0 & 0 & 20 & 72 & 159 \\ 0 & 0 & 0 & -556 & -1112 \end{bmatrix}.$$

●

It can be shown that the divisor used in (9.6) is the largest possible such divisor. If the entries $a_{ij}^{(0)}$ are polynomials of degree δ, then the degrees of the entries $a_{ij}^{(k)}$ are bounded by $(k+1)\delta$. (A similar result holds when the coefficients are integers; see Exercise 9.1.) Since the entry $a_{nn}^{(n-1)}$ produced in the above method is $\det(A)$ (see Theorem 9.1 below), we have a polynomial-time determinant algorithm. Moreover, with the addition of an analogous fraction-free ''back-solving'' scheme we will have a complete solving algorithm. However, we postpone discussion of these details. Instead, we will prove (following Bareiss [2]) that the exact division in (9.6) always works; in so doing, a worthwhile improvement to this formula follows.

Let us now *define* elements $a_{ij}^{(k)}$ as subdeterminants of $A^{(0)}$ by

$$a_{00}^{(-1)} = 1, \ [a_{ij}^{(0)}] = A^{(0)}, \tag{9.7}$$

$$a_{ij}^{(k)} = \det \begin{bmatrix} a_{11} & a_{12} & \ldots & a_{1k} & a_{1j} \\ a_{21} & a_{22} & \ldots & a_{2k} & a_{2j} \\ \ldots & \ldots & \ldots & \ldots & \ldots \\ a_{k1} & a_{k2} & \ldots & a_{kk} & a_{kj} \\ a_{i1} & a_{i2} & \ldots & a_{ik} & a_{ij} \end{bmatrix}, \tag{9.8}$$

$$k+1 \le i \le n, \ k+1 \le j \le n+1,$$

for $1 \le k < n$. (The relationship with our previous use of "$a_{ij}^{(k)}$" will soon become clearer.) We then have the following:

Theorem 9.1 (Sylvester's Identity). For $a_{ij}^{(k)}$ defined in (9.7) - (9.8) we have

$$\det(A)\,[a_{ll}^{(l-1)}]^{n-l-1} = \det \begin{bmatrix} a_{l+1,l+1}^{(l)} & \cdots & a_{l+1,n}^{(l)} \\ \cdots & \cdots & \cdots \\ a_{n,l+1}^{(l)} & \cdots & a_{nn}^{(l)} \end{bmatrix} \tag{9.9}$$

for $1 \le l \le n-1$.

Proof: Partition and factor the matrix A as

$$A = \begin{bmatrix} A_{11} & A_{12} \\ A_{21} & A_{22} \end{bmatrix} = \begin{bmatrix} A_{11} & 0 \\ A_{21} & I \end{bmatrix} \begin{bmatrix} I & A_{11}^{-1}A_{12} \\ 0 & A_{22}-A_{21}A_{11}^{-1}A_{12} \end{bmatrix},$$

where A_{11} is a nonsingular $k \times k$ matrix. Then

$$\det(A) = \det(A_{11})\det(A_{22} - A_{21}A_{11}^{-1}A_{12});$$

in fact, since the last determinant on the right hand side is of order $n-k$,

$$\det(A_{11})^{n-k-1}\det(A) = \det(\det(A_{11})\cdot(A_{22} - A_{21}A_{11}^{-1}A_{12})).$$

Since $A_{11}^{-1}\cdot\det(A_{11}) = A_{11}^{adj}$, we also have, for $k \le i \le n$ and $k \le j \le n$,

$$\begin{aligned} \det(A_{11})\,(A_{22} - A_{21}A_{11}^{-1}A_{12}) &= \det(A_{11})[\,a_{ij} - \sum_{r=1}^{k} a_{ir} \sum_{s=1}^{k} (A_{11}^{-1})_{rs} a_{sj}\,] \\ &= \det(A_{11})a_{ij} - \sum_{r=1}^{k} a_{ir} \sum_{s=1}^{k} \tilde{A}_{sr} a_{sj} \\ &= a_{ij}^{(k)}, \end{aligned}$$

where $\tilde{A}_{sr}$ is the cofactor of $(A_{11})_{sr}$. (The last line follows by expanding the determinant (9.8) along the last row from right to left.) Then (9.9) follows by taking determinants of both sides. It is noted by Bareiss [1] that this holds even when A_{11} is singular.

●

If Theorem 9.1 is applied to (9.8) (i.e. view $a_{ij}^{(k)}$ as a determinant), we obtain

$$a_{ij}^{(k)}\,[a_{ll}^{(l-1)}]^{k-l} = \det \begin{bmatrix} a_{l+1,l+1}^{(l)} & \cdots & a_{l+1,k}^{(l)} & a_{l+1,j}^{(l)} \\ \cdots & \cdots & \cdots & \cdots \\ a_{k,l+1}^{(l)} & \cdots & a_{kk}^{(l)} & a_{kj}^{(l)} \\ a_{i,l+1}^{(l)} & \cdots & a_{ik}^{(l)} & a_{ij}^{(l)} \end{bmatrix}, \tag{9.10}$$

for $1 \le l \le k-1$. Hence the right hand side above is divisible by the factor $[a_{ll}^{(l-1)}]^{k-l}$. If we let $l = k-1$ in the above, we obtain

$$a_{ij}^{(k)} = \frac{1}{a_{k-1,k-1}^{(k-2)}} \det \begin{bmatrix} a_{kk}^{(k-1)} & a_{kj}^{(k-1)} \\ a_{ik}^{(k-1)} & a_{ij}^{(k-1)} \end{bmatrix}, \tag{9.11}$$

which is precisely (9.6)! This correspondence is made clearer by the observations that

$$a_{kj}^{(k-1)} = 0, \quad j < k$$

(since the corresponding determinant has two identical columns), and

$$a_{nn}^{(n-1)} = \det(A).$$

The former means that the transformed (augmented) matrix may be written

$$A^{(n-1)} = \begin{bmatrix} a_{11}^{(0)} & a_{12}^{(0)} & \cdots & & \cdots & a_{1n}^{(0)} & a_{1,n+1}^{(0)} \\ & a_{22}^{(1)} & \cdots & & \cdots & a_{2n}^{(1)} & a_{2,n+1}^{(1)} \\ & & \cdots & & \cdots & \cdots & \cdots \\ & & & a_{kk}^{(k-1)} & \cdots & a_{kn}^{(k-1)} & a_{k,n+1}^{(k-1)} \\ & & & & \cdots & \cdots & \cdots \\ & & & & & a_{nn}^{(n-1)} & a_{n,n+1}^{(n-1)} \end{bmatrix}, \tag{9.12}$$

where the entries not shown are zero. Then, it can be shown that the solutions of the system with augmented matrix (9.12) are precisely those of (9.1), in the form

$$\sum_{j=1}^{n} a_{kj}^{(k-1)} x_j^* = \det(A)\, a_{k,n+1}^{(k-1)}$$

for $k = 1, 2, \ldots, n$ where $\mathbf{x}^* = \det(A) \cdot \mathbf{x}$; this corresponds to the *adjoint form* of (9.1),

$$A\,\mathbf{x}^* = \det(A) \cdot \mathbf{b}. \tag{9.13}$$

(See Exercise 9.2.) Bareiss [1] describes how (9.10) may be used to derive fraction-free schemes which eliminate over $k-l$ columns at once. Of particular interest is the formula obtained by setting $l = k-2$:

$$a_{ij}^{(k)} = \frac{1}{[a_{k-2,k-2}^{(k-3)}]^2} \det \begin{bmatrix} a_{k-1,k-1}^{(k-2)} & a_{k-1,k}^{(k-2)} & a_{k-1,j}^{(k-2)} \\ a_{k,k-1}^{(k-2)} & a_{kk}^{(k-2)} & a_{kj}^{(k-2)} \\ a_{i,k-1}^{(k-2)} & a_{ik}^{(k-2)} & a_{ij}^{(k-2)} \end{bmatrix}.$$

It turns out that the quantity $a_{k-2,k-2}^{(k-3)}$ divides all of the cofactors of $a_{k-1,j}^{(k-2)}$, $a_{kj}^{(k-2)}$, $a_{ij}^{(k-2)}$, in the above. This follows for $a_{ij}^{(k-2)}$ by (9.11), and for the others because row/column interchanges (which could make them corner elements) would not affect the value of the determinant except possibly in sign. Hence, we have the *two-step fraction-free* scheme:

$$a_{00}^{(-1)} = 1\,, \quad [a_{ij}^{(0)}] = A^{(0)}\,, \tag{9.14}$$

$$c_0^{(k-2)} = (a_{k-1,k-1}^{(k-2)}\, a_{kk}^{(k-2)} - a_{k-1,k}^{(k-2)}\, a_{k,k-1}^{(k-2)}) \,/\, a_{k-2,k-2}^{(k-3)}\,, \tag{9.15}$$

$$c_{i1}^{(k-2)} = (a_{k-1,k}^{(k-2)}\, a_{i,k-1}^{(k-2)} - a_{k-1,k-1}^{(k-2)}\, a_{ik}^{(k-2)}) \,/\, a_{k-2,k-2}^{(k-3)}\,, \tag{9.16}$$

$$c_{i2}^{(k-2)} = (a_{k,k-1}^{(k-2)}\, a_{ik}^{(k-2)} - a_{kk}^{(k-2)}\, a_{i,k-1}^{(k-2)}) \,/\, a_{k-2,k-2}^{(k-3)}\,, \tag{9.17}$$

$$a_{ij}^{(k)} = (c_0^{(k-2)}\, a_{ij}^{(k-2)} + c_{i1}^{(k-2)}\, a_{kj}^{(k-2)} + c_{i2}^{(k-2)}\, a_{k-1,j}^{(k-2)}) \,/\, a_{k-2,k-2}^{(k-3)}\,, \tag{9.18}$$

for $k+1 \le i \le n,\ k+1 \le j \le n+1$;

$$a_{kk}^{(k-1)} = c_0^{(k-2)}\,, \tag{9.19}$$

$$a_{kl}^{(k-1)} = a_{kl}^{(k)} = (a_{k-1,k-1}^{(k-2)}\, a_{kl}^{(k-2)} - a_{k-1,l}^{(k-2)}\, a_{k,k-1}^{(k-2)}) \,/\, a_{k-2,k-2}^{(k-3)}\,, \tag{9.20}$$

for $k+1 \le l \le n+1$,

which we apply for $k = 2, 4, \ldots, 2\left\lfloor \frac{n-1}{2} \right\rfloor$.

We note that when n is even, the last elimination is performed by the single-step method. In effect, the formulas (9.14) - (9.20) are obtained by applying the single-step scheme to itself. This causes some terms to cancel, and thus results in fewer computations (Exercise 9.4).

Example 9.4. For the system of the previous example, the augmented matrix transforms according to the two-step scheme as follows:

$$A^{(0)} = \begin{bmatrix} 3 & 4 & -2 & 1 & -2 \\ 1 & -1 & 2 & 2 & 7 \\ 4 & -3 & 4 & -3 & 2 \\ -1 & 1 & 6 & -1 & 1 \end{bmatrix}; \quad A^{(2)} = \begin{bmatrix} 3 & 4 & -2 & 1 & -2 \\ 0 & -7 & 8 & 5 & 23 \\ 0 & 0 & 20 & 72 & 159 \\ 0 & 0 & -56 & -7 & -56 \end{bmatrix};$$

$$A^{(3)} = \begin{bmatrix} 3 & 4 & -2 & 1 & -2 \\ 0 & -7 & 8 & 5 & 23 \\ 0 & 0 & 20 & 72 & 159 \\ 0 & 0 & 0 & -556 & -1112 \end{bmatrix}.$$

•

Analyses of the relative efficiency of the one and two-step methods by Bareiss [2] and Lipson [12] (for the cases of integer and multivariate polynomial coefficients, respectively) show that the latter is about 50% faster asymptotically.

We remark that in practice, some pivoting (i.e. row interchanges) may be necessary in the course of any of the single-step schemes (when $a_{kk}^{(k-1)} = 0$) or the two-step scheme (when $c_0^{(k-2)} = a_{kk}^{(k-1)} = 0$). In the latter case, for example, it is necessary to switch one (or both) of the "active" rows ($k-1$, k) with later rows of $A^{(k-2)}$ until $c_0^{(k-2)} \ne 0$. (Note that we also require $a_{k-1,k-1}^{(k-2)} \ne 0$.) Such an exchange will be impossible if $\det(A) = 0$; but, it is often possible to continue elimination from a later column (e.g. for a homogeneous system). We also mention that when the entries of $A^{(k)}$ are not of uniform size, it may be worthwhile to interchange rows in order to obtain a smaller pivot for the next step. A version of the single-step fraction-free scheme (which doubles as a determinant algorithm) is presented as Algorithm 9.1. We leave the task of justifying the extension to general $m \times n$ matrices as an exercise for the reader.

Algorithm 9.1. Fraction-Free Gaussian Elimination.

```
procedure FractionFreeElim(A)

  # Given an m × n matrix A (with entries a_ij),
  # reduce it to upper echelon form via formulas (9.5), (9.6).

  sign ← 1 ; divisor ← 1 ; r ← 1

  # Eliminate below row r, with pivot in column k.
  for k from 1 to n while r ≤ m do {

    # Find a nonzero pivot.
    for p from r to m while a_pk = 0 do { }
    if p ≤ m then {

      # Pivot is in row p, so switch rows p and r.
      for j from k to n do { interchange a_pj and a_rj }

      # Keep track of sign changes due to row exchange.
      if r ≠ p then sign ← −sign
      for i from r+1 to m do {
        for j from k+1 to n do {
          a_ij ← (a_rk a_ij − a_rj a_ik) / divisor }
        a_ik ← 0 }
      divisor ← a_rk ; r ← r + 1 } }

  # Optionally, compute the determinant for square
  # or augmented matrices.
  if r = m + 1 then det ← sign·divisor else det ← 0
  return(A)
end
```

It also turns out that fraction-free back-solving is possible. Since we know that $x_i^* \in D$ for each x_i^* in (9.13), and that the x_i^* are determined by (9.12), we have

$$x_n^* = a_{n,n+1}^{(n-1)}, \tag{9.21}$$

$$x_k^* = (a_{nn}^{(n-1)}\, a_{k,n+1}^{(k-1)} - \sum_{j=k+1}^{n} a_{kj}^{(k-1)} x_j^*) \,/\, a_{kk}^{(k-1)}, \tag{9.22}$$

for $k = n-1, n-2, \ldots, 1$.

Example 9.5. In Example 9.4 we computed the matrix

$$A^{(3)} = \begin{bmatrix} 3 & 4 & -2 & 1 & -2 \\ 0 & -7 & 8 & 5 & 23 \\ 0 & 0 & 20 & 72 & 159 \\ 0 & 0 & 0 & -556 & -1112 \end{bmatrix}.$$

Then $\det(A) = a_{44}^{(3)} = -556$, and according to (9.21) and (9.22) we have

$$x_4^* = -1112\,,\ x_3^* = -417\,,\ x_2^* = 556\,,\ x_1^* = -278\,.$$

Therefore the solutions to the (original) linear system of Example 9.1 are

$$x_1 = \frac{1}{2}\,,\ x_2 = -1\,,\ x_3 = \frac{3}{4}\,,\ x_4 = 2\,.$$

•

Bareiss [1] develops back-solving strategies (for both the one and two-step schemes) which proceed, in effect, by applying the respective elimination schemes to the *rows* of $A^{(n-1)}$.

Once an upper echelon form of the augmented matrix is known, it is easily determined whether the associated linear system is inconsistent, under-determined or uniquely determined. The details of formulating the ''solution'' procedure in the latter cases is left as a (straightforward) exercise.

9.4. ALTERNATIVE METHODS FOR SOLVING LINEAR EQUATIONS

The algorithms for fraction-free Gaussian elimination described in the previous section should be considered reasonably good ones, since their asymptotic complexity is polynomial of low order. And (once the rather complicated derivations are complete) they are fairly simple to implement. However, a variety of other possible algorithms exist. These methods deserve mention not just out of completeness, but because some of them may be particularly well-suited to certain types of problems.

It is fairly easy to see that our fraction-free elimination methods will be less efficient for problems with multivariate polynomial coefficients than for ones with integer or univariate coefficients. This is due to the growth which occurs when, for example, the products are formed in an elimination step (9.6) (i.e. before division by a previous pivot). Given two univariate polynomials with s, t terms, their product has at most $s+t$ terms. However, if such polynomials are multivariate the product may contain as many as $s \cdot t$ terms. Therefore an elimination such as the single- (or double-) step fraction-free scheme is prone to some intermediate expression swell.

Minor Expansion

As pointed out in the previous section, when a unique solution to (9.1) exists it may be computed via Cramer's rule in terms of determinants. Typically, one imagines a determinant in terms of the *cofactor definition*:

$$\det(A) = \sum_{j=1}^{n} a_{ij} \tilde{A}_{ij}, \quad \text{for any } 1 \le i \le n,$$

$$= \sum_{i=1}^{n} a_{ij} \tilde{A}_{ij}, \quad \text{for any } 1 \le j \le n,$$

where $\tilde{A}_{ij}$ is the cofactor of a_{ij}. These are referred to as *minor expansions* since each cofactor is, in turn, defined as $\tilde{A}_{ij} = (-1)^{i+j} \det(M_{ij})$ where M_{ij} is the (i, j) minor of matrix A. The evaluation of a determinant in this form requires $O(n!)$ multiplications, i.e. exponential computing time (even when such a scheme is implemented efficiently, i.e. there are no redundant computations of sub-determinants). As such, this may be considered a poor algorithm in view of the results of the previous section. However, the reader should be aware that in many instances sub-optimal algorithms may outperform optimal (or merely "better", in terms of asymptotic complexity) ones. A surprising case study by Gentleman and Johnson [8] provides evidence of this in the present context. Their analysis compared the computing time requirements for the single-step, fraction-free elimination scheme and the minor expansion method, applied to the computation of determinants of dense matrices with polynomial entries. They found (using particular computational models) that when the matrix entries are dense univariate polynomials with t terms, the minor expansion method is superior unless:

(a) $n = 6, t > 1$; (b) $n = 7, t > 3$; (c) $n \ge 8$.

Further, when the matrix entries are (totally) sparse multivariate polynomials the minor expansion method *always* requires less computation.

The above results suggest that the solution of linear systems using minor expansion may indeed be feasible for many practical problems. This would seem to apply particularly to sparse systems, since Cramer's rule after all requires the computation of $n+1$ determinants. Elimination methods tend to "fill in" the upper triangular portion of an initially sparse matrix; however, in such cases minor expansion may be extremely effective if (at each level) the expansion proceeds along a near-empty row or column (or equivalently, if the rows are "suitably" permuted). The combination of the minor expansion method with heuristics for selecting advantageous row orderings is examined by Griss [9].

The Method of Sasaki and Murao

The growth problem described above may be alleviated to some extent using a modification of fraction-free elimination due to Sasaki and Murao [14], which we sketch very briefly. Suppose that we re-name the diagonal entries of the augmented matrix $[a_{ij}^{(0)}]$, i.e. write $a_{ii}^{(0)} = X_i$ for $1 \le i \le n$. Then the subdeterminant defined in (9.8) becomes

$$a_{ij}^{(k)} = \det \begin{bmatrix} X_1 & a_{12} & \dots & a_{1k} & a_{1j} \\ a_{21} & X_2 & \dots & a_{2k} & a_{2j} \\ \dots & \dots & \dots & \dots & \dots \\ a_{k1} & a_{k2} & \dots & X_k & a_{kj} \\ a_{i1} & a_{i2} & \dots & a_{ik} & a_{ij} \end{bmatrix}. \tag{9.23}$$

As before we may apply the single-step fraction-free formulas (9.5) and (9.6), say, to obtain $a_{ij}^{(k)}$ from $a_{ij}^{(k-1)}$ and $a_{k-1,k-1}^{(k-2)}$. However, we note from the above formula that the quantity $a_{k-1,k-1}^{(k-2)}$ is actually a polynomial in the variables X_i; it is of degree $k-1$, and has $P = X_1 X_2 \cdots X_{k-1}$ as its leading term. Therefore we may divide P^2 by $a_{k-1,k-1}^{(k-2)}$, i.e. write

$$P^2 = Q_{k-2} \cdot a_{k-1,k-1}^{(k-2)} + R_{k-2}, \tag{9.24}$$

where R_{k-2} contains no terms which are multiples of P and Q_{k-2} is linear in each of $X_1, \dots,$ X_{k-1}. By multiplying (9.6) and (9.24) we may obtain

$$a_{ij}^{(k)} P^2 = (a_{kk}^{(k-1)} a_{ij}^{(k-1)} - a_{kj}^{(k-1)} a_{ik}^{(k-1)}) Q_{k-2} + a_{ij}^{(k)} R_{k-2}.$$

Now, by (9.23) we know that $a_{ij}^{(k)}$ is linear in each of the X_i. Hence, the second term of the right hand side (in view of the definition of R_{k-2}) has no terms which are multiples of P^2. This means that $a_{ij}^{(k)}$ may be determined from the *first* term of the right hand side, simply by taking all terms which are multiples of P^2. In fact, since Q_{k-2} is linear in the X_i we need only take those terms of the parenthesized part which are multiples of P before examining the product. The X_i are replaced by the original diagonal values when $a_{ij}^{(n-1)}$ is obtained. Consider, as a very simple example, the matrix

$$A^{(0)} = \begin{bmatrix} a & b & c & \dots \\ e & f & g & \dots \\ p & q & r & \dots \\ \dots & \dots & \dots & \dots \end{bmatrix} = \begin{bmatrix} X_1 & b & c & \dots \\ e & X_2 & g & \dots \\ p & q & X_3 & \dots \\ \dots & \dots & \dots & \dots \end{bmatrix}.$$

After one iteration of the fraction-free formula (9.6) we obtain

$$A^{(1)} = \begin{bmatrix} X_1 & b & c & \cdots \\ 0 & (X_1X_2-be) & (X_1g-ce) & \cdots \\ 0 & (X_1q-bp) & (X_1X_3-cp) & \cdots \\ \cdots & \cdots & \cdots & \cdots \\ 0 & \cdots & \cdots & \cdots \end{bmatrix}.$$

Now for $k = 2$ we have $P^2 = X_1^2$ and $Q_0 = X_1$; so to find (for example) $a_{33}^{(2)}$ at the next iteration we compute the coefficient of X_1 in

$$(X_1X_2 - be)(X_1X_3 - cp) \; - \; (X_1q - bp)(X_1g - ce) .$$

Hence, only the terms

$$X_1X_2X_3 - X_3be - X_2cp - X_1gq + ceq + bgp$$

need be formed.

The key point in the above procedure is that the required terms may be computed efficiently *without* fully expanding all products. For this purpose, it is possible to define a special (associative) multiplication operation for polynomials in X_i which, in essence, only expands products which contribute essential terms. We will not attempt to describe this operation here, but refer the reader to Sasaki and Murao [14] for further information.

Homomorphism Methods

We noted above that minor expansion may be superior to Gaussian elimination when a system is sparse, has sparse polynomial coefficients, or is of small order. We now consider the opposite extremes: namely, dense systems with dense polynomial coefficients which may also be of large order. In such cases, the use of modular/evaluation homomorphisms combined with the CRA/interpolation techniques of Chapter 5 may be especially appropriate. In fact, the use of homomorphisms for inverting matrices with integer entries seems to have been one of the seminal applications of the congruence method (cf. Takahasi and Ishibashi [15]). In view of the fact that Chapter 5 deals with homomorphism methods in great detail, we will not completely specify algorithms here. Instead we will examine those details which are specific to the problem at hand.

As in the previous section, we consider the adjoint form (9.13) of the system (9.1). When A is nonsingular, we may determine the solutions to (9.1) as $\mathbf{x} = \mathbf{x}^* / d$ where

$$\mathbf{x}^* = A^{adj}\,\mathbf{b}\,, \quad d = \det(A)\,. \tag{9.25}$$

Note, however, that these quantities (which exist even when $\mathbf{x}$ does not) lie in the coefficient (integral) domain; hence, they may be computed by homomorphism techniques. We recall that an example of such a computation has already been provided in Section 5.5.

Let us first assume that the system (9.1) has integer coefficients. In this case we may apply to it a modular homomorphism of the form ϕ_m: $\mathbf{Z} \rightarrow \mathbf{Z}_m$ where m is prime. Since the solutions to (9.1) may contain both signs, the symmetric representation of $\mathbf{Z}_m$ is usually chosen. Further assume (for now) that A is nonsingular. Then, unless the modulus m divides $\det(A)$, the adjoint solution $(\mathbf{x}^*, d)$ may be easily computed in $\mathbf{Z}_m$ using the ordinary Gaussian elimination scheme (9.2). (Of course, we may now use field inverses, rather than reciprocals, of the pivot elements.) That is, we first determine x_i for $1 \le i \le n$ from the transformed matrix $A^{(n-1)}$; then we use the fact that the determinant is the product of the pivot elements to obtain $x_i^* \equiv d \cdot x_i \pmod m$.

Example 9.6. Consider the linear system of Example 9.1. If we apply formula (9.2) while working over $\mathbf{Z}_{11}$, then we obtain

$$A^{(3)} \equiv \begin{bmatrix} 3 & 4 & -2 & 1 & -2 \\ 0 & 5 & -1 & -2 & 4 \\ 0 & 0 & 5 & -4 & 4 \\ 0 & 0 & 0 & 3 & -5 \end{bmatrix}.$$

Then (modulo 11), we solve to obtain

$$x_4 \equiv 3^{-1} \cdot (-5) \equiv 2\,,\ x_3 \equiv -2\,,\ x_2 \equiv -1\,,\ x_1 \equiv -5\,.$$

We also see from $A^{(3)}$ that

$$d \equiv \det(A) \equiv 3 \cdot 5 \cdot 5 \cdot 3 \equiv 5\,.$$

Therefore

$$x_1^* \equiv 5 \cdot (-5) \equiv -3\,,\ x_2^* \equiv -5\,,\ x_3^* \equiv 1\,,\ x_4^* \equiv -1\,.$$

•

Those prime moduli m which divide d may be rejected as "unlucky", since $\mathbf{x}^*$ may not be determined by the above method. We mention, though, that Cabay [3] provides a method for computing $\mathbf{x}^*$ in case $d \equiv 0 \bmod m$. (Note that this may only be desirable, though, if $d \not\equiv 0$ for other moduli; indeed, if $d \equiv 0 \bmod m$, it is very likely that $d = 0$.) When sufficiently many images of $\mathbf{x}^*$, d have been computed these quantities may be reconstructed in terms of their mixed-radix representations (cf. 5.15)

$$\mathbf{x}^{*(k)} = \mathbf{u}_1 + \mathbf{u}_2(m_1) + \cdots + \mathbf{u}_k\left(\prod_{i=1}^{k-1} m_i\right), \tag{9.26}$$

$$d^{(k)} = v_1 + v_2(m_1) + \cdots + v_k\left(\prod_{i=1}^{k-1} m_i\right) \tag{9.27}$$

via Algorithm 5.1. In general, we require that the moduli satisfy

$$2 \cdot \max(|d|, |x_1^*|, \ldots, |x_n^*|) \le \prod_{i=1}^{k} m_i - 1\,.$$

Since the x_i^* are determined by (9.3) we may apply Hadamard's inequality to obtain (cf. Howell and Gregory [10])

$$\max_{1 \le i \le n} (|d|, |x_i^*|) \le \prod_{i=1}^{n} [\sum_{j=1}^{n} a_{ij}^2]^{1/2} \sum_{k=1}^{n} |b_k| . \qquad (9.28)$$

We mention that this may often yield too conservative a bound; see Exercise 9.8. It turns out that there are other ways to determine how many images are required. Suppose that we update the mixed-radix representations (9.26), (9.27) as each new image is computed. If, at some point, the new mixed-radix coefficients of x_i^*, d all vanish, we may check to see if (9.13) is satisfied by the current representations; we then continue with additional moduli only if it is not. In fact, it is shown by Cabay [3] that if the update coefficients for x_i^*, d all vanish for sufficiently many consecutive moduli (after, say, m_k) then the current representations $\mathbf{x}^{*(k)}$, $d^{(k)}$ satisfy (9.13) even if they are not complete! We thereby obtain the solution $\mathbf{x}$ if $d^{(k)} \ne 0$, or a proof that $d=0$ when $d^{(k)}=0$ but $\mathbf{x}^{*(k)} \ne 0$. In general, we may prove that $d=0$ only by continuing (9.26) and (9.27) until a bound such as (9.28) is reached.

When the coefficients of (9.1) are multivariate polynomials over $\mathbf{Z}$ (i.e. elements of $\mathbf{Z}[y_1, \ldots, y_l]$), we proceed in an analogous manner using (composite) evaluation homomorphisms of the form

$$\phi_{y_1-c_1} \phi_{y_2-c_2} \cdots \phi_{y_l-c_l} : D[y_1, \ldots, y_l] \to D .$$

If the entries a_{ij}, b_i are of degree (at most) δ_k in y_k, then $\mathbf{x}^*$, d are of degree at most $n\,\delta_k$; hence these quantities may be reconstructed by recursively applying Algorithm 5.2, using $n\,\delta_k + 1$ images to invert $\phi_{y_k - c_k}$. If we first apply a modular homomorphism

$$\phi_m : \mathbf{Z}[y_1, \ldots, y_l] \to \mathbf{Z}_m[y_1, \ldots, y_l]$$

(cf. Figure 5.1) to the original system, then the interpolation proceeds over $\mathbf{Z}_m$. This requires also a bound on the size of the integers involved, which is easily obtained. If on the other hand, one is willing to involve rational numbers (in spite of the disadvantages of such an approach), then *any* method may be used to solve the image problems over $\mathbf{Z}$.

Example 9.7. Consider the simultaneous linear equations

$$\begin{aligned} cx + (c+1)y + \quad\; z &= 1 , \\ x + \quad cy + (c+1)z &= 2 , \\ (c+1)x + \quad y + \quad cz &= -1 , \end{aligned}$$

in x, y, z. If we apply the evaluation homomorphism ϕ_{c-0} (corresponding to $c=0$), the augmented matrix for this system becomes $A^{(0)} = [A \mid \mathbf{b}]$ where

$$A \equiv \begin{bmatrix} 0 & 1 & 1 \\ 1 & 0 & 1 \\ 1 & 1 & 0 \end{bmatrix}, \ \mathbf{b} \equiv \begin{bmatrix} 1 \\ 2 \\ -1 \end{bmatrix}$$

with the obvious notation $\mathbf{x} = [x, y, z]^T$. By applying fraction-free elimination, we obtain the solution image

$$(\mathbf{x}^*, d) \equiv ([0, -2, 4]^T, 2) .$$

Similarly, after applying ϕ_{c-1}, ϕ_{c-2}, and ϕ_{c+2}, we obtain

$$([-6, 2, 6]^T, 4), ([-16, 14, 8]^T, 18), ([0, 14, 0]^T, -14),$$

respectively. (The homomorphism ϕ_{c+1} is rejected since $\phi_{c+1}(d) = 0 \neq \phi_{c-0}(d)$.) By Cramer's rule, we know that $\deg(\det(A)) \leq 3$, so four images will suffice. Therefore we can reconstruct $(\mathbf{x}^*, d)$ using the interpolation algorithm to obtain

$$d = 2 + 2(c) + 6c(c-1) + 2c(c-1)(c-2) = 2c^3 + 2 ,$$

$$x^* = -2c^2 - 4c ,$$

$$y^* = 4c^2 - 2 ,$$

$$z^* = 2c + 4 .$$

●

The interested reader is referred to Bareiss [2] and Cabay and Lam [4] for comparisons of the relative efficiencies (in the integer coefficient case) of the modular and fraction-free elimination methods. We also mention that McClellan [13] has investigated such details as coefficient and degree bounds, choosing homomorphisms, termination, and computing time analysis in detail for the polynomial coefficient case. These studies reveal that the homomorphism methods are indeed superior in terms of computing time bounds. However, as one finds with the GCD problem, the crucial assumption of dense coefficients is not always realistic.

9.5. NONLINEAR EQUATIONS AND RESULTANTS

In this section, we examine the problem of solving (or, at least somehow representing the solutions of) systems of nonlinear equations over a field F. As in the previous section, we pay particular attention to the solution of equations with coefficients in **Q** or $\mathbf{Q}(c_1, \ldots, c_l)$. We therefore consider (certainly without loss of generality) systems of equations over an integral domain D; some of the results of this section are in fact valid for commutative rings with identity. However, we bear in mind that in the present context, the solutions will lie in an extension field of the quotient field F_D.

We recall that a *solution* (or *root*) of a system of k polynomial equations in r variables

$$p_i(x_1, x_2, \ldots, x_r) = 0 , \ 1 \leq i \leq k \tag{9.29}$$

over D is an r-tuple $(\alpha_1, \alpha_2, \ldots, \alpha_r)$ from a suitably chosen extension field of F_D such that

$$p_i(\alpha_1, \alpha_2, \ldots, \alpha_r) = 0 \,, \quad 1 \le i \le k \,.$$

It is well known that even single univariate equations of degree greater than four cannot (in general) be explicitly solved in terms of radicals. Hence, the solution problem for (9.29) inevitably leads to the related problem of computing in algebraic extensions. Here, we will proceed as though the roots of a single univariate equation are known (i.e. in principle); we therefore concentrate on the problem of reducing a given system to an equivalent form in which the roots are easily obtained, modulo this assumption.

Example 9.8. Consider the following systems:

$$\text{(i)}\quad xy - x + 2 = 0 \,,$$
$$xy - x + 3 = 0 \,;$$

$$\text{(ii)}\quad x^2 - y^2 = 0 \,,$$
$$x^2 - x + 2y^2 - y - 1 = 0 \,;$$

$$\text{(iii)}\quad x^2 c + xy - yc - 1 = 0 \,,$$
$$2xy^2 + yc^2 - c^2 - 2 = 0 \,,$$
$$x + y^2 - 2 = 0 \,.$$

The first system in x, y has no solutions, as is easily verified by subtracting the equations. The system (ii) in x, y has precisely four roots, namely

$$\{\, x = y = 1 \,\}, \quad \{\, x = y = -\frac{1}{3} \,\},$$

$$\{\, x = \frac{1}{\sqrt{3}}, y = \frac{-1}{\sqrt{3}} \,\}, \quad \{\, x = \frac{-1}{\sqrt{3}}, y = \frac{1}{\sqrt{3}} \,\}.$$

If the system (iii) is solved over $\mathbf{Q}(c)[x,y]$ (i.e. for *variables* x, y in terms of the *parameter* c), it has only the solution

$$\{\, x = y = 1 \,\} \,.$$

If, however, we consider the equations over $\mathbf{Q}[x,y,c]$, then the above set represents a *family* of solutions. That is, there are infinitely many since c may be given any value. Furthermore, there are also 9 complex-number solutions corresponding to the roots of

$$c^9 - 14c^7 - 10c^6 + 10c^5 - 2c^4 - 56c^3 + 64c^2 - 24c - 8 = 0$$

which do not belong to this family.

●

It is well known from linear algebra that when the p_i are linear, the existence and uniqueness of solutions to (9.29) may be neatly characterized in terms of the rank of the corresponding coefficient matrix. The above example illustrates that even when there are as many independent equations as variables, these issues are not so simple. In general, when $k = r$ the number of solutions will be an exponential function of the degrees of the p_i. (See, for example, Bézout's theorem in Vol. 2 of van der Waerden [16].) However, such matters are better treated by the Gröbner basis theory/methods of the next chapter. We will instead proceed by describing a method for solving nonlinear equations (in the sense described above) which is (roughly) analogous to Gaussian elimination. Of course, the nonlinear methods will also be more difficult computationally. We see that just as determinants are fundamental to linear elimination, the basic tool of this section is the polynomial resultant (cf. Section 7.3). While resultant theory seems to have emerged primarily through study of the GCD problem (and has been applied most often to algebraic elimination), we mention that applications to quantifier elimination (cf. Collins [7]) and computing in algebraic extensions also exist. After presenting some additional theory of resultants, we will outline some algorithms for their computation. Finally, we show how they may be used to solve systems of polynomial equations.

Properties of Resultants

We recall from Definitions 7.2 and 7.3 that for polynomials $f, g \in R[x]$ (where R is a commutative ring with identity), res(f, g) is defined as the determinant of the Sylvester matrix (7.5) of f, g when this matrix is defined. It is also typical to let $\mathrm{res}(0, g) = 0$ for nonzero $g \in R[x]$, and $\mathrm{res}(\alpha, \beta) = 1$ for nonzero constants $\alpha, \beta \in R$.

Example 9.9. For the polynomials

$$f = 3yx^2 - y^3 - 4\,, \quad g = x^2 + y^3x - 9\,,$$

considered as elements of $\mathbf{Z}[y][x]$, we compute the resultant (with respect to x) as

$$\mathrm{res}_x(f, g) = \det \begin{bmatrix} 3y & 0 & -y^3-4 & 0 \\ 0 & 3y & 0 & -y^3-4 \\ 1 & y^3 & -9 & 0 \\ 0 & 1 & y^3 & -9 \end{bmatrix}$$

$$= -3y^{10} - 12y^7 + y^6 - 54y^4 + 8y^3 + 729y^2 - 216y + 16\,.$$

•

We note the obvious but important fact that $\mathrm{res}(f, g) \in R$; that is, it does not contain the indeterminate x. Several additional properties of resultants may be deduced almost directly from the properties of determinants.

Theorem 9.2. Let R be a commutative ring with identity, and let $f = \sum_{i=0}^{m} a_i x^i$ and $g = \sum_{i=0}^{n} b_i x^i$ be polynomials in R[x] of nonzero degrees m, n, respectively. Let $c \in$ R be a nonzero constant. Then:

(i) $\mathrm{res}(c, g) = c^n$;

(ii) $\mathrm{res}(f, f) = 0$;

(iii) $\mathrm{res}(f, g) = (-1)^{mn}\, \mathrm{res}(g, f)$;

(iv) $\mathrm{res}(cf, g) = c^n\, \mathrm{res}(f, g)$;

(v) $\mathrm{res}(x^k f, g) = b_0^k\, \mathrm{res}(f, g)$, $k > 0$;

(vi) if $\phi: \mathrm{R} \to \tilde{\mathrm{R}}$ is a homomorphism between commutative rings with identity, $\deg(\phi(f)) = m$ and $\deg(\phi(g)) = k$, $0 \le k \le n$, then

$$\phi(\,\mathrm{res}(f, g)\,) = \phi(a_m)^{n-k} \mathrm{res}(\,\phi(f), \phi(g)\,)\,.$$

Proof: Parts (i)-(iv) follow directly from Definition 7.3 and the properties of determinants. The proof of (v) is left as an exercise (see Exercise 9.13). Finally, we consider (vi). If $\deg(\phi(g)) = n$, the result holds since $\det \phi(M) = \phi(\det M)$, where M is the Sylvester matrix of f, g. If $\deg(\phi(g)) = k < n$, the first $n-k$ columns of the "g" entries become 0. So, the Sylvester matrix becomes upper-triangular for the first $n-k$ columns, with $\phi(a_m)$ on the diagonal. Reducing the determinant by cofactor expansion (along the first $n-k$ columns), we obtain the desired result. (Note that the resultant on the right hand side is obtained by deleting $n-k$ rows and columns from that on the left hand side.)

●

As we showed in Section 7.3, resultants have a basic connection with polynomial GCD's which is specified by Sylvester's criterion. In addition to this result and the useful relation (7.6), the resultant may be characterized by the following theorem.

Theorem 9.3. Let $f(x) = a_m \prod_{i=1}^{m} (x-\alpha_i)$ and $g(x) = b_n \prod_{i=1}^{n} (x-\beta_i)$ be polynomials over an integral domain D with indeterminates α_i, β_i. Then

(i) $$\mathrm{res}(f, g) = a_m^n\, b_n^m \prod_{i=1}^{m} \prod_{j=1}^{n} (\alpha_i - \beta_j) ;$$

(ii) $$\mathrm{res}(f, g) = a_m^n \prod_{i=1}^{m} g(\alpha_i) ;$$

(iii) $$\mathrm{res}(f, g) = (-1)^{mn}\, b_n^m \prod_{i=1}^{n} f(\beta_i)$$

Proof: We first observe that if we write $f(x)$ (respectively, $g(x)$) in the form $f(x) = \sum_{i=0}^{m} a_i x^i$ (respectively, $g(x) = \sum_{j=0}^{n} b_j x^j$), we may express the a_i (respectively, b_j) as symmetric functions of the α_i (respectively, β_j), multiplied by a_m (respectively, b_n). Since the resultant is homogeneous of degree n in the a_i and of degree m in the b_j, it must equal $a_m^n b_n^m$ times a symmetric function of the α_i and β_j. By Sylvester's criterion, res(f, g) vanishes when $\alpha_k = \beta_l$ since f, g have a common factor. Hence it is divisible by $(\alpha_i - \beta_j)$ for $1 \le i \le m$, $1 \le j \le n$. It is therefore also divisible by the product on the right hand side of (i), since all of its factors are relatively prime. Since this quantity and res(f, g) are of the same degree, they must be equal up to a constant multiple. They are in fact equal since they both contain the term

$$a_m^n\, b_n^m\, (-1)^{mn} \left(\prod_{j=1}^{n} \beta_j\right)^m = a_m^n\, b_0^m$$

(Here we have used the fact that $b_0 = (-1)^n b_n \prod_{j=1}^{n} \beta_j$.) The relations (ii) and (iii) follow directly from (i). ●

We mention that the above formulae are of particular interest when the indeterminates α_i, β_j are the roots of f, g. In fact, (i) is often presented as the definition of res(f, g).

Bézout's Determinant

The most obvious means of computing the resultant of two polynomials is by direct evaluation of Sylvester's determinant (which is of order $m+n$). But since Sylvester's matrix has a rather special structure, it is actually possible to express the determinant in terms of one of lower order. Specifically, we consider a method given by Bézout in the eighteenth century for computing the resultant as a determinant of order max(m, n). Assume (without loss of generality) that $m > n$, and consider the equations

$$f = a_m x^m + a_{m-1} x^{m-1} + \cdots + a_1 x + a_0 = 0,$$

$$x^{m-n} g = b_n x^m + b_{n-1} x^{m-1} + \cdots + b_1 x^{m-n+1} + b_0 x^{m-n} = 0.$$

Suppose we eliminate the leading terms by multiplying by b_n, a_m, respectively, and subtracting. If we repeat this using the multipliers $b_n x + b_{n-1}$, $a_m x + a_{m-1}$, etc., we obtain n new equations, namely (in reverse order):

$$
\begin{array}{l}
(a_m b_0 - a_{m-n} b_n) x^{m-1} + (a_{m-1} b_0 - a_{m-n} b_{n-1} - a_{m-n-1} b_n) x^{m-2} + \cdots - a_0 b_1 = 0 , \\
\qquad \vdots \\
(a_m b_{n-1} - a_{m-1} b_n) x^{m-1} + (a_m b_{n-2} - a_{m-2} b_n) x^{m-2} + \quad \cdots \quad - a_0 b_n = 0 .
\end{array}
\tag{9.30}
$$

Then, construct an additional $m-n$ equations as

$$
\begin{array}{rcl}
x^{m-n-1} g & = & b_n x^{m-1} + b_{n-1} x^{m-2} + \cdots + b_0 x^{m-n-1} = 0 , \\
x^{m-n-2} g & = & b_n x^{m-2} + \quad \cdots \quad + b_0 x^{m-n-2} = 0 , \\
\vdots & & \vdots \\
g & = & b_n x^n + \quad \cdots \quad + b_0 = 0 .
\end{array}
\tag{9.31}
$$

Now, *Bézout's determinant* is that of the coefficient matrix for the system given by (9.30), (9.31) in the unknowns $x^{m-1}, x^{m-2}, \ldots, 1$. Note that we need not explicitly subtract the terms which cancel when forming (9.30).

Example 9.10. Consider polynomials f, g of degree 3, 2 respectively. Following the method described above, Bézout's determinant is

$$
\bar{r} = \det \begin{bmatrix} a_3 b_0 - a_1 b_2 & a_2 b_0 - a_1 b_1 - a_0 b_2 & -a_0 b_1 \\ a_3 b_1 - a_2 b_2 & a_3 b_0 - a_1 b_2 & -a_0 b_2 \\ b_2 & b_1 & b_0 \end{bmatrix} .
$$

Note then that we might also pre-multiply the Sylvester matrix of f, g by a suitable matrix to obtain

$$
\begin{bmatrix} 1 & 0 & 0 & 0 & 0 \\ 0 & 1 & 0 & 0 & 0 \\ -b_2 & -b_1 & a_3 & a_2 & 0 \\ 0 & -b_2 & 0 & a_3 & 0 \\ 0 & 0 & 0 & 0 & 1 \end{bmatrix}
\begin{bmatrix} a_3 & a_2 & a_1 & a_0 & 0 \\ 0 & a_3 & a_2 & a_1 & a_0 \\ b_2 & b_1 & b_0 & 0 & 0 \\ 0 & b_2 & b_1 & b_0 & 0 \\ 0 & 0 & b_2 & b_1 & b_0 \end{bmatrix}
$$

$$= \begin{bmatrix} a_3 & a_2 & a_1 & a_0 & 0 \\ 0 & a_3 & a_2 & a_1 & a_0 \\ 0 & 0 & (a_3b_0-a_1b_2) & (a_2b_0-a_1b_1-a_0b_2) & -a_0b_1 \\ 0 & 0 & (a_3b_1-a_2b_2) & (a_3b_0-a_1b_2) & -a_0b_2 \\ 0 & 0 & b_2 & b_1 & b_0 \end{bmatrix} .$$

If we take determinants on both sides, we obtain

$$(a_3)^2 res(f, g) = (a_3)^2 \bar{r} ,$$

which proves the equivalence of Bézout's determinant to the resultant. (This also provides a clue to the proof in general.)

●

Resultants and the Division Algorithm

The above approach may be effective if the polynomials involved are sparse. However, it may be shown (cf. Exercise 9.6) that the (worst-case) behavior of the Bézout algorithm is exponential in the degree of the input. It is possible, though, to use the connection between resultants and GCD's to devise algorithms of polynomial complexity. Such a method may be obtained using the following result:

Theorem 9.4. Let D be an integral domain and let $p_1, p_2, p_3 \in$ D[x] be polynomials in x such that $p_1(x) = p_2(x)q(x) + p_3(x)$ and $\deg(p_1) = l$, $\deg(p_2) = m$, $\text{lcoeff}(p_2) = a_m$, $\deg(p_3) = n$. Then

$$\text{res}(p_2, p_1) = a_m^{l-n}\, \text{res}(p_2, p_3) .$$

Proof: Let α_i be the roots of p_2. Then using Theorem 9.3 (ii), we have

$$\begin{aligned} \text{res}(p_2, p_3) &= a_m^n \prod_{i=1}^{m} p_3(\alpha_i) = a_m^n \prod_{i=1}^{m} (p_1(\alpha_i) - p_2(\alpha_i) q(\alpha_i)) \\ &= a_m^n \prod_{i=1}^{m} p_1(\alpha_i) \\ &= a_m^{n-l}\, \text{res}(p_2, p_1) . \end{aligned}$$

●

Now, given $p_1 = f, p_2 = g \in$ D[x] such that GCD(p_1,p_2) $\in$ D (so that res(f, g) $\neq 0$), we use the Euclidean algorithm to construct a remainder sequence (cf. Chapter 7) for f, g (over F_D, if D is not a field), say $p_1, p_2, \ldots, p_k$. We then define $n_i = \deg(p_i)$, and note that

$$n_k = 0 \implies \text{res}(p_{k-1}, p_k) = \text{lcoeff}(p_k)^{n_{k-1}} ,$$

to obtain the formula

$$\operatorname{res}(p_1, p_2) = \operatorname{lcoeff}(p_k)^{n_{k-1}} \prod_{i=1}^{k-2} (-1)^{n_i n_{i+1}} \operatorname{lcoeff}(p_{i+1})^{n_i - n_{i+2}} . \tag{9.32}$$

Example 9.11. The polynomials of Example 9.9, viewed as elements of $\mathbf{Z}[y][x]$, are

$$f = (3y)x^2 - (y^3 + 4) , \quad g = x^2 + (y^3)x - 9 .$$

We may compute a remainder sequence over $\mathbf{Q}(y)$ via the Euclidean algorithm as follows:

$$p_1 = f , \; p_2 = g ,$$

$$p_3 = \operatorname{rem}_x(p_1, p_2) = (-3y^4)x + (-y^3 + 27y - 4) ,$$

$$p_4 = \operatorname{rem}_x(p_2, p_3) = \frac{-3y^{10} - 12y^7 + y^6 - 54y^4 + 8y^3 + 729y^2 - 216y + 16}{9y^8} .$$

Then by equation (9.32), we have

$$\operatorname{res}_x(f, g) = (p_4) \operatorname{lcoeff}_x(p_3)^2$$

$$= -3y^{10} - 12y^7 + y^6 - 54y^4 + 8y^3 + 729y^2 - 216y + 16 .$$

●

This has the now familiar disadvantage of requiring computations over the quotient field of D. However, it turns out that similar formulae may be derived for the superior reduced and subresultant PRS algorithms. These offer an obvious advantage when the coefficient domain D is, in turn, a multivariate polynomial domain.

The Modular Resultant Algorithm

In what follows we will be particularly interested in computing resultants of polynomials in $\mathbf{Z}[x_1, \ldots, x_s]$, viewed as univariate polynomials in

$$\mathbf{Z}[x_1, \ldots, x_{j-1}, x_{j+1}, \ldots, x_s][x_j]$$

for some $1 \le j \le s$. In such instances, we will write $\operatorname{res}_j(f, g)$ to denote the resultant with respect to the (main) variable x_j. It also happens that in many practical situations, it is necessary to compute resultants of large, dense polynomials. We will therefore briefly discuss a further improvement to the above methods, namely the *modular* resultant algorithm of Collins [6]. This scheme, in essence, uses homomorphisms and the CRA/interpolation (cf. Chapter 5) to reduce the multivariate problem over $\mathbf{Z}$ to a series of univariate ones over finite fields. Consider first a modular homomorphism ϕ_m: $\mathbf{Z} \to \mathbf{Z}_m$ where m is prime. We recall that by property (vi) of Theorem 9.2, the CRA allows us to compute $\operatorname{res}(f, g)$ if $\phi_{m_i}(\operatorname{res}(f, g))$ is known for sufficiently many prime moduli m_i such that $\deg(\phi_{m_i}(f)) = \deg(f)$ and

$\deg(\phi_{m_i}(g)) = \deg(g)$. The reconstruction via Algorithm 5.1 requires a bound on the coefficients of the resultant, which we derive as follows. For a multivariate polynomial in s variables,

$$q = \sum_{i=0}^{l} c_i(x_1, \ldots, x_{s-1}) x_s^i \in \mathbf{Z}[x_1, \ldots, x_s],$$

we define a norm by

$$\| q \|_+ = \begin{cases} | q | & , s = 0 \ (q \in \mathbf{Z}) \\ \sum_{i=0}^{l} \| c_i \|_+ & , s \geq 1 . \end{cases}$$

Then it is easily seen that for the polynomials $f = \sum_{i=0}^{m} a_i(x_1, \ldots, x_{s-1}) x_s^i$ and $g = \sum_{i=0}^{n} b_i(x_1, \ldots, x_{s-1}) x_s^i$, the resultant in x_s satisfies

$$\| \mathrm{res}_s(f, g) \|_+ \leq (m+n)! \, A^n B^m , \tag{9.33}$$

where

$$A = \max \{ \| a_i \|_+ : 0 \leq i \leq m \} , \quad B = \max \{ \| b_i \|_+ : 0 \leq i \leq n \} .$$

Next, each multivariate problem over a finite field $\mathbf{Z}_p$ is reduced to a series of univariate ones using evaluation homomorphisms of the form

$$\phi_{x_1 - c_1} \cdots \phi_{x_{s-1} - c_{s-1}} : \mathbf{Z}_p[x_1, \ldots, x_{s-1}] \rightarrow \mathbf{Z}_p$$

for which the leading coefficients of the $\phi_p(f)$, $\phi_p(g)$ do not vanish. In the Euclidean domain $\mathbf{Z}_p[x_s]$, the resultant may be easily computed by the formula (9.32). This result is then lifted back to $\mathbf{Z}_p[x_1, \ldots, x_s]$ by interpolation, with the degree bound

$$\deg_r(\mathrm{res}_s(f, g)) \leq \deg_r(f) \deg_s(g) + \deg_r(g) \deg_s(f) , \tag{9.34}$$

which holds for $1 \leq r < s$. Collins shows that the maximum time required to compute the resultant of two s-variate polynomials of maximum degree n and whose coefficients are at most δ digits is $O(n^{2s+1} \delta + n^{2s} \delta^2)$.

Resultants and Nonlinear Elimination

We have already seen that the resultant of two polynomials f, $g \in R[x]$ is an eliminant (i.e. $\mathrm{res}(f, g) \in R$) with a connection to the GCD of f, g. The following result shows precisely how resultants may be used to solve systems of algebraic equations.

Theorem 9.5 (Fundamental Theorem of Resultants). Let $\tilde{F}$ be an algebraically closed field, and let

$$f = \sum_{i=0}^{m} a_i(x_2, \ldots, x_r)x_1^i \ , \quad g = \sum_{i=0}^{n} b_i(x_2, \ldots, x_r)x_1^i$$

be elements of $\tilde{F}[x_1, \ldots, x_r]$ of positive degrees in x_1. Then if $(\alpha_1, \ldots, \alpha_r)$ is a common zero of f and g, their resultant with respect to x_1 satisfies

$$\mathrm{res}_1(f, g)(\alpha_2, \ldots, \alpha_r) = 0 . \tag{9.35}$$

Conversely, if the above resultant vanishes at $(\alpha_2, \ldots, \alpha_r)$, then at least one of the following holds:

(i) $a_m(\alpha_2, \ldots, \alpha_r) = \cdots = a_0(\alpha_2, \ldots, \alpha_r) = 0$;

(ii) $b_n(\alpha_2, \ldots, \alpha_r) = \cdots = b_0(\alpha_2, \ldots, \alpha_r) = 0$;

(iii) $a_m(\alpha_2, \ldots, \alpha_r) = b_n(\alpha_2, \ldots, \alpha_r) = 0$;

(iv) $\exists\, \alpha_1 \in \tilde{F}$ such that $(\alpha_1, \alpha_2, \ldots, \alpha_r)$ is a common zero of f and g.

Proof: The first part of the result is obvious using Theorem 7.1. Now assume that (9.35) holds, and that $a_m(\alpha_2, \ldots, \alpha_r) \neq 0$. Denote by ϕ the homomorphism corresponding to evaluation at $(\alpha_2, \ldots, \alpha_r)$. Then by Theorem 9.2 (vi), we have $\mathrm{res}_1(\phi(f), \phi(g)) = 0$. If $\deg(\phi(g)) = 0$, this implies (by the definition of the resultant) that

$$(\phi(g))^m = 0 \implies b_0(\alpha_2, \ldots, \alpha_r) = 0 ,$$

i.e. that (ii) holds. If $\deg(\phi(g)) > 0$, then (by Sylvester's criterion) $\phi(f)$ and $\phi(g)$ have a non-constant common divisor $h \in \tilde{F}[x_1]$. Since $\tilde{F}$ is algebraically closed, this has a root which we denote by α_1. It follows that f, g have a common root at $(\alpha_1, \alpha_2, \ldots, \alpha_r)$, i.e. that (iv) holds.

Similarly, if we assume that (9.35) holds and $b_n(\alpha_2, \ldots, \alpha_r) \neq 0$, we find that either (i) or (iv) holds. The case (iii) is the remaining possibility.

●

Let us assume for now that our system (9.29) has finitely many solutions. Then, as a consequence of Theorem 9.5, we may eliminate x_1 from this system to obtain new equations in $r-1$ variables without losing information about the common roots of (9.29). We may similarly proceed to eliminate $x_2, x_3, \ldots, x_{r-1}$ (in turn) and obtain a univariate polynomial $q_r(x_r)$. The set of roots of this polynomial then contains all possible values of x_r which may appear in roots of the original system. We also note that if we were able to eliminate *all* of the variables to obtain a nonzero constant (as the final resultant), there would be *no* solutions to the given system. This follows from Theorem 7.1, since each resultant along the way may be expressed as a combination of the original polynomials p_i. Hence there would exist polynomials $a_i \in \tilde{F}[x_1, \ldots, x_r]$ such that

$$\sum_{i=1}^{k} a_i(x_1, \ldots, x_r)\, p_i(x_1, \ldots, x_r) = 1 ,$$

which implies that the simultaneous vanishing of the p_i is impossible. This does not mean that our final resultant *must* be a constant when no common roots exist, however.

Example 9.12. Consider the system from Example 9.8 (i):

$$p_1 = xy - x + 2 = 0 ,$$

$$p_2 = xy - x + 3 = 0 .$$

We find that

$$\mathrm{res}_x(p_1, p_2) = y - 1 .$$

However, p_1 and p_2 have no common roots.

●

The above example shows that while we lose no information about the roots by computing resultants, we may gain some spurious information. The value $y = 1$ associated with the resultant above is called an *extraneous root*; in this case it arises from part (iii) of Theorem 9.5.

There is no particular significance to the order of the variables when we specify the polynomial ring as $D[x_1, x_2, \ldots, x_r]$. So, we may repeat the elimination process for different permutations of variables (e.g. consider our polynomials as elements of $D[x_1, \ldots, x_r, x_{r-1}]$ instead of $D[x_1, \ldots, x_{r-1}, x_r]$), to similarly obtain univariate polynomials $q_i(x_i)$ for $1 \le i \le r-1$. Then the (finite) set of r-tuples of roots of the q_i contains the set of roots of (9.29).

Example 9.13. For the system of Example 9.8 (ii),

$$p_1 = x^2 - y^2 = 0 ,$$

$$p_2 = x^2 - x + 2y^2 - y - 1 = 0$$

we find

$$q_2 = \mathrm{res}_x(p_1, p_2) = 9y^4 - 6y^3 - 6y^2 + 2y + 1$$
$$= (y - 1)(3y + 1)(3y^2 - 1) ,$$

$$q_1 = \mathrm{res}_y(p_1, p_2) = 9x^4 - 6x^3 - 6x^2 + 2x + 1 .$$

●

While the above certainly constitutes a ''fairly explicit'' representation of the solutions, several shortcomings of this approach are obvious. First, if the system has infinitely many solutions then no such finite inclusion of the roots will exist. Second, there remains the problem of somehow determining which r-tuples are in fact roots of the original system. For this purpose, a decision procedure of Tarski is suggested by Collins [6]. Finally, we observe that when there are many variables ($r \geq 3$), the elimination process may be quite time-consuming. Given r polynomials in r variables, we normally must compute $r(r-1)/2$ resultants to obtain a univariate polynomial. However, as each variable is eliminated, the degrees of the intermediate polynomials grow according to (9.34). In view of this exponential degree growth, the computational cost of constructing the univariate polynomial depends on the permutation of variables used, *including* which variable is not eliminated.

For the above reasons, it is more practical to obtain (or represent) the solutions in terms of only one elimination process. Formally, we require the following:

Definition 9.1. Let $P = \{ p_1, \ldots, p_k \} \subset \tilde{F}[x_1, \ldots, x_r]$ be a set of polynomials over an algebraically closed field $\tilde{F}$, where $k \geq r$. A *reduced system* for P is a list of r sets of polynomials $G = \{E_1, \ldots, E_r\}$, such that the polynomials of each $E_i \subset \tilde{F}[x_i, \ldots, x_r]$ for $1 \leq i \leq r$ contain all roots in the variables $x_i, \ldots, x_r$ which are possible values for common roots in E_{i-1}, where $E_0 = P$.

●

We note that G is the nonlinear analogue of a triangulation of the original system, in that the set E_i (cf. row/equation i in a reduced linear system) contains only the last $r-i+1$ variables. Once G is known, the problem of finding the roots of P is reduced to a series of univariate problems. Each common root, say α_r, of the univariate polynomial(s) in E_r may be substituted into the polynomials in E_{r-1}. The GCD of the resulting univariate polynomials will, in turn, yield a number of roots. Each of these roots, say α_{r-1}, corresponds to a pair (α_{r-1}, α_r) which is a root of $\{E_{r-1}, E_r\}$. This "back-solving" process continues until either an r-tuple $(\alpha_1, \ldots, \alpha_r)$ is obtained (i.e. a root of P), or a partial root cannot be extended for some reason. (The reader should compare the implications of cases (i)-(iii) of Theorem 9.5.)

The sets E_i are computed as sets of resultants (with respect to x_{i-1}) as follows. Suppose that initially each $p_i \in P$ contains all variables, and let $E_1 = P$. Then we construct E_2 by eliminating x_1:

$$E_2 = \{ \mathrm{res}_1(p_1, p_2), \mathrm{res}_1(p_1, p_3), \mathrm{res}_1(p_1, p_4), \ldots \} , \tag{9.36}$$

until (at least) $r-1$ independent nonzero elements are obtained. If there is a subset $\tilde{P} \subset P$ of polynomials which do not contain x_1, we simply set $E_1 = P - \tilde{P}$ and $E_2 = \tilde{P}$. The remainder of E_2 is then computed from E_1 as above. We similarly obtain E_i from E_{i-1} for $3 \leq i \leq r$. A possible algorithm to compute a reduced system for an arbitrary set of polynomials appears as Algorithm 9.2. Note that if the system is under-determined (i.e. $k < r$) or has infinitely many solutions, it may still be useful to carry out the elimination as far as possible, in spite of the fact that a reduced system will not exist.

Algorithm 9.2. Nonlinear Elimination Algorithm.

```
procedure NonlinearElim(P)

    # Given a set of polynomials P = { p_1, ..., p_k }
    # ⊂ F̃[x_1, ..., x_r] (none of which are constants or multiples
    # of any others), construct a reduced system if possible.
    # Note that we always remove the (multivariate) content,
    # because of exponential coefficient growth.

    # Distribute the p_i into subsets (by domains); also note
    # all pairs of (distinct) polynomials in each subset.

    for i from 1 to r do { E_i ← ∅ ; B_i ← ∅ }
    E_{r+1} ← ∅
    foreach p ∈ P do {
        if p ∈ F̃[x_j, ..., x_r] − F̃[x_{j+1}, ..., x_r] then {
            B_j ← B_j ∪ { [q, pp(p)] : q ∈ E_j }
            E_j ← E_j ∪ { pp(p) }      } }

    # Compute resultants until each E_i has a member, or
    # no more resultants are left.

    do until (E_i ≠ ∅, 1 ≤ i ≤ r) or (B_i = ∅, 1 ≤ i ≤ r−1) {
        k ← max { i : B_i ≠ ∅ }
        [f, g] ← an element of B_k ; B_k ← B_k − { [f, g] }
        p ← pp(res_k(f, g))
        if p ≠ 0 then {
            if p = 1 then return( no solutions )
            else if p ∈ F̃[x_j, ..., x_r] − F̃[x_{j+1}, ..., x_r] and
                    (no s ∈ E_j divides p) then {
                B_j ← B_j ∪ { [s, p] : s ∈ E_j } ; E_j ← E_j ∪ { p }   } } }
    return( { E_1, ..., E_r } )
end
```

Example 9.14. For the system

$$p_1 = 43 - 52x - 96y + 4z + 5yz + 26xz + 2xy = 0 ,$$

$$p_2 = -69 - 35y - 8yz - 14xy + 3xz - 75z^2 = 0 ,$$

$$p_3 = -44 - 3xz - 78xy - 8y^2 + 8z^2 = 0 ,$$

we set $E_1 = \{ p_1, p_2, p_3 \}$ and compute

$$e_{21} = \mathrm{res}_x(p_1, p_2)$$

$$= 3588 - 1923z + 2284y - 150yz - 1414y^2 + 3888z^2 - 373z^2y + 54y^2z - 1950z^3 ,$$

$$e_{22} = \mathrm{res}_x(p_1, p_3)$$

$$= 2288 - 1015z + 3266y + 24yz - 7072y^2 - 404z^2 + 31z^2y + 182y^2z - 16y^3 + 208z^3 .$$

Then $E_2 = \{ e_{21}, e_{22} \}$. Finally, $E_3 = \{ e_{31} \}$ where

$$e_{31} = \mathrm{res}_y(e_{21}, e_{22})$$

$$= 1048220182880z^9 - 587251755239544z^8 + 27783779484105880z^7 - 384968091539943616z^6 + 1241155878703492504z^5 - 1738563821260717568z^4 + 2173581545971252448z^3 - 2265902857293468008z^2 + 900061675251400176z - 80343644009944O576$$

$$= 8\,(2015808044z^7 - 1086471118279z^6 + 30254224198765z^5 - 560701530255559z^4 + 529526368775 92z^3 - 1015361249394 05z^2 + 22799235683318z - 40941522630424)(65z^2 - 1382z + 2453) .$$

●

We reiterate that the variables x_i may be eliminated in *any* order, and that the overall difficulty of this process is much greater for some orders than for others. Therefore, in practice one usually first selects x_1 as the variable which "appears to the lowest degree" in the set P (in some sense), and considers the choice of x_2 only *after* computing resultants in x_1 (see Exercise 9.15). (We further add that although the resultant pairs may be chosen in any order within each set B_i, it is important to take account of (9.34) here as well.) Ideally, for a system of r equations in r unknowns it should be possible to compute E_{i+1} after obtaining the $r-i+1$ polynomials of E_i. However, extra resultants may have to be computed if some of the first $r-i$ resultants computed above are identically zero. Indeed, this will happen when the polynomials involved are not relatively prime. The approach taken in Algorithm 9.2 is to consider new pairs of polynomials until the required number of nonzero elements is obtained. For example, if the polynomials of E_2 do not yield $r-2$ elements for E_3, we may augment (9.36) using

$$\mathrm{res}_1(p_2, p_3),\ \mathrm{res}_1(p_2, p_4),\ \ldots$$

and hope that some of the new resultants in x_2 are not zero. In general, this procedure cannot be avoided since we do not know in advance if P admits a reduced system or not (even in the case of r equations in r unknowns; e.g. Example 9.8 (iii)). Still, this alone is not an entirely satisfactory solution, since some of the resultants at the *next* stage may be zero. Therefore, a better approach is to ensure that the polynomials of each E_i are relatively prime. When polynomials $f, g \in E_i$ have a common divisor h, the system $\{f = g = 0\}$ is clearly equivalent to

$$\{\, h = 0 \,\} \quad \text{or} \quad \{\, \frac{f}{h} = \frac{g}{h} = 0 \,\}\,.$$

Yun [17] notes that a division into subsystems offers a further advantage, in that the subsequent resultants will be of lower degree. In fact, since

$$\mathrm{res}(fg, h) \;=\; \mathrm{res}(f, h)\,\mathrm{res}(g, h) \tag{9.37}$$

(Exercise 9.14), we may reduce the growth of intermediate results by complete system subdivision at each resultant step.

Example 9.15. For the polynomials of Example 9.8 (iii),

$$\begin{aligned} p_1 &= x^2c + xy - yc - 1 = 0\,, \\ p_2 &= 2xy^2 + yc^2 - c^2 - 2 = 0\,, \\ p_3 &= x + y^2 - 2 = 0\,, \end{aligned}$$

where c is considered a parameter, we proceed by temporarily viewing the $p_i \in \mathbf{Z}[c][x,y]$ as elements of $\mathbf{Z}[x,y,c]$. (Note that we might also use $\mathbf{Z}[y,x,c]$.) Then we find $E_2 = \{e_{21}, e_{22}\}$, where

$$\begin{aligned} e_{21} &= \mathrm{res}_x(p_3, p_1) \\ &= y^4c - y^3 - 4y^2c - yc + 2y + 4c - 1 \\ &= (y-1)(y^3c + y^2c - y^2 - 3yc - y - 4c + 1)\,, \end{aligned}$$

$$\begin{aligned} e_{22} &= \mathrm{res}_x(p_3, p_2) \\ &= -2y^4 + 4y^2 + yc^2 - c^2 - 2 \\ &= -(y-1)(2y^3 + 2y^2 - 2y - c^2 - 2)\,. \end{aligned}$$

Since these are actually univariate polynomials in $\mathbf{Z}[c][y]$, we need only consider $\mathrm{GCD}(e_{21}, e_{22}) = y - 1$. Hence, we find that the only solutions are $\{x = 1,\ y = 1\}$. Note that, although $E_2 = \{e_{21}\}$ would have completed a reduced system, the extraneous roots are more easily detected using the "extra" polynomial e_{22}.

If we wish to solve the above system for x, y, c, we compute E_2 as before except that its elements are now polynomials in $\mathbf{Z}[y,c]$. Then

$$\mathrm{res}_y(e_{21}, e_{22}) = 0\,,$$

since e_{21} and e_{22} are not relatively prime. However, we may use the factorizations above to derive the subsystems

$$p_1 = 0\,,\ p_2 = 0\,,\ p_3 = 0\,,\ y - 1 = 0\,,$$

and

$$\begin{aligned} &p_1 = 0\,,\ p_2 = 0\,,\ p_3 = 0\,, \\ &\tilde{e}_{21} = y^3c + y^2c - y^2 - 3yc - y - 4c + 1 = 0\,, \\ &\tilde{e}_{22} = 2y^3 + 2y^2 - 2y - c^2 - 2 = 0\,. \end{aligned}$$

The former is not a reduced system, but quickly yields the family of solutions $\{x = 1,\ y = 1\}$ nonetheless. For the latter, we compute

$$\begin{aligned} \tilde{e}_{31} &= \mathrm{res}_y(\tilde{e}_{21}, \tilde{e}_{22}) \\ &= -c^9 + 14c^7 + 10c^6 - 10c^5 + 2c^4 + 56c^3 - 64c^2 + 24c + 8 = 0\,, \end{aligned}$$

and hence the reduced (sub-)system

$$\tilde{G} = \{\ \{p_1, p_2, p_3\}, \{\tilde{e}_{21}, \tilde{e}_{22}\}, \{\tilde{e}_{31}\}\ \}\,.$$

It may be shown that the roots of $\tilde{G}$ do not belong to the family above.

●

Of course, it is not generally possible to exactly solve univariate equations (i.e. in terms of radicals). However the back-solving process used for triangular linear systems may in principle be generalized to the nonlinear case. The resulting algorithm appears as Algorithm 9.3.

We conclude this section with a brief discussion of some important practical considerations. First, it seems that the non-existence of a reduced system for a given set may only be established after all possible resultant pairs have been tried (at considerable computational expense). Even then, Algorithm 9.2 may arrive at a collection of polynomials which is not

Algorithm 9.3. Solution of a Nonlinear System of Equations.

```
procedure NonlinearSolve(P)

  # Given a set P ⊂ F̃[x_1, ..., x_r] corresponding to a
  # system of k nonlinear equations (k ≥ r) with finitely
  # many solutions, find the common roots of P.

  G ← NonlinearElim(P) ; roots ← ∅

  # If we obtain a nonzero constant, there are no solutions.

  if no solutions exist then return(roots)
  else {

    # The reduced system G has the form {E_1, ..., E_r}.

    # Find the roots of the univariate polynomials.

    q ← GCD( polynomials in E_r )
    roots ← roots ∪ { (α) : q(α) = 0 }

    # Now extend each partial root by back-solving.

    for j from r−1 by −1 to 1 do {
      R ← ∅
      foreach (α_{j+1}, ..., α_r) ∈ roots do {
        U_j ← {e(x_j,α_{j+1}, ...,α_r) : e ∈ E_j} − {0}
        q ← GCD( polynomials in U_j )

        # Note that q may sometimes be constant.

        R ← R ∪ { (α,α_{j+1}, ...,α_r) : q(α) = 0 }  }
    roots ← R  }
  return(roots)
end
```

really a reduced system. However, such sets will be detected only when the back-solving procedure fails. (See Exercise 9.22, for example.) We shall find a much nicer solution to this problem in the next chapter, where a more refined counterpart to the reduced system (i.e. a Gröbner basis) is examined. Still, in such cases the back-solving process may often be continued by solving for x_{i-1} in terms of x_i when E_i cannot be completed by Algorithm 9.2 (or when U_i is empty in Algorithm 9.3). This reformulation is analogous to the treatment of a linear, homogeneous system, except that the result here will be an algebraic function of x_i.

Second, we recall that the resultants are computed from *pairs* of polynomials; but, the common roots of a pair f, g are not necessarily those of a triple f, g, h. Therefore, it is possible that the introduction at some point of extraneous roots (i.e. those which cannot be extended to complete common roots) to the reduced system will cause unnecessary expression growth. (Note that some of the GCD's computed in Algorithm 9.3 may be nonzero constants.) We can estimate the magnitude of this growth if we consider a system of r homogeneous polynomials in r variables of degree d. By Bézout's theorem, the number of solutions of such a system (and so, the required degree of a univariate polynomial in a reduced system) is d^r. On the other hand, it may be shown that the use of resultants will yield a final univariate polynomial of degree $d^{2^{r-1}}$. Once such a univariate polynomial is obtained, extraneous factors may often be removed by computing extra resultants in each E_i. Then, the set E_r may be reduced to the GCD of any polynomials it contains. In fact if some extra resultants are computed in the early stages of the elimination, perhaps less complicated resultants may be found later on (since there are more pairs of polynomials to choose from). Nonetheless, some intermediate expression swell is unavoidable. Finally, even when a univariate polynomial of degree less than 5 is found, it may not be possible to carry out the back-solving process (i.e. exactly) in practice when nested radicals occur (or when the coefficient domain involves extra parameters).

Exercises

1. Suppose that a matrix with integer entries $a_{ij}^{(0)}$ of length at most δ digits is transformed according to (9.5) and (9.6). Derive a bound on the size of the entries of $a_{ij}^{(k)}$.

2. Prove that if $a_{ij}^{(k)}$ is defined by (9.7) - (9.8), the solutions of
$$\sum_{j=1}^{n} a_{kj}^{(k-1)} x_j^* = \det(A)\, a_{k,n+1}^{(k-1)}$$
are the same as those of (9.13) for $1 \le k \le n$.

3. Using your favorite computer algebra system, implement the single-step fraction-free elimination scheme (9.5) - (9.6). You should include pivoting (row interchanges) in case $a_{kk}^{(k-1)} = 0$, or in case $a_{kk}^{(k-1)}$ is larger than other possible pivots. Test your code on Example 9.3, and on the following matrices:

(a) $$A = \mathrm{LCM}(2,3,4,5,6,7)\begin{bmatrix} 1 & \frac{1}{2} & \frac{1}{3} & \frac{1}{4} \\ \frac{1}{2} & \frac{1}{3} & \frac{1}{4} & \frac{1}{5} \\ \frac{1}{3} & \frac{1}{4} & \frac{1}{5} & \frac{1}{6} \\ \frac{1}{4} & \frac{1}{5} & \frac{1}{6} & \frac{1}{7} \end{bmatrix};$$

(b) $$A = \begin{bmatrix} 1 & 1+x & 1+x+x^2 & 1+x+\cdots+x^3 \\ 1+x & 1+x+x^2 & 1+x+\cdots+x^3 & 1+x+\cdots+x^4 \\ 1+x+x^2 & 1+x+\cdots+x^3 & 1+x+\cdots+x^4 & 1+x+\cdots+x^5 \\ 1+x+\cdots+x^3 & 1+x+\cdots+x^4 & 1+x+\cdots+x^5 & 1+x+\cdots+x^6 \end{bmatrix};$$

(c) $$A = \begin{bmatrix} 1 & x & y & z \\ x & 1 & x & y \\ y & x & 1 & x \\ z & y & x & 1 \end{bmatrix};$$

(d) $$A = \begin{bmatrix} 1 & t & t^2 & t^3 \\ 1 & x & x^2 & x^3 \\ 1 & y & y^2 & y^3 \\ 1 & z & z^2 & z^3 \end{bmatrix}.$$

4. Obtain formulas (9.14) - (9.20) by applying the single-step scheme to itself. How many additions/multiplications are saved in this manner?

5. Formulate a pivoting strategy for the two-step fraction-free elimination scheme. Implement this scheme and test your code on the matrices of Exercise 3.

6. Show that the number of multiplications necessary for the evaluation of an n-th order determinant by minor expansion is $n(2^{n-1}-1)$. (This shows that the time complexity of Bézout's resultant scheme is exponential in the degrees of the input.) *Hint:* During an expansion along, say, rows how many distinct minors of order k are there?

7. Implement the minor expansion algorithm for computing determinants, avoiding redundant computation of subdeterminants (cf. Exercise 6). Test your code and compare it with Gaussian elimination on the matrices of Exercise 3, (a)-(d). Repeat this comparison (including the respective storage requirements) as each matrix is extended to order 5, 6, 7, 8 (with a suitable limit on the time for each computation).

8. Derive a bound for the adjoint solution in the case of integer coefficients, i.e. $\max(|d|, \|\mathbf{x}^*\|_\infty)$, which is tighter than (9.28). *Hint:* Apply Hadamard's inequality directly to (9.3).

9. By repeating Example 9.6 for a suitable number of prime moduli (see Exercise 8), obtain the solutions of Example 9.5 (over **Q**) via the CRA.

10. Use (single-step) fraction-free elimination and back-solving to solve linear systems $A\mathbf{x} = \mathbf{b}$ for $\mathbf{b} = [1,2,3,4]^T$ and for the following coefficient matrices:

    (a) the (scaled) Hilbert matrix of Exercise 3(a);

    (b) the univariate band matrix whose nonzero entries are defined by

    $$a_{ii} = c^2 \, , \quad a_{i,i+1} = 1 + c \, , \quad a_{i+1,i} = 1 - c \, ;$$

    (c) the multivariate band matrix whose nonzero entries are defined by

    $$a_{ii} = c \, , \quad a_{i,i+1} = d \, , \quad a_{i,i+2} = e \, , \quad a_{i+1,i} = f \, , \quad a_{i+2,i} = g \, .$$

11. Compare the solution by minor expansion (and Cramer's rule) to the fraction-free elimination method as the systems of Exercise 10 are extended to orders 5, 6, 7, 8. (The vector $\mathbf{b}$ should be extended to $[1, \ldots, n]^T$.)

12. Use the CRA/interpolation method to solve the system of Exercise 10(c).

13. Prove part (v) of Theorem 9.2. *Hint:* First prove that $\mathrm{res}(xf, g) = b_0 \mathrm{res}(f, g)$ directly from Definition 7.3.

14. Derive formula (9.37).

15. Consider the system of Example 9.14.

    (a) Which permutation(s) of variables will yield the univariate polynomial of smallest degree? How might this be predicted?

    (b) Show that the polynomial "e_{31}" contains an extraneous factor.

16. Consider the system of nonlinear equations over $\mathbf{Q}$

    $$3yx^2 - y^3 - 4 = 0 \, , \quad x^2 + y^3 x - 9 = 0$$

    from Example 9.9. How many solutions (in x, y) does this system possess? How many *real* solutions are there? How many of these real solutions can you approximate (i.e. not using nonlinear elimination) using a fixed-point iteration technique (e.g. Newton's method)?

17. Solve the following system (explicitly!) for x, y, z in terms of parameter c:

    $$x^2 + y^2 + 2cz = 0 \, ,$$

    $$cxy - z^2 = 0 \, ,$$

    $$x + y + z - c = 0 \, .$$

18. For the polynomials

$$f = 5x^2y^2 + x^2 - 2xy + 3x + 2y^2 + 3y - 7\,,$$
$$g = x^2y - 2x^2 + xy^2 + 9xy + 3x - 6y^2 - 2y - 9\,,$$

compute $\mathrm{res}_x(f, g)$:

(a) directly;

(b) by Bézout's determinant;

(c) using the Euclidean algorithm and (9.32).

19. Use the results of Section 7.3 and the fact that $\mathrm{res}(f,g) = S(0,f,g)$ to derive the following formula: if $p_1, p_2, \ldots, p_k$ is a reduced PRS for $p_1, p_2 \in D[x]$, $n_i = \deg(p_i)$, $\delta_i = n_i - n_{i+1}$, and $n_k = 0$, then

$$\mathrm{res}(p_1, p_2) = (-1)^{\sigma_k}\, \mathrm{lcoeff}(p_k)^{\delta_{k-1}} \left[\prod_{i=2}^{k-1} \mathrm{lcoeff}(p_i)^{-\delta_{i-1}(\delta_i - 1)}\right],$$

where $\sigma_k = \sum_{i=1}^{k-1} n_i n_{i+1}$.

20. Using your favorite computer algebra system, implement the following resultant algorithms:

(a) Bézout's method;

(b) the reduced PRS method;

(c) the modular method.

Test your implementations on the polynomials of Example 9.14.

21. Compare the schemes implemented in Exercise 20 by computing the resultant with respect to x of the following pairs of polynomials:

$$\text{(a)}\ f = 43x^{25} + 11x^{24} - 7x^{21} + 5x^{20} - 83x^{19} + 53x^{17} + 3x^{16} + 84x^{15} - 93x^{14} + 4x^{13} - 178x^{12} + 9x^{11} + 83x^{10} + 7x^9 - x^8 - 712x^7 + 3x^6 + 93x^5 - 27x^4 + 110x^3 - 61x^2 + 3x + 301\,,$$

$$g = 15x^{23} + 8x^{22} - 16x^{21} + 832x^{20} - 3x^{19} + 93x^{17} + 47x^{15} - 15x^{14} + 48x^{13} + 53x^{12} - 116x^{11} + 9x^{10} + 127x^9 + 67x^8 + 2x^7 + 10x^6 + 108x^5 - 93x^4 - x^3 + 17x^2 + 55x - 19\,;$$

(b) $f = 4 + 9y - 11y^2 - 11y^3 + 8y^4 - 4y^5 - 3x + 6xy + 2xy^2$
$- 9xy^3 - 10xy^4 + 5x^2 - 2x^2y - 4x^2y^2 - x^2y^3 + 10x^3$
$- 7x^3y + 6x^3y^2 - 2x^4 + 11x^4y - 6x^5$,

$g = 7 + y - 2y^2 + 6y^3 - 10y^4 - 9y^5 - 4x - 9xy - 5xy^2$
$- 10xy^3 - 2xy^4 + 11x^2 - 9x^2y + 8x^2y^2 + 11x^2y^3 - 3x^3$
$- 2x^3y - 2x^3y^2 - 3x^4 - 3x^4y + 3x^5$;

(c) $f = -10 + 4x + 11x^3 + 10x^2 - 3v + 7v^2y - 5v^2x + 7v^2z$
$+ 3vx^2 + 7v^3 + 10v^2 - 9vz - 6vz^2 - 7vy - 6vy^2 - 7vx$
$- 3xz - 6xz^2 - 3x^2z - 2y^2z + 3yz + 3yz^2 + 6z^2 - 8z^3 - 6xyz$
$- 8vyz + 4vxy - vxz + 3xy - 5x^2y + 9y + 11y^2 + 9y^3 + 8z$,

$g = -3 + 4x + 7x^3 + 3v + 7v^2y - 7v^2x + 10v^2z - 10vx^2 - 6v^3$
$+ v^2 - 9vz + vz^2 + 6vy - 6vy^2 - 7vx + 11xz + xz^2 + 2x^2z$
$- 3y^2z - 2yz + 4yz^2 + 3z^2 + 2z^3 + 3xyz - vyz$
$+ 4vxy - 4vxz - 10xy^2 - 2x^2y + 3y - 11y^2 - 7y^3 - 9z$.

22. Apply Algorithm 9.2 to the polynomials of Example 9.8 (iii), considered as elements of $\mathbf{Q}[c,x,y]$. Is the result a reduced system? Suggest how Algorithm 9.3 may be generalized to treat this case.

23. Using the resultant scheme of your choice, solve the following systems:

(a) $yz + 19wx + 5w^3 + 45 = 0$,
$z - 7y + 9x - w + 44 = 0$,
$53yz + 2wx + 11xy + 454 = 0$,
$3w^2 + wyz - 6y + 30 = 0$,

for x, y, z, w. *Hint:* You will not be able to find the solutions explicitly, but your reduced system should contain a univariate polynomial of degree 12.

(b) $27y^2z^2 + 9xy^2 - 36xz^2 - 45z^2 - 7x - 8 = 0$,
$2x^2z + y^2z - 3z + x + y + 1 = 0$,
$9xy^2 + 9y^3 - 12xy - 7x - 15y - 8 = 0$,

for x, y, z. *Hint:* There are infinitely many solutions, consisting of a one-parameter family plus exactly 9 numerical roots.

References

1. E.H. Bareiss, ''Sylvester's Identity and Multistep Integer-Preserving Gaussian Elimination,'' *Math. Comp.*, **22**(103) pp. 565-578 (1968).
2. E.H. Bareiss, ''Computational Solutions of Matrix Problems Over an Integral Domain,'' *J. Inst. Maths Applcs*, **10** pp. 68-104 (1972).
3. S. Cabay, ''Exact Solution of Linear Equations,'' pp. 392-398 in *Proc. SYMSAM '71*, ed. S.R. Petrick, ACM Press (1971).
4. S. Cabay and T.P.L. Lam, ''Congruence Techniques for the Exact Solution of Integer Systems of Linear Equations,'' *ACM TOMS*, **3**(4) pp. 386-397 (1977).
5. J.F. Canny, E. Kaltofen, and L. Yagati, ''Solving Systems of Non-Linear Polynomial Equations Faster,'' pp. 121-128 in *Proc. ISSAC '89*, ed. G.H. Gonnet, ACM Press (1989).
6. G.E. Collins, ''The Calculation of Multivariate Polynomial Resultants,'' *J. ACM*, **18**(4) pp. 515-532 (1971).
7. G. E. Collins, ''Quantifier Elimination for Real Closed Fields: A Guide to the Literature,'' pp. 79-81 in *Computer Algebra - Symbolic and Algebraic Computation (Second Edition)*, ed. B. Buchberger, G.E. Collins and R. Loos, Springer-Verlag, Wien - New York (1983).
8. W.M. Gentleman and S.C. Johnson, ''Analysis of Algorithms, A Case Study: Determinants of Matrices with Polynomial Entries,'' *ACM TOMS*, **2**(3) pp. 232-241 (1976).
9. M.L. Griss, ''The Algebraic Solution of Sparse Linear Systems via Minor Expansion,'' *ACM TOMS*, **2**(1) pp. 31-49 (1976).
10. J.A. Howell and R.T. Gregory, ''An Algorithm for Solving Linear Algebraic Equations using Residue Arithmetic I, II,'' *BIT*, **9** pp. 200-224, 324-337 (1969).
11. D. Lazard, ''Systems of Algebraic Equations,'' pp. 88-94 in *Proc. EUROSAM '79, Lecture Notes in Computer Science* **72**, ed. W. Ng, Springer-Verlag (1979).
12. J.D. Lipson, ''Symbolic methods for the computer solution of linear equations with applications to flowgraphs,'' pp. 233-303 in *Proc. of the 1968 Summer Inst. on Symb. Math. Comp.*, ed. R. G. Tobey, (1969).
13. M.T. McClellan, ''The Exact Solution of Systems of Linear Equations with Polynomial Coefficients,'' *J. ACM*, **20**(4) pp. 563-588 (1973).
14. T. Sasaki and H. Murao, ''Efficient Gaussian Elimination Method for Symbolic Determinants and Linear Systems,'' *ACM TOMS*, **8**(3) pp. 277-289 (1982).
15. H. Takahasi and Y. Ishibashi, ''A New Method for 'Exact Calculation' by a Digital Computer,'' *Inf. Processing in Japan*, **1** pp. 28-42 (1961).

16. B.L. van der Waerden, *Modern Algebra (Vols. I and II),* Ungar (1970).

17. D.Y.Y. Yun, ''On Algorithms For Solving Systems of Polynomial Equations,'' *ACM SIGSAM Bull.*, **27** pp. 19-25 (1973).

CHAPTER 10

GRÖBNER BASES FOR POLYNOMIAL IDEALS

10.1. INTRODUCTION

We have already seen that, among the various algebraic objects we have encountered, polynomials play a central role in symbolic computation. Indeed, many of the (higher-level) algorithms discussed in Chapter 9 (and later in Chapters 11 and 12) depend heavily on computation with multivariate polynomials. Hence, considerable effort has been devoted to improving the efficiency of algorithms for arithmetic, GCD's and factorization of polynomials. It also happens, though, that a fairly wide variety of problems involving polynomials (among them, simplification and the solution of equations) may be formulated in terms of *polynomial ideals*. This should come as no surprise, since we have already used particular types of ideal bases (i.e. those derived as kernels of homomorphisms) to obtain algorithms based on interpolation and Hensel's lemma. Still, satisfactory *algorithmic* solutions for many such problems did not exist until the fairly recent development of a special type of ideal basis, namely the Gröbner basis.

We recall that, given a commutative ring with identity R, a non-empty subset $I \subseteq R$ is an *ideal* when:

(i) $p, q \in I \;\Rightarrow\; p - q \in I$;

(ii) $p \in I, r \in R \;\Rightarrow\; r\,p \in I$.

Every (finite) set of polynomials $P = \{ p_1, \ldots, p_k \} \subset F[x_1, \ldots, x_n]$ generates an ideal

$$<P> \;=\; <p_1, \ldots, p_k> \;=\; \{ \sum_{i=1}^{k} a_i p_i \;:\; a_i \in F[x_1, \ldots, x_n] \} .$$

The set P is then said to form a *basis* for this ideal. Unfortunately, while P generates the (infinite) set $<P>$, the polynomials p_i in P may not yield much insight into the nature of this ideal. For example, a set of simple polynomials over **Q** such as

$$p_1 = x^3yz - xz^2, \;\; p_2 = xy^2z - xyz, \;\; p_3 = x^2y^2 - z^2$$

generates a polynomial ideal in $\mathbf{Q}[x,y,z]$; namely,

$$< p_1, p_2, p_3 > = \{ a_1 \cdot p_1 + a_2 \cdot p_2 + a_3 \cdot p_3 : a_1, a_2, a_3 \in \mathbf{Q}[x,y,z] \} .$$

It is not difficult to show that $q = x^2yz - z^3$ is a member of this ideal since one can find polynomials a, b, c such that

$$q = ap_1 + bp_2 + cp_3 .$$

In this case, one could eventually determine these a, b, c by trial-and-error. However, it is generally a difficult problem to decide whether a given q is in the ideal $< p_1, \ldots, p_k >$ for arbitrary polynomials p_i. We mention that the ''ideal membership'' problem (which was considered, but not fully solved by Hermann [23] in 1926) may be viewed as an instance of the ''zero-equivalence'' problem studied in Chapter 3. For example, deciding if $q \in < p_1, p_2, p_3 >$ in the previous problem is the same as deciding if q simplifies to 0 with respect to the side relations

$$x^3yz - xz^2 = 0 , \quad xy^2z - xyz = 0 , \quad x^2y^2 - z^2 = 0 .$$

It is easy to show that for a fixed set of polynomials P, the relation ~ defined by

$$q_1 \sim q_2 \quad \Leftrightarrow \quad q_1 - q_2 \in < P >$$

is an equivalence relation. Hence, both of these problems will be solved *if* we can find a normal function (i.e. a zero-equivalence simplifier) for $F[x_1, \ldots, x_n]$ with respect to ~.

Consider also the problem of solving a system of nonlinear equations

$$p_1 = 0 , \; p_2 = 0 , \; \ldots , \; p_k = 0 ,$$

where each $p_i \in F[x_1, \ldots, x_n]$ and F is a field. In the previous chapter we used resultants to transform a set of polynomials $P = \{ p_1, \ldots, p_k \}$ into an equivalent set (i.e. one with all of the original common zeros) from which the roots could be more easily obtained. For example, the nonlinear system of equations

$$\{ x^2y - x^2 + 5xy - 2y + 1 = 0 , \; xy^2 - 2xy + x - 4y^3 - 7 = 0 \}$$

may be ''reduced'' into the system

$$\{ x^2y - x^2 + 5xy - 2y + 1 = 0 , \; xy^2 - 2xy + x - 4y^3 - 7 = 0 ,$$
$$16y^7 + 4y^6 - 42y^5 + 85y^4 - 37y^3 - 56y^2 + 78y - 48 = 0 \} ,$$

which is then solved. However, we noted in Chapter 9 that such a reduced system will not always exist; moreover one cannot always tell from a reduced system whether a given system of equations is solvable or not. In hindsight, it should be clear that a reduced system for P is simply an alternate (but more useful) basis for the ideal $< P >$. What we would like, however, is an alternate ideal basis which always exists and from which the existence and uniqueness of solutions (as well as the solutions themselves) may easily be determined.

It is reasonable to wonder if the above problems might be solvable, if only an arbitrary ideal basis could be transformed into a sufficiently potent form. In fact, Hironaka [24] established the *existence* of such a basis (which he called a ''standard basis'') for ideals of formal power series in 1964. However it was Buchberger [5] who, in his Ph.D. thesis, first presented an *algorithm* to perform the required transformation in the context of polynomial ideals. He soon named these special bases *Gröbner bases* (after his supervisor, W. Gröbner), and refined both the concept and algorithm further. Hence, most of the concepts (and, in fact, many of the proofs) we present are due to Buchberger. Today, most modern computer algebra systems include an implementation of Buchberger's algorithm.

In this chapter, we will first present the concepts of reduction and Gröbner bases, in terms of the ideal membership problem. We develop Buchberger's algorithm for computing Gröbner bases, and consider its practical improvement. Various extensions of the algorithm, and its connection with other symbolic algorithms are (briefly) discussed. Finally, we examine some of the applications of Gröbner bases, including solving systems of algebraic equations.

10.2. TERM ORDERINGS AND REDUCTION

For univariate polynomials, the zero-equivalence problem is easily solved since F[x] is a Euclidean domain. Hence, we can simplify with respect to univariate polynomials using ordinary polynomial division (i.e. the ''rem'' function). For multivariate domains, however, the situation is much less clear, as our previous example shows. Still, it was pointed out in Chapter 5 that a multivariate polynomial domain over a field (while not a Euclidean domain, or even a principal ideal domain) is a Noetherian ideal domain; that is, every ideal in such a domain has a finite basis. Fortunately, this is almost enough to allow us to solve the above problems - and more. The missing (but easily supplied) element is a small amount of additional structure on the polynomial ring, which will permit a more algorithmic treatment of multivariate polynomials. As in earlier chapters, we will denote the polynomial ring by F[**x**] when the (ordered) set of variables $\mathbf{x} = (x_1, x_2, \ldots, x_n)$ is understood.

Orderings of Multivariate Terms

We begin by defining the set of *terms in* **x** by

$$T_{\mathbf{x}} = \{ x_1^{i_1} \cdots x_n^{i_n} : i_1, \ldots, i_n \in \mathbf{N} \},$$

where **N** is the set of non-negative integers. Note that this constitutes a (vector space) basis for F[**x**] over the field (coefficient domain) F. We will require that these terms be ordered as follows.

Definition 10.1. An *admissible total ordering* $<_T$ for the set $T_{\mathbf{x}}$ is one such that:

(i) $1 \le_T t$;

(ii) $s <_T t \;\Rightarrow\; s \cdot u <_T t \cdot u$

for all $s, t, u \in T_{\mathbf{x}}$, where $1 = x_1^0 \cdots x_n^0$.

•

A wide variety of admissible orderings are possible. (See, for example, Exercise 10.17.) However, we will discuss the two which are most common in the literature (and which seem to be the most useful in practice).

Definition 10.2. The *(pure) lexicographic* term ordering is defined by

$$s = x_1^{i_1} \cdots x_n^{i_n} <_L x_1^{j_1} \cdots x_n^{j_n} = t \quad \Leftrightarrow$$

$$\exists\, l \text{ such that } i_l < j_l \text{ and } i_k = j_k,\ 1 \le k < l\ .$$

•

Note that by specifying the polynomial ring as $F[x_1, \ldots, x_n]$, the precedence

$$x_1 >_L x_2 >_L \cdots >_L x_n$$

is implied.

Example 10.1. The trivariate terms in (x, y, z) are lexicographically ordered

$$1 <_L z <_L z^2 <_L \cdots <_L y <_L yz <_L yz^2 <_L \cdots$$

$$<_L y^2 <_L y^2 z <_L \cdots <_L x <_L xz <_L \cdots <_L xy <_L \cdots .$$

•

Definition 10.3. The *(total) degree* (or *graduated*) term ordering is defined by

$$s = x_1^{i_1} \cdots x_n^{i_n} <_D x_1^{j_1} \cdots x_n^{j_n} = t \quad \Leftrightarrow$$

$$\deg(s) < \deg(t)\ , \text{ or}$$

$$\{ \deg(s) = \deg(t) \text{ and } \exists\, l \text{ such that } i_l > j_l \text{ and } i_k = j_k,\ l < k \le n \}\ .$$

•

We note that terms of equal total degree are ordered using an *inverse* lexicographic ordering, which is admissible within these graduations. Obviously, a different term ordering results from using the regular lexicographic ordering for this purpose. Both types are referred to as "total degree" orderings in the literature; however, we will use Definition 10.3 exclusively.

Example 10.2. The trivariate terms in (x, y, z) are degree-ordered

$$1 <_D z <_D y <_D x <_D$$
$$<_D z^2 <_D yz <_D xz <_D y^2 <_D xy <_D x^2$$
$$<_D z^3 <_D yz^2 <_D xz^2 <_D y^2z <_D xyz <_D \cdots .$$

●

Clearly, any polynomial in F[**x**] contains a monomial whose term is maximal with respect to a given term ordering $<_T$. We will adopt the following notation.

Definition 10.4. The *leading monomial* of $p \in$ F[**x**] with respect to $<_T$ is the monomial appearing in p whose term is maximal among those in p. We denote this by $\mathrm{M}_T(p)$, or simply by $\mathrm{M}(p)$ if the term ordering $<_T$ is understood. Also define hterm(p) to be the maximal ("head") term, and hcoeff(p) to be the corresponding coefficient, so that

$$\mathrm{M}(p) = \mathrm{hcoeff}(p)\ \mathrm{hterm}(p) .$$

We adopt the convention that hcoeff(0) = 0 and hterm(0) = 1.

●

Example 10.3. Suppose we consider

$$p = -2x^2yz + x^2y^2 + x^2z^2 + x^2y + 2xy^2z^2 - 3xyz^3 - xy + yz + z^2 + 5$$

as an element of **Q**[x,y,z]. We may write p so that its terms are in descending order with respect to $<_D$, as

$$p = 2xy^2z^2 - 3xyz^3 + x^2y^2 - 2x^2yz + x^2z^2 + x^2y - xy + yz + z^2 + 5 .$$

Clearly, then, we have

$$\mathrm{M}(p) = 2xy^2z^2 ,\ \ \mathrm{hterm}(p) = xy^2z^2 ,\ \ \mathrm{hcoeff}(p) = 2 .$$

If p is considered as an element of **Q**(z)[x,y], then we write

$$p = x^2y^2 - (2z-1)x^2y + (2z^2)xy^2 + (z^2)x^2 - (3z^3+1)xy + (z)y + (z^2+5) ;$$

hence

$$\mathrm{M}(p) = x^2y^2 ,\ \ \mathrm{hterm}(p) = x^2y^2 ,\ \ \mathrm{hcoeff}(p) = 1 .$$

We note finally that under the *lexicographic* ordering for $\mathrm{T}_{(x,y)}$ we would write the terms in descending order as

$$p = x^2y^2 - (2z-1)x^2y + (z^2)x^2 + (2z^2)xy^2 - (3z^3+1)xy + (z)y + (z^2+5) .$$

●

Reduction in Multivariate Domains

The above structure on F[**x**] now permits a certain type of simplification.

Definition 10.5. For nonzero $p, q \in$ F[**x**] we say that p *reduces modulo* q (with respect to a fixed term ordering) if there exists a monomial in p which is divisible by hterm(q). If $p = \alpha t + r$ where $\alpha \in \mathrm{F} - \{0\}$, $t \in \mathrm{T}_{\mathbf{x}}$, $r \in \mathrm{F}[\mathbf{x}]$ and

$$\frac{t}{\mathrm{hterm}(q)} = u \in \mathrm{T}_{\mathbf{x}},$$

then we write

$$p \mapsto_q p - \frac{\alpha t}{\mathrm{M}(q)} \cdot q = p - \frac{\alpha}{\mathrm{hcoeff}(q)} u \cdot q = p'$$

to signify that p *reduces to* p' *(modulo* q*)*. If p reduces to p' modulo some polynomial in $Q = \{q_1, q_2, \ldots, q_m\}$, we say that p *reduces modulo* Q and write $p \mapsto_Q p'$; otherwise, we say that p is *irreducible* (or *reduced*) *modulo* Q. We adopt the convention that 0 is always irreducible.

●

It is apparent that the process of reduction involves subtracting an appropriate multiple of one polynomial from another, to obtain a result which is (in a sense) smaller. As such, it may be viewed as one step in a generalized division.

Example 10.4. For the polynomials

$$p = 6x^4 + 13x^3 - 6x + 1, \quad q = 3x^2 + 5x - 1$$

we have

$$p \mapsto_q p - 2x^2 \cdot q = 3x^3 + 2x^2 - 6x + 1$$

if we reduce the leading term. We might also compute

$$p \mapsto_q p - \frac{13}{3} x \cdot q = 6x^4 - \frac{65}{3} x^2 - \frac{5}{3} x + 1$$

if we instead reduce the term of degree 3. We note that in either case, we could continue reducing to eventually obtain 0, since in fact $q \mid p$. (Note that in this case, reduction and polynomial division are equivalent.)

●

Example 10.5. Consider the polynomials

$$p = 2y^2 z - xz^2, \quad q = 7y^2 + yz - 4, \quad r = 2yz - 3x + 1,$$

and impose the ordering $<_D$ on $\mathrm{T}_{(x,y,z)}$. (As before, we have written the terms of these polynomials in descending order with respect to $<_D$.) Then we have

$$p \mapsto_q p - \frac{2}{7}z \cdot q = -xz^2 - \frac{2}{7}yz^2 + \frac{8}{7}z ,$$

which is irreducible modulo q, and

$$p \mapsto_r p - y \cdot r = -xz^2 + 3xy - y ,$$

which is irreducible modulo r. Hence, p reduces modulo $Q = \{q, r\}$, but the result is not uniquely defined.

●

A fundamental property of reduction is the following.

Theorem 10.1. For a fixed set Q and ordering $<_T$, there is no infinite sequence of reductions

$$p_0 \mapsto_Q p_1 \mapsto_Q p_2 \mapsto_Q \cdots \qquad (10.1)$$

Proof: We proceed by induction on i, the number of variables in p_0. It is clear that there is no infinite sequence for $i = 0$, since either Q contains elements of headterm 1 ($p_0 \mapsto_Q 0$), or p_0 is irreducible modulo Q. (We may now ignore the possibility that $<Q> = <1>$.)

Now consider $i = 1$. Assume there is no infinite sequence of reductions of a polynomial in $F[x_1]$ of degree $k-1$, and suppose that $p_0 \in F[x_1]$ is of degree k. (The previous point treats the case $k = 0$.) By assumption, there is no infinite sequence of reductions on the lower order terms of p_0. Hence, an infinite sequence of reductions on p_0 requires that the term of degree k eventually be reduced. However, this would yield a polynomial of lower degree.

Similarly, suppose that there is no infinite sequence of reductions of a polynomial in $F[x_1, x_2]$ of degree $l-1$ in x_2, and write $p_0 \in F[x_1, x_2]$ with $\deg_2(p_0) = l$ as an element of $F[x_1][x_2]$. (The previous paragraph treats the case $l = 0$.) The terms in p_0 of degree l in x_2 consist of a polynomial of fixed degree (say, m) in x_1, times x_2^l. These must eventually be reduced as part of an infinite sequence of reductions of p_0, since there is no infinite sequence for terms of degree less than l in x_2. Then, by the argument used in the previous paragraph, there can be no infinite sequence of reductions of *these* terms.

The above argument may be extended for $i = 3, 4, \ldots$ to establish the result for arbitrarily many variables. Note, though, that it is independent of the term ordering $<_T$.

●

Let $\overset{+}{\mapsto}_Q$ denote the reflexive, transitive closure of $\mapsto_Q$. That is, $p \overset{+}{\mapsto}_Q q$ if and only if there is a sequence (possibly trivial) of polynomials such that

$$p = p_0 \mapsto_Q p_1 \mapsto_Q \cdots \mapsto_Q p_n = q .$$

If $p \overset{+}{\mapsto}_Q q$ and q is irreducible, we will write $p \overset{*}{\mapsto}_Q q$. By Theorem 10.1, we may construct an algorithm which, given a polynomial p, finds a q such that $p \mapsto_Q^* q$. While Example 10.5 shows that such a q is not uniquely defined, we will (temporarily) ignore this shortcoming and examine some of the details of the reduction process. For the sake of *efficiency*, it makes the most sense to organize the algorithm so that the *largest* monomials are reduced first, since these reductions affect the lower order monomials anyway. (Compare, for example, this and the opposite strategy on Example 10.4.) Therefore, we formulate our scheme to first reduce the leading monomial M(p) (as part of p), and then $p - \mathrm{M}(p)$ (as a distinct polynomial). Since we will only need to find reducers for leading monomials, it is convenient to adopt the following notation. Noting that 0 is irreducible, we define its *reducer set* by $R_{0,Q} = \emptyset$; for nonzero p, define

$$R_{p,Q} = \{ q \in Q - \{0\} \text{ such that } \mathrm{hterm}(q) \mid \mathrm{hterm}(p) \} .$$

We note that if several reducers exist for M(p), *any* one may be chosen. However, this choice will again affect the efficiency of the algorithm. In practice, the optimal selection depends on the term ordering used. (See Exercise 10.3, for example.) We will therefore write "selectpoly($R_{p,Q}$)" to denote that some reducer (e.g. the first one) is chosen. A possible reduction algorithm is presented below as Algorithm 10.1.

Algorithm 10.1. Full Reduction of p Modulo Q.

```
procedure Reduce(p, Q)

   # Given a polynomial p and a set of polynomials Q
   # from the ring F[x], find a q such that p ↦* q.

   # Start with the whole polynomial.

   r ← p ; q ← 0

   # If no reducers exist, strip off the leading monomial;
   # otherwise, continue to reduce.

   while r ≠ 0 do {
      while R_{r,Q} ≠ ∅ do {
         f ← selectpoly(R_{r,Q})
         r ← r − M(r) f / M(f) }
      q ← q + M(r) ; r ← r − M(r) }
   return( q )
end
```

There are several ways in which the efficiency of this procedure may be further improved. For example, it should actually terminate when all terms in "r" as large as the smallest headterm in Q have been reduced. This is a small point, however, since no significant amount of arithmetic is performed in this phase. It is far more important to economize, where possible, on the amount of (coefficient) arithmetic performed in the innermost loop. One approach is to first divide each of the polynomials in Q by its head coefficient. Another approach is possible when the coefficient field is the fraction field of some integral domain D. Namely, as in the previous chapter it is possible to (temporarily) perform most of the computations in the domain D (essentially, in the manner of the primitive PRS algorithm). (See Czapor [19] for the details.)

Example 10.6. Consider the set $P = \{ p_1, p_2 \} \subset \mathbf{Q}[x,y]$, where

$$p_1 = x^2y + 5x^2 + y^2 \,,\; p_2 = 7xy^2 - 2y^3 + 1 \,;$$

impose the lexicographic term ordering (where $x >_L y$). Then for the polynomial

$$q = 3x^3y + 2x^2y^2 - 3xy + 5x \,,$$

we have (using Algorithm 10.1 and the "first available" reducer)

$$\begin{aligned} q &\mapsto_{p_1} q - 3x \cdot p_1 \\ &= -15x^3 + 2x^2y^2 - 3xy^2 - 3xy + 5x \\ &\mapsto_{p_1} -15x^3 - 10x^2y - 3xy^2 - 3xy + 5x - 2y^3 \\ &\mapsto_{p_1} -15x^3 + 50x^2 - 3xy^2 - 3xy + 5x - 2y^3 + 10y^2 \\ &\mapsto_{p_2} -15x^3 + 50x^2 - 3xy + 5x - \frac{20}{7}y^3 + 10y^2 + \frac{3}{7} \,. \end{aligned}$$

The final result is the fully reduced form of q modulo P, Reduce(q, P). Note that we have written the terms of each polynomial in descending order with respect to $<_L$.

●

The following example illustrates another shortcoming of the reduction process.

Example 10.7. Suppose we adopt the degree ordering for $T_{(x,y,z)}$, and consider again the set of polynomials $P = \{ p_1, p_2, p_3 \} \subset \mathbf{Q}[x,y,z]$ where

$$p_1 = x^3yz - xz^2 \,,\; p_2 = xy^2z - xyz \,,\; p_3 = x^2y^2 - z^2 \,.$$

Also, let

$$q = x^2y^2z - z^3 \,,\; r = -x^2y^2z + x^2yz \,.$$

Then

$$q \mapsto_{p_3} x^2y^2z - z^3 - z(x^2y^2 - z^2) = 0\,,$$

and similarly $r \mapsto_{p_2} 0$. However, $q + r = x^2yz - z^3$ is irreducible modulo P.

●

The fact that, in the above example, $q+r$ is irreducible when q, r each reduce to 0 suggests that reduction (as it stands) is of limited usefulness. The following theorems, while more modest than one would like, illustrate some of the less obvious properties of reduction and will be used in the next section to overcome the current difficulties.

Theorem 10.2. Consider $p, q, r \in \mathrm{F}[\mathbf{x}]$ and $S \subset \mathrm{F}[\mathbf{x}]$. If $p - q \mapsto_S r$, then there exist $\tilde{p}$, $\tilde{q}$ such that

$$p \overset{*}{\mapsto}_S \tilde{p}\,,\ q \overset{*}{\mapsto}_S \tilde{q}\,,\ r = \tilde{p} - \tilde{q}\,.$$

Proof: Let $s \in S$, $\alpha \in \mathrm{F}$, $v \in T_{\mathbf{x}}$ be such that

$$r = (p - q) - \alpha\, v \cdot \frac{s}{\mathrm{M}(s)}\,.$$

(Then v is the term eliminated in the reduction.) Suppose that v has coefficient β_1 in p, and coefficient β_2 in q. Assume that $\beta_1 \neq \beta_2$, since v actually appears in $p - q$ (with coefficient $\alpha = \beta_1 - \beta_2$) and at least one of p or q. Now let $u = v\,/\,\mathrm{hterm}(s)$, and choose

$$\tilde{p} = p - \frac{\beta_1}{\mathrm{hcoeff}(s)}\, u \cdot s\,,\ \tilde{q} = q - \frac{\beta_2}{\mathrm{hcoeff}(s)}\, u \cdot s\,.$$

●

Theorem 10.3. Suppose $p, q \in \mathrm{F}[\mathbf{x}]$ are such that $p - q \overset{*}{\mapsto}_S 0$ for $S \subset \mathrm{F}[\mathbf{x}]$. Then there exists $r \in \mathrm{F}[\mathbf{x}]$ such that $p \overset{*}{\mapsto}_S r$ and $q \overset{*}{\mapsto}_S r$, i.e. p, q have a "common successor" when reduced modulo S.

Proof: As with many of the results of this chapter, we proceed by induction (in this case, on the number of steps necessary to reduce $p - q$ to 0). Clearly, if $p = q$ the result is true. Now assume that the result holds for $n-1$ reduction steps, and suppose that

$$p - q \mapsto_S h_1 \mapsto_S h_2 \mapsto_S \cdots \mapsto_S h_n = 0\,.$$

By Theorem 10.2, $\exists\ \tilde{p}$, $\tilde{q}$ such that $p \overset{*}{\mapsto}_S \tilde{p}$, $q \overset{*}{\mapsto}_S \tilde{q}$, and $\tilde{p} - \tilde{q} = h_1$. But then, by hypothesis, $\tilde{p}$ and $\tilde{q}$ (and hence p, q) have a common successor.

●

Theorem 10.4. If p_1, p_2 are polynomials such that $p_1 \mapsto_Q p_2$, then for any polynomial r, there exists s such that

$$p_1 + r \overset{*}{\mapsto}_Q s \,,\; p_2 + r \overset{*}{\mapsto}_Q s \,.$$

Proof: Let $\alpha \in$ F, $u \in \mathrm{T}_{\mathbf{x}}$, $q \in Q$ be such that $p_2 = p_1 - \alpha u \cdot q \,/\, \mathrm{hcoeff}(q)$, and let $t = u \cdot \mathrm{hterm}(q)$ be the term cancelled in the reduction. For arbitrary r, suppose that t has coefficient β in r (or in $p_2 + r$); then t has coefficient $\alpha + \beta$ in $p_1 + r$. Now, for $\tilde{q} = q \,/\, \mathrm{hcoeff}(q)$ we have

$$p_1 + r \overset{*}{\mapsto}_q s_1 = (p_1 + r) - (\alpha+\beta) u \cdot \tilde{q} \,,$$

$$p_2 + r \overset{*}{\mapsto}_q s_2 = (p_2 + r) - \beta u \cdot \tilde{q} \,,$$

and

$$s_1 - s_2 = [\alpha - (\alpha+\beta) + \beta]\, u \cdot \tilde{q} = 0 \,.$$

Therefore, $s = s_1 = s_2$ is the required polynomial.

●

10.3. GRÖBNER BASES AND BUCHBERGER'S ALGORITHM

While it is certainly true that $p \in \langle Q \rangle$ if $p \overset{*}{\mapsto}_Q 0$, Example 10.7 shows that the converse is not true. Hence, the process of reduction will not solve the zero-equivalence problem as it stands. It turns out that this is not, strictly speaking, due to a deficiency of Algorithm 10.1, but rather the structure of the ideal basis Q. We therefore propose the following:

Definition 10.6. An ideal basis $G \subset$ F[**x**] is called a *Gröbner basis* (with respect to a fixed term ordering $<_T$ and the implied permutation of variables) if

$$p \in \langle G \rangle \;\Leftrightarrow\; p \overset{*}{\mapsto}_G 0 \,.$$

●

Equivalently, G is a Gröbner basis when the only irreducible polynomial in $\langle G \rangle$ is $p = 0$. (Whenever G is a Gröbner basis and $p \in \langle G \rangle$ is irreducible, we can have $p \overset{*}{\mapsto}_G 0$ only if $p = 0$. Conversely, for any $p \in \langle G \rangle$ compute a q such that $p \overset{*}{\mapsto}_G q$. Clearly $q \in \langle G \rangle$ and q is irreducible; so if q must be 0, then $p \overset{*}{\mapsto}_G 0$.) This, in turn, implies that G is a Gröbner basis precisely when reduction modulo G (in any formulation) is a normal simplifier for F[**x**]$/\langle G \rangle$.

Example 10.8. For the polynomials $P = \{p_1, p_2, p_3\}$ and q, r of the previous example,

$$G = \{p_1,\, p_2,\, p_3,\, x^2yz - z^3,\, xz^3 - xz^2, yz^3 - z^3,\, xyz^2 - xz^2,\, x^2z^2 - z^4,\, z^5 - z^4\}$$

is a Gröbner basis (with respect to the degree ordering for $T_{(x,y,z)}$) such that $<P> = <G>$. Note that $q \mapsto^*_G 0$, $r \mapsto^*_G 0$, and $q + r \mapsto^*_G 0$, irrespective of the sequence of reductions that is followed.

●

Unfortunately, we do not yet have a means to actually prove that the above set is a Gröbner basis. Thus we require an algorithm for their construction.

Alternate Characterizations of Gröbner Bases

We have already seen that an arbitrary ideal basis P does not, in general, constitute a Gröbner basis for $<P>$. The idea behind Buchberger's method is to "complete" the basis P by adding (a finite number of) new polynomials to it. Buchberger's primary contribution was to show that this completion only requires consideration of the following quantity, for finitely many pairs of polynomials from P.

Definition 10.7. The *S-polynomial of* $p, q \in F[\mathbf{x}]$ is

$$\mathrm{Spoly}(p,q) = \mathrm{LCM}(\mathrm{M}(p), \mathrm{M}(q)) \left[\frac{p}{\mathrm{M}(p)} - \frac{q}{\mathrm{M}(q)} \right] . \tag{10.2}$$

●

Example 10.9. For the polynomials $p_1, p_2 \in \mathbf{Q}[x,y]$ defined by

$$p_1 = 3x^2y - y^3 - 4 , \quad p_2 = xy^3 + x^2 - 9 ,$$

using the degree ordering on $T_{(x,y)}$, we have

$$\begin{aligned} \mathrm{Spoly}(p_1, p_2) &= y^2(3x^2y - y^3 - 4) - 3x(xy^3 + x^2 - 9) \\ &= -y^5 - 3x^3 - 4y^2 + 27x . \end{aligned}$$

●

It is useful to view the S-polynomial (which generalizes the operation of reduction) as the difference between reducing LCM(M(p), M(q)) modulo p and reducing it modulo q. This plays a crucial role in the following (fundamental) theorem of Buchberger [8], which leads almost directly to an algorithm for computing Gröbner bases.

Theorem 10.5 (Alternate Characterizations of Gröbner Bases). The following are equivalent:

(i) G is a Gröbner basis;

(ii) $\mathrm{Spoly}(p, q) \mapsto^*_G 0$ for all $p, q \in G$;

(iii) If $p \mapsto^*_G q$ and $p \mapsto^*_G r$, then $q = r$.

Proof: Although the proof is rather involved, we present the details in order to further acquaint the reader with the subtleties of reduction. We proceed in three stages.

(i) $\Longrightarrow$ (ii): This is clear, since $\mathrm{Spoly}(p, q) \in \,<G>$ implies that

$$\mathrm{Spoly}(p,q) \;\mapsto^{*}_{G}\; 0 .$$

(ii) $\Longrightarrow$ (iii): We proceed by induction on the headterm of p. First, consider the case $\mathrm{hterm}(p) = 1$. Clearly the assertion is true, since either p is irreducible (i.e. it is already reduced) or reduces in one step to 0. Suppose, then, that (iii) holds for all p such that $\mathrm{hterm}(p) <_T t$ for some fixed $t \in T_{\mathbf{x}}$ (the "main" induction hypothesis); consider p such that $\mathrm{hterm}(p) = t$. If t is irreducible (modulo G), $p \mapsto^{*}_{G} q$ and $p \mapsto^{*}_{G} r$, the result is fairly clear. This is because the reductions may involve only the lower order terms; i.e. if

$$p \;=\; \mathrm{M}(p) + p - \mathrm{M}(p) \;\mapsto^{*}_{G}\; \mathrm{M}(p) + p_1 \;=\; q$$

and hence

$$p \;\mapsto^{*}_{G}\; \mathrm{M}(p) + p_2 \;=\; r ,$$

the induction hypothesis (applied to $p-\mathrm{M}(p)$) implies $p_1 = p_2$ and hence $q = r$. We therefore assume that t is reducible, and write

$$R_{p,G} \;=\; \{\, g_1, \ldots, g_m \,\} ,$$

where the order is fixed but arbitrary. Take p_1, p_2, q such that

$$\mathrm{M}(p) \;\mapsto_{g_1}\; p_1 \,,\; p - \mathrm{M}(p) \;\mapsto^{*}_{G}\; p_2 \,,\; p_1 + p_2 \;\mapsto^{*}_{G}\; q \,, \tag{10.3}$$

and hence also

$$p \;\mapsto^{*}_{G}\; \mathrm{M}(p) + p_2 \;\mapsto_{G}\; p_1 + p_2 \;\mapsto^{*}_{G}\; q \,. \tag{10.4}$$

(This is always possible by reducing the lower order terms in $p - \mathrm{M}(p)$ first, since $\mathrm{hterm}(p_1)$ $<_T t$.) Now suppose that there is also an r such that $p \mapsto^{*}_{G} r$. We consider two cases.

(a) For the time being, assume that the latter reduction is of the form

$$\mathrm{M}(p) \;\mapsto_{g_1}\; p_1 \,,\; p - \mathrm{M}(p) \;\mapsto^{*}_{G}\; p_3 \,,\; p_1 + p_3 \;\mapsto^{*}_{G}\; r \,, \tag{10.5}$$

where once again the middle reductions (if there are any) are carried out first. We claim that, under the present conditions, $p_1 + p_2$ and $p_1 + p_3$ have a common successor. This is established by induction on k, the number of steps in the reduction $p_3 \mapsto^{*}_{G} p_2$ (which is always possible in view of the induction hypothesis). If $k = 0$, this is trivial. Let us assume that, say, $p_1 + f$ and $p_1 + p_3$ have a common successor if $p_3 \mapsto^{*}_{G} f$ in l steps. Now let $\tilde{f}$ be such that $p_3 \mapsto^{*}_{G} \tilde{f}$ in l steps, and $\tilde{f} \mapsto_{G} p_2$. By Theorem 10.4, $\exists\, g$ such that

$$p_1 + \tilde{f} \mapsto^{*}_{G} g \,,\ p_1 + p_2 \mapsto^{*}_{G} g \,.$$

Since, by hypothesis, $\exists\, h$ such that

$$p_1 + p_3 \mapsto^{*}_{G} h \,,\ p_1 + \tilde{f} \mapsto^{*}_{G} h \,,$$

it follows that for some $\tilde{g}$, $\tilde{h}$,

$$p_1 + \tilde{f} \mapsto^{*}_{G} \tilde{g} \,,\ p_1 + p_2 \mapsto^{*}_{G} \tilde{g} \,,$$

$$p_1 + p_3 \mapsto^{*}_{G} \tilde{h} \,,\ p_1 + \tilde{f} \mapsto^{*}_{G} \tilde{h} \,.$$

Since the headterms of all these polynomials are smaller than t, the main induction hypothesis implies $\tilde{g} = \tilde{h}$; i.e. $p_1 + p_2$ and $p_1 + p_3$ have a common successor. Together with (10.3), (10.5) and the main induction hypothesis, this implies that $r = q$.

(b) Assume that we have $\tilde{p}_1, p_3$ such that

$$\mathrm{M}(p) \mapsto_{g_n} \tilde{p}_1 \,,\ p - \mathrm{M}(p) \mapsto^{*}_{G} p_3 \,,\ \tilde{p}_1 + p_3 \mapsto^{*}_{G} r \,, \tag{10.6}$$

where $2 \le n \le m$, and hence

$$p \mapsto^{*}_{G} \mathrm{M}(p) + p_3 \mapsto_{G} \tilde{p}_1 + p_3 \mapsto^{*}_{G} r \,. \tag{10.7}$$

Consider also the reductions

$$\mathrm{M}(p) \mapsto_{g_1} p_1 \,,\ p - \mathrm{M}(p) \mapsto^{*}_{G} p_3 \,,\ p_1 + p_3 \mapsto^{*}_{G} \tilde{r} \,; \tag{10.8}$$

that is,

$$p \mapsto^{*}_{G} \mathrm{M}(p) + p_3 \mapsto_{G} p_1 + p_3 \mapsto^{*}_{G} \tilde{r} = q \tag{10.9}$$

(noting the result of case (a)). Now, we find that

$$(\tilde{p}_1 + p_3) - (p_1 + p_3) = \tilde{p}_1 - p_1$$

$$= \mathrm{M}(p) \left[\frac{g_1}{\mathrm{M}(g_1)} - \frac{g_n}{\mathrm{M}(g_n)} \right] . \tag{10.10}$$

Since $g_1, g_n \in R_{p,G}$, the above quantity is the product of Spoly(g_1, g_n) times a monomial. Applying (ii) and Theorem 10.3, $\exists f$ such that

$$\tilde{p}_1 + p_3 \mapsto^{*}_{G} f \,,\ p_1 + p_3 \mapsto^{*}_{G} f \,.$$

Therefore, by (10.6)-(10.9), and the main hypothesis, $r = q$.

(iii) $\Rightarrow$ (i): If $p \in \langle G \rangle$, then $\exists\, h_i \in F[\mathbf{x}]$ such that

$$p = \sum_{i=1}^{l} h_i \, g_i \,. \tag{10.11}$$

We proceed by induction on the maximal term t among the headterms of $h_1g_1, h_2g_2, \ldots, h_lg_l$. First, if $t = 1$ the result is trivial. (Either $p = 0$, or $p \in$ F and $p \mapsto_G 0$.) Now assume that for some t, we have $p \mapsto^*_G 0$ whenever (10.11) holds with hterm$(h_ig_i) <_T t$ for $1 \le i \le l$; then consider a polynomial p with hterm$(h_ig_i) \le_T t$ for some $1 \le i \le l$. We suppose (without loss of generality) that $\{h_1g_1, \ldots, h_mg_m\}$ are the (nonzero) polynomials in (10.11) which have headterm t. We will show that $p \mapsto^*_G 0$ by induction on m. If $m = 1$, then

$$p = h_1g_1 + \sum_{i=2}^{l} h_ig_i \ \mapsto_{g_1}\ \tilde{p} \ = \ (h_1 - \mathrm{M}(h_1))g_1 + \sum_{i=2}^{l} h_ig_i\,,$$

and by the main hypothesis $\tilde{p} \mapsto^*_G 0$. Now assume that $p \mapsto^*_G 0$ when $m \le k$ and consider $m = k+1$. (That is, the representation (10.11) of p has $k+1$ components with headterm t.) Now, for $\alpha \in$ F write

$$p = h_1g_1 + h_2g_2 + \sum_{i=3}^{l} h_ig_i \ = \ \tilde{p} + p'\,,$$

where

$$\tilde{p} = \mathrm{M}(h_1)g_1 + [\,\mathrm{M}(h_2) + \alpha\cdot\mathrm{hterm}(h_2)\,]\cdot g_2\,,$$

$$p' = (h_1 - \mathrm{M}(h_1))g_1 + [\,h_2 - \mathrm{M}(h_2) - \alpha\cdot\mathrm{hterm}(h_2)\,]\cdot g_2 + \sum_{i=3}^{l} h_ig_i\,, \tag{10.12}$$

and choose α such that

$$\tilde{p} = \beta\, u\cdot\mathrm{Spoly}(g_1, g_2)$$

for some $\beta \in$ F, $u \in \mathrm{T}_{\mathbf{x}}$ (Exercise 10.4). On one hand, the representation (10.12) has at most k components of headterm t; hence by hypothesis $p' \mapsto^*_G 0$. On the other hand, we can show that $\tilde{p} \mapsto^*_G 0$, as follows. We note that

$$\frac{\mathrm{LCM}(\mathrm{M}(g_1), \mathrm{M}(g_2))}{\mathrm{M}(g_1)}\cdot g_1 \ \mapsto_{g_2}\ \mathrm{Spoly}(g_1, g_2) \ \mapsto^*_G\ q$$

for some q. But also

$$\frac{\mathrm{LCM}(\mathrm{M}(g_1), \mathrm{M}(g_2))}{\mathrm{M}(g_1)}\cdot g_1 \ \mapsto_{g_1}\ 0\,,$$

which in view of (iii) implies $q = 0$. It follows that $\tilde{p} = p - p' \mapsto^*_G 0$ as well. Therefore, by Theorem 10.3 and (iii), $\exists\, r$ such that $p \mapsto^*_G r$ and $p' \mapsto^*_G r$. But since $p' \mapsto^*_G 0$, we conclude that $r = 0$.

●

Corollary 10.6. G is a Gröbner basis if and only if $\forall f, g \in G$ either

(1) $\text{Spoly}(f, g) \mapsto^{*}_{G} 0$, or

(2) $\exists\, h \in G,\ f \neq h \neq g$, such that

$$\text{hterm}(h) \mid \text{LCM}(\,\text{hterm}(f), \text{hterm}(g)\,)\,, \tag{10.13}$$

$$\text{Spoly}(f, h) \mapsto^{*}_{G} 0\,,\ \ \text{Spoly}(h, g) \mapsto^{*}_{G} 0\,. \tag{10.14}$$

Proof: If we replace (ii) in Theorem 10.5 with the above condition, we need only extend part (b) of the proof that (ii) $\Rightarrow$ (iii) when (2) holds; we therefore resume the proof up to (10.9). We first note that by (10.13), $h \in R_{p,G}$. As before, we let p_1', s be such that

$$\text{M}(p) \mapsto_{h} p_1'\,,\ p - \text{M}(p) \mapsto^{*}_{G} p_3\,,\ p_1' + p_3 \mapsto^{*}_{G} s\,, \tag{10.15}$$

$$p \mapsto^{*}_{G} \text{M}(p) + p_3 \mapsto_{G} p_1' + p_3 \mapsto^{*}_{G} s\,. \tag{10.16}$$

Also (as before!),

$$(p_1' + p_3) - (p_1 + p_3) \;=\; \text{M}(p)\,[\,\frac{g_1}{\text{M}(g_1)} - \frac{h}{\text{M}(h)}\,] \mapsto^{*}_{G} 0\,,$$

$$(\tilde{p}_1 + p_3) - (p_1' + p_3) \;=\; \text{M}(p)\,[\,\frac{h}{\text{M}(h)} - \frac{g_n}{\text{M}(g_n)}\,] \mapsto^{*}_{G} 0\,,$$

by (10.14). Thus, by Theorem 10.3 and the induction hypothesis, we conclude that $r = s = q$. ●

In view of Theorem 10.5 (iii), the result of reduction modulo a Gröbner basis is always unique. Therefore, we may write $q = \text{Reduce}(p, G)$ instead of $p \mapsto^{*}_{G} q$ since the details of the reduction algorithm (cf. Algorithm 10.1) will not affect the outcome.

Corollary 10.7. If G is a Gröbner basis, then

$$\text{Reduce}(p, G) = \text{Reduce}(q, G) \quad \Leftrightarrow \quad p - q \in \,<G>\,.$$

Proof:

$\Rightarrow$: Suppose $r = \text{Reduce}(p, G) = \text{Reduce}(q, G)$. Then $p - r \in \,<G>$ and $q - r \in \,<G>$. Therefore,

$$(p - r) - (q - r) = p - q \in \,<G>\,.$$

$\Leftarrow$: Apply Theorem 10.3 (noting $p - q \in \,<G>$), and then part (iii) of Theorem 10.5. ●

The need for Corollary 10.6 will become apparent in the next section. Corollary 10.7 shows that if G is a Gröbner basis, then its reduction algorithm is not only a normal simplifier, but also a *canonical* simplifier (cf. Chapter 3). Decision procedures follow for a host of related problems in polynomial ideal theory, including ideal inclusion (Exercise 10.5) and computing in the quotient ring $F[\mathbf{x}]/<G>$. We postpone discussion of these, and other applications, until later sections. Instead, we will now fulfill our promise to present an algorithm for the computation of Gröbner bases.

Buchberger's Algorithm

Characterization (ii) of Theorem 10.5 suggests how we may transform an arbitrary ideal basis into a Gröbner basis. Given a finite set $P \subset F[\mathbf{x}]$, we may immediately test P by checking whether

$$\mathrm{Spoly}(p,q) \mapsto^*_P 0 \quad \text{for all } p,q \in P,\ p \neq q\ .$$

If we find a pair (p, q) such that

$$\mathrm{Spoly}(p,q) \mapsto^*_P r \neq 0\,,$$

then $<P> = <P,r>$ and $\mathrm{Spoly}(p, q) \mapsto^*_{P\cup\{r\}} 0$. That is, we may add the nonzero result to the basis, and begin testing of the *augmented* set. To see that such a process will terminate, let H_i be the set of headterms of the basis after the i-th new polynomial is added. Since new headterms are not multiples of old ones, the inclusions

$$<H_1> \subset <H_2> \subset \cdots$$

are proper. Given that $F[\mathbf{x}]$ is a Noetherian integral domain, such a chain of ideals must terminate by Hilbert's "divisor chain condition" (see van der Waerden [35] for example). The resulting algorithm appears below as Algorithm 10.2. As in Algorithm 10.1, we have used a procedure "selectpair" to denote that some selection is made from a non-empty set "B". Since the particular selection is of no *theoretical* importance, the reader may assume for now that the first element is chosen. (The reason "G" appears as an argument to selectpair will be given later.)

Example 10.10. Consider the set $P \subset \mathbf{Q}[x,y,z]$ defined by

$$P = \{ x^2 + yz - 2,\ y^2 + xz - 3,\ xy + z^2 - 5 \}\,,$$

using the degree ordering $<_D$. (As usual, we write all terms in $<_D$-descending order.) We first set $G = P$, $k = 3$, and $B = \{[1,2], [1,3], [2,3]\}$. Then

$$\begin{aligned}
\mathrm{Spoly}(G_1, G_2) &= y^2\cdot(x^2 + yz - 2) - x^2\cdot(y^2 + xz - 3) = -x^3z + y^3z + 3x^2 - 2y^2 \\
&\mapsto_{G_1} y^3z + xyz^2 + 3x^2 - 2y^2 - 2xz \\
&\mapsto_{G_2} 3x^2 - 2y^2 - 2xz + 3yz \\
&\mapsto_{G_1} -2y^2 - 2xz + 6 \\
&\mapsto_{G_2} 0\ ,
\end{aligned}$$

Algorithm 10.2. Buchberger's Algorithm for Gröbner Bases.

```
procedure Gbasis(P)
   # Given a set of polynomials P, compute G such
   # that < G > = < P > and G is a Gröbner basis.

   G ← P ; k ← length(G)

   # We denote the i-th element of the ordered set
   # G by G_i.

   B ← { [i, j] : 1 ≤ i < j ≤ k }
   while B ≠ ∅ do {
      [i, j] ← selectpair(B, G)
      B ← B − {[i, j]}
      h ← Reduce(Spoly(G_i, G_j), G)
      if h ≠ 0 then {
         G ← G ∪ {h} ; k ← k + 1
         B ← B ∪ { [i, k] : 1 ≤ i < k }   }}
   return( G )
end
```

whereupon $B = \{[1,3],[2,3]\}$. Then

$$\begin{aligned}\mathrm{Spoly}(G_1, G_3) &= y^2z - xz^2 + 5x - 2y \\ &\mapsto_{G_2} -2xz^2 + 5x - 2y + 3z \ ,\end{aligned}$$

which is irreducible. We therefore set $G_4 = -2xz^2 + 5x - 2y + 3z$, ($k = 4$,) and $B = \{[2,3], [1,4],[2,4],[3,4]\}$. Continuing in this manner:

$$\mathrm{Spoly}(G_2, G_3) \mapsto_{G_1} G_5 = -2yz^2 - 3x + 5y + 2z \ ,$$

$$B = \{ [1,4],[2,4],[3,4],[1,5],[2,5],[3,5],[4,5] \} ;$$

$$\mathrm{Spoly}(G_1, G_4) \mapsto^*_G 0 \ ,$$

$$B = \{ [2,4],[3,4],[1,5],[2,5],[3,5],[4,5] \} ;$$

$$\mathrm{Spoly}(G_2, G_4) \mapsto^*_G 0 \ ,$$

$$B = \{ [3,4],[1,5],[2,5],[3,5],[4,5] \} ;$$

$$\mathrm{Spoly}(G_3, G_4) \mapsto^*_G G_6 = -2z^4 - 2xz - 3yz + 15z^2 - 19 \ ,$$

$$B = \{ [1, 5], [2, 5], [3, 5], [4, 5], [1, 6], [2, 6], [3, 6], [4, 6], [5, 6] \} \ ,$$

after which all further S-polynomial reductions lead to 0.

●

When applied to linear polynomials, Algorithm 10.2 specializes to a Gaussian elimination algorithm. When applied to univariate polynomials, it specializes to Euclid's algorithm for several polynomials. The relationship with polynomial division processes is, in the bivariate case, fully specified by Lazard [31]. It has also been shown that Algorithm 10.2 and the Knuth-Bendix [28] algorithm for rewrite rules are both instances of a more general "critical pair/completion" algorithm. (See Buchberger [12] or Le Chenadec [17], for example.) This connection has been exploited by Bachmair and Buchberger [2] to shorten the proof of Theorem 10.5, and by Winkler [36] to carry over improvements to Algorithm 10.2 to the Knuth-Bendix procedure. We mention also that the algorithm has been generalized to various Euclidean domains (e.g. **Z**); see Buchberger [13] or Kandri-Rody and Kapur [26], for example.

10.4. IMPROVING BUCHBERGER'S ALGORITHM

It must be noted that if, in Example 10.10, we had used a different permutation of variables, or another term ordering, we would have obtained a completely different basis. It should also be pointed out that, these issues aside, Gröbner bases are by no means unique.

Example 10.11. Consider the set P and corresponding Gröbner basis G of Example 10.8. It may be shown (Exercise 10.7) that $G - \{p_1\}$ is also a Gröbner basis for $<P>$.

●

Reduced Gröbner Bases

Fortunately, the problem of non-uniqueness is very easily remedied, as we now illustrate.

Definition 10.8. A set $G \subset F[\mathbf{x}]$ is *reduced* if $\forall\, g \in G,\ g = \mathrm{Reduce}(g, G-\{g\})$; it is *monic* if $\forall\, g \in G,\ \mathrm{hcoeff}(g) = 1$.

●

Theorem 10.8 (Buchberger [7]). If G, H are reduced, monic Gröbner bases such that $<G> = <H>$, then $G = H$.

●

We see that if the polynomials are scaled in *any* consistent manner, a Gröbner basis may be made unique by ensuring that each element is reduced modulo the others. A possible algorithm to perform such a transformation appears as Algorithm 10.3. A proof that Algorithm 10.3 terminates is given by Buchberger [5]. When applied at the end of Algorithm 10.2, it is easy to see that only a subset of G must be reduced. Namely, for any G_i, G_j in Algorithm

Algorithm 10.3. Construction of a Reduced Ideal Basis.

```
procedure ReduceSet(E)
    # Given a set E (not necessarily a Gröbner basis),
    # compute Ẽ such that <E> = <Ẽ> and Ẽ is reduced.
    # First, remove any redundant elements.
    R ← E ; P ← ∅
    while R ≠ ∅ do {
        h ← selectpoly(R) ; R ← R − {h}
        h ← Reduce(h, P)
        if h ≠ 0 then {
            Q ← { q ∈ P such that hterm(h) | hterm(q) }
            R ← R ∪ Q
            P ← P − Q ∪ {h} } }
    # Ensure each element is reduced modulo the others.
    Ẽ ← ∅ ; S ← P
    foreach h ∈ P do {
        h ← Reduce(h, S − {h})
        Ẽ ← Ẽ ∪ {h} }
    return(Ẽ)
end
```

10.2 such that $\mathrm{hterm}(G_j) \mid \mathrm{hterm}(G_i)$, $\mathrm{Spoly}(G_i, G_j)$ is equal (up to a rescaling) to the reduced form of G_i modulo G_j; hence, G_i may be discarded at the end of the algorithm. And, although the result will not be unique if the input set "E" is not a Gröbner basis, Algorithm 10.3 may also be applied before Algorithm 10.2. In fact, a reformulation of Algorithm 10.2 is possible in which the partial basis G is reduced after each new polynomial is added. Still, it is not clear how much pre- or inter-reduction is best in practice. (See Czapor [19], for example.)

The Problem of Unnecessary Reductions

It should be apparent from Example 10.10 that Algorithm 10.2 is capable of producing extremely complex calculations from (apparently) modest input polynomials. Note, for example, that as "k" (the number of polynomials) grows, the number of S-polynomials in "B" grows rapidly. It turns out that when applied to polynomials of the form $s - t$, where s, $t \in T_{\mathbf{x}}$, Algorithm 10.2 specializes to one for the "uniform word problem" for commutative semigroups (see Ballantyne and Lankford [3]). This relationship is used by Mayr and Meyer

[32] to demonstrate that the congruence problem for polynomial ideals is exponentially space complete. Hence, the problem of constructing Gröbner bases is intrinsically hard. This does not mean that Algorithm 10.2 is of no practical use; however, it is well worth considering some refinements which will improve its performance.

It is also clear that most of the computational cost of the algorithm is in the polynomial arithmetic of the reduction step. Now, it is easy to see that *full* reduction of each S-polynomial is not actually necessary; a partially reduced form $\tilde{h} \neq 0$ will suffice as long as $\mathrm{M}(\tilde{h})$ is irreducible (i.e. it is not possible that $\tilde{h} \mapsto^{*}_{G} 0$). However, it may happen that the fully reduced form leads to simpler polynomials later in the algorithm; so, the actual benefits of this approach are difficult to assess. Quite typically, though, only a relatively small proportion of the S-polynomials which are reduced will yield new (nonzero) results. Therefore, a great deal of computation is wasted. Fortunately, Buchberger has shown that many of these 0-reductions may be detected *a priori*, without a significant amount of computation. This is accomplished, in part, using the following result.

Theorem 10.9. If $\mathrm{LCM}(\mathrm{hterm}(p), \mathrm{hterm}(q)) = \mathrm{hterm}(p)\cdot\mathrm{hterm}(q)$, then

$$\mathrm{Spoly}(p, q) \mapsto^{*}_{\{p,q\}} 0 .$$

Proof: We obtain

$$\begin{aligned}\mathrm{Spoly}(p, q) &= \alpha\,(\,\mathrm{M}(q)\cdot p - \mathrm{M}(p)\cdot q\,) \\ &= \alpha\,[\,\mathrm{M}(q)\cdot(p - \mathrm{M}(p)) - \mathrm{M}(p)\cdot(q - \mathrm{M}(q))\,],\end{aligned}$$

where $\alpha = \mathrm{GCD}(\mathrm{hcoeff}(p), \mathrm{hcoeff}(q))^{-1}$. No terms cancel in the subtraction above, since the terms of the two polynomials $p - \mathrm{M}(p)$ and $q - \mathrm{M}(q)$ are distinct. (This is an easy consequence of the fact that $\mathrm{M}(p)$, $\mathrm{M}(q)$ must contain distinct sets of variables.) Then note that

$$\mathrm{M}(p) \mapsto_{p} p - \mathrm{M}(p), \quad \mathrm{M}(q) \mapsto_{q} q - \mathrm{M}(q) .$$

●

The above result provides a condition under which certain S-polynomials (i.e. pairs $[i, j]$) may be skipped. Namely, $[i, j]$ may be safely ignored if it does not satisfy the function

$$\begin{aligned}&\mathrm{criterion1}([i, j], G) \iff \\ &\quad \mathrm{LCM}(\,\mathrm{hterm}(G_i), \mathrm{hterm}(G_j)\,) \neq \mathrm{hterm}(G_i)\,\mathrm{hterm}(G_j) .\end{aligned}$$

In addition, Corollary 10.6 implies that we may skip $\mathrm{Spoly}(i, j)$ if $[i, j]$ does not satisfy

$$\begin{aligned}&\mathrm{criterion2}([i, j], B, G) \iff \\ &\quad \neg\ \exists\, k,\ 1 \le k \le \mathrm{length}(G), \text{ such that} \\ &\qquad \{i \neq k \neq j, \\ &\qquad \mathrm{hterm}(G_k) \mid \mathrm{LCM}(\mathrm{hterm}(G_i), \mathrm{hterm}(G_j)), \\ &\qquad [i, k] \notin B,\ [k, j] \notin B\} .\end{aligned}$$

The reader is referred to Buchberger [9] and Buchberger and Winkler [10], which together supply greater insight into the derivation of criterion2. In practical terms, the effect of using these criteria is dramatic. According to Buchberger, for example, criterion2 results (roughly speaking) in a reduction of the number of S-polynomial reductions from $O(K^2)$ to $O(K)$, where K is the final length of the basis. The improved form of Buchberger's algorithm appears below as Algorithm 10.4.

Algorithm 10.4. Improved Construction of Reduced Gröbner Basis.

```
procedure Gbasis(P)
  # Given polynomials P, find the corresponding reduced
  # Gröbner basis G.
  # First, pre-reduce the raw input set;
  # optionally, just set G ← P.
  G ← ReduceSet(P) ;  k ← length(G)
  B ← { [i, j] : 1 ≤ i < j ≤ k }
  while B ≠ ∅ do {
    [i, j] ← selectpair(B, G) ;  B ← B − {[i, j]}
    if criterion1([i, j], G) and criterion2([i, j], B, G) then {
      h ← Reduce(Spoly(G_i, G_j), G)
      if h ≠ 0 then {
        G ← G ∪ {h} ;  k ← k + 1
        B ← B ∪ { [i, k] : 1 ≤ i < k }   }}}
  # Discard redundant elements and inter-reduce.
  R ← { g ∈ G such that R_{g,G} − {g} ≠ ∅ }
  return( ReduceSet( G − R ) )
end
```

Buchberger and Winkler [10] also present an important argument regarding the procedure selectpair. It can be shown that if we always select $[i, j]$ such that

$$\text{LCM}(\text{hterm}(G_i), \text{hterm}(G_j)) =$$

$$\min_{<_T} \{ \text{LCM}(\text{hterm}(G_u), \text{hterm}(G_v)) : [u, v] \in B \} \tag{10.17}$$

(the "normal" selection strategy), then criterion1, criterion2 are "good" in the sense that *all* possible reductions (i.e. not just one particular reduction) of $\text{Spoly}(G_i, G_j)$ will yield 0. Moreover, the likelihood that criterion2 can even be applied is increased. Finally, if the degree ordering is used, this strategy would seem to lead to simpler polynomials than other choices. (Although, it has recently become apparent that the same cannot be said when the lexicographic ordering is used; see Czapor [19].)

Computational Complexity

We conclude this section with some brief remarks on the complexity of Buchberger's algorithm. It is useful to determine bounds on the maximum degree of any polynomial produced by the algorithm; this, in turn, may bound the number of polynomials and the (maximum) number of reduction steps required for each. Although this is difficult in general, Buchberger [9] has shown the following: in the bivariate case, when a criterion similar to criterion2 is used (in conjunction with the normal selection strategy) the polynomials produced by the algorithm with the degree ordering are bounded by $4D(P)$, where

$$D(P) = \max \{ \deg(P_i) : 1 \le i \le \mathrm{length}(P) \} ;$$

the number of computational steps is then bounded by

$$2 (\mathrm{length}(P) + 16 D(P)^2)^4 .$$

Of course, the actual computational cost depends on the coefficient field as well. Recent results (see Winkler [37], for example) show progress with regard to development of a version of Algorithm 10.2 which uses a homomorphism/lifting approach similar to the EZ-GCD scheme (cf. Chapter 7). Since algorithms involving polynomial division (e.g. PRS algorithms) are plagued by the problem of coefficient growth, this is an important area of study. The role of the term ordering used is illustrated by the following result of Buchberger [11]: for every natural number n, $\exists\, P \subset F[x,y]$ with $n = D(P)$ such that

(a) for all Gröbner bases for P with respect to $<_D$, $D(G) \ge 2n - 1$;

(b) for all Gröbner bases for P with respect to $<_L$, $D(G) \ge n^2 - n + 1$.

Apparently, the complexity of the algorithm is lower when using $<_D$ than when using $<_L$. Lazard [30] shows that for $<_D$ (or similar orderings), the maximum degree of the reduced basis is usually below $\Sigma \deg(P_i) - n + 1$, where n is the number of variables. (See also Möller and Mora [33] for other interesting results.)

10.5. APPLICATIONS OF GRÖBNER BASES

We have seen that Buchberger's algorithm completely solves the simplification problem for polynomials modulo side relations. That is, when $G \subset F[\mathbf{x}]$ is a Gröbner basis the corresponding reduction algorithm Reduce($\cdot$, G) is a canonical function for $[F[\mathbf{x}]; \sim]$, where $\sim$ is the "equivalence modulo G" relation used in Section 1. In view of the central role of polynomial domains in symbolic computation, this alone establishes the importance of Gröbner bases. However, a survey of some of the applications of this powerful technique suggests that it is indeed one of the fundamental algorithms of symbolic computation. We will not attempt to list all such applications here; this is an active area of research, and any such list would soon be incomplete. Moreover, a discussion of the recent use of Gröbner bases in such fields as bifurcation theory (Armbruster [1]) and spline theory (Billera and Rose [4]) is beyond the scope of this book. We instead restrict our attention to a few simple, but important, examples.

Computing in Quotient Rings

The close connection between the simplification problem and arithmetic in the quotient ring $F[\mathbf{x}]/<G>$ is illustrated by the following theorem.

Theorem 10.10. Suppose G is a Gröbner basis, and define

$$U = \{ [u], \text{ where } u \in T_{\mathbf{x}} \text{ is such that } \neg\exists\, g \in G \text{ with hterm}(g) \mid u \}, \qquad (10.18)$$

where $[u]$ is the congruence class of u modulo G. Then U is a linearly independent (vector space) basis for $F[\mathbf{x}]/<G>$.

Proof: Suppose we have a dependence

$$a_1[u_1] + \cdots + a_m[u_m] = 0,$$

where $a_i \in F$, $u_i \in U$ for $1 \le i \le m$. Since we now know that for $p \in F[\mathbf{x}]$,

$$[p] = 0 \iff p \in <G>,$$

and that reduction modulo G is a canonical simplifier, there must be a polynomial $g = a_1u_1 + \cdots + a_mu_m \in <G>$. But it is only possible that $\text{Reduce}(g, G) = 0$ if we have $a_i = 0,\ 1 \le i \le m$.

●

The reader should compare the above result to the well known fact that an extension field of F of the form

$$F[\mathbf{x}]/<p> \; = \; \{ a_0 + a_1x + \cdots + a_{n-1}x^{n-1} : a_i \in F \}$$

where $p \in F[\mathbf{x}]$ is an irreducible polynomial of degree n, is a vector space of dimension n with basis $[1], [x], \ldots, [x^{n-1}]$.

Theorem 10.10 allows us to easily decide if the quotient ring is finite dimensional (when considered as a vector space), since this is so if and only if the set U has finitely many elements. This observation will prove useful later on in this section. However, its immediate importance is that it guarantees we can perform arithmetic in the quotient ring.

Example 10.12. We recall that the set

$$G = \{ x^2 + yz - 2,\ y^2 + xz - 3,\ xy + z^2 - 5,\ -2xz^2 + 5x - 2y + 3z,$$
$$-2yz^2 - 3x + 5y + 2z,\ -2z^4 - 2xz - 3yz + 15z^2 - 19 \}$$

computed in Example 10.10 is a Gröbner basis in $\mathbf{Q}[x,y,z]$ with respect to $<_D$. In fact, it is also a reduced Gröbner basis. Then

$$U = \{ [1], [x], [y], [z], [xz], [yz], [z^2], [z^3] \}$$

is a basis for $\mathbf{Q}[x,y,z]/<G>$. To compute $[xz]\cdot[yz]$, for example, we merely find

$$\text{Reduce}(xz\cdot yz, G) = xz + \frac{3}{2}yz - \frac{5}{2}z^2 + \frac{19}{2} ;$$

then

$$[xz]\cdot[yz] = 1[xz] + \frac{3}{2}[yz] - \frac{5}{2}[z^2] + \frac{19}{2}[1] .$$

•

In addition to the basic arithmetic operations, Theorem 10.10 allows us to compute inverses, when they exist, in $F[\mathbf{x}]/<G>$.

Example 10.13. Consider again the sets G, U of Example 10.12. Since U has finitely many entries, it may be possible to compute ring inverses for some of those entries. For example, if $[x]$ has an inverse, it must be of the form

$$[x]\cdot(a_0[1] + a_1[x] + a_2[y] + a_3[z] + a_4[xz] + a_5[yz] + a_6[z^2] + a_7[z^3]) = 1 .$$

Then Theorem 10.10 implies that the reduced form of the polynomial

$$p = x(a_0 + a_1x + a_2y + a_3z + a_4xz + a_5yz + a_6z^2 + a_7z^3) - 1$$

vanishes. Since we find that

$$\text{Reduce}(p, G) = (-1 + 2a_1 + 5a_2) + (a_0 + \frac{3}{2}a_4 + \frac{5}{2}a_6)x + (-\frac{5}{2}a_4 - a_6)y + (a_4 + 5a_5 + \frac{3}{2}a_6)z$$
$$+ (-a_1 - a_7)yz + (a_3 + \frac{5}{2}a_7)xz + (-a_2 + \frac{3}{2}a_7)z^2 - a_5z^3 ,$$

we obtain the system of linear equations

$$2a_1 + 5a_2 = 1 , \quad a_0 + \frac{3}{2}a_4 + \frac{5}{2}a_6 = 0 ,$$
$$-\frac{5}{2}a_4 - a_6 = 0 , \quad a_4 + 5a_5 + \frac{3}{2}a_6 = 0 , \quad -a_1 - a_7 = 0 ,$$
$$a_3 + \frac{5}{2}a_7 = 0 , \quad -a_2 + \frac{3}{2}a_7 = 0 , \quad -a_5 = 0 .$$

If we solve this system (e.g. by the one of the methods of Chapter 9), we find the solution

$$a_0 = a_4 = a_5 = a_6 = 0, \ a_1 = -\frac{2}{11}, a_2 = \frac{3}{11}, a_3 = -\frac{5}{11}, a_7 = \frac{2}{11} ;$$

hence

$$[x]^{-1} = \frac{2}{11}[z^3] - \frac{2}{11}[x] + \frac{3}{11}[y] - \frac{5}{11}[z] .$$

•

This type of construction turns out to be very useful in the next subsection.

Solution of Systems of Polynomial Equations

We now turn our attention to the more common problem of solving systems of polynomial equations. To this end, we will take a somewhat more modern approach than that of Section 9.5. Namely, we will view a set of equations over a field F

$$p_i(x_1, x_2, \ldots, x_n) = 0, \quad 1 \le i \le k,$$

in terms of the ideal $<p_1, p_2, \ldots, p_k>$. It is easily established that if $<P> = <G>$, then the sets of common zeros of the sets P, $G \subset$ F[**x**] are identical. (Exercise 10.10.) If G is a Gröbner basis for $<P>$, then one expects (by now!) to be able to obtain more information about these zeros from G than from P. This is indeed the case, as the following results of Buchberger [6] show.

Theorem 10.11. Let G be a monic Gröbner basis for $<P> = <p_1, \ldots, p_k> \subseteq$ F[**x**]. Then P, viewed as a system of algebraic equations, is solvable if and only if $1 \notin G$.

Proof: It is well known from (modern) algebra (see for example Hilbert's "Nullstellensatz", in van der Waerden [35]) that P is unsolvable if and only if there exists a combination of the p_i (over F[**x**]) which equals a nonzero constant, say 1. Since $<P> = <G>$, this is equivalent to $1 \in <G>$. Since G is a Gröbner basis, this implies that Reduce(1, G) = 0; this, in turn, means that $1 \in G$.

●

We note that a system P is unsolvable if and only if a Gröbner basis for $<P>$ contains an element of headterm 1. In such a case, the reduced monic Gröbner basis will simply be {1}.

Example 10.14. The reduced, monic Gröbner basis (over **Q**[x,y]) for the ideal $<p_1, p_2, p_3>$ where

$$p_1 = x^2y + 4y^2 - 17, \quad p_2 = 2xy - 3y^3 + 8, \quad p_3 = xy^2 - 5xy + 1,$$

is {1}, irrespective of the term ordering used. Therefore, the corresponding system of algebraic equations

$$p_1 = 0, \quad p_2 = 0, \quad p_3 = 0$$

has no solutions.

●

Theorem 10.12. Let G be a Gröbner basis for $<P> \subseteq \mathrm{F}[\mathbf{x}]$, and let H be the set

$$H = \{ \mathrm{hterm}(g) : g \in G \} .$$

Then the system of equations corresponding to P has finitely many solutions if and only if for all $1 \le i \le n$, there is an $m \in \mathbf{N}$ such that $(x_i)^m \in H$.

Proof: The headterms of G have the required "separation property" iff the set U defined in Theorem 10.10 has finitely many entries; i.e. $\mathrm{F}[\mathbf{x}]/<G>$ is finite dimensional as a vector space. This, however, is true if and only if the set G (or P) has finitely many solutions. (This is plausible in view of our earlier remark on algebraic extension fields of F. However, the reader is referred to Gröbner [22], or van der Waerden [35] for more details.) ●

It must be noted that these powerful results do not depend on the term ordering chosen to construct the Gröbner basis. Neither do they require that the solutions themselves be produced. The latter fact may be important in practice, since the construction of solutions may (for some reason) be impractical when Algorithm 10.4 is not.

Example 10.15. We found in Example 10.10 that the reduced Gröbner basis for

$$<P> = <x^2 + yz - 2,\ y^2 + xz - 3,\ xy + z^2 - 5> \subset \mathbf{Q}[x,y,z]$$

with respect to $<_D$ has 6 polynomials with headterms

$$H = \{ x^2, y^2, xy, xz^2, yz^2, z^4 \} .$$

Since $1 \notin H$, the system corresponding to P is solvable; also, by Theorem 10.12, there are finitely many solutions. ●

Example 10.16. Consider the set of polynomials (and system of equations corresponding to)

$$<P> = < zx + yx - x + z^2 - 2,\ xy^2 + 2zx - 3x + z + y - 1,$$
$$2z^2 + zy^2 - 3z + 2zy + y^3 - 3y > .$$

If we order $\mathrm{T}_{(x,y,z)}$ with the lexicographic order $<_L$, we may obtain a reduced Gröbner basis for $<P>$,

$$\{ xz^2 - 2x - z^4 + 4z^2 - 4,$$
$$y + z^4 + 2z^3 - 5z^2 - 3z + 5,$$
$$z^6 + 2z^5 - 7z^4 - 8z^3 + 15z^2 + 8z - 10 \} .$$

Again, we see that the system corresponding to P is solvable; however, in this case there are infinitely many solutions. ●

The above examples illustrate an important distinction between "total degree" and "lexicographic" Gröbner bases. The degree basis shown in Example 10.12 offers no direct insight into the solutions of the system; however, a quick inspection of the lexicographic basis of Example 10.16 suggests a more powerful result. Apparently, it will be more difficult to obtain solutions from some types of Gröbner bases than from others. Since the choice of term ordering affects the complexity (and practical behaviour) of Algorithm 10.4, it is well worth developing solution methods for both $<_D$ and $<_L$.

We consider first the use of the degree ordering. In Example 10.13, we exploited the fact that if a polynomial (with indeterminate coefficients) $p = \Sigma a_j t_j$ is in $<G>$, the requirement that Reduce(p, G) = 0 yields conditions on the indeterminates a_i. The difficulty lies in determining which $t_i \in T_{\mathbf{x}}$ to include in the representation of p. But, if G has finitely many solutions, certain types of polynomials are guaranteed to exist. Namely, for each x_i, $1 \le i \le n$, there must exist a univariate polynomial $p_i = \Sigma a_{ij}(x_i)^j$ whose roots contain all possible values of x_i which may appear in solutions of G. For a set $P \subset F[\mathbf{x}]$ and $\tilde{x} \in \mathbf{x}$, the polynomial of least degree in $<P> \cap F[\tilde{x}]$ may be constructed by Algorithm 10.5 below. This algorithm is clearly valid for *any* admissible term ordering, although we will soon see that for $<_L$ it is unnecessary.

Example 10.17. Consider the set $P \subset \mathbf{Q}[x,y,z]$ defined in Example 10.10, along with the corresponding total degree basis G. In order to find the polynomial $p \in <P> \cap \mathbf{Q}[z]$ of least degree, we note that 1, z, z^2, z^3 are irreducible modulo G. Therefore, we first let $p = a_0 + a_1 z + a_2 z^2 + a_3 z^3 + a_4 z^4$, and set

$$\text{Reduce}(p, G) = (a_0 - \frac{19}{2}a_4) + a_1 z + (a_2 + \frac{15}{2}a_4)z^2 - \frac{3}{2}a_4 yz - a_4 xz + a_3 z^3$$
$$= 0 .$$

This implies that $a_0 = \cdots = a_4 = 0$. In like manner, the polynomials of degrees 5, 6, and 7 all vanish identically. When we try $p = \sum_{k=0}^{8} a_k z^k$,

$$\begin{aligned}\text{Reduce}(p, G) = {} & (a_0 - \frac{19}{2}a_4 - \frac{3325}{8}a_8 - \frac{285}{4}a_6) + (a_1 - \frac{25}{2}a_5 - \frac{775}{8}a_7)z \\ & + (-\frac{109}{4}a_7 - \frac{11}{4}a_5)y + (\frac{13}{8}a_7 - \frac{1}{4}a_5)x \\ & + (a_2 + \frac{175}{4}a_6 + \frac{925}{4}a_8 + \frac{15}{2}a_4)z^2 + (-\frac{743}{8}a_8 - 14a_6 - \frac{3}{2}a_4)yz \\ & + (-\frac{337}{8}a_8 - \frac{31}{4}a_6 - a_4)xz + (a_3 + \frac{175}{4}a_7 + \frac{15}{2}a_5)z^3 .\end{aligned}$$

The resulting system of linear, homogeneous equations has the nontrivial solution

$$a_0 = \frac{361}{8}a_8,\ a_1 = 0,\ a_2 = -95a_8,\ a_3 = 0,$$

Algorithm 10.5. Solution of System P in Variable $\tilde{x}$.

procedure Solve1(P, $\tilde{x}$)

```
# Given a system P with finitely many solutions, find
# the smallest polynomial containing the solutions in x̃.

G ← Gbasis(P)

# Assume a polynomial of form Σ a_k x̃^k;
# then require that
#    Reduce(Σ a_k x̃^k, G) = Σ a_k Reduce(x̃^k, G) = 0.

k ← 0

# If G does not satisfy Theorem 10.12, the following loop
# may be infinite!

do {
   p_k ← Reduce(x̃^k, G)
   if ∃ (a_0, ..., a_k) ≠ (0, ..., 0) such that Σ_{j=0}^{k} a_j p_j = 0 then
      return( a_k^{-1} · Σ_{j=0}^{k} a_j x̃^j )
   else k ← k + 1 }
end
```

$$a_4 = \frac{219}{4}a_8,\ a_5 = 0,\ a_6 = -\frac{25}{2}a_8,\ a_7 = 0\,.$$

Without loss of generality, we choose $a_8 = 1$ to obtain

$$p = z^8 - \frac{25}{2}z^6 + \frac{219}{4}z^4 - 95z^2 + \frac{361}{8}\,.$$

•

A complete set of n univariate polynomials obtained in the above manner constitutes a finite inclusion of the roots of the original system. As noted in Section 9.5, though, not all n-tuples so defined are roots. One can do better if one of the univariate polynomials splits into factors over F. For example, if $p_1 \in \langle P\rangle \cap \mathrm{F}[x_1]$ admits a factorization $p_1 = q_1^{e_1} q_2^{e_2} \cdots q_m^{e_m}$, then Gbasis($P$) may be refined to Gbasis($P \cup \{q_i\}$) with respect to each irreducible, distinct factor q_i. Thereafter, each component basis will yield different (smaller) univariate polynomials in $x_2, \ldots, x_n$. Carried even further, this approach suggests a scheme (specified by Algorithm 10.6) to explicitly determine the roots of P.

Algorithm 10.6. Complete Solution of System P.

```
procedure GröbnerSolve(P)

  # Given system P ⊂ F[x] with finitely many solutions,
  # find these solutions over an ''appropriate'' extension of F.

  # We store partially refined bases and partial roots in Q.

  Q ← { [P, ()] }
  for k from n by −1 to 1 do {
    S ← ∅

    # Refine/extend each element of Q one more level.

    foreach [G, (α_{k+1}, ..., α_n)] ∈ Q do {
      G̃ ← { g(x_1, ..., x_k, α_{k+1}, ..., α_n) : g ∈ G }
      G̃ ← Gbasis(G̃)
      p ← Solve1(G̃, x_k)

      # The roots of p in x_k yield several new partial roots.

      if p ≠ 1 then
        S ← S ∪ { [G̃, (α, α_{k+1}, ..., α_n)] : p(α) = 0 }  }
    Q ← S  }
  roots ← ∅
  foreach [G, (α_1, ..., α_n)] ∈ Q do {
    roots ← roots ∪ { (α_1, ..., α_n) }  }
  return( roots )
end
```

Of course, it will not always be possible to solve all univariate polynomials exactly. Moreover, the successive refinement of each Gröbner basis may be impractical if complicated extensions of F are involved. (Note that F may be a rational function field!) Still, Algorithm 10.6 provides a complete solution in theory when P has finitely many solutions.

Lexicographic Bases and Elimination

We now recall from Example 10.16 that Gröbner bases with respect to $<_L$ seem to provide more information, in a way, than total degree bases. So, in spite of the increased difficulty of computing such bases, their use may offer a valuable alternative to Algorithm 10.6. The basis for such a method is the following theorem.

Theorem 10.13. Let $<_T$ be an admissible ordering on $T_{\mathbf{x}}$ which is such that $s <_T t$ whenever $s \in T_{(x_k, \ldots, x_n)}$ and $t \in T_{(x_1, \ldots, x_{k-1})}$; let G be a Gröbner basis over F with respect to $<_T$. Then

$$<G> \cap F[x_k, \ldots, x_n] = <G \cap F[x_k, \ldots, x_n]>,$$

where the ideal on the right hand side is formed in $F[x_k, \ldots, x_n]$.

Proof: For convenience, we define $G^{(k)} = G \cap F[x_k, \ldots, x_n]$. First, suppose that $p \in <G> \cap F[x_k, \ldots, x_n]$. Since G is a Gröbner basis, $p \mapsto^{*}_{G} 0$. But since p contains only the variables $x_k, \ldots, x_n$ this means that there exist polynomials $p_i \in F[x_k, \ldots, x_n]$, $g_i \in G^{(k)}$ such that $p = \sum_{i=1}^{m} p_i g_i$; this implies that $p \in <G^{(k)}>$.

We remark that if G is a Gröbner basis with respect to $<_T$, then $G^{(k)}$ must also be a Gröbner basis, since Theorem 10.5 (ii) requires that

$$\mathrm{Spoly}(p, q) \mapsto^{*}_{G^{(k)}} 0$$

for all $p, q \in G^{(k)} \subseteq G$. It follows that if $p \in <G^{(k)}>$, we also have $p \in <G> \cap F[x_k, \ldots, x_n]$.

●

Now, consider specifically the ordering $<_L$, which satisfies the requirements of the above result. Then Theorem 10.13 says that the polynomials in G which only depend on the last $n-k+1$ variables are a Gröbner basis for the "k-th elimination ideal" of G (i.e. the subset of $<G>$ which depends only on these variables). Suppose that G is the lexicographic Gröbner basis of a set $P \subset F[\mathbf{x}]$ which has finitely many solutions. Then by Theorem 10.12, G must contain a single univariate polynomial in x_n; namely, the polynomial in $<P> \cap F[x_n]$ of least degree. In addition, G must contain at least one polynomial in each elimination ideal in which the "highest" variable is separated.

Example 10.18. The reduced, monic Gröbner basis with respect to $<_L$ for

$$<P> = <x^2 + yz - 2,\ y^2 + xz - 3,\ xy + z^2 - 5> \subset \mathbf{Q}[x,y,z]$$

is

$$\{\, x - \frac{88}{361}z^7 + \frac{872}{361}z^5 - \frac{2690}{361}z^3 + \frac{125}{19}z\,,$$

$$y + \frac{8}{361}z^7 + \frac{52}{361}z^5 - \frac{740}{361}z^3 + \frac{75}{19}z\,,$$

$$z^8 - \frac{25}{2}z^6 + \frac{219}{4}z^4 - 95z^2 + \frac{361}{8}\,\}.$$

Hence, in order to solve the nonlinear system associated with P we may solve instead the reduced equations

$$\begin{aligned}
x \qquad - \frac{88}{361}z^7 + \frac{872}{361}z^5 - \frac{2690}{361}z^3 + \frac{125}{19}z &= 0\,, \\
y + \frac{8}{361}z^7 + \frac{52}{361}z^5 - \frac{740}{361}z^3 + \frac{75}{19}z &= 0\,, \\
z^8 - \frac{25}{2}z^6 + \frac{219}{4}z^4 - 95z^2 + \frac{361}{8} &= 0\,.
\end{aligned}$$

Note how closely this resembles a triangular linear system.

●

A lexicographic basis may, of course, contain other polynomials whose headterms are not separated. But since each subset $G \cap F[x_k, \ldots, x_n]$ is also a Gröbner basis, no simpler (i.e. "more separated") basis may exist for the given permutation of variables. A (simpler) counterpart to Algorithm 10.6 appears below as Algorithm 10.7. We note that the basis refinements (i.e. the additional Gröbner basis computations) in Algorithm 10.7 are univariate subproblems, and therefore amount to GCD computations. In fact, it has been shown (e.g. see Kalkbrener [25]) that these calculations are unnecessary; it suffices to select the element of "G_k" of minimal degree in x_k whose leading coefficient (in x_k) does not vanish under the current evaluation. Thus, the above process is indeed simpler than Algorithm 10.6. Still, it may not be possible to carry out in practice. We mention that, as before, the Gröbner basis may be decomposed into irreducible components if any of its elements factor. (In fact, these components can be computed much more efficiently by factoring *during* Algorithm 10.4; see Czapor [19].)

It should also be pointed out that even when a given system has *infinitely* many solutions, the lexicographic Gröbner basis will be as "triangular" as possible. Therefore, it is still possible to obtain the solutions directly from the basis.

Example 10.19. In Example 10.16, we computed the lexicographic basis

$$\begin{aligned}
G = \{\, & xz^2 - 2x - z^4 + 4z^2 - 4\,, \\
& y + z^4 + 2z^3 - 5z^2 - 3z + 5\,, \\
& z^6 + 2z^5 - 7z^4 - 8z^3 + 15z^2 + 8z - 10 \,\}\,,
\end{aligned}$$

over $\mathbf{Q}[x,y,z]$. If we factor the final, univariate polynomial we obtain

$$(z^2 - 2)(z^4 + 2z^3 - 5z^2 - 4z + 5)\,.$$

The roots of the larger factor,

$$p_1(z) \;=\; z^4 + 2z^3 - 5z^2 - 4z + 5 \;=\; 0\,,$$

may be extended using Algorithm 10.7 to four complete roots for x, y, z; however, the roots of $p_2(z) = z^2 - 2 = 0$ yield only the solutions $\{y = 1 \mp \sqrt{2},\ z = \pm\sqrt{2}\}$, in which x may take any value. Alternatively, we may refine the basis with respect to each of the univariate polynomials p_1, p_2 to obtain

Algorithm 10.7. Solution of P using Lexicographic Gröbner Basis.

```
procedure LexSolve(P)
    # First, find a reduced Gröbner basis with respect to <_L
    # for the ideal generated by P ⊂ F[x].
    G ← Gbasis(P) ;  roots ← ∅

    # If P has finitely many solutions, we proceed to
    # solve the univariate polynomial in x_n.
    p ← selectpoly( G ∩ F[x_n] )
    roots ← roots ∪ { (α) : p(α) = 0 }

    # Now, back-solve (cf. Section 9.5).
    for k from n − 1 by −1 to 1 do {
        S ← ∅
        G_k ← G ∩ F[x_k, ..., x_n] − F[x_{k+1}, ..., x_n]
        foreach (α_{k+1}, ..., α_n) ∈ roots do {
            G̃ ← { g(x_k, α_{k+1}, ..., α_n) : g ∈ G_k }
            G̃ ← Gbasis(G̃)
            p ← selectpoly(G̃ ∩ F[x_k])
            if p ≠ 1 then
                S ← S ∪ { (α, α_{k+1}, ..., α_n ) : p(α) = 0 }  }
        roots ← S  }
    return( roots )
end
```

$$\text{Gbasis}(G \cup \{p_1\}) = \{ x - z^2 + 2,\ y + z,\ z^4 + 2z^3 - 5z^2 - 4z + 5 \},$$

$$\text{Gbasis}(G \cup \{p_2\}) = \{ y + z - 1,\ z^2 - 2 \}.$$

Note that, in the irreducible components of the lexicographic basis, it becomes clear when specific variables must be viewed as parameters in the corresponding solutions. Moreover, it is clear *before* we begin back-solving (or basis decomposition) that some variables will be parameters. (The reader should compare this with the solution of a reduced system in Section 9.5.)

●

Obviously, there is a strong connection between lexicographic Gröbner bases and the resultant techniques of the previous chapter. However, the Gröbner basis is clearly a more powerful and elegant result than a reduced system (cf. Definition 9.1). For example, the "final" univariate polynomial is of minimal degree, and therefore contains no extraneous roots. Also, the degree of the polynomial which must be solved at each phase of back-solving will be no larger than the number of roots.

On the other hand, for some types of input polynomials the computation of a reduced system via resultants may be much faster than the computation of a lexicographic Gröbner basis (via Algorithm 10.4). Hence, Pohst and Yun [34] proposed the combined use of resultants, pseudo-division and S-polynomial reduction. Also, the speed of either scheme (for a given problem) depends very strongly on the permutation of variables $x_1, \ldots, x_n$ used for the elimination. It is much easier to choose a good permutation for the resultant method (one variable at a time) than to choose a good permutation (a priori) for Algorithm 10.4, when $<_L$ is used. Although the problem of determining the *optimal* permutation is difficult, Böge et al. [14] have proposed a simple heuristic for choosing a "reasonable" permutation. However, it is important to note that when a degree ordering is used, the algorithm is not nearly as sensitive to this choice (cf. Exercise 10.14). Therefore the lexicographic algorithm is "unstable" in this sense. We recall also that the complexity of Algorithm 10.4 is greater when using $<_L$ than when using $<_D$. Consider a system of n polynomial equations in n variables of degree at most d, which has only finitely many solutions. The results of Lazard[30] (mentioned at the end of Section 10.4) imply that the computation of a total degree basis is of complexity $O(d^{nk})$ for some $k \in \mathbf{N}$ (i.e. polynomial in d^n). On the other hand, Caniglia et al. [15, 16] have recently shown that the complexity of the lexicographic calculation is $O(d^{n^2})$. Hence it is clear that (in general) the computation of a lexicographic Gröbner basis via Algorithm 10.4 is *much* more difficult than the corresponding total degree computation.

In spite of the above, a lexicographic basis is more useful than a degree basis in that the corresponding solutions are easily obtained from it via Algorithm 10.7. Recently, Faugère et al. [20] presented an important generalization of Algorithm 10.5 which computes a lexicographic basis from (say) a degree basis by an efficient "change of basis" transformation. However, for systems with infinitely many solutions no alternative to Algorithm 10.4 currently exists.

10.6. ADDITIONAL APPLICATIONS

Along with the applications discussed above, Buchberger's algorithm provides constructive solutions for a great many problems in polynomial ideal theory such as computation of Hilbert functions for polynomial ideals, free resolution of polynomial ideals and syzigies, and determination of algebra membership. However, most such topics require more algebraic background than we wish to discuss here. We consider instead two very basic problems which, perhaps surprisingly, may be solved in terms of Gröbner bases.

Geometry Theorem Proving

In the past few years, the automated proving of elementary geometry theorems has become a topic of great interest in symbolic computation. This is primarily due (it seems) to the recent work of Wu [38]. We will not dwell here on the foundations of the subject; rather, we will attempt to present some of the basic ideas with emphasis on the possible role of Gröbner bases. The main idea is that often a theorem (i.e. a set of hypotheses implying a conclusion), for which the geometric relationships may be expressed as polynomials, can be proven algebraically. In Wu's method, one attempts to show that the set of common zeros (in an algebraically closed field) of the hypothesis polynomials is contained in the set of zeros of the conclusion polynomial. Unfortunately, in elementary geometry one is concerned with real (rather than complex) zeros; so, the method is not *complete* in the sense that not all valid theorems may be proven. In spite of this, Wu and also Chou [18] have succeeded in proving a large number of such theorems.

It is not surprising that, in the above problem, Gröbner basis techniques have been successfully applied. We will sketch one such approach due to Kapur [27]. (A different approach presented by Kutzler and Stifter [29] appears to be faster, but less powerful.) Details of the equivalence between Wu's formulation of the problem and Kapur's (and a comparison of the various methods) are given by Kapur [27].

Let F be a field of characteristic zero, and let $\tilde{F}$ be an algebraically closed field containing F. Suppose we can represent hypotheses as polynomials $h_i \in F[\mathbf{x}]$, the conclusion as a polynomial $c \in F[\mathbf{x}]$, and any subsidiary hypotheses (to be explained later) as a polynomial $s_i \in F[\mathbf{x}]$. Then we will consider statements of the form

$$\forall\, x_1, \ldots, x_n \in \tilde{F},\ \{\, h_1=0, h_2=0, \ldots, h_k=0, s_1 \neq 0, \ldots, s_l \neq 0 \,\} \Rightarrow \quad c = 0\,. \tag{10.19}$$

The above statement is a theorem if the zeros in $\tilde{F}$ of c include the admissible common zeros of the h_i. This form is actually quite general because any (quantifier-free) formula involving boolean connectives may also be expressed as a (finite) set of polynomial equations. Namely, Kapur shows in [27] that:

(a) $p_1 = 0$ and $p_2 = 0 \quad \Leftrightarrow \quad \{\, p_1 = 0, p_2 = 0 \,\}$;

(b) $p_1 = 0$ or $p_2 = 0 \quad \Leftrightarrow \quad \{\, p_1 p_2 = 0 \,\}$;

(c) $p_1 \neq 0 \quad \Leftrightarrow \quad \{\, p_1 z - 1 = 0 \,\}$,

where z in the above is a new indeterminate. He then proposes the following, which (as in Theorem 10.11) is based on Hilbert's Nullstellensatz:

Theorem 10.14 ([27]). The validity of a (geometry) statement of the form (10.19) is equivalent to the validity of

$$< h_1, \ldots, h_k, s_1 z_1 - 1, \ldots, s_l z_l - 1, cz - 1> \; = \; <1>\,,$$

where z, $\{z_i\}$ are additional indeterminates.

●

That is, the problem reduces to that of showing that a related system (which includes the contradiction of the conclusion) is not solvable over $\tilde{F}$.

Example 10.20. Consider the problem of proving the following simple proposition: if the right bisector of the hypotenuse of a right triangle intersects the right vertex, then the triangle is isosceles.

Without loss of generality, we set up a plane coordinate system in which the right vertex is at the origin, and the triangle sits in the first quadrant. Suppose the other two vertices are at $(y_1, 0)$ and $(0, y_2)$, and the midpoint of the hypotenuse is (y_3, y_4). Then $y_4 = y_2 / 2$ and $y_3 = y_1 / 2$ (since we have a midpoint), and $y_4 / y_3 = -(-y_2 / y_1)^{-1}$ (since the bisector is perpendicular to the hypotenuse). Furthermore, the triangle will be isosceles if and only if $| y_1 | = | y_2 |$. Since the reduced, monic Gröbner basis of

$$< y_1 - 2y_3,\ y_2 - 2y_4,\ y_1y_3 - y_2y_4,\ (y_1^2 - y_2^2)z - 1 >$$

(over $\mathbf{Q}[y_1, y_2, y_3, y_4, z]$) is $\{1\}$, the theorem is valid.

●

We must, finally, mention the role of the subsidiary hypotheses $\{s_i\}$ in the above. It may happen that some theorems may only be established in the above manner when certain degenerate cases are ruled out. For example, in Example 10.20 it might have been necessary to specify that $y_1 \neq 0 \neq y_2$; thus we would have added the polynomials

$$s_1 = y_1z_1 - 1\,,\quad s_2 = y_2z_2 - 1$$

to the set above. An example of a case in which such extra conditions are necessary is provided in Exercise 10.20. (Methods for detecting such cases are discussed by Kapur [27] and Wu [38].)

Polynomial GCD Computation

Recently, Gianni and Trager [21] outlined how Gröbner basis calculations may be used to compute multivariate GCD's. While not especially practical, such methods do serve to illustrate the significance of Gröbner bases. They propose the following method:

Theorem 10.15. Let $f_1, \ldots, f_m, g \in \mathrm{F}[y, \mathbf{x}]$ be primitive with respect to y and I be a maximal ideal in $\mathrm{F}[\mathbf{x}]$ (i.e. I is contained in no other ideal). Suppose that

$$< f_1, \ldots, f_m, \mathrm{I} > \; = \; < 1 > ,$$

$$< \mathrm{lcoeff}_y(f_i \cdot g), \mathrm{I} > \; = \; < 1 > \quad \text{for some } 1 \le i \le m\,,$$

and let $G_{(k)}$ be a reduced Gröbner basis for the ideal

$$< f_1 \cdot g,\ \ldots,\ f_m \cdot g,\ \mathrm{I}^k >$$

with respect to $<_D$. Then for $k > [\deg(g)]^2$, the unique polynomial $\tilde{g}$ in $G_{(k)}$ of least total degree is an associate of g.

●

The idea in the above is to produce an ideal in which the GCD is the element of least degree. This depends, in part, on the observation that

$$< f_1, \ldots, f_m, \mathrm{I}> = < 1> \Rightarrow < f_1 \cdot g, \ldots, f_m \cdot g, \mathrm{I}^k> = < g, \mathrm{I}^k>$$

for $k > 0$. (The proof of this is left as an exercise for the reader.) In practice, the ideal I is chosen to be of the form

$$\mathrm{I} = < y_1 - a_1,\ y_2 - a_2,\ \ldots,\ y_m - a_m>$$

for $a_i \in$ F. The reader should compare the above requirements for the $\{f_i\}$, g and I with those imposed on homomorphisms in Section 7.4.

Example 10.21. Consider the problem of finding the GCD of

$$p_1 = 2yxz - 2y^3 + 4y - 7x^2z + 7xy^2 - 14x + xz^2 - zy^2 + 2z\ ,$$

$$p_2 = 3x^2z - 3xy^2 + 6x - xz^2 + zy^2 - 2z - xz + y^2 - 2\ .$$

We first note that both of these polynomials are primitive in x. If we choose $\mathrm{I} = < y, z-1>$, for example, then a basis for I^5 is

$$Q = \{ y^5, y^4(z-1), y^3(z-1)^2, y^2(z-1)^3, y(z-1)^4, (z-1)^5 \}\ .$$

It is easily verified that the remaining conditions of Theorem 10.15 are met. By means of Algorithm 10.4, we may compute a reduced basis (with respect to $<_D$) for $< p_1, p_2, Q>$, namely

$$\{ yz^4 - 4yz^3 + 6yz^2 - 4yz + y,\ z^5 - 5z^4 + 10z^3 - 10z^2 + 5z - 1,$$
$$xyz^2 - 2yz^3 - 2xyz + 8yz^2 + xy - 10yz + 4y,$$
$$x^3 - 12xz^2 + 16z^3 + 6x^2 + 24xz - 72z^2 + 96z - 32,$$
$$xz^3 - 2z^4 - 3xz^2 + 10z^3 + 3xz - 18z^2 - x + 14z - 4,$$
$$x^2y - 4xyz + 4yz^2 + 8xy - 16yz + 16y,$$
$$x^2z - 4xz^2 + 4z^3 - x^2 + 12xz - 20z^2 - 8x + 32z - 16,$$
$$y^2 - xz - 2 \}\ .$$

The polynomial of least degree in this basis is

$$g = y^2 - xz - 2\ .$$

Since we have

$$p_1 = (2y - 7x + z)(xz - y^2 + 2),\ \ p_2 = (3x - z - 1)(xz - y^2 + 2)\ ,$$

g is indeed the required GCD. •

We mention that, in addition, a Gröbner basis method is given by Gianni and Trager [21] for performing multivariate factorization.

Exercises

(Those exercises marked with an * may require a significant amount of computer time; time limits should be set at "appropriate" values.)

1. Consider arbitrary $p, q, r \in F[\mathbf{x}]$ and $P \subset F[\mathbf{x}]$ such that $p \mapsto_P q$. Is it true that $p+r \mapsto_P q+r$? Is it true that $p+r \mapsto^*_P q+r$?

2. Prove that if $p, q, r \in F[\mathbf{x}]$ are such that $p = q \cdot r$, then $p \mapsto^*_{\{q\}} 0$ for any term ordering $<_T$.

3. Formulate a strategy for the procedure selectpoly, which selects the "best" of several reducers in Algorithm 10.1 when the degree ordering is used. *Hint:* Show that the number of distinct n-variate terms of total degree k is
$$\binom{k+n-1}{n-1} \; ;$$
hence the number of distinct terms of degree less than or equal to d is
$$\sum_{k=0}^{d} \binom{k+n-1}{n-1} = \binom{d+n}{n} \; .$$

4. For arbitrary $g_1, g_2, h_1, h_2 \in F[\mathbf{x}]$, find $\alpha, \beta \in F$ and $u \in T_{\mathbf{x}}$ such that
$$\mathrm{M}(h_1)g_1 + [\,\mathrm{M}(h_2) + \alpha \cdot \mathrm{hterm}(h_2)\,] \cdot g_2 = \beta\, u \cdot \mathrm{Spoly}(g_1, g_2) \; .$$

5. Devise an algorithm to decide, given $P_1, P_2 \subset F[\mathbf{x}]$, if $<P_1> \subseteq <P_2>$.

6. Using your favorite computer algebra system, implement Algorithm 10.2 using both term orderings $<_D$ and $<_L$ over $\mathbf{Q}[\mathbf{x}]$. (Note: you need only construct different leading monomial functions M_D and M_L.) You should use the "first available" pair selection strategy, and make all polynomials monic as they are added to the partial basis. Test your code for $<_D$ on Examples 10.8 and 10.10, and for $<_L$ on Examples 10.16 and 10.18.

7. Implement Algorithm 10.3, and hence modify the code from Exercise 6 to yield a reduced, monic Gröbner basis. Compare the results of the new code to that of the old on Examples 10.16, 10.18 using $<_L$; repeat this comparison for Example 10.8 using $<_D$. Assuming the implementation of Exercise 6 is correct, can you devise a procedure which verifies the correctness of the new code?

8. Improve the implementation of Exercise 7 by adding criterion1, criterion2 as in Algorithm 10.4. (Note: the efficiency of criterion2 depends on a fast means of testing if $[u, v] \in B$.) Compare this code and that of Exercise 7 on:

   (a) the polynomials of Example 10.10, for $<_D$ and $<_L$;

   (b)* the set

   $$\{ x^2z + xz^2 + yz^2 - z^2 - 2z,\ zy^2 + 2xz - 3z + x + y - 1,$$
   $$2x^2 + xy^2 - 3x + 2xy + y^3 - 3y \},$$

   using the same orderings on $T_{(x,y,z)}$.

9. Further (and finally!) improve the implementation of Exercise 8 by modifying the procedure selectpair to use the "normal" selection strategy (10.17). (Note: a careful choice of data structure for the set B will help.) Carefully compare this and the code of Exercise 8 on the following:

   (a) the polynomials of Exercise 8(b), using $<_D$;

   (b) the same set of polynomials, using $<_L$;

   (c) the polynomials of Example 10.10 using $<_L$.

10. Prove that if sets $P, Q \subset F[\mathbf{x}]$ are such that $<P> = <Q>$, then the roots of P, Q are identical.

11. Implement Algorithm 10.5 in your favorite computer algebra system, i.e. whichever one was used in Exercises 6-9. (Note: in order to save some trouble, you may use whatever system routines are available for the solution of systems of linear equations.) Test your implementation by computing the polynomial $p(z)$ found in Example 10.17. Then, for the set P defined in Example 10.10, use your implementation to compute the counterparts $q(y)$, $r(x)$ to p.

12. Show that the set $\{p, q, r\}$ computed in Exercise 11 is a reduced, monic Gröbner basis. (Note that this does not require computation of this set.) Why does this not contradict the uniqueness of the basis computed in Example 10.10?

13. Is it possible for a reduced ideal basis, *not* composed entirely of univariate polynomials, to be a Gröbner basis with respect to more than one term ordering? Give an example (if possible) of such a set which is a Gröbner basis with respect to an *arbitrary* admissible ordering.

14. Compute monic, reduced Gröbner bases with respect to $<_L$ for the set

$$\{ y_3y_4 + 19y_1y_2 + 5y_1^3 + 45,\ y_4 - 7y_3 + 9y_2 - y_1 + 44,$$
$$53y_3y_4 + 2y_1y_2 + 11y_2y_3 + 454,\ y_1y_3y_4 + 3y_1^2 - 6y_3 + 30 \},$$

using the following permutations of variables:

(a) $\mathbf{x} = (y_4, y_3, y_2, y_1)$;

(b) $\mathbf{x} = (y_2, y_4, y_1, y_3)$;

(c)* $\mathbf{x} = (y_1, y_2, y_3, y_4)$.

Compare the times required for the above computations with those required using $<_D$. Suggest a procedure for choosing a permutation for which the lexicographic computation is relatively simple.

15. Explain how the lexicographic basis of Exercise 14(c) could be computed by first computing its degree basis, assuming the lexicographic basis takes the simplest possible form. (*Hint:* Guess the likely form for the lexicographic basis, and then use the fact that the reduced form of each polynomial must vanish.) Can you generalize this approach to the case where the form of the lexicographic basis is not known in advance, assuming that the corresponding system of equations has only finitely many solutions?

16. Using Gröbner bases, solve the following systems of equations (or the systems corresponding to given sets of polynomials) as explicitly as possible:

(a) the set of Exercise 9.23(b) for x, y, z;

(b) the system of Exercise 9.17 (noting that your implementation of Buchberger's algorithm may have to be modified for coefficients in $\mathbf{Q}(c)$);

(c)* the set

$$\{ x_3x_4 + 19x_1x_2 + 5x_1^2 + 45,\ 33x_3x_4 + 2x_2^2 + 11x_2x_3 + 454,$$
$$x_2x_3^2 - 2x_3x_4 - 2x_1x_4 - 14,\ x_4 - 7x_3 + 9x_2 - x_1 + 44,\ x_1x_3x_4 - 6x_3 + 30 \},$$

for x_1, x_2, x_3, x_4.

17. Prove that an ordering $<_M$ on $T_{\mathbf{x}}$ defined for $1 \le m \le n$ by

$$s = x_1^{i_1} \cdots x_n^{i_n} <_M x_1^{j_1} \cdots x_n^{j_n} = t \quad \Leftrightarrow$$
$$\{\exists\, l,\ 1 \le l \le m \text{ such that } i_l < j_l \text{ and } i_k = j_k,\ 1 \le k < l\}, \text{ or}$$
$$\{i_k = j_k,\ 1 \le k \le m\ , \text{ and } x_{m+1}^{i_{m+1}} \cdots x_n^{i_n} <_D x_{m+1}^{j_{m+1}} \cdots x_n^{j_n}\}$$

is admissible according to Definition 10.1. Suggest two possible uses for such an ordering.

18. Implement (i.e. modify one of your previous implementations of) Buchberger's algorithm for Gröbner bases over $\mathbf{Z}_p$, where p is prime. Compute Gröbner bases over $\mathbf{Z}_{17}[x,y,z]$ with respect to $<_D$ for the following:

    (a) $\{ x^4y + 2x^3y + 5x^3, x^2y^2 - 3xy, xy^4 + xy^2 \}$;

    (b) $\{ 2xy^2 + 7y^2 + 9x + 2, xy + 4x - 5y + 11 \}$.

    Compare both results above to the corresponding bases over $\mathbf{Q}[x,y,z]$. What conclusions may be drawn?

19. Devise a modification of Buchberger's algorithm which, given $p \in \langle f_1, \ldots, f_m \rangle \subseteq$ F[**x**], finds $a_i \in$ F[**x**] such that $p = \sum_{i=1}^{m} a_i f_i$. Implement your scheme, and use it to find a_1, a_2, a_3 such that

$$a_1 \cdot (x^3yz - xz^2) + a_2 \cdot (xy^2z - xyz) + a_3 \cdot (x^2y^2 - z^2) = xz^4 - xyz^3 .$$

20. Use Gröbner bases to prove that a parallelogram is a square iff its diagonals are perpendicular and equal in length. *Hint:* Does the problem make sense if your "arbitrary points" do not really define a parallellogram?

References

1. D. Armbruster, "Bifurcation Theory and Computer Algebra: An Initial Approach," pp. 126-137 in *Proc. EUROCAL '85, Vol. 2, Lecture Notes in Computer Science* **204**, ed. B. F. Caviness, Springer-Verlag (1985).

2. L. Bachmair and B. Buchberger, "A Simplified Proof of the Characterization Theorem for Gröbner Bases," *ACM SIGSAM Bull.*, **14**(4) pp. 29-34 (1980).

3. A.M. Ballantyne and D.S. Lankford, "New Decision Algorithms for Finitely Presented Commutative Semigroups," *Comp. Math. Appl.*, **7** pp. 159-165 (1981).

4. L.J. Billera and L.L. Rose, "Gröbner Basis Methods for Multivariate Splines," RRR # 1-89, Rutgers Univ. Department of Mathematics and Center for Operations Research (1989).

5. B. Buchberger, "An Algorithm for Finding a Basis for the Residue Class Ring of a Zero-Dimensional Polynomial Ideal (German)," Ph.D. Thesis, Univ. of Innsbruck, Math. Inst. (1965).

6. B. Buchberger, "An Algorithmical Criterion for the Solvability of Algebraic Systems of Equations (German)," *Aequationes math.*, **4**(3) pp. 374-383 (1970).

7. B. Buchberger, "Some Properties of Gröbner-Bases for Polynomial Ideals," *ACM SIGSAM Bull.*, **10**(4) pp. 19-24 (1976).

8. B. Buchberger, ''A Theoretical Basis for the Reduction of Polynomials to Canonical Forms,'' *ACM SIGSAM Bull.*, **10**(3) pp. 19-29 (1976).

9. B. Buchberger, ''A Criterion for Detecting Unnecessary Reductions in the Construction of Gröbner Bases,'' pp. 3-21 in *Proc. EUROSAM '79, Lecture Notes in Computer Science* **72**, ed. W. Ng, Springer-Verlag (1979).

10. B. Buchberger and F. Winkler, ''Miscellaneous Results on the Construction of Gröbner Bases for Polynomial Ideals,'' Tech. Rep. 137, Univ. of Linz, Math. Inst. (1979).

11. B. Buchberger, ''A Note on the Complexity of Constructing Gröbner Bases,'' pp. 137-145 in *Proc. EUROCAL '83, Lecture Notes in Computer Science* **162**, ed. H. van Hulzen, Springer-Verlag (1983).

12. B. Buchberger and R. Loos, ''Algebraic Simplification,'' pp. 11-43 in *Computer Algebra - Symbolic and Algebraic Computation (Second Edition)*, ed. B. Buchberger, G. Collins and R. Loos, Springer-Verlag, Wein - New York (1983).

13. B. Buchberger, ''Gröbner Bases: An Algorithmic Method in Polynomial Ideal Theory,'' pp. 184-232 in *Progress, directions and open problems in multidimensional systems theory*, ed. N.K. Bose, D. Reidel Publishing Co. (1985).

14. W. Böge, R. Gebauer, and H. Kredel, ''Some Examples for Solving Systems of Algebraic Equations by Calculating Gröbner Bases,'' *J. Symbolic Comp.*, **2**(1) pp. 83-98 (1986).

15. L. Caniglia, A. Galligo, and J. Heintz, ''Some New Effectivity Bounds in Computational Geometry,'' pp. 131-152 in *Proc. AAECC-6, Lecture Notes in Computer Science* **357**, Springer-Verlag (1989).

16. L. Caniglia, A. Galligo, and J. Heintz, ''How to Compute the Projective Closure of an Affine Algebraic Variety in Subexponential Time,'' in *Proc. AAECC-7 (to appear)*, (1989).

17. P. Le Chenadec, ''Canonical Forms in Finitely Presented Algebras (French),'' Ph.D. Thesis, Univ. of Paris-Sud, Centre d'Orsay (1983).

18. S.C. Chou, ''Proving Elementary Geometry Theorems Using Wu's Algorithm,'' *Contemporary Math.*, **29** pp. 243-286 (1984).

19. S.R. Czapor, ''Gröbner Basis Methods for Solving Algebraic Equations,'' Ph.D. Thesis, University of Waterloo, Dept. of Applied Math. (1988).

20. J.C. Faugère, P. Gianni, D. Lazard, and T. Mora, ''Efficient Computation of Zero-Dimensional Gröbner Bases by Change of Ordering,'' Preprint (1990).

21. P. Gianni and B. Trager, ''GCD's and Factoring Multivariate Polynomials Using Gröbner Bases,'' pp. 409-410 in *Proc. EUROCAL '85, Vol. 2, Lecture Notes in Computer Science* **204**, ed. B.F. Caviness, Springer-Verlag (1985).

22. W. Gröbner, *Modern Algebraic Geometry* (German), Springer-Verlag, Wien-Innsbruck (1949).

23. G. Hermann, "The Question of Finitely Many Steps in Polynomial Ideal Theory (German)," *Math. Ann.*, **95** pp. 736-788 (1926).

24. H. Hironaka, "Resolution of Singularities of an Algebraic Variety over a Field of Characteristic Zero I, II," *Ann. Math.*, **79** pp. 109-326 (1964).

25. M. Kalkbrener, "Solving Systems of Algebraic Equations by Using Buchberger's Algorithm," pp. 282-297 in *Proc. EUROCAL '87, Lecture Notes in Computer Science* **378**, ed. J.H. Davenport, Springer-Verlag (1989).

26. A. Kandri-Rody and D. Kapur, "Algorithms for Computing Gröbner Bases of Polynomial Ideals over Various Euclidean Rings," pp. 195-206 in *Proc. EUROSAM '84, Lecture Notes in Computer Science* **174**, ed. J. Fitch, Springer-Verlag (1984).

27. D. Kapur, "Geometry Theorem Proving Using Hilbert's Nullstellensatz," pp. 202-208 in *Proc. SYMSAC '86*, ed. B.W. Char, ACM Press (1986).

28. D.E. Knuth and P.B. Bendix, "Simple Word Problems in Universal Algebras," pp. 263-298 in *Proc. OXFORD '67*, ed. J. Leech, Pergamon Press, Oxford (1970).

29. B. Kutzler and S. Stifter, "Automated Geometry Theorem Proving Using Buchberger's Algorithm," pp. 209-214 in *Proc. SYMSAC '86*, ed. B. W. Char, ACM Press (1986).

30. D. Lazard, "Gröbner Bases, Gaussian Elimination, and Resolution of Systems of Algebraic Equations," pp. 146-156 in *Proc. EUROCAL '83, Lecture Notes in Computer Science* **162**, ed. H. van Hulzen, Springer-Verlag (1983).

31. D. Lazard, "Ideal Bases and Primary Decomposition: Case of Two Variables," *J. Symbolic Comp.*, **1**(3) pp. 261-270 (1985).

32. E. Mayr and A. Meyer, "The Complexity of the Word Problems for Commutative Semigroups and Polynomial Ideals," Report LCS/TM-199, M.I.T. Lab. of Computer Science (1981).

33. H.M. Möller and F. Mora, "Upper and Lower Bounds for the Degree of Gröbner Bases," pp. 172-183 in *Proc. EUROSAM '84, Lecture Notes in Computer Science* **174**, ed. J. Fitch, Springer-Verlag (1984).

34. M.E. Pohst and D.Y.Y. Yun, "On Solving Systems of Algebraic Equations via Ideal Bases and Elimination Theory," pp. 206-211 in *Proc. SYMSAC '81*, ed. P.S. Wang, ACM Press (1981).

35. B.L. van der Waerden, *Modern Algebra (Vols. I and II)*, Ungar (1970).

36. F. Winkler, "Reducing the Complexity of the Knuth-Bendix Completion Algorithm: A Unification of Different Approaches," pp. 378-389 in *Proc. EUROCAL '85, Vol. 2, Lecture Notes in Computer Science* **204**, ed. B.F. Caviness, Springer-Verlag (1985).

37. F. Winkler, "A p-adic Approach to the Computation of Gröbner Bases," *J. Symbolic Comp.*, **6** pp. 287-304 (1988).

38. W. Wu, "Basic Principles of Mechanical Theorem Proving in Elementary Geometries," *J. Syst. Sci. and Math. Sci.*, **4**(3) pp. 207-235 (1984).

CHAPTER 11

INTEGRATION OF

RATIONAL FUNCTIONS

11.1. INTRODUCTION

The problem of indefinite integration is one of the easiest problems of mathematics to describe: given a function $f(x)$, find a function $g(x)$ such that

$$g'(x) = f(x) .$$

If such a function can be found then one writes

$$\int f(x)\, dx = g(x)$$

(or $g(x) + c$ where c denotes a constant). As presented in most introductory courses, the indefinite integration problems are commonly solved by a collection of heuristics: substitution, trigonometric substitution, integration by parts, etc. Only when the integrand is a rational function is there any appearance of an algorithmic approach (via the method of partial fractions) to solving the integration problem.

Interestingly enough, most computer algebra systems begin their integration routines by following the same simple heuristic methods used in introductory calculus. As an example, we can consider the first stage of the integration routine used by MAPLE. In this first stage integrals of polynomials (or for that matter finite Laurent series) are readily determined. Next, MAPLE tries to integrate using a simple table lookup process; that is, it checks to see if an integrand is one of approximately 35 simple functions (e.g. cosine or tangent), and, if so, then the result is looked up in a table and returned. MAPLE then looks (as any calculus student would) for other specific types of integrands and uses appropriate methods. Thus, for example, integrands of the form

$$e^{ax+b} \cdot \sin(cx + d) \cdot p(x)$$

for constants a, b, c, d and a polynomial $p(x)$ are solved using the standard integration by parts technique. When the above methods fail, MAPLE uses a form of substitution called the "derivative-divides" method. This method examines the integrand to see if it has a composite function structure. If this is the case, it then attempts to substitute for any composite functions, $f(x)$, by dividing its derivative into the integrand and checking if the result is independent of x after the substitution $u = f(x)$ occurs.

The heuristic methods used by MAPLE obtain the correct answer for a surprisingly large percentage of integral problems. It also obtains the answer relatively quickly. In particular, the heuristic methods solve a trivial problem in trivial time, a highly desirable feature. However, heuristics by their very nature do not solve all problems and so we are still left with a significant class of integrals that remain unsolved at this point. As such we must use deterministic algorithms to obtain our answers.

It is when deterministic algorithms are required that the methods used by computer algebra systems diverge sharply from those found in calculus books. The algorithmic approach used by these systems is the subject of both this and the next chapter. In this chapter we describe the simplest example where a deterministic algorithm can be applied, namely the case where we are integrating a rational function. This is a case where the integral can always be determined. We show that the methods normally taught in introductory calculus courses need to be altered in order to avoid their computational inefficiencies. Interestingly enough, the methods presented in this chapter will be the basis for integration (or for that matter deciding if an integral even exists) of a much wider class of functions than just rational functions. This is the Risch algorithm for integrating elementary functions which is the topic of Chapter 12.

11.2. BASIC CONCEPTS OF DIFFERENTIAL ALGEBRA

The concepts of differentiation and integration are usually defined in the context of analysis, employing the process of taking the *limit* of a function. However, it is possible to develop these concepts in an *algebraic* setting. The calculation of the derivative of a given function, or the integral of a given function, is a process of algebraic manipulation. If one considers the problem of finding the indefinite integral of a rational function $r = a / b$, where a and b are polynomials in the integration variable x, the mathematical tools required include: polynomial division with remainder, GCD computation, polynomial factorization, and solving equations; in short, the polynomial manipulation algorithms which have been the central topic of this book. If one considers the more general problem of finding the indefinite integral of an elementary function f (which may involve exp, log, sin, cos, arctan, n-th roots, etc.), it is not so obvious that the mathematical tools required are the algorithms of multivariate polynomial manipulation discussed in earlier chapters. However, the various non-rational subexpressions appearing in f are ultimately treated as independent symbols, so that f is viewed as a multivariate rational function. The computational steps in the integration algorithm for a non-rational function f are then remarkably similar to the case of a rational function.

In order to proceed with this algebraic description, we require some basic terminology from differential algebra. These concepts date back to the work of J.F. Ritt [9] from the early 1940's.

Definition 11.1. A *differential field* is a field F of characteristic 0 on which is defined a mapping $D : \mathrm{F} \rightarrow \mathrm{F}$ satisfying, for all $f, g \in \mathrm{F}$:

$$D(f+g) = D(f) + D(g), \tag{11.1}$$

$$D(f \cdot g) = f \cdot D(g) + g \cdot D(f). \tag{11.2}$$

The mapping D is called a *derivation* or *differential operator*.

●

Theorem 11.1. If D is a differential operator on a differential field F then the following properties hold:

(i) $D(0) = D(1) = 0$;

(ii) $D(-f) = -D(f)$, for all $f \in \mathrm{F}$;

(iii) $D(\frac{f}{g}) = \frac{g \cdot D(f) - f \cdot D(g)}{g^2}$, for all $f, g \in \mathrm{F}$ $(g \neq 0)$;

(iv) $D(f^n) = nf^{n-1}D(f)$, for all $n \in \mathbf{Z}, f \in \mathrm{F}$ $(f \neq 0)$.

Proof: Equation (11.1) along with the property of the additive identity in F, imply that for any $f \in \mathrm{F}$

$$D(f) = D(f+0) = D(f) + D(0)$$

from which it follows that $D(0) = 0$. Similarly, equation (11.2) combined with the property of the multiplicative identity in F gives

$$D(1) = D(1 \cdot 1) = 1 \cdot D(1) + 1 \cdot D(1) = D(1) + D(1)$$

from which it follows that $D(1) = 0$. This proves (i).

To prove (ii), equation (11.1) and part (i) of this Theorem imply that for any $f \in \mathrm{F}$, we have

$$0 = D(0) = D(f + (-f)) = D(f) + D(-f)$$

implying that $D(-f) = -D(f)$.

Letting $g \in \mathrm{F}$ be nonzero and using part (i) above as well as equation (11.2), gives

$$0 = D(1) = D(g \cdot g^{-1}) = g \cdot D(g^{-1}) + g^{-1} \cdot D(g)$$

from which it follows that

$$D(g^{-1}) = -\frac{D(g)}{g^2} .$$

Applying equation (11.2) once again, gives

$$D(\frac{f}{g}) = f \cdot D(g^{-1}) + g^{-1} D(f) = \frac{-f \cdot D(g)}{g^2} + \frac{D(f)}{g}$$

which gives part (iii) when taken over a common denominator.

Part (iv) follows similar arguments and is left as an exercise for the reader (cf. Exercise 11.1).

●

Notice that the proofs in Theorem 11.1 follow those that would be encountered in an introductory calculus course. Namely, introductory courses invariably use the definition of the derivative (via limits) to prove that the linearity property and the product rule hold. Subsequent rules such as the quotient rule (iii) and the power rule (iv) are then usually proved via simple algebraic manipulations of these first two rules.

Definition 11.2. Let F and G be differential fields with differential operators D_F and D_G, respectively. Then G is a *differential extension field* of F if G is an extension field of F and

$$D_F(f) = D_G(f) \text{ for all } f \in \mathrm{F} \ .$$

●

Definition 11.3. Let F and G be differential fields with differential operators D_F and D_G, respectively. The mapping $\phi: \mathrm{F} \to \mathrm{G}$ is called a *differential homomorphism* if ϕ is a field homomorphism (see Definition 5.2) and

$$\phi(D_F(f)) = D_G(\phi(f)) \text{ for all } f \in \mathrm{F} \ .$$

●

Definition 11.4. Let F be a differential field with differential operator D. The field of constants (or constant field) of F is the subfield of F defined by

$$K = \{\ c \in \mathrm{F} : D(c) = 0\ \}\ .$$

●

Consider for a moment the particular field $\mathbf{Q}(x)$ of rational functions in the variable x over the coefficient field of rational numbers. Let D be a differential operator defined on $\mathbf{Q}(x)$ such that

$$D(x) = 1\ . \tag{11.3}$$

Then $\mathbf{Q}(x)$ is a differential field. The following theorem proves that D is the familiar differential operator defined on the subdomain $\mathbf{Q}[x]$ of polynomials.

Theorem 11.2. If $p \in \mathbf{Q}[x]$ then $D(p) \in \mathbf{Q}[x]$, where D is a differential operator satisfying (11.3). More specifically, if

$$p = \sum_{k=0}^{n} a_k x^k \in \mathbf{Q}[x], \text{ with } a_n \neq 0$$

then

$$D(p) = \begin{cases} 0, & \text{if } \deg(p) = n = 0 \\ \sum_{k=0}^{n-1} (k+1) \cdot a_{k+1} x^k, & \text{if } \deg(p) = n > 0 . \end{cases}$$

In particular, $\deg(D(p)) = \deg(p) - 1$ when $\deg(p) > 0$.

Proof: First consider the case $\deg(p) = n = 0$; i.e. $p = a_0 \in \mathbf{Q}$. We must prove that $D(p) = 0$ for all $p \in \mathbf{Q}$. From Theorem 11.1 we have that $D(0) = 0$ and $D(1) = 0$. For any positive integer n, we have

$$D(n) = D(1 + (n-1)) = D(1) + D(n-1) = D(n-1)$$

and hence, by induction, $D(n) = D(0) = 0$ for all positive integers n. Since $D(-n) = -D(n)$ the result also holds for all negative integers. Finally, using the quotient rule gives

$$D(m/n) = \frac{nD(m) - mD(n)}{n^2} = 0$$

and hence $D(p) = 0$ for all $p \in \mathbf{Q}$.

Turning now to the case $\deg(p) = n > 0$, for the polynomial p given in the statement of the theorem, we have

$$D(p) = D(a_0 + \sum_{k=1}^{n} a_k x^k) = \sum_{k=1}^{n} D(a_k x^k)$$

using the summation rule for D and the fact that $D(a_0) = 0$. Further, applying the product rule for D, Theorem 11.1 (iv), and (11.3), yields

$$D(p) = \sum_{k=1}^{n} a_k D(x^k) = \sum_{k=1}^{n} k a_k x^{k-1}$$

which is the desired result (with a shift in the index of summation).

•

Using the quotient rule from Theorem 11.1, it is clear that D is the familiar differential operator on all rational functions $r \in \mathbf{Q}(x)$.

Example 11.1. Let $r \in \mathbf{Q}(x)$ be the rational function

$$r = \frac{1}{x+1} \; .$$

Then

$$D(r) = \frac{-1}{x^2 + 2x + 1} \; .$$

●

Theorem 11.3. For the differential field $\mathbf{Q}(x)$ with differential operator D satisfying (11.3), the constant field is $\mathbf{Q}$.

Proof: If $c \in \mathbf{Q}$ then $D(c) = 0$, by Theorem 11.2. Conversely, suppose that $D(r) = 0$ for $r \in \mathbf{Q}(x)$. We must prove that $r \in \mathbf{Q}$. Now any rational function $r \in \mathbf{Q}(x)$ may be expressed in canonical form $r = p/q$ for $p, q \in \mathbf{Q}[x]$, with $q \neq 0$, and $\mathrm{GCD}(p,q) = 1$. We have

$$D(r) = \frac{q \cdot D(p) - p \cdot D(q)}{q^2} = 0$$

so

$$q \cdot D(p) - p \cdot D(q) = 0, \quad \text{that is,} \quad D(p) = \frac{p \cdot D(q)}{q} \; .$$

From Theorem 11.2, $D(p) \in \mathbf{Q}[x]$, and, since p and q have no common factors, we conclude that there exists a polynomial $s \in \mathbf{Q}[x]$ such that

$$\frac{D(q)}{q} = s \; .$$

But then $D(q) = s \cdot q$ and taking degrees (under the assumption that $\deg(q) > 0$) yields

$$\deg(q) - 1 = \deg(s) + \deg(q) \; .$$

This is impossible; hence $\deg(q) = 0$ (i.e. $D(q) = 0$ and $s = 0$). A previous relationship then yields $D(p) = 0$. We have thus proved that $p, q \in \mathbf{Q}$ as desired.

●

The problem of indefinite integration is to compute the inverse D^{-1} of the differential operator D. Clearly, for any function $r \in \mathbf{Q}(x)$, $D(r) \in \mathbf{Q}(x)$ by definition of a differential operator. One might pose the question: Given any function $s \in \mathbf{Q}(x)$, does there exist a function $r \in \mathbf{Q}(x)$ such that $D(r) = s$? In other words, is the differential field $\mathbf{Q}(x)$ closed under the inverse operator D^{-1}? Note that it is clear from Theorem 11.2 that for any polynomial

$$q = \sum_{k=0}^{m} b_k x^k \in \mathbf{Q}[x]$$

there exists a polynomial $p = D^{-1}(q) \in \mathbf{Q}[x]$; specifically,

$$p = \sum_{k=1}^{m+1} \frac{1}{k} b_{k-1} x^k .$$

In the non-polynomial case, consider for example the rational function

$$s = \frac{-1}{x^2 + 2x + 1} \in \mathbf{Q}(x) .$$

From Example 11.1, we see that in this case there exists a rational function $r \in \mathbf{Q}(x)$ such that $D(r) = s$. However, as is well known, the general answer is negative: $\mathbf{Q}(x)$ is not closed under the inverse operator D^{-1}. The following theorem proves the classical counterexample.

Theorem 11.4. For the rational function $1/x \in \mathbf{Q}(x)$, there does not exist a rational function $r \in \mathbf{Q}(x)$ such that $D(r) = 1/x$.

Proof: Suppose that $r = p/q \in \mathbf{Q}(x)$ satisfies $D(p/q) = 1/x$, with $p, q \in \mathbf{Q}[x]$ and $\mathrm{GCD}(p, q) = 1$. Then

$$\frac{q \cdot D(p) - p \cdot D(q)}{q^2} = \frac{1}{x}$$

so

$$x \cdot q \cdot D(p) - x \cdot p \cdot D(q) = q^2. \tag{11.4}$$

Write

$$q = x^n \cdot \hat{q}, \text{ where } \hat{q} \in \mathbf{Q}[x] \text{ and } \mathrm{GCD}(\hat{q}, x) = 1. \tag{11.5}$$

By equation (11.4), x divides into q^2, and hence also into q. Therefore $n \ge 1$. Substituting equation (11.5) into (11.4) gives

$$x^{n+1} \cdot \hat{q} \cdot D(p) - nx^n \cdot p \cdot \hat{q} - x^{n+1} \cdot p \cdot D(\hat{q}) = x^{2n} \cdot \hat{q}^2$$

which simplifies to

$$n \cdot p \cdot \hat{q} = x \cdot (\hat{q} \cdot D(p) - p \cdot D(\hat{q}) - x^{n-1} \cdot \hat{q}^2).$$

Since $\mathrm{GCD}(\hat{q}, x) = 1$, we must have that $x \mid p$. But then p and q have a common factor, a contradiction.

●

It follows from Theorem 11.4 that in order to express the indefinite integral of a rational function, it may be necessary to extend the field $\mathbf{Q}(x)$ with new functions. It will be seen that the only new functions required are logarithms. The concept of a logarithm will now be defined algebraically.

Definition 11.5. Let F be a differential field and let G be a differential extension field of F. If, for a given $\theta \in$ G, there exists an element $u \in$ F such that

$$D(\theta) = \frac{D(u)}{u}$$

then θ is called *logarithmic over* F and we write $\theta = \log(u)$.

●

It should be noted that the concept of "multiple branches" of the logarithm function which arises in an analytic definition of logarithms over the complex number field, is completely avoided in the above algebraic definition. Such an issue should be avoided in the content of indefinite integration since we will only determine an indefinite integral to within an arbitrary additive constant. Thus "any branch" of the analytic logarithm function will suffice.

Given a differential field F, the process of indefinite integration for a given element $f \in$ F is to determine a differential extension field

$$\mathrm{G} = \mathrm{F}(\theta_1, \ldots, \theta_n)$$

in which an element $g \in$ G exists such that $D(g) = f$, or else to determine that no such extension field exists within the context of a well-defined class of "allowable extensions". For the differential field $\mathrm{F} = \mathbf{Q}(x)$ of rational functions, the indefinite integral of a function $r \in \mathbf{Q}(x)$ can always be expressed in an extension field requiring only two types of extensions: logarithmic extensions and algebraic number extensions. The latter concept relates to the field of constants; if the base field of rational functions was expressed as $\mathrm{K}(x)$ over a field of constants K which is algebraically closed (e.g. $\mathrm{K} = \mathbf{C}$, the field of complex numbers) then only logarithmic extensions would be required. However, for the fields of most interest in a symbolic computation system where every element has a finite exact representation, the property of algebraic closure is not present. Notationally, and conceptually in the algorithm for rational function integration, we will separate these two types of extensions. Rather than expressing the extension field in the form

$$\mathrm{G} = \mathbf{Q}(x, \theta_1, \ldots, \theta_m)$$

where each θ_i $(1 \le i \le m)$ may be either a logarithm or an algebraic number extension, we choose instead to express the extension field in the form

$$\mathrm{G} = \mathbf{Q}(\alpha_1, \ldots, \alpha_k)(x, \theta_1, \ldots, \theta_n)$$

where each θ_i $(1 \le i \le n)$ is a logarithm and each α_i $(1 \le i \le k)$ is an algebraic number. Thus, the field of constants of the differential extension field G will be $\mathbf{Q}(\alpha_1, \ldots, \alpha_k)$.

Adopting conventional notation, we will usually express the derivative $D(g)$ by g' and we will express the indefinite integral D^{-1} by $\int f$. We will not use the conventional dx symbol to denote the particular variable because this will be clear from the context. (x will commonly be the variable of differentiation and integration.) Furthermore, we choose not to express the conventional "plus an arbitrary constant" when expressing the result of indefinite integration. (In other words, we will be expressing a *particular* indefinite integral.) Thus we will write

$$\int f = g \text{ if } g' = f \ .$$

Hence, from Example 11.1 we have

$$\int \frac{-1}{x^2+2x+1} = \frac{1}{x+1} \in \mathbf{Q}(x)$$

where we note that no extensions of the field $\mathbf{Q}(x)$ were required to express the integral. In order to express the integral of the function $1/x$ considered in Theorem 11.4, we introduce the new function $\theta_1 = \log(x)$ defined by the condition $\theta_1{}' = 1/x$. Then we may write

$$\int \frac{1}{x} = \log(x) \in \mathbf{Q}(x, \log(x))$$

where one logarithmic extension to the field $\mathbf{Q}(x)$ was required to express the integral. Example 11.2 shows a case where two logarithmic extensions are introduced; Example 11.3 shows a case where the constant field must be extended by an algebraic number.

Example 11.2.

$$\int \frac{1}{x^3+x} = \log(x) - \frac{1}{2}\log(x^2+1) \in \mathbf{Q}(x, \log(x), \log(x^2+1)).$$

●

Example 11.3.

$$\int \frac{1}{x^2-2} = \frac{1}{4}\sqrt{2}\log(x-\sqrt{2}) - \frac{1}{4}\sqrt{2}\log(x+\sqrt{2})$$

which requires the extension $\mathbf{Q}(\sqrt{2})(x, \log(x-\sqrt{2}), \log(x+\sqrt{2}))$.

●

We have not yet stated the integration algorithm which was used to compute the integrals in the above two examples, but for the moment they can be verified by differentiation and rational function normalization. Note that there are other forms in which these integrals could be expressed. In Example 11.2, a commonly-used method would proceed by completely factoring the denominator as

$$x^3 + x = x(x+i)(x-i) \ ,$$

where i denotes the algebraic number satisfying $i^2+1=0$ (i.e. the complex square root of -1). Then a partial fraction expansion would yield

$$\frac{1}{x^3+x} = \frac{1}{x} - \frac{1/2}{x+i} - \frac{1/2}{x-i}$$

from which we obtain

$$\int \frac{1}{x^3+x} = \log(x) - \frac{1}{2}\cdot\log(x+i) - \frac{1}{2}\cdot\log(x-i)$$

which lies in the extension $\mathbf{Q}(i)(x, \log(x), \log(x+i), \log(x-i))$. The algorithm which computed the results specified in Example 11.2 has the property that it avoided any algebraic number extensions. This has two computational advantages: (i) it avoids the cost of computing a complete factorization of the denominator, and (ii) it avoids the cost of manipulating algebraic numbers. Example 11.3 shows that this algorithm cannot always avoid introducing algebraic number extensions. Indeed we will show later that the integral in Example 11.3 cannot be expressed without introducing the number $\sqrt{2}$. An important property of the algorithm developed in the next section is that it expresses the integral using the minimal number of algebraic extensions. Note that this is not to say that it will use the minimal number of logarithmic extensions; the integral in Example 11.3 could also be expressed in the form

$$\frac{1}{4}\sqrt{2}\log\left(\frac{x-\sqrt{2}}{x+\sqrt{2}}\right).$$

The algorithm presented in the next section uses strictly polynomial operations and the arguments to the log functions will always be polynomials.

One of the tools used in the next section is the method commonly known as *integration by parts*. The following theorem proves its validity for any differential field.

Theorem 11.5. If F is a differential field with differential operator D then for any elements $u, v \in$ F we have

$$\int u \cdot D(v) = u \cdot v - \int v \cdot D(u).$$

Proof: Applying the operator D to the right hand side expression, and applying the sum rule and the product rule for D, we get

$$u \cdot D(v) + v \cdot D(u) - v \cdot D(u)$$

which simplifies to $u \cdot D(v)$. This proves the result.

●

11.3. RATIONAL PART OF THE INTEGRAL: HERMITE'S METHOD

Throughout this chapter we will work with a differential field of rational functions K(x) over an arbitrary constant field K with characteristic 0, and with a differential operator satisfying $x'=1$. Thus K could be **Q**, or **C**, or it could be a field of rational functions $\mathbf{Q}(\alpha_1, \ldots, \alpha_k)(y_1, \ldots, y_v)$ involving algebraic numbers α_i $(1 \le i \le k)$ and other variables y_i $(1 \le i \le v)$ independent of the integration variable x. The latter type of field will arise in the more general algorithm for integrating elementary functions, considered in the next chapter, where a fundamental sub-algorithm will be the algorithm of this section.

A method was presented by Hermite [1] more than a century ago which, by using only polynomial operations and without introducing any algebraic extensions, reduces the problem to

$$\int \frac{p}{q} = \frac{c}{d} + \int \frac{a}{b} \tag{11.6}$$

where $p, q, a, b, c, d \in$ K[x], deg(a) < deg(b), and b is monic and square-free. (Recall from Chapter 8 that b is square-free if and only if GCD(b, b') = 1.) In this form, c/d is called the *rational part* of the integral because the remaining integral (if it is not zero at this point) can be expressed only by introducing logarithmic extensions. Thus the unevaluated integral on the right hand side of (11.6) is called the *logarithmic part* of the integral.

Hermite's method proceeds as follows. Let $p/q \in K(x)$ be normalized such that $GCD(p,q)=1$ and q is monic. Apply Euclidean division to p and q yielding polynomials $s, r \in K[x]$ such that $p = q \cdot s + r$ with $r = 0$ or $\deg(r) < \deg(q)$. We then have

$$\int \frac{p}{q} = \int s + \int \frac{r}{q}.$$

Integrating the polynomial s is trivial and its integral (called the *polynomial part)* is one contribution to the term c/d appearing in equation (11.6). To integrate the proper fraction r/q, compute the square-free factorization (cf. Section 8.2) of the denominator

$$q = \prod_{i=1}^{k} q_i^i$$

where each q_i $(1 \le i \le k)$ is monic and square-free, GCD (q_i, q_j)=1 for $i \ne j$, and $\deg(q_k) > 0$. Compute the partial fraction expansion of the integrand $r/q \in K(x)$ in the form

$$\frac{r}{q} = \sum_{i=1}^{k} \sum_{j=1}^{i} \frac{r_{ij}}{q_i^j}$$

where for $1 \le i \le k$ and $1 \le j \le i$, $r_{ij} \in K[x]$ and

$$\deg(r_{ij}) < \deg(q_i) \text{ if } \deg(q_i) > 0,\ r_{ij} = 0 \text{ if } q_i = 1. \tag{11.7}$$

The integral of r/q can then be expressed in the form

$$\int \frac{r}{q} = \sum_{i=1}^{k} \sum_{j=1}^{i} \int \frac{r_{ij}}{q_i^j}. \tag{11.8}$$

The task now will be to apply reductions on the integrals appearing in the right hand side of (11.8) until each integral that remains has a denominator which is square-free (rather than a power $j > 1$ of a square-free q_i). The main tools in this process will be integration by parts and application of the extended Euclidean algorithm.

Consider a particular nonzero integrand r_{ij}/q_i^j with $j > 1$. Since q_i is square-free, GCD $(q_i, q_i') = 1$ so we may apply the method of Theorem 2.6 to compute polynomials $s, t \in K[x]$ such that

$$s \cdot q_i + t \cdot q_i' = r_{ij} \tag{11.9}$$

where $\deg(s) < \deg(q_i) - 1$ and $\deg(t) < \deg(q_i)$. (The latter inequality holds because of the inequality in (11.7).) Dividing by q_i^j in equation (11.9) yields

$$\int \frac{r_{ij}}{q_i^j} = \int \frac{s}{q_i^{j-1}} + \int \frac{t\, q_i'}{q_i^j}.$$

Now apply integration by parts to the second integral on the right with, in the notation of

Theorem 11.5,

$$u = t, \quad v = \frac{-1}{(j-1)q_i^{\,j-1}};$$

we get

$$\int \frac{t \cdot q_i'}{q_i^{\,j}} = \frac{-t}{(j-1)q_i^{\,j-1}} + \int \frac{t'}{(j-1)q_i^{\,j-1}} .$$

Thus we have achieved the reduction

$$\int \frac{r_{ij}}{q_i^{\,j}} = \frac{-t/(j-1)}{q_i^{\,j-1}} + \int \frac{s + t'/(j-1)}{q_i^{\,j-1}} .$$

Note that this process has produced a rational function which contributes to the term c/d in equation (11.6), and a remaining integral with the power of q_i reduced by 1. It may happen that the numerator of the new integrand is zero in which case the reduction process terminates. Otherwise, if $j-1 = 1$ then this integral contributes to the logarithmic part to be considered in the next subsection, and if $j-1 > 1$ then the same reduction process may be applied again. Note that the numerator of the new integrand satisfies the degree constraint

$$\deg(s + t'/(j-1)) \le \max\{\deg(s), \deg(t')\} < \deg(q_i) - 1 \tag{11.10}$$

which is consistent with the original numerator degree constraint expressed in (11.7). Therefore the degree constraints associated with equation (11.9) will still hold. By repeated application of this reduction process until the denominators of all remaining integrands are square-free, we obtain the complete rational part of the integral.

●

It is interesting to note a particular situation where there will be no logarithmic part. Namely, in the integrals appearing on the right hand side of equation (11.8) suppose that $r_{i1} = 0$ $(1 \le i \le k)$. In other words for each term that appears, the denominator is a power $j > 1$ of a square-free polynomial q_i. Suppose further that $\deg(q_i) = 1$ $(1 \le i \le k)$, in other words, each square-free part is a linear polynomial q_i. In such a case, for each integral appearing in (11.8) the reduction process will terminate in one step with no remaining logarithmic part because inequality (11.10) becomes

$$\deg(s + t'/(j-1)) < 0$$

which implies that $s + t'/(j-1) = 0$. Looking at it another way, in this situation each numerator r_{ij} appearing in (11.8) is a constant (due to condition (11.7)) and equation (11.9) becomes trivial with $s = 0$ and t a constant, whence $s + t'/(j-1) = 0$. Example 11.4 shows an example of this situation. Of course, we may relax the condition $\deg(q_i) = 1$ $(1 \le i \le k)$ and it is still possible to have $s + t'/(j-1) = 0$. However, this will occur only if $r_{ij} = c \cdot q_i'$ for some constant $c \in K$ (cf. Exercise 11.7), in which case equation (11.9) again has the simple solution $s = 0$, $t = c$. In such cases, the integral is

$$\int \frac{r_{ij}}{q_i^{\,j}} = \int \frac{c \cdot q_i{}'}{q_i^{\,j}} = \frac{-c/(j-1)}{q_i^{\,j-1}} \quad .$$

Algorithm 11.1. Hermite's Method for Rational Functions.

procedure HermiteReduction(p, q, x)

  # Given a rational function p/q in x, this algorithm
  # uses Hermite's method to reduce $\int p/q$.

  # Determine polynomial part of integral
  poly_part ← quo(p,q); r ← rem(p,q)

  # Calculate the square-free factorization of q, returning a list
  # $q[1], \ldots, q[k]$ of polynomials.
  ($q[1], \ldots, q[k]$) ← SquareFree(q)

  # Calculate the partial fraction decomposition for r/q, returning
  # numerators $r[i,j]$ for $q[i]^j$
  r ← PartialFractions($r, q[1], \ldots, q[k]$)
  rational_part ← 0; integral_part ← 0

  for i **from** 1 **to** k **do** {
    integral_part ← integral_part + $r[i,1]/q[i]$
    for j **from** 2 **to** i **do** {
      $n \leftarrow j$
      while $n > 1$ **do** {
        solve($s \cdot q[i] + t \cdot q[i]' = r[i,n]$) for s and t
        $n \leftarrow n - 1$
        rational_part ← rational_part − $t/n/q[i]^n$
        $r[i,n] \leftarrow s + t'/n$ }
      integral_part ← integral_part + $r[i,1]/q[i]$ } }

  return(rational_part + $\int$poly_part + $\int$integral_part)

end

Example 11.4. Consider $\int f$ where $f \in \mathbf{Q}(x)$ is

$$f = \frac{441x^7 + 780x^6 - 2861x^5 + 4085x^4 + 7695x^3 + 3713x^2 - 43253x + 24500}{9x^6 + 6x^5 - 65x^4 + 20x^3 + 135x^2 - 154x + 49}.$$

Normalizing to make the denominator monic and then applying Euclidean division yields

$$f = P + \frac{r}{q}$$

where

$$P = 49x + 54\ ,\ \ r = 735x^4 + 441x^2 - \frac{12446}{3}x + \frac{21854}{9}\ ,$$

and

$$q = x^6 + \frac{2}{3}x^5 - \frac{65}{9}x^4 + \frac{20}{9}x^3 + 15x^2 - \frac{154}{9}x + \frac{49}{9}.$$

The square-free factorization of the denominator q is

$$q = (x + 7/3)^2 \cdot (x - 1)^4\ .$$

Partial fraction expansion yields

$$\frac{r}{q} = \frac{294}{(x + 7/3)^2} + \frac{441}{(x - 1)^2} - \frac{49}{(x - 1)^4}\ .$$

To integrate the first term appearing here, equation (11.9) takes the form

$$s \cdot (x + 7/3) + t = 294$$

which trivially has the solution $s = 0$, $t = 294$. Thus

$$\int \frac{294}{(x + 7/3)^2} = \frac{-t}{x + 7/3} + \int \frac{s + t'}{x + 7/3} = \frac{-294}{x + 7/3}.$$

The other terms are equally trivial to integrate, yielding

$$\int \frac{r}{q} = -\frac{294}{x + 7/3} - \frac{441}{x - 1} + \frac{49/3}{(x - 1)^3}$$

or, if expressed as a single rational function,

$$\int \frac{r}{q} = \frac{-735x^3 + 735x^2 + 2254/3\ x - 6272/9}{(x + 7/3) \cdot (x - 1)^3}.$$

Finally, adding the integral of P yields the complete result:

$$\int f = \frac{49}{2}x^2 + 54x + \frac{-735x^3 + 735x^2 + 2254/3\ x - 6272/9}{(x + 7/3) \cdot (x - 1)^3} \in \mathbf{Q}(x).$$

●

Example 11.5. Consider the problem of computing the integral of $g \in \mathbf{Q}(x)$ where

$$g = \frac{36x^6 + 126x^5 + 183x^4 + 13807/6\ x^3 - 407\ x^2 - 3242/5\ x + 3044/15}{(x^2 + 7/6\ x + 1/3)^2 (x - 2/5)^3}\ .$$

Note that this is already in the form

$$g = \frac{p}{q}$$

with $\deg(p) < \deg(q)$ and q monic. Moreover, the denominator of g is expressed in its square-free factorization. Note that, although the quadratic term factors over $\mathbf{Q}$ into

$$x^2 + \frac{7}{6}x + \frac{1}{3} = (x + \frac{1}{2})\cdot(x + \frac{2}{3})\,,$$

the square-free factorization leaves the quadratic unfactored. Partial fraction expansion yields

$$g = \frac{-\frac{36875}{16}x - \frac{346625}{96}}{x^2 + \frac{7}{6}x + \frac{1}{3}} + \frac{-\frac{4425}{2}x - \frac{5225}{4}}{(x^2 + \frac{7}{6}x + \frac{1}{3})^2} + \frac{\frac{37451}{16}}{x - \frac{2}{5}} + \frac{\frac{354}{5}}{(x - \frac{2}{5})^2} + \frac{\frac{864}{25}}{(x - \frac{2}{5})^3}\,.$$

Consider the integral of the second term here. Equation (11.9) takes the form

$$s\cdot(x^2 + \frac{7}{6}x + \frac{1}{3}) \;+\; t\cdot(2x + \frac{7}{6}) \;=\; -\frac{4425}{2}x - \frac{5225}{4}$$

which has the solution

$$s = 2250,\;\; t = -1125x - 3525/2\;.$$

Thus we have the reduction

$$\int \frac{-\frac{4425}{2}x - \frac{5225}{4}}{(x^2 + \frac{7}{6}x + \frac{1}{3})^2} = \frac{1125x + \frac{3525}{2}}{x^2 + \frac{7}{6}x + \frac{1}{3}} + \int \frac{1125}{x^2 + \frac{7}{6}x + \frac{1}{3}}\,.$$

The reduction of the fourth and fifth terms in the partial fraction expansion of g are trivial (as in Example 11.4) yielding rational functions as the integrals. The first and third terms contribute directly to the logarithmic part. Therefore, we have split the integral into its rational and logarithmic parts as

$$\int g = \frac{1125x + \frac{3525}{2}}{x^2 + \frac{7}{6}x + \frac{1}{3}} + \frac{\frac{-354}{5}}{x - \frac{2}{5}} + \frac{\frac{-432}{25}}{(x - \frac{2}{5})^2}$$
$$+ \int \left(\frac{\frac{-36875}{16}x - \frac{346625}{96}}{x^2 + \frac{7}{6}x + \frac{1}{3}} + \frac{1125}{x^2 + \frac{7}{6}x + \frac{1}{3}} + \frac{\frac{37451}{16}}{x - \frac{2}{5}} \right).$$

Or, expressing each part as a single rational function,

$$\int g = \frac{\frac{5271}{5}x^3 + \frac{39547}{50}x^2 - \frac{31018}{25}x + \frac{7142}{25}}{(x^2 + \frac{7}{6}x + \frac{1}{3})(x - \frac{2}{5})^2} + \int \frac{36x^2 + 1167x + \frac{3549}{2}}{(x^2 + \frac{7}{6}x + \frac{1}{3})(x - \frac{2}{5})}\,.$$

•

We note that there is a variation of Hermite's method which does not require a partial fraction decomposition of r/q. As before, let

$$q = \prod_{i=1}^{k} q_i^{\,i}$$

be a square-free factorization of q. If $k = 1$, then q is square-free; otherwise set

$$f = q_k \,, \quad g = \frac{q}{q_k^{\,k}} .$$

Then $\mathrm{GCD}(g \cdot f', f) = 1$, so that there are polynomials s and t such that

$$s \cdot g \cdot f' + t \cdot f = r \,.$$

Dividing both sides by $g \cdot f^k$, integrating, and using integration by parts gives

$$\int \frac{r}{g \cdot f^k} = \frac{\hat{s}}{f^{k-1}} + \int \frac{t - g \cdot \hat{s}'}{g \cdot f^{k-1}}$$

with $\hat{s} = s/(1-k)$. As before, this process may be repeated until the denominator of the integral is square-free.

11.4. RATIONAL PART OF THE INTEGRAL: HOROWITZ' METHOD

There is a simple relationship among the various denominator polynomials $q, b, d \in \mathrm{K}[x]$ appearing in equation (11.6), as stated in the next theorem. This leads to an alternate method of splitting the integral of a rational function into its rational and logarithmic parts. We need only consider the case where the integrand is a proper fraction since determining the polynomial part is trivial.

Theorem 11.6. Let $p/q \in \mathrm{K}(x)$ be such that $\mathrm{GCD}(p,q) = 1$, q monic, and $\deg(p) < \deg(q)$. Let the rational part of its integral be c/d and a/b be the integrand appearing in the logarithmic part, as expressed in equation (11.6). Then

$$d \;=\; \mathrm{GCD}(q, q')$$

and

$$b \;=\; q/d \,.$$

Furthermore, $\deg(a) < \deg(b)$ and $\deg(c) < \deg(d)$.

Proof: Let the square-free factorization of q be

$$q \;=\; \prod_{i=1}^{k} q_i^{\,i} \,. \tag{11.11}$$

Then the integral of p/q can be expressed in the form of equation (11.8) where $r = p$ in the present case. The proof consists of noting the form of the contribution to the rational part and to the logarithmic part from each term on the right hand side of equation (11.8), when the reduction process of Hermite's method is applied. One finds that the result can be expressed in the form

$$\int \frac{p}{q} = \frac{c}{d} + \int \frac{a}{b}$$

where

$$b = \prod_{i=1}^{k} q_i, \quad d = \prod_{i=2}^{k} q_i^{i-1},$$

$$\deg(a) < \deg(b) \quad \text{and} \quad \deg(c) < \deg(d)\,. \tag{11.12}$$

Moreover, as noted in Chapter 8, it follows from the square-free factorization (11.11) that

$$\mathrm{GCD}(q, q') = \prod_{i=2}^{k} q_i^{i-1} = d$$

and

$$\frac{q}{\mathrm{GCD}(q, q')} = \prod_{i=1}^{k} q_i = b\ .$$

●

It follows from Theorem 11.6 that the following method can be used to split the integral of a rational function into the form of equation (11.6). This alternate method is usually called Horowitz' method since it was studied by Horowitz[3]. First, the polynomial part will be determined by Euclidean division as before, so assume that the integrand is $p/q \in K(x)$ where $\mathrm{GCD}(p, q) = 1$, q is monic, and $\deg(p) < \deg(q)$. Let the two denominators $b, d \in K(x)$ be computed by the formulas in Theorem 11.6. (Note that this method does not require the computation of the complete square-free factorization of q.) If $d = 1$ (i.e. $\mathrm{GCD}(q, q') = 1$ so that q is already square-free) then there is no rational part. Otherwise, b and d are each polynomials of positive degree. Let $m = \deg(b)$, $n = \deg(d)$, and let a and c be polynomials with degrees $m - 1$ and $n - 1$, respectively, with undetermined coefficients:

$$a = a_{m-1}x^{m-1} + \cdots + a_1 x + a_0,$$

$$c = c_{n-1}\, x^{n-1} + \cdots + c_1 x + c_0\,.$$

Substituting these polynomials into equation (11.6) and differentiating both sides gives

$$\frac{p}{q} = \frac{d \cdot c' - c \cdot d'}{d^2} + \frac{a}{b}$$

or, multiplying through by the denominator $q = b \cdot d$

$$p = b \cdot c' - c\left(\frac{b \cdot d'}{d}\right) + d \cdot a\ . \tag{11.13}$$

This is a polynomial identity. (It can be verified from the square-free factorizations (11.12) that $b \cdot d'$ is divisible by d.) Note that on the left hand side of (11.13)

$$\deg(p) < \deg(q) = \deg(b) + \deg(d) = m + n$$

while on the right hand side the degree is

$$\max(m + n - 2, m + n - 2, m + n - 1) = m + n - 1\ .$$

By equating coefficients on the left and right of (11.13), we have a system of $m + n$ linear equations over the field K in the $m + n$ unknowns

$a_i \ (0 \le i \le m-1)$ and $c_i \ (0 \le i \le n-1)$.

Algorithm 11.2. Horowitz' Reduction for Rational Functions.

```
procedure HorowitzReduction(p, q, x)

    # For a given rational function p/q in x, this algorithm calculates
    # the reduction of ∫p/q into a polynomial part and logarithmic part
    # via Horowitz' algorithm.

    poly_part ← quo(p,q); p ← rem(p, q)
    d ← GCD(q, q'); b ← quo(q, d)
    m ← deg(b); n ← deg(d)
    a ← Σ_{i=0}^{m-1} a[i]·x^i; c ← Σ_{i=0}^{n-1} c[i]·x^i
    r ← b·c' − c·quo(b·d', d) + d·a

    for i from 0 to m + n − 1 do
        eqns[i] ← coeff(p, i) = coeff(r, i)

    solve(eqns, {a[0], . . . , a[m−1], c[0], . . . , c[n−1]})
    return(c/d + ∫poly_part + ∫a/b)

end
```

Example 11.6. Let us apply Horowitz' method to $\int f$ where f is the rational function in Example 11.4. We first make the denominator monic and apply Euclidean division as before, so that

$$\int f = \frac{49}{2}x^2 + 54x + \int \frac{r}{q}$$

where

$$\frac{r}{q} = \frac{735x^4 + 441x^2 - 12446/3\,x + 21854/9}{x^6 + 2/3\,x^5 - 65/9\,x^4 + 20/9\,x^3 + 15x^2 - 154/9\,x + 49/9} .$$

Applying the formulas of Theorem 11.6, the denominator of the rational part of the integral is

$$d = \mathrm{GCD}(q, q') = x^4 - \frac{2}{3}x^3 - 4x^2 + 6x - \frac{7}{3}$$

and the denominator appearing in the logarithmic part of the integral is

$$b = \frac{q}{d} = x^2 + \frac{4}{3}x - \frac{7}{3}.$$

We define $m = \deg(b) = 2,\ n = \deg(d) = 4$, and

$$a = a_1x + a_0, \quad c = c_3x^3 + c_2x^2 + c_1x + c_0.$$

Expanding out the right hand side of equation (11.13) which is formed from the polynomials a, b, c, d and

$$\frac{b \cdot d'}{d} = 4x + 6,$$

and then equating coefficients with the numerator polynomial r yields the equations:

$$\begin{array}{rcl}
a_1 & = & 0 \\
-2/3a_1 + a_0 - c_3 & = & 735 \\
-4a_1 - 2/3a_0 - 2c_3 - 2c_2 & = & 0 \\
6a_1 - 4a_0 - 7c_3 - 10/3c_2 - 3c_1 & = & 441 \\
-7/3a_1 + 6a_0 - 14/3c_2 - 14/3c_1 - 4c_0 & = & -12446/3 \\
-7/3a_0 - 7/3c_1 - 6c_0 & = & 21854/9
\end{array}$$

Solving these six linear equations yields:

$$a_1 = 0,\ a_0 = 0,\ c_3 = -735,\ c_2 = 735,\ c_1 = 2254/3,\ c_0 = -6272/9.$$

Thus the logarithmic part of the integral vanishes and we have determined that

$$\int \frac{r}{q} = \frac{-735x^3 + 735x^2 + 2254/3\,x - 6272/9}{x^4 - 2/3\,x^3 - 4\,x^2 + 6\,x - 7/3}$$

which is consistent with the result obtained in Example 11.4.

•

Example 11.7. Let us apply Horowitz' method to $\int g$ where $g = p/q$ is the rational function in Example 11.5, which is (with q expanded)

$$g = \frac{36x^6 + 126x^5 + 183x^4 + \frac{13807}{6}x^3 - 407x^2 - \frac{3242}{5}x + \frac{3044}{15}}{x^7 + \frac{17}{15}x^6 - \frac{263}{900}x^5 - \frac{1349}{2250}x^4 + \frac{2}{1125}x^3 + \frac{124}{1125}x^2 + \frac{4}{1125}x - \frac{8}{1125}}.$$

We already have q monic and $\deg(p) < \deg(q)$. Applying the formulas of Theorem 11.6, the denominator of the rational part is

$$d = \mathrm{GCD}(q, q') = x^4 + \frac{11}{30}x^3 - \frac{11}{25}x^2 - \frac{2}{25}x + \frac{4}{75}$$

and the denominator appearing in the logarithmic part is

$$b = \frac{q}{d} = x^3 + \frac{23}{30}x^2 - \frac{2}{15}x - \frac{2}{15}.$$

We define $m = \deg(b) = 3,\ n = \deg(d) = 4$, and let a and c be polynomials of degree 2 and 3, respectively. Expanding out the right hand side of equation (11.13) which is formed from the polynomials a, b, c, d and

$$\frac{b \cdot d'}{d} = 4x^2 + \frac{27}{10}x + \frac{1}{5}$$

and then equating coefficients with the numerator polynomial p yields the equations:

$$\begin{array}{rrrrrrrrr}
a_2 & & & & & & & = & 36 \\
11/30a_2 + & a_1 & & - & c_3 & & & = & 126 \\
-11/25a_2 + & 11/30a_1 + & a_0 - & 2/5c_3 - & 2c_2 & & & = & 183 \\
-2/25a_2 - & 11/25a_1 + & 11/30a_0 - & 3/5c_3 - & 7/6c_2 - & 3c_1 & & = & 13807/6 \\
4/75a_2 - & 2/25a_1 - & 11/25a_0 - & 2/5c_3 - & 7/15c_2 - & 29/15c_1 - & 4c_0 & = & -407 \\
& 4/75a_1 - & 2/25a_0 & - & 4/15c_2 - & 1/3c_1 - & 27/10c_0 & = & -3242/5 \\
& & 4/75a_0 & & - & 2/15c_1 - & 1/5c_0 & = & 3044/15
\end{array}$$

Solving these seven linear equations yields:

$$a_2 = 36,\ a_1 = 1167,\ a_0 = \frac{3549}{2},$$

$$c_3 = \frac{5271}{5},\ c_2 = \frac{39547}{50},\ c_1 = \frac{31018}{25},\ c_0 = \frac{7142}{25}.$$

Thus we have split the integral into its rational and logarithmic parts as follows:

$$\int g = \frac{5271/5\,x^3 + 39547/50\,x^2 - 31018/25\,x + 7142/25}{x^4 + 11/30\,x^3 - 11/25\,x^2 - 2/25\,x + 4/75} + \int \frac{36x^2 + 1167x + 3549/2}{x^3 + 23/30\,x^2 - 2/15\,x - 2/15}.$$

●

11.5. LOGARITHMIC PART OF THE INTEGRAL

The Rothstein/Trager Method

Consider now the problem of expressing the logarithmic part of the integral of a rational function, which is an integral of the form

$$\int \frac{a}{b}$$

where $a, b \in K[x]$, $\deg(a) < \deg(b)$, and b is monic and square-free. As noted in Examples 11.2 and 11.3 and the discussion following those examples, it may be necessary to extend the constant field K to $K(\alpha_1, \ldots, \alpha_k)$ where α_i $(1 \le i \le k)$ are algebraic numbers over K. However, we would like to avoid algebraic number extensions whenever possible and, in general, to express the integral using the minimal algebraic extension field. Two different methods for achieving this were discovered independently by Rothstein[11] and by Trager[13].

First, let us note the general form in which the integral can be expressed. Ignoring any concern about the number of algebraic extensions, let the denominator $b \in K[x]$ be completely factored over its splitting field K_b into the form

$$b = \prod_{i=1}^{m} (x - \beta_i)$$

where β_i ($1 \le i \le m$) are m distinct elements of K_b, an algebraic extension of K. Then the integrand can be expressed in a partial fraction expansion of the form

$$\frac{a}{b} = \sum_{i=1}^{m} \frac{\gamma_i}{x - \beta_i} \text{ where } \gamma_i, \beta_i \in K_b \tag{11.14}$$

and so

$$\int \frac{a}{b} = \sum_{i=1}^{m} \gamma_i \cdot \log(x - \beta_i) \tag{11.15}$$

with the result of the integration expressed in the extension field $K_b(x, \log(x - \beta_1), \ldots, \log(x - \beta_m))$. In the traditional analytic setting where K is the algebraically closed field **C** of complex numbers, the problem of rational function integration is completely solved at this point.

When K is a field which is not algebraically closed, such as **Q**, then the above method has serious practical difficulties. In the worst case, the splitting field of a degree-m polynomial $b \in K[x]$ is of degree $m!$ over K (cf. Herstein [2]). This exponential degree growth means that the computation of the splitting field is, in general, impossible even with today's computing power. Fortunately, the splitting field of b is not always required in order to express the integral. A very simple example of this was seen in Example 11.2 where the denominator is

$$b = x^3 + x \in \mathbf{Q}[x] .$$

The splitting field of b is $\mathbf{Q}(i)$ where $i^2 + 1 = 0$, yielding the complete factorization

$$b = x(x + i)(x - i) \in \mathbf{Q}(i)[x] .$$

However, the integral in Example 11.2 is expressed without requiring the extension of the constant field from **Q** to $\mathbf{Q}(i)$. A more significant example is presented in Example 11.11. For the simple case of Example 11.2, let us note the relationship between the integral expressed in the form (11.15) which is

$$\int \frac{1}{x^3 + x} = \log(x) - \frac{1}{2} \cdot \log(x + i) - \frac{1}{2} \cdot \log(x - i)$$

and the form of the integral presented in Example 11.2. Namely, noting that the coefficients of the second and third log terms above are identical, these two terms may be combined using the product rule for logarithms:

$$-\frac{1}{2}\cdot\log(x+i)-\frac{1}{2}\cdot\log(x-i)=-\frac{1}{2}\cdot\log((x+i)(x-i))=-\frac{1}{2}\cdot\log(x^2+1).$$

In this way we obtain the form of the integral expressed in Example 11.2, in which we discover that it was not necessary to split the factor x^2+1 into linear factors.

Clearly, if $\gamma_i=\gamma_j$, then the corresponding logarithms appearing in (11.15) can be combined using the addition law of logarithms (cf. Exercise 11.3). Thus we are led to searching for only the distinct γ_i in equation (11.14). Let β be one of the roots of b, and consider the Laurent power series expansion of a/b about a pole $x=\beta$

$$\frac{a}{b}=\frac{\gamma}{(x-\beta)}+c_0+c_1(x-\beta)+c_2(x-\beta)^2+\cdots \quad (11.16)$$

The value γ is called the *residue* of the rational function a/b at the pole β. It is not hard to show (cf. Exercise 11.8) that one can calculate a residue at a pole β via

$$\gamma=\frac{a(\beta)}{b'(\beta)}. \quad (11.17)$$

Integrating (11.16) term by term, one obtains

$$\int\frac{a}{b}=\gamma\log(x-\beta)+c_0(x-\beta)+\frac{c_1}{2}(x-\beta)^2+\cdots$$

hence one can see that the γ_i appearing in (11.15) are nothing more than the residues of a/b at the poles $\beta_1,\ldots,\beta_m$. Equation (11.15) can thus be given as

$$\int\frac{a}{b}=\sum_{\beta\,|\,b(\beta)=0}\frac{a(\beta)}{b'(\beta)}\log(x-\beta). \quad (11.18)$$

Thus, we desire to find the distinct residues of a/b, that is, the distinct solutions of (11.17) as β varies over the poles of a/b. Writing (11.17), as

$$0=a(\beta)-\gamma\cdot b'(\beta)$$

we see that this is the same as finding the distinct roots of

$$R(z)=\prod(a(\beta)-z\cdot b'(\beta))$$

as β runs over all the distinct roots of b. From Theorem 9.3, we have (up to multiplication by a nonzero constant)

$$R(z)=\mathrm{res}_x(a(x)-z\cdot b'(x),\,b(x)). \quad (11.19)$$

Any repeated root of (11.19) will result in a reduction in the number of log terms on the right of (11.15). Thus, we may write

$$\int\frac{a}{b}=\sum_{i=1}^{k}c_i\log(v_i) \quad (11.20)$$

where the c_i are the distinct roots of (11.19) and the v_i are monic, square-free, and pairwise relatively prime polynomials. Theorem 11.7 shows that all these distinct roots are necessary. In addition it provides a simple mechanism for constructing the polynomials v_i appearing in

(11.20) once the distinct residues have been calculated.

Theorem 11.7 (Rothstein/Trager Method - Rational Function Case). Let $K^*(x)$ be a differential field over some constant field K^*. Let $a, b \in K^*[x]$ be such that $\mathrm{GCD}(a, b) = 1$, with b monic and square-free, and $\deg(a) < \deg(b)$. Suppose that

$$\int \frac{a}{b} = \sum_{i=1}^{n} c_i \cdot \log(v_i) \tag{11.21}$$

where $c_i \in K^*$ $(1 \le i \le n)$ are distinct nonzero constants and $v_i \in K^*[x]$ $(1 \le i \le n)$ are monic, square-free, pairwise relatively prime polynomials of positive degree. Then c_i $(1 \le i \le n)$ are the distinct roots of the polynomial

$$R(z) = \mathrm{res}_x(a - zb', b) \in K^*[z]$$

and v_i $(1 \le i \le n)$ are the polynomials

$$v_i = \mathrm{GCD}(a - c_i b', b) \in K^*[x].$$

Proof: From the form of the integral assumed in the statement of the theorem, differentiating both sides yields

$$\frac{a}{b} = \sum_{i=1}^{n} c_i \cdot \frac{v_i'}{v_i} . \tag{11.22}$$

Setting

$$u_i = \prod_{j \ne i}^{n} v_j, \quad 1 \le i \le n ,$$

and multiplying both sides of (11.22) by $b \cdot \prod_{j=1}^{n} v_j$ yields

$$a \cdot \prod_{j=1}^{n} v_j = b \cdot \sum_{i=1}^{n} c_i v_i' \cdot u_i . \tag{11.23}$$

We now claim that

$$b = \prod_{j=1}^{n} v_j . \tag{11.24}$$

To prove (11.24), first note that since $\mathrm{GCD}(a, b) = 1$ it follows from equation (11.23) that $b \mid \prod_{j=1}^{n} v_j$. In the other direction, for each j we have from equation (11.23) that

$$v_j \mid b \cdot \sum_{i=1}^{n} c_i \cdot v_i' \cdot u_i .$$

Since v_j is explicitly a factor of u_i for each $i \ne j$, this implies that

$$v_j \mid b \cdot v_j' \cdot u_j .$$

Now $\mathrm{GCD}(v_j, v_j') = 1$ since v_j is assumed to be square-free, and $\mathrm{GCD}(v_j, u_j) = 1$ since

v_i $(1 \le i \le n)$ are pairwise relatively prime, hence $v_j \mid b$. Since this holds for each j, we have $\prod_{j=1}^{n} v_j \mid b$. Finally, since b and v_i $(1 \le i \le n)$ are all assumed to be monic, we have (11.24).

Equations (11.23) and (11.24) imply

$$a = \sum_{i=1}^{n} c_i \, v_i' \cdot u_i \ .$$

Our next claim is that for each j,

$$v_j \mid (a - c_j \cdot b') \ . \tag{11.25}$$

To see this, first note from (11.24) that

$$b' = \sum_{i=1}^{n} v_i' \cdot u_i \ .$$

Hence,

$$a - c_j \cdot b' = \sum_{i=1}^{n} c_i \cdot v_i' \cdot u_i - c_j \cdot \sum_{i=1}^{n} v_i' \cdot u_i = \sum_{i=1}^{n} (c_i - c_j) v_i' u_i . \tag{11.26}$$

In the latter sum, for each term with $i \ne j$, $v_j \mid u_i$ while when $i = j$ the term vanishes. Thus (11.25) is true.

It follows from (11.24) and (11.25) that v_j is a common divisor of $a - c_j \cdot b'$ and b, for each j. We must show that it is the greatest common divisor. It suffices to show that for $h \ne j$, $\mathrm{GCD}(a - c_j \cdot b', v_h) = 1$. To this end we have, using (11.26),

$$\begin{aligned} \mathrm{GCD}(a - c_j \cdot b', v_h) &= \mathrm{GCD}(\sum_{i=1}^{n} (c_i - c_j) \cdot v_i' \cdot u_i, v_h) \\ &= \mathrm{GCD}((c_h - c_j) v_h' \cdot u_h, v_h) \end{aligned}$$

where the last equality holds because v_h is a factor of each u_i $(i \ne h)$. But for $h \ne j$, the above GCD is 1 because $c_h \ne c_j$, $\mathrm{GCD}(v_h, v_h') = 1$, and $\mathrm{GCD}(v_h, u_h) = 1$. We have thus proved that

$$v_j = \mathrm{GCD}(a - c_j \cdot b', b) \ \text{ for } \ j = 1, \ldots, n \ .$$

The above fact implies that $\mathrm{res}_x(a - c_j \cdot b', b) = 0$ (since there is a nontrivial common factor) and hence c_j is a root of the polynomial $R(z)$ defined in the statement of the theorem. Conversely, let c be any root of $R(z)$. (At this point, we must assume that $c \in K_R^*$, the splitting field of $R(z)$, since we have not yet proved that $R(z)$ splits over K^*. The following argument uses operations in the domain $K_R^*[x]$). Then $\mathrm{res}_x(a - c \cdot b', b) = 0$ which implies that

$$\mathrm{GCD}(a - c \cdot b', b) = G$$

with $\deg(G) > 0$. Let g be an irreducible factor of G. Then, since $g \mid b$ and using (11.24), there exists one and only v_j such that $g \mid v_j$. Now,

$$g \mid (a - c \cdot b')$$

implies, using the form (11.26), that

$$g \mid \sum_{i=1}^{n} (c_i - c) \cdot v_i' \cdot u_i \ .$$

But $g \mid u_i$ for each $i \neq j$ (because $g \mid v_j$) and therefore we have

$$g \mid (c_j - c) \cdot v_j' \cdot u_j$$

which can be true only if $c_j - c = 0$. Therefore c is one of the constants c_j appearing in the form assumed for the integral. We have thus proved, under the hypotheses of the theorem, that the polynomial $R(z) \in K^*[z]$ completely splits over K^* and that c_j $(1 \le j \le n)$ are all the distinct roots of $R(z)$.

●

Theorem 11.7 shows how c_i $(1 \le i \le n)$ and v_i $(1 \le i \le n)$ appearing in the form (11.20) can be computed if the conditions of the theorem hold. We now prove that the method arising from the theorem is completely general and, moreover, it leads to the minimal constant field.

Theorem 11.8. Let K(x) be a differential field over a constant field K. Let $a, b \in K[x]$ be such that GCD$(a,b) = 1$, b monic and square-free, and deg$(a) <$ deg(b). Let K^* be the minimal algebraic extension field of K such that the integral can be expressed in the form

$$\int \frac{a}{b} = \sum_{i=1}^{n^*} c_i^* \cdot \log(v_i^*) \qquad (11.27)$$

where $c_i^* \in K^*$, $v_i^* \in K^*[x]$. Then

$$K^* = K(c_1, \ldots, c_n)$$

where c_i $(1 \le i \le n)$ are the distinct roots of the polynomial

$$R(z) = \text{res}_x(a - zb', b) \in K[z] \ .$$

In other words, K^* is the splitting field of $R(z) \in K[z]$. Moreover the formulas in Theorem 11.7 may be used to calculate the integral using the minimal constant field.

Proof: Let the integral be expressed in the form (11.27). If the c_i^* and v_i^* do not satisfy the conditions stated in Theorem 11.7 then it is possible to rearrange formula (11.27) so that they do, as follows. First, if for some i, v_i^* is not square-free then let its square-free factorization be

$$v_i^* = \prod_{j=1}^{k} v_j^{\,j}$$

and use the replacement

$$\log(v_i^*) = \sum_{j=1}^{k} j\cdot\log(v_j) .$$

This can be done for each term until the arguments to the log functions are all square-free. Also, each argument can be made monic by simply dividing out the leading coefficient, noting that any term with a constant argument to the log function may be discarded and we still have a valid equation of the form (11.27). So we may assume that the v_i^* in (11.27) are monic square-free polynomials of positive degree. Next, if for some i and j, $\mathrm{GCD}(v_i^*, v_j^*) = v$ with $\deg(V) > 1$ then use the replacements

$$\log(v_i^*) = \log(v) + \log(v_i^*/v) , \quad \log(v_j^*) = \log(v) + \log(v_j^*/v) .$$

Note that all arguments to the log functions remain in $K^*[x]$. We may continue applying such replacements, collecting into a single term any terms with identical log arguments, until the arguments to the log functions are pairwise relatively prime. Note that this process maintains the property that the polynomials are monic, square-free, and of positive degree. Finally, if in the form (11.27) we have $c_i^* = c_j^*$ for $i \neq j$ then we may use the replacement

$$c_i^*\cdot\log(v_i^*) + c_j^*\cdot\log(v_j^*) \; = \; c_i^*\cdot\log(v_i^*\cdot v_j^*) ,$$

noting that this transformation maintains the property that the arguments to the log functions are monic, square-free, pairwise relatively prime polynomials of positive degree. Thus we have proved that an expression of the form (11.27) can be rearranged into an expression of the same form for which the hypotheses of Theorem 11.7 hold.

Now, Theorem 11.7 tells us that c_i^* $(1 \le i \le n^*)$ are the distinct roots of the resultant polynomial $R(z)$ and v_i^* $(1 \le i \le n^*)$ are as defined in that theorem. It follows that K^* is the splitting field of the polynomial $R(z) \in K[z]$.

●

We now have a method to express the logarithmic part of the integral of a rational function in the form (11.27) using the minimal algebraic extension to the constant field. The following examples illustrate the method and a complete rational function integration algorithm is presented in Algorithm 11.3.

Example 11.8. Let us apply the Rothstein/Trager method to compute the integral given in Example 11.2, where the integrand is

$$\frac{a}{b} = \frac{1}{x^3 + x} \in \mathbf{Q}(x) .$$

Since b is square-free and $\deg(a) < \deg(b)$, the integral has only a logarithmic part. First compute the resultant

$$R(z) = \mathrm{res}_x(a - z\cdot b', b) = \mathrm{res}_x((1 - z) - (3z)\cdot x^2, x + x^3)$$

Algorithm 11.3. Rothstein/Trager Method.

```
procedure LogarithmicPartIntegral(a, b, x)
  # Given a rational function a/b in x with deg(a) < deg(b),
  # b monic and square-free, we calculate ∫ a/b .

  R(z) ← pp_z(res_x(a − z·b′, b))
  (r_1(z), . . . , r_k(z)) ← factors(R(z))
  integral ← 0
  for i from 1 to k do {
     d ← deg(r_i(z))
     if d = 1 then {
        c ← solve(r_i(z)=0, z)
        v ← GCD(a − c·b′, b); v ← v/lcoeff(v)
        integral ← integral + c·log(v) }
     else {
        # Need to do GCD over algebraic number field
        v ← GCD(a − α·b′, b); v ← v/lcoeff(v)
        #           (where α = RootOf(r_i(z)))
        if d = 2 then {
           # Give answer in terms of radicals
           c ← solve(r_i(z)=0, z)
           for j from 1 to 2 do {
              v[j] ← substitute(α=c[j],v)
              integral ← integral + c[j]·log(v[j]) } }
        else {
           # Need answer in RootOf notation
           for j from 1 to d do {
              v[j] ← substitute(α=c[j],v)
              integral ← integral + c[j]·log(v[j])
              # (where c[j] = RootOf(r_i(z)) ) } } } }
  return(integral)

end
```

$$= \det \begin{bmatrix} -3z & 0 & 1-z & 0 & 0 \\ 0 & -3z & 0 & 1-z & 0 \\ 0 & 0 & -3z & 0 & 1-z \\ 1 & 0 & 1 & 0 & 0 \\ 0 & 1 & 0 & 1 & 0 \end{bmatrix} = -4z^3 + 3z + 1 \in \mathbf{Q}[z] .$$

Next computing the complete factorization of $R(z)$ in the domain $\mathbf{Q}[z]$ gives

$$R(z) = -4\,(z-1)(z+1/2\,)^2 \in \mathbf{Q}[z] .$$

In this case, $R(z)$ completely splits over the constant field $\mathbf{Q}$ and therefore no algebraic number extensions are required to express the integral. The distinct roots of $R(z)$ are

$$c_1 = 1\,,\ \ c_2 = -1/2\ .$$

The corresponding log arguments are computed via GCD computations in the domain $\mathbf{Q}[x]$

$$v_1 = \mathrm{GCD}(a - c_1 \cdot b', b) = x \in \mathbf{Q}[x]\,,$$

$$v_2 = \mathrm{GCD}(a - c_2 \cdot b', b) = x^2 + 1 \in \mathbf{Q}[x]\,.$$

Hence,

$$\int \frac{1}{x^3 + x} = c_1 \cdot \log(v_1) + c_2 \cdot \log(v_2)$$

$$= \log(x) - 1/2\,\log(x^2+1) \in \mathbf{Q}(x, \log(x), \log(x^2+1))\ .$$

●

Example 11.9. Let us apply the Rothstein/Trager method to compute the integral given in Example 11.3, where the integrand is

$$\frac{a}{b} = \frac{1}{x^2 - 2} \in \mathbf{Q}(x).$$

Since b is square-free and $\deg(a) < \deg(b)$, the integral has only a logarithmic part. Computing the resultant gives

$$R(z) = \mathrm{res}_x(a - z \cdot b', b) = -8z^2 + 1 \in \mathbf{Q}[z].$$

The complete factorization of $R(z)$ in the domain $\mathbf{Q}[z]$ is

$$R(z) = -8\,(z^2 - 1/8\,) \in \mathbf{Q}[z].$$

In this case $R(z)$ fails to split over the constant field $\mathbf{Q}$ so it is necessary to adjoin algebraic numbers to the constant field. The splitting field of $R(z)$ is $\mathbf{Q}(\alpha)$ where[1] $\alpha = \mathrm{RootOf}(z^2 - 1/8)$ and the complete factorization of $R(z)$ over its splitting field is

$$R(z) = -8(z - \alpha)(z + \alpha) \in \mathbf{Q}(\alpha)[z].$$

The distinct roots of $R(z)$ are

---

[1] As in Section 1.4, we use this notation to denote any α such that $\alpha^2 - 1/8 = 0$.

$$c_1 = \alpha,\ c_2 = -\alpha.$$

The log argument corresponding to $c_1 = \alpha$ is computed via a GCD computation in the domain $\mathbf{Q}(\alpha)[x]$

$$v_1 = \mathrm{GCD}(a - c_1 \cdot b', b) = x - 4\alpha \in \mathbf{Q}(\alpha)[x].$$

Since c_2 is a conjugate of c_1, the corresponding GCD computation defining v_2 is essentially the same computation in the domain $\mathbf{Q}(\alpha)[x]$ as above and need not be repeated. We simply substitute to obtain

$$v_2 = x + 4\alpha.$$

Therefore we have determined the integral as

$$\int \frac{1}{x^2-2} = c_1 \cdot \log(v_1) + c_2 \cdot \log(v_2) = \alpha \cdot \log(x - 4\alpha) - \alpha \cdot \log(x + 4\alpha)$$

in the domain $\mathbf{Q}(\alpha)(x, \log(x-4\alpha), \log(x+4\alpha))$, where $\alpha = \mathrm{RootOf}(z^2 - 1/8)$. Now if the algebraic number α could not be expressed in terms of radicals then this would be the best possible closed-form solution. (Note that, in practice, one could proceed to calculate a numerical approximation for the algebraic number α and substitute this value into the expression if numerical calculation is the ultimate objective.) However, in this example, we are able to express α in terms of radicals by

$$\alpha = 1/4 \cdot \sqrt{2},$$

yielding the final result

$$\int \frac{1}{x^2-2} = \frac{1}{4} \cdot \sqrt{2} \cdot \log(x - \sqrt{2}) - \frac{1}{4} \cdot \sqrt{2} \cdot \log(x + \sqrt{2})$$

with the answer in the extension field $\mathbf{Q}(\sqrt{2})(x, \log(x - \sqrt{2}), \log(x + \sqrt{2}))$.

•

Example 11.10. Let us apply the Rothstein/Trager method to complete the computation of the integral in Example 11.7, where the integrand is

$$g = \frac{36x^6 + 126x^5 + 183x^4 + \frac{13807}{6}x^3 - 407x^2 - \frac{3242}{5}x + \frac{3044}{15}}{x^7 + \frac{17}{15}x^6 - \frac{263}{900}x^5 - \frac{1349}{2250}x^4 + \frac{2}{1125}x^3 + \frac{124}{1125}x^2 + \frac{4}{1125}x - \frac{8}{1125}}$$

from the domain $\mathbf{Q}(x)$. We had reduced the problem to the computation of the logarithmic part which has the integrand

$$\frac{a}{b} = \frac{36x^2 + 1167x + 3549/2}{x^3 + 23/30\,x^2 - 2/15\,x - 2/15}.$$

Computing the resultant gives

$$R(z) = \mathrm{res}_x(a - z\cdot b', b)$$
$$= \frac{16}{625}z^3 - \frac{576}{625}z^2 - \frac{20872009}{16}z + 2730177900 \in \mathbf{Q}[z]\,.$$

The complete factorization of $R(z)$ in the domain $\mathbf{Q}[z]$ is

$$R(z) = \frac{16}{625}\,(z - \frac{37451}{16})(z + 8000)(z - \frac{91125}{16})$$

which has completely split into linear factors and therefore no algebraic number extensions are required to express the integral. We get

$$c_1 = \frac{37451}{16},\ c_2 = -8000,\ c_3 = \frac{91125}{16},$$

and performing the required GCD computations in the domain $\mathbf{Q}[x]$ yields

$$v_1 = x - \frac{2}{5},\ v_2 = x + \frac{1}{2},\ v_3 = x + \frac{2}{3}\,.$$

This determines the logarithmic part of the integral and putting it together with the rational part determined in Example 11.7, we have the complete result

$$\int g = \frac{\frac{5271}{5}x^3 + \frac{39547}{50}x^2 - \frac{31018}{25}x + \frac{7142}{25}}{x^4 + \frac{11}{30}x^3 - \frac{11}{25}x^2 - \frac{2}{25}x + \frac{4}{75}}$$
$$+ \frac{37451}{16}\log(x - \frac{2}{5}) - 8000\log(x + \frac{1}{2}) + \frac{91125}{16}\log(x + \frac{2}{3})$$

with the answer in $\mathbf{Q}(x, \log(x - 2/5), \log(x + 1/2), \log(x + 2/3))$.

●

Example 11.11. The final example in this section was presented by Tobey [12] in his 1967 thesis as an example where the splitting field of the denominator polynomial is an algebraic number field of extremely high degree over $\mathbf{Q}$ and yet the integral can be expressed using a single algebraic extension of degree 2. Tobey posed the problem of finding an algorithm which would compute the result using the minimal algebraic extension field. The Rothstein/Trager method is such an algorithm and it computes the integral very easily, whereas a traditional method based on computing the splitting field of the denominator is hopeless in this case. The integrand is

$$\frac{a}{b} = \frac{7x^{13} + 10x^8 + 4x^7 - 7x^6 - 4x^3 - 4x^2 + 3x + 3}{x^{14} - 2x^8 - 2x^7 - 2x^4 - 4x^3 - x^2 + 2x + 1} \in \mathbf{Q}(x)\,.$$

Since b is square-free and $\deg(a) < \deg(b)$, the integral has only a logarithmic part. Computing the resultant gives

$$
\begin{aligned}
R(z) &= \mathrm{res}_x(a - zb', b) \\
&= -2377439676624535552z^{14} + 16642077363717488864z^{13} \\
&\quad - 45765713775022309376z^{12} + 58247272077301121024z^{11} \\
&\quad - 23922986746034388992z^{10} - 17682207594894983168z^9 \\
&\quad + 15861980342479323136z^8 + 3417569535147769856z^7 \\
&\quad - 3965495085619830784z^6 - 1105137974680936448z^5 \\
&\quad + 373796667906787328z^4 + 227528406551957504z^3 \\
&\quad + 44693079858420224z^2 + 4063007259856384z + 145107402137728.
\end{aligned}
$$

The complete factorization of $R(z)$ in the domain $\mathbf{Q}[z]$ is

$$R(z) = -2377439676624535552\,(z^2 - z - 1/4)^7 .$$

Although $R(z)$ does not completely split over the constant field $\mathbf{Q}$, we see that its splitting field is simply $\mathbf{Q}(\alpha)$ where $\alpha = \mathrm{RootOf}(z^2 - z - 1/4)$. The complete factorization of $R(z)$ over its splitting field is

$$R(z) = -2377439676624535552\,(z - \alpha)^7(z + \alpha - 1)^7 \in \mathbf{Q}(\alpha)[z] .$$

The distinct roots of $R(z)$ are

$$c_1 = \alpha,\ c_2 = 1 - \alpha .$$

The log argument corresponding to $c_1 = \alpha$ is

$$v_1 = \mathrm{GCD}(a - c_1{\cdot}b', b) = x^7 + (1 - 2\alpha)x^2 - 2\alpha\, x - 1 \in \mathbf{Q}(\alpha)[x]$$

and since c_2 is a conjugate of c_1, substitution gives

$$v_2 = x^7 + (2\alpha - 1)x^2 - (2 - 2\alpha)x - 1 .$$

Therefore, the integral can be expressed in the form

$$\int \frac{a}{b} = \alpha \log(x^7 + (1 - 2\alpha)x^2 - 2\alpha x - 1) + (1 - \alpha)\log(x^7 + (2\alpha - 1)x^2 - (2 - 2\alpha)x - 1)$$

where $\alpha = \mathrm{RootOf}\,(z^2 - z - 1/4)$. Since α may be expressed in terms of radicals

$$\alpha = 1/2\ (1 + \sqrt{2}),$$

the final form of the result is

$$
\begin{aligned}
\int \frac{a}{b} &= \frac{1}{2}(1 + \sqrt{2}){\cdot}\log(x^7 - \sqrt{2}x^2 - (1 + \sqrt{2})x - 1) \\
&\quad + \frac{1}{2}(1 - \sqrt{2}){\cdot}\log(x^7 + \sqrt{2}x^2 - (1 - \sqrt{2})x - 1) .
\end{aligned}
$$

•

The Lazard/Rioboo/Trager Improvement

The Rothstein/Trager algorithm successfully avoids computing with extraneous algebraic numbers, that is, it only uses those algebraic numbers $c_1, \ldots, c_k$ which must appear in the final answer of the integral. However, there is still a significant amount of computation that takes place using algebraic extension fields. For instance, in Example 11.11 v_1 and v_2 are calculated via GCD computations of polynomials of degrees 13 and 14. The GCD's are calculated, however, over the coefficient field $\mathbf{Q}(\sqrt{2})$, rather than over $\mathbf{Q}$, a considerable increase in complexity. In general, the polynomials v_i appearing inside the log terms of our result involve GCD computations over the field $K(c_1, \ldots, c_k)$. In this section we present a simple observation by Lazard and Rioboo [6] that allows us to avoid the use of algebraic numbers when computing these v_i's. This improvement was also discovered independently by Trager (unpublished) during the process of implementing the Rothstein/Trager algorithm in SCRATCHPAD II.

Suppose we calculate the resultant $R(z)$ (11.19) by using the subresultant PRS algorithm from Chapter 7 (cf. Theorem 9.4). Let $S_i(x, z)$ denote the remainder (i.e. subresultant) of degree i in x appearing in the computation of the resultant via this algorithm. Then, rather than computing GCD's over algebraic extension fields, the v_i appearing in Theorem 11.7 can be determined by simple substitution into the $S_i(x, z)$. Specifically, we have

Theorem 11.9. Let $S_i(x,z)$ be the remainder of degree i in x appearing in the computation of $R(z)$ via the subresultant algorithm. Let

$$R(z) = \prod_{i=1}^{k} R_i(z)^i \tag{11.28}$$

be the square-free factorization of $R(z)$ over $K[z]$. Then

$$\int \frac{a}{b} = \sum_{i=1}^{k} \sum_{c\,:\,R_i(c)=0} c \cdot \log(S_i(x, c)). \tag{11.29}$$

Proof: Let c be a root of $R(z)$ of multiplicity i. We will show that

$$\mathrm{GCD}(a(x) - c \cdot b'(x), b(x)) = S_i(x, c). \tag{11.30}$$

Theorem 11.9 then follows directly from equation (11.21) in Theorem 11.7.

From Theorem 9.3, we have the formula

$$R(z) = m \cdot \prod_{\beta\,:\,b(\beta)=0} (a(\beta) - z \cdot b'(\beta)),$$

with m a constant independent of x and z. Since c is a root of multiplicity i, there are i values $\beta_1, \ldots, \beta_i$ such that

$$b(\beta_j) = 0, \text{ and } a(\beta_j) - c \cdot b'(\beta_j) = 0 \text{ for } j = 1, \ldots, i.$$

Each β_j is a root of the GCD on the left of equation (11.30). Since b is square-free, these account for all the roots of the GCD, hence

$$\deg(\mathrm{GCD}(a(x) - c \cdot b'(x), b(x))) = i \quad \text{(multiplicity of } c\text{)}. \tag{11.31}$$

When c is a root of $R(z)$ then the subresultant algorithm applied to $a(x) - c \cdot b'(x)$ and $b(x)$ results in the GCD of the two polynomials. When c has multiplicity i, the previous paragraph shows that this GCD has degree i. Equation (11.30) (and hence Theorem 11.9) then follows from a well-known property of the subresultant algorithm: for any constant c, the PRS resulting from applying the subresultant algorithm to the pair $a(x) - c \cdot b'(x)$ and $b(x)$ as polynomials in $K(c)[x]$, is the same as the PRS resulting from using the subresultant algorithm for the pair $a(x) - z \cdot b'(x)$ and $b(x)$ as polynomials in $K(z)[x]$ and then substituting $z = c$.

•

Example 11.12. Let us determine

$$\int \frac{a}{b} = \int \frac{6x^5 + 6x^4 - 8x^3 - 18x^2 + 8x + 8}{x^6 - 5x^4 - 8x^3 - 2x^2 + 2x + 1}.$$

Calculating $R(z) = \mathrm{res}_x(a - z \cdot b', b)$ via the subresultant algorithm, and computing a square-free factorization of the result gives

$$R(z) = -1453248 \cdot (z^2 - 2z - 2)^3.$$

$R_3(z) = z^2 - 2z - 2$ does not split over the constant field $\mathbf{Q}$, but its splitting field is just $\mathbf{Q}(\alpha)$ where $\alpha = \mathrm{RootOf}(z^2 - 2z - 2)$. The distinct roots of $R_3(z)$ in its splitting field are

$$c_1 = \alpha, \;\; c_2 = 2 - \alpha.$$

The remainder of degree 3 in x in the PRS defined by the subresultant algorithm is given by

$$S_3(x, z) = (-976z^3 + 2640z^2 - 120z - 680)\,x^3 + (648z^3 - 1032z^2 - 1824z + 1560)\,x^2$$
$$(1480z^3 - 3168z^2 - 1848z + 1808)\,x + (664z^3 - 1776z^2 + 264z + 200).$$

We wish to evaluate this polynomial at α, a root of $z^2 - 2z - 2$ and then normalize the result to obtain a polynomial that is monic in x. We can do this as follows. Applying the extended Euclidean algorithm to the minimal polynomial of α and the leading coefficient of $S_3(x, z)$, considered as a polynomial in x, gives

$$1 = s(z) \cdot (-976z^3 + 2640z^2 - 120z - 680) + t(z) \cdot (z^2 - 2z - 2)$$

with

$$s(z) = -\frac{1}{2088} z + \frac{1}{2088}$$

(there must be a solution; see Exercise 11.15). Multiplying $S_3(x,z)$ by $s(z)$ and taking the remainder on division by $z^2 - 2z - 2$ gives

$$x^3 + (1 - z)x^2 - zx - 1.$$

This is equivalent to evaluating at α and normalizing coefficients in the domain $\mathbf{Q}(\alpha)[x]$. Therefore, the integral can be expressed in the form

Algorithm 11.4. Lazard/Rioboo/Trager Improvement.

```
procedure LogarithmicPartIntegral(a, b, x)

  # Given a rational function a/b in x, with deg(a) < deg(b)
  # b monic and square-free, we calculate ∫a/b .

  # Calculate (via the subresultant algorithm)
  R(z) = res_x(a − z·b', b)
  S_i(x,z) = remainder of degree i in x in this computation
  (R_1(z), . . . , R_k(z)) ← SquareFree(R(z))

  # Process nontrivial square-free factors of R(z)
  integral ← 0
  for i from 1 to k with R_i(z) ≠ 1 do {
    # Normalize to make results monic
    w(z) = lcoeff_x(S_i(x, z))
    EEA(w(z), R_i(z); s(z), t(z))
    S_i(x, z) = pp_z( rem(s(z)·S_i(x,z), R_i(z)) )

    # Convert the S_i(x,c) for c a root of R_i(z) into a simpler form
    # (cf. Algorithm 11.3)
    (r_{i,1}, . . . , r_{i,k_i}) ← factors(R_i(z))
    for j from 1 to k_i do {
      d_j ← deg_z(r_{i,j}(z))
      if d_j = 1 then {
        c ← solve(r_{i,j}(z) = 0, z)
        integral ← integral + c·log(S_i(x, c))}
      elseif d_j = 2 then { #  Give answer in terms of radicals
        c ← solve(r_{i,j}(z) = 0, z)
        for n from 1 to 2 do {
          integral ← integral + c[n]·log(S_i(x, c[n])) } }
      else { #  Need answer in RootOf notation
        for n from 1 to d_j do {
            integral ← integral + c[n]·log(S_i(x, c[n]))
            # (where c[n] = RootOf(r_{i,j}(z))) } } } }
  return(integral)
end
```

$$\int \frac{a}{b} = \alpha \log(x^3 + (1-\alpha)x^2 - \alpha x - 1)$$
$$+ (2-\alpha)\log(x^3 - (1-\alpha)x^2 - (2-\alpha)x - 1)$$

where $\alpha = \mathrm{RootOf}\,(z^2 - 2z - 2)$. Since α may be expressed in terms of radicals by

$$\alpha = 1 + \sqrt{3}\,,$$

the final form of the result is

$$\int \frac{a}{b} = (1+\sqrt{3})\cdot\log(x^3 - \sqrt{3}x^2 - (1+\sqrt{3})x - 1)$$
$$+ (1-\sqrt{3})\cdot\log(x^3 + \sqrt{3}x^2 - (1-\sqrt{3})x - 1)\,.$$

•

Example 11.13. Consider the integral

$$\int \frac{a}{b} = \int \frac{2x^5 - 19x^4 + 60x^3 - 159x^2 + 50x + 11}{x^6 - 13x^5 + 58x^4 - 85x^3 - 66x^2 - 17x + 1}\,.$$

Calculating $R(z) = \mathrm{res}_x(a - z\cdot b', b)$ via the subresultant algorithm, and computing a square-free factorization of the result gives

$$R(z) = -190107645728000\cdot(z^3 - z^2 + z + 1)^2.$$

$R_2(z) = z^3 - z^2 + z + 1$ does not split over the constant field **Q**.

The remainder of degree 2 in x in the PRS defined by the subresultant algorithm is given by

$$S_2(x,z) = 880\cdot\{\ (-197244z^4 + 321490z^3 - 323415z^2 - 66508z + 144376)\,x^2$$
$$+ (-73478z^4 + 84426z^3 - 65671z^2 - 50870z - 85852)\,x$$
$$+ (-11516z^4 + 1997z^3 + 24623z^2 - 23675z - 14084)\ \}.$$

Making this monic in x gives

$$x^2 + (2z-5)x + z^2$$

hence the integral becomes

$$\int \frac{a}{b} = \sum_{\alpha \mid \alpha^3 - \alpha^2 + \alpha + 1 = 0} \alpha\cdot\log(x^2 + (2\alpha - 5)x + \alpha^2).$$

•

Exercises

1. Prove part (iv) (the power rule) of Theorem 11.1.

2. Let F be a differential field with differential operator D. Prove that the constant field

$$K = \{ c \in \mathrm{F} : D(c) = 0 \}$$

is indeed a subfield of F.

3. Using the algebraic definition of logarithms (Definition 11.5) prove that

(a) $\log(f \cdot g) = \log(f) + \log(g)$;

(b) $\log(f^n) = n \cdot \log(f)$, where $n \in \mathbf{Z}$.

4. Calculate, using Hermite reduction,

$$\int \frac{x^5 - x^4 + 4x^3 + x^2 - x + 5}{x^4 - 2x^3 + 5x^2 - 4x + 4} \, .$$

5. Repeat Exercise 11.4, but using Horowitz' method.

6. There is another reduction method (cf. Mack [7]) that avoids partial fraction decompositions when calculating $\int p/q$ with $\deg(p) < \deg(q)$. Let the square-free decomposition of q be $q_1 q_2^2 \cdots q_k^k$. Write $q = c \cdot q_k^k$. Show that one can find polynomials d and e such that

$$\int \frac{p}{q} = \frac{d}{c \cdot q_k^{k-1}} + \int \frac{e}{c \cdot q_k^{k-1}} \quad \text{and} \quad \deg(d) < \deg(q_k).$$

By then applying the same reduction process to the integral on the right side, we obtain a method for determining the rational part of the integral.

7. Let K(x) be a differential field over a constant field K and let $p/q^j \in \mathrm{K}(x)$, with j a positive integer, q square-free and $\deg(p) < \deg(q)$. Prove that

$$\int \frac{p}{q^j} \in \mathrm{K}(x)$$

if and only if $p = c \cdot q'$ for some constant $c \in \mathrm{K}$.

8. Show that, if b is a square-free polynomial and β is a finite pole of a/b, then the corresponding residue can be determined by equation (11.17).

9. Suppose that a/b is a rational function of x over a field K, with b irreducible over K. Prove that, if all the residues of a/b are contained in K, then the nonzero residues are all equal.

10. Solve Example 11.9 using the Lazard/Rioboo/Trager improvement. Is there an advantage in using this improvement in this case?

11. Solve Example 11.11 using the Lazard/Rioboo/Trager improvement. Is there an advantage in using this improvement in this case?

12. Calculate
$$\int \frac{8x^9 + x^8 - 12x^7 - 4x^6 - 26x^5 - 6x^4 + 30x^3 + 23x^2 - 2x - 7}{x^{10} - 2x^8 - 2x^7 - 4x^6 + 7x^4 + 10x^3 + 3x^2 - 4x - 2}.$$

13. Calculate
$$\int \frac{6x^7 + 7x^6 - 38x^5 - 53x^4 + 40x^3 + 96x^2 - 38x - 39}{x^8 - 10x^6 - 8x^5 + 23x^4 + 42x^3 + 11x^2 - 10x - 5}.$$

14. Solve Example 11.12 without using the Lazard/Rioboo/Trager improvement.

15. Let $S_i(x,z)$ and $R_i(z)$ be as in the last section. Why is it true that
$$\mathrm{GCD}(R_i(z), \mathrm{lcoeff}(S_i(x,z),x)) = 1 \ ?$$

The next three questions are concerned with the problem of *definite* integration. They are from the Ph.D. thesis of R. Rioboo

16. Compute
$$\int_1^2 \frac{x^4 - 3x^2 + 6}{x^6 - 5x^4 + 5x^2 + 4}$$
by first computing the indefinite integral using the methods from this chapter. Comment on the correctness of the result.

17. Let $u \in \mathbf{Q}(x)$. Show that
$$\frac{d}{dx}\log\left(\frac{u+i}{u-i}\right) = -2i \cdot \frac{d}{dx}\arctan(u).$$
Using this, repeat Exercise 11.16. Comment of the correctness of the result.

18. Let u and v be relatively prime polynomials in $\mathbf{Q}[x]$. Let s and t be polynomials with $\deg(s) < \deg(v)$ and $\deg(t) < \deg(u)$ satisfying $u \cdot s - v \cdot t = 1$.

a) Show that
$$\frac{d}{dx}\log\left(\frac{u+i\cdot v}{u-i\cdot v}\right) = -2 \cdot i \cdot \frac{d}{dx}\arctan(u \cdot t + v \cdot s) + \frac{d}{dx}\log\left(\frac{t+i\cdot s}{t-i\cdot s}\right).$$

b) Use part a) to give an algorithm for rewriting expressions of the form $\log\left(\frac{u+i\cdot v}{u-i\cdot v}\right)$ into a sum of arctangents having only polynomial arguments.

c) Use the method of part b) to recompute the integral from Exercise 11.16. Again comment of the correctness of the result.

References

1. E. Hermite, ''Sur l'Integration des Fractions Rationelles,'' *Nouvelles Annales de Mathematiques*, pp. 145-148 (1872).

2. I.N. Herstein, *Topics in Algebra,* Blaisdell (1964).

3. E. Horowitz, ''Algorithms for Partial Fraction Decomposition and Rational Integration,'' pp. 441-457 in *Proc. SYMSAM '71*, ed. S.R. Petrick, ACM Press (1971).

4. I. Kaplansky, ''An Introduction to Differential Algebra (second edition),'' Publications de l'Institut de Mathematique de l'Universite de Nancago V, Herman, Paris (1976).

5. E.R. Kolchin, *Differential Algebra and Algebraic Groups,* Academic Press, London (1973).

6. D. Lazard and R. Rioboo, ''Integration of Rational Functions: Rational Computation of the Logarithmic Part,'' *J. Symbolic Comp.*, **9**(2) pp. 113-116 (1990).

7. D. Mack, ''On Rational Function Integration,'' UCP-38, Computer Science Dept., University of Utah (1975).

8. J. Moses, ''Symbolic Integration: The Stormy Decade,'' *Comm. ACM*, **14** pp. 548-560 (1971).

9. J.F. Ritt, *Integration in Finite Terms,* Columbia University Press, New York (1948).

10. J.F. Ritt, ''Differential Algebra,'' *AMS Colloquium Proceedings*, **33**(1950).

11. M. Rothstein, ''Aspects of Symbolic Integration and Simplification of Exponential and Primitive Functions,'' Ph.D. Thesis, Univ. of Wisconsin, Madison (1976).

12. R.G. Tobey, *Algorithms for Antidifferentiation of Rational Functions,* Ph.D. Thesis, Harvard University (1967).

13. B. Trager, ''Algebraic Factoring and Rational Function Integration,'' pp. 219-226 in *Proc. SYMSAC '76*, ed. R.D. Jenks, ACM Press (1976).

CHAPTER 12

THE RISCH

INTEGRATION ALGORITHM

12.1. INTRODUCTION

When solving for an indefinite integral, it is not enough simply to ask to find an antiderivative of a given function $f(x)$. After all, the fundamental theorem of integral calculus gives the area function

$$A(x) = \int_a^x f(t)\, dt$$

as an antiderivative of $f(x)$. One really wishes to have some sort of closed expression for the antiderivative in terms of well-known functions (e.g. $\sin(x)$, e^x, $\log(x)$) allowing for common function operations (e.g. addition, multiplication, composition). This is known as the problem of integration in closed form or integration in finite terms. Thus, one is given an elementary function $f(x)$, and asks to find if there exists an elementary function $g(x)$ which is the antiderivative of $f(x)$ and, if so, to determine $g(x)$.

The abilities of symbolic computation systems to solve indefinite integrals are often met with equal doses of mystery and amazement. Amazement because such seemingly complex integrals as

$$\int \frac{x(x+1)\{(x^2e^{2x^2}-\log^2(x+1))^2+2xe^{3x^2}(x-(2x^3+2x^2+x+1)\log(x+1))\}}{((x+1)\log^2(x+1)-(x^3+x^2)e^{2x^2})^2}\, dx$$

can be worked out very quickly by systems using the Risch algorithm to yield

$$x - \log(x+1) - \frac{xe^{x^2}\log(x+1)}{\log^2(x+1) - x^2(e^{x^2})^2}$$
$$+ \frac{1}{2}\log(\log(x+1) + xe^{x^2}) - \frac{1}{2}\log(\log(x+1) - xe^{x^2})\,;$$

mystery in that these systems can invariably compute

$$\int \frac{1}{1+e^x}\, dx = x - \log(1+e^x)$$

yet

$$\int \frac{x}{1+e^x}\,dx$$

returns unevaluated (or else the answer is in terms of a non-elementary special function known as a logarithmic integral). Clearly, the first integral is so complicated that heuristics will probably not succeed. Therefore, an algorithmic approach would need to be used, and, since the above is not a rational function, the approach would require methods other than those from the previous chapter. However, it is confusing that any algorithm which can determine an answer in the first two cases cannot obtain an answer in the third case. The confusion stems from the fact that in the third case an unevaluated integral is an acceptable answer; namely, no closed form in terms of common elementary functions exists for the integral.

The problem of integration in finite terms (or integration in closed form) has a long history. It was studied extensively about 150 years ago by the French mathematician Joseph Liouville. The contribution of another nineteenth-century French mathematician, Charles Hermite, to the case of rational function integration is reflected in computational methods used today. For the case of transcendental elementary functions, apart from the sketch of an integration algorithm presented in G.H. Hardy's 1928 treatise, the constructive (computational, algebraic) approach to the problem received little attention beyond Liouville's work until the 1940's when J.F. Ritt [18] started to develop the topic of Differential Algebra. With the advent of computer languages for symbolic mathematical computation, there has been renewed interest in the topic since 1960 and the mathematics of the indefinite integration problem has evolved significantly. The modern era takes as its starting point the fundamental work by Risch [14] in 1968, where a complete decision procedure was described for the first time.

12.2. ELEMENTARY FUNCTIONS

The class of functions commonly referred to as the *elementary functions* includes rational functions, exponentials, logarithms, algebraic functions (e.g. n-th roots, and more generally, the solution of a polynomial equation whose coefficients are elementary functions), as well as the trigonometric, inverse trigonometric, hyperbolic, and inverse hyperbolic functions. Any finite nesting (composition) of the above functions is again an elementary function. Given an elementary function f, the problem of finding another elementary function g such that $g' = f$ (i.e. $g = \int f$), if such a g exists, is the indefinite integration problem of calculus. As commonly taught in calculus courses for the past century, the process of indefinite integration has been seen as a heuristic process employing a ''bag of tricks'' and a table of standard integrals. Only the case of rational function integration had the form of a finite algorithm; for non-rational functions, if one could not find an integral g for f then one still had not proved the nonexistence of an elementary integral for f. Indeed, it was not generally believed that there could exist a finite decision procedure for this problem. Consider, for example, the variety of different functions appearing in the following cases:

$$\int \frac{1}{1+x^2} = \arctan(x)\,; \tag{12.1}$$

$$\int \cos(x) = \sin(x)\,; \tag{12.2}$$

$$\int \frac{1}{\sqrt{1-x^2}} = \arcsin(x)\,; \tag{12.3}$$

$$\int \operatorname{arccosh}(x) = x \operatorname{arccosh}(x) - \sqrt{x^2-1}\,. \tag{12.4}$$

In these examples, there does not appear to be a regular relationship between the input (the integrand) and the output (the resulting integral). It would seem that for a given integrand f, in order to find an integral g (if it exists) one must search among a vast array of different functions.

As it happens, there does exist a reasonably simple relationship between an integrand f and its integral g if g exists as an elementary function. The lack of a discernible relationship in the above examples is mainly due to an unfortunate choice of mathematical notation. Recall that the rational function integration algorithm presented in the preceding chapter is based on the fact that if $f \in \mathrm{K}(x)$ then $\int f$ can be expressed using only logarithmic extensions of $\mathrm{K}(x)$ (in cases where the constant field K is algebraically closed). Moreover, precisely which logarithmic extensions are required is determined by a relatively simple algorithm. However, note that in case (12.1) above we expressed the integral of a rational function in a very different form using an inverse trigonometric function. Using Algorithm 11.3, case (12.1) would be expressed instead in the form

$$\int \frac{1}{1+x^2} = \frac{1}{2} i \log(x+i) - \frac{1}{2} i \log(x-i) \tag{12.5}$$

using logarithms and introducing the complex number i (an algebraic number satisfying $i^2+1=0$). Now most calculus students (and calculus instructors) would argue that the form (12.1) is "simpler" than the form (12.5) . However, for the purpose of obtaining a precise algorithm for rational function integration, the form (12.5) is much preferable because it fits into a simple framework in which we know that the only extensions required are logarithms (and algebraic number extensions to the constant field), and there is an algorithm to compute the result. Surprisingly, there exists a similar simple framework for the general case of elementary function integration: the only extensions required are logarithms (and algebraic numbers). Furthermore, the general integration algorithm is remarkably similar to the rational function integration algorithm.

In order to achieve this "simple framework" for the elementary functions, we discard the special notation for trigonometric, inverse trigonometric, hyperbolic, and inverse hyperbolic functions, noting that they all may be expressed using only exponentials, logarithms, and square roots (and allowing into the constant field the algebraic number i satisfying $i^2+1=0$). In this new notation, the integrals in (12.1) – (12.4) take the forms shown in (12.5) – (12.8).

$$\int \left(\frac{1}{2} \exp(ix) + \frac{1}{2} \exp(-ix) \right) = -\frac{1}{2} i \exp(ix) + \frac{1}{2} i \exp(-ix) \tag{12.6}$$

$$\int \frac{1}{\sqrt{1-x^2}} = -i \log(\sqrt{1-x^2} + ix) \tag{12.7}$$

$$\int \log(x + \sqrt{x^2-1}) = x \log(x + \sqrt{x^2-1}) - \sqrt{x^2-1}\,. \tag{12.8}$$

In these new forms, it can be seen that whatever functions appear in the integrand generally appear also in the expression for the integral, plus new logarithmic extensions may appear. There is a more regular relationship between the integrand and its integral than was apparent in formulas (12.1) - (12.4) .

It must be remarked that, in the context of placing these algorithms into a computer algebra system for practical use, it is still possible to use the more familiar notation appearing in formulas (12.1) - (12.4) . We are simply moving the choice of notation to a different level. By adopting the exp-log notation of the following definitions, we achieve a finite decision procedure for the integration of a significant class of functions. One could imagine, in a computer algebra system, an input transformation algorithm which takes the user's input and converts it into the exp-log notation for the integration algorithm, and an output transformation algorithm which converts the result back into the more familiar notation. However, the latter transformation process encounters the difficulties of the general *simplification problem* (see Chapter 3). A more practical approach generally adopted in computer algebra systems is to invoke initially a heuristic integration procedure which uses some standard transformations and table look-up (in the spirit of a classical first-year calculus student) to obtain the result in "familiar form" if possible. If the heuristic method fails then the problem is converted into the exp-log notation and the finite decision procedure is invoked. The result from the latter procedure will be either the integral expressed in the exp-log notation or an indication that there does not exist an elementary integral.

Definition 12.1. Let F be a differential field and let G be a differential extension field of F.

(i) For an element $\theta \in$ G, if there exists an element $u \in$ F such that

$$\theta' = \frac{u'}{u}$$

then θ is called *logarithmic over* F and we write $\theta = \log(u)$.

(ii) For an element $\theta \in$ G, if there exists an element $u \in$ F such that

$$\frac{\theta'}{\theta} = u'$$

then θ is called *exponential over* F and we write $\theta = \exp(u)$.

(iii) For an element $\theta \in$ G, if there exists a polynomial $p \in$ F$[z]$ such that

$$p(\theta) = 0$$

then θ is called *algebraic* over F.

•

Definition 12.2. Let F be a field and let G be an extension field of F. An element $\theta \in G$ is called *transcendental over* F if θ is not algebraic over F.

●

In some of the proofs of theorems, and indeed in the integration algorithm itself, it is necessary to distinguish exponential and logarithmic extensions which are transcendental from extensions which are algebraic. The manipulations which are valid for transcendental symbols (in proofs and also in algorithms) are quite different from the manipulations of symbols which satisfy an algebraic relationship.

Definition 12.3. Let F be a differential field and let G be a differential extension field of F. G is called a *transcendental elementary extension* of F if it is of the form

$$G = F(\theta_1, \ldots, \theta_n)$$

where for each $i = 1, \ldots, n$, θ_i is transcendental and either logarithmic or exponential over the field $F_{i-1} = F(\theta_1, \ldots, \theta_{i-1})$. G is called an *elementary extension* of F if it is of the form

$$G = F(\theta_1, \ldots, \theta_n)$$

where for each $i = 1, \ldots, n$, θ_i is either logarithmic, or exponential, or algebraic over the field $F_{i-1} = F(\theta_1, \ldots, \theta_{i-1})$. (In this notation, $F_0 = F$.)

●

Definition 12.4. Let $K(x)$ be a differential field of rational functions over a constant field K which is a subfield of the field of complex numbers. If F is a transcendental elementary extension of $K(x)$ then F is called a *field of transcendental elementary functions*. Similarly, if F is an elementary extension of $K(x)$ then F is called a *field of elementary functions*.

●

A Structure Theorem

Before proceeding, it is important to note that the three extensions defined in Definition 12.1 are not mutually exclusive. Indeed, according to Definition 12.1, an element $\theta \in G$ could be logarithmic over F (or exponential over F) when in fact $\theta \in F$, in which case θ is trivially "algebraic over F". The less trivial cases shown in the following examples must be recognized in order to proceed correctly.

Example 12.1. The function

$$f = \exp(x) + \exp(2x) + \exp(x/2)$$

could be represented as

$$f = \theta_1 + \theta_2 + \theta_3 \in \mathbf{Q}(x, \theta_1, \theta_2, \theta_3)$$

where $\theta_1 = \exp(x)$, $\theta_2 = \exp(2x)$, and $\theta_3 = \exp(x/2)$. Taking derivatives gives $\theta_1' / \theta_1 = x' \in \mathbf{Q}(x)$, so θ_1 is exponential over $\mathbf{Q}(x)$. Also, θ_2 is exponential over $\mathbf{Q}(x,\theta_1)$

since $\theta_2' / \theta_2 = (2x)' \in \mathbf{Q}(x,\theta_1)$. Similarly, θ_3 is exponential over $\mathbf{Q}(x,\theta_1,\theta_2)$ since $\theta_3' / \theta_3 = (x/2)' \in \mathbf{Q}(x,\theta_1,\theta_2)$. However

$$\mathbf{Q}(x,\theta_1,\theta_2) = \mathbf{Q}(x,\theta_1)$$

since

$$\theta_2 = \theta_1^2 \in \mathbf{Q}(x,\theta_1).$$

Hence a simpler representation for the function f is

$$f = \theta_1 + \theta_1^2 + \theta_3 \in \mathbf{Q}(x,\theta_1,\theta_3)$$

where $\theta_1 = \exp(x)$ and $\theta_3 = \exp(x/2)$.

Given the field $\mathbf{Q}(x,\theta_1)$, the function θ_3 is not only exponential over this field but it is also algebraic over this field since

$$\theta_3^2 - \theta_1 = 0.$$

In other words, $\theta_3 = \theta_1^{1/2} \in \mathbf{Q}(x,\theta_1,\theta_1^{1/2})$. Thus the function f could be represented in the form

$$f = \theta_1 + \theta_1^2 + \theta_1^{1/2} \in \mathbf{Q}(x,\theta_1,\theta_1^{1/2}).$$

Alternatively, the simplest representation for f would be

$$f = \theta_3^2 + \theta_3^4 + \theta_3 \in \mathbf{Q}(x,\theta_3)$$

where $\theta_3 = \exp(x/2)$.

●

Example 12.2. The function

$$g = \sqrt{\log(x^2 + 3x + 2)\,(\log(x + 1) + \log(x + 2))}$$

could be represented as

$$g = \theta_4 \in \mathbf{Q}(x, \theta_1, \theta_2, \theta_3, \theta_4)$$

where $\theta_1 = \log(x^2 + 3x + 2)$, $\theta_2 = \log(x + 1)$, $\theta_3 = \log(x + 2)$, and θ_4 satisfies the algebraic equation

$$\theta_4^2 - \theta_1(\theta_2 + \theta_3) = 0.$$

In this view, θ_1 is logarithmic over $\mathbf{Q}(x)$, θ_2 is logarithmic over $\mathbf{Q}(x,\theta_1)$, θ_3 is logarithmic over $\mathbf{Q}(x,\theta_1,\theta_2)$, and θ_4 is algebraic over $\mathbf{Q}(x,\theta_1,\theta_2,\theta_3)$. Thus g is viewed as an algebraic function at the "top level". However, g is also logarithmic since

$$g' = \frac{(x^2 + 3x + 2)'}{x^2 + 3x + 2};$$

i.e. $g = \log(x^2 + 3x + 2)$. This is easily seen by applying the rule of logarithms. Thus the simplest representation for g is:

$$g = \theta_1 \in \mathbf{Q}(x, \theta_1)$$

where $\theta_1 = \log(x^2 + 3x + 2)$.

•

Example 12.3. The function

$$h = \exp(\log(x)/2)$$

could be represented as

$$h = \theta_2 \in \mathbf{Q}(x, \theta_1, \theta_2)$$

where $\theta_1 = \log(x)$ and $\theta_2 = \exp(\theta_1/2)$. In this view, θ_1 is logarithmic over $\mathbf{Q}(x)$ and θ_2 is exponential over $\mathbf{Q}(x,\theta_1)$. However, θ_2 is also algebraic over $\mathbf{Q}(x)$ since

$$\theta_2^2 - x = 0 .$$

Therefore a simpler representation for h is

$$h = \theta_2 \in \mathbf{Q}(x, \theta_2)$$

where $\theta_2^2 - x = 0$. In other words, $h = x^{1/2}$.

•

Definition 12.5. An element θ is *monomial* over a differential field F if

(i) $F(\theta)$ and F have the same constant field,

(ii) θ is transcendental over F,

(iii) θ is either exponential or logarithmic over F.

•

Thus we want to determine when a new element is a monomial over F. It turns out that this can be found by checking a set of linear equations for a solution. The following theorem gives explicit requirements for new extensions to be ''independent'' of the previous elementary extensions. Because of the large quantity of algebraic machinery required to prove this result, we do not prove Theorem 12.1, but rather refer the reader to Risch [17].

Theorem 12.1 (Structure Theorem). Let F be a field of constants, and $F_n = F(x, \theta_1, \ldots, \theta_n)$ an extension of $F(x)$ having F as its field of constants. Suppose further that each θ_j is either

(a) algebraic over $F_{j-1} = F(x, \theta_1, \ldots, \theta_{j-1})$,

(b) w_j with $w_j = \log(u_j)$ and $u_j \in F_{j-1}$, or

(c) u_j with $u_j = \exp(w_j)$ and $w_j \in F_{j-1}$.

Then:

(i) $g = \log(f)$ with $f \in \mathrm{F}_n - \mathrm{F}$ is a monomial over F_n if and only if there is no product combination

$$f^k \cdot \prod u_j^{k_j} \in \mathrm{F} \quad (k, k_j \in \mathbf{Z}, \text{ and } k \neq 0).$$

(ii) $f = \exp(g)$ with $g \in \mathrm{F}_n - \mathrm{F}$ is a monomial over F_n if and only if there is no linear combination

$$g + \sum c_i w_i \in \mathrm{F}, \quad (c_i \in \mathbf{Q}).$$

●

If we are given a new exponential, then Theorem 12.1 implies that, to see if it is a monomial, we need only check that its argument can be written as

$$c + \sum c_i w_i$$

with $c \in \mathrm{F}$ and $c_i \in \mathbf{Q}$. By differentiating this we obtain a linear system of equations in the c_i. Once the c_i are known we can determine c.

A similar approach can be used when we are given a new logarithm and wish to determine if it is a monomial. In this case, we need only check whether $f^k \cdot \Pi u_j^{k_j}$ can be made to lie in F, for some suitable choice of integers k and k_j. If we write

$$h = f^k \cdot \prod u_j^{k_j}$$

then $h \in \mathrm{F}$ if and only if $h' = 0$. This is identical to determining if

$$0 = \frac{h'}{h} = \frac{k \cdot f'}{f} + \sum k_j \cdot \frac{u_j'}{u_j}$$

has a solution. Thus, as before, determining if an element is monomial is equivalent to determining if a particular system of linear equations has a solution.

Example 12.4. Let

$$g = \log(\sqrt{x^2+1} + x) + \log(\sqrt{x^2+1} - x).$$

If we set $\theta_1 = \sqrt{x^2+1}$, then g can be considered as belonging to the extension $\mathbf{Q}(x, \theta_1, \log(\theta_1 + x), \log(\theta_1 - x))$. If $\theta_2 = \log(\theta_1 + x)$ and $\theta_3 = \log(\theta_1 - x)$, we check to see if θ_3 is a monomial over $\mathbf{Q}(x, \theta_1, \theta_2)$.

By Theorem 12.1 θ_3 is a monomial iff there exist no integers $k \neq 0$ and k_1 such that

$$h = (\theta_1 - x)^k (\theta_1 + x)^{k_1} \in \mathbf{Q}.$$

Differentiating with respect to x and dividing both sides by h gives this as

$$0 = \frac{h'}{h} = k\frac{(\theta_1 - x)'}{(\theta_1 - x)} + k_1\frac{(\theta_1 + x)'}{(\theta_1 + x)}.$$

Differentiating and clearing denominators gives

$$0 = k(\frac{x}{\sqrt{x^2+1}} - 1)(\sqrt{x^2+1} + x) + k_1(\frac{x}{\sqrt{x^2+1}} + 1)(\sqrt{x^2+1} - x)$$

$$0 = k - k_1.$$

One particular solution is therefore $k = k_1 = 1$.

Thus θ_3 would not add a new independent transcendental logarithmic extension onto $\mathbf{Q}(x, \theta_1, \theta_2)$. Indeed, since

$$(\theta_1 - x)\cdot(\theta_1 + x) = 1 \in \mathbf{Q}$$

we have

$$\theta_3 = \log(\theta_1 - x) = -\log(\theta_1 + x) = -\theta_2 \in \mathbf{Q}(x, \theta_1, \theta_2).$$

●

12.3. DIFFERENTIATION OF ELEMENTARY FUNCTIONS

We now have a precise definition of what we mean by a field of elementary functions. Specifically, it is any finitely generated extension field of a field $K(x)$ of rational functions such that each extension is one of three types: logarithmic, exponential, or algebraic. It would be mathematically convenient to let the constant field K in Definition 12.4 always be the algebraically closed field $\mathbf{C}$ of complex numbers. However, we wish to operate in domains with exact arithmetic (where operations such as polynomial GCD computation will be well-defined) so that the constant field K will be of the form $\mathbf{Q}(\alpha_1, \ldots, \alpha_k)$ where $\mathbf{Q}$ is the field of rational numbers and α_i $(1 \le i \le k)$ are algebraic number extensions of $\mathbf{Q}$ required by the problem at hand. The problem of elementary function integration can be stated in the following terms. Given an elementary function f, first determine a specification for an elementary function field F such that $f \in$ F :

$$\mathrm{F} = \mathbf{Q}(\alpha_1, \ldots, \alpha_k)(x, \theta_1, \ldots, \theta_n).$$

Then determine the additional extensions required so that $g = \int f$ lies in the *new* elementary function field

$$\mathbf{Q}(\alpha_1, \ldots, \alpha_k, \ldots, \alpha_{k+h})(x, \theta_1, \ldots, \theta_n, \ldots, \theta_{n+m})$$

and explicitly determine g, or else prove that no such elementary function g exists.

In order to obtain a complete decision procedure, we first need to investigate the possible forms for such an integral. This requires that we know how the differentiation operator behaves in these elementary extensions. In this section we present some basic properties of the differential operator in simple elementary extensions, that is, those fields given as a single logarithmic, exponential or algebraic extension.

Theorem 12.2 (Differentiation of logarithmic polynomials). Let F be a differential field and let F(θ) be a differential extension field of F having the same subfield of constants. Suppose that θ is transcendental and logarithmic over F (with say $\theta' = u'/u$ for $u \in$ F). For any $a(\theta) \in$ F[θ] with deg($a(\theta)$) > 0 the following properties hold:

(i) $a(\theta)' \in$ F[θ];

(ii) if the leading coefficient of the polynomial $a(\theta)$ is a constant then $\deg(a(\theta)') = \deg(a(\theta)) - 1$;

(iii) if the leading coefficient of the polynomial $a(\theta)$ is not a constant then $\deg(a(\theta)') = \deg(a(\theta))$.

Proof: Write $a(\theta)$ in the form

$$a(\theta) = \sum_{i=0}^{n} a_i \theta^i$$

where $a_n \neq 0$ and $n > 0$. Then

$$a(\theta)' = \sum_{i=0}^{n-1} (a_i' + (i+1)a_{i+1}\theta')\,\theta^i + a_n'\,\theta^n .$$

Since $\theta' = u'/u \in$ F, property (i) is obvious. If $a_n' \neq 0$ then

$$\deg(a(\theta)') = \deg(a(\theta)) = n$$

proving (iii). Property (ii) is the case where $a_n' = 0$ in which case it is clear that $\deg(a(\theta)') < n$. To prove the more precise statement, suppose that the degree $n-1$ coefficient vanishes

$$a_{n-1}' + n\,a_n\,\theta' = 0 .$$

In this case, it follows that

$$(n\,a_n\,\theta + a_{n-1})' = n\,a_n'\,\theta + n\,a_n\,\theta' + a_{n-1}' = 0 .$$

Now $n\,a_n\,\theta + a_{n-1} \in$ F(θ) and it is a constant so, in particular, $n\,a_n\,\theta + a_{n-1} \in$ F. This contradicts the assumption that θ is transcendental over F. Thus property (ii) is proved.

●

Theorem 12.3 (Differentiation of exponential polynomials). Let F be a differential field and let F(θ) be a differential extension field of F having the same subfield of constants. Suppose that θ is transcendental and exponential over F (with say $\theta'/\theta = u'$ for $u \in$ F). For any $a(\theta) \in$ F[θ] with deg($a(\theta)$) > 0 the following properties hold:

(i) if $h \in$ F, $h \neq 0$, $n \in \mathbf{Z}$, $n \neq 0$, then $(h\cdot\theta^n)' = \bar{h}\cdot\theta^n$ for some $\bar{h} \in$ F with $\bar{h} \neq 0$;

(ii) $a(\theta)' \in$ F[θ] and $\deg(a(\theta)') = \deg(a(\theta))$;

(iii) $a(\theta)$ divides $a(\theta)'$ if and only if $a(\theta)$ is a monomial (i.e. $a(\theta) = h\cdot\theta^n$ for some $h \in$ F, $n \in \mathbf{Z}$).

Proof: Since

$$(h\,\theta^n)' = h'\,\theta^n + nh\theta^{n-1}\theta' = (h'+nhu')\theta^n,$$

we must show that $\bar{h} = h' + nhu' \neq 0$ if $h \neq 0$ and $n \neq 0$. But if $\bar{h} = 0$ then $h\theta^n$ is a constant in F(θ) so, in particular, $h\theta^n \in$ F. This contradicts the assumption that θ is transcendental over F. This proves (i).

To prove (ii), write $a(\theta)$ in the form

$$a(\theta) = \sum_{i=0}^{n} a_i\,\theta^i$$

where $a_n \neq 0$ and $n > 0$. Then

$$a(\theta)' = \sum_{i=0}^{n} (a_i\theta^i)' = \sum_{i=0}^{n} \bar{a}_i\,\theta^i$$

for $\bar{a}_i \in$ F and, by property (i), $\bar{a}_n \neq 0$ hence the degrees are equal.

Suppose that $a(\theta)$ is a monomial $a(\theta) = h\theta^n$ for $h \in$ F and $n > 0$. Then by property (i), $a(\theta)' = \bar{h}\,\theta^n$ for $\bar{h} \in$ F and clearly $a(\theta)$ divides $a(\theta)'$. Now suppose that $a(\theta)$ is not a monomial but that $a(\theta)$ divides $a(\theta)'$. Then we have the divisibility relationship

$$a(\theta)' = g\,a(\theta)$$

for some $g \in$ F[θ]. By property (ii) $\deg(a(\theta)') = \deg(a(\theta))$ so we conclude that $g \in$ F. Writing $a(\theta)$ in descending powers of θ, $a(\theta)$ contains at least two terms so it takes the form

$$a(\theta) = a_n\,\theta^n + a_m\,\theta^m + b(\theta)$$

where $n, m \in \mathbf{Z}$, $n > m \geq 0$, $a_n \neq 0$, $a_m \neq 0$, and either $b(\theta) = 0$ or $\deg(b(\theta)) < m$. By properties (i) and (ii),

$$a(\theta)' = \bar{a}_n\,\theta^n + \bar{a}_m\,\theta^m + b(\theta)'$$

where $\bar{a}_n \neq 0$, $\bar{a}_m \neq 0$, and either $b(\theta)' = 0$ or $\deg(b(\theta)') < m$. (There is one possibility not covered by property (i), namely the case $m = 0$. In this case, $a(\theta) = a_n\,\theta^n + a_0$ and $a(\theta)' = \bar{a}_n\,\theta^n + a_0{}'$ where $\bar{a}_n \neq 0$ but perhaps $a_0{}' = 0$. However, this cannot happen under the present circumstances because the divisibility relationship would become

$$\bar{a}_n\,\theta^n = g\,a_n\,\theta^n + g\,a_0$$

with $n > 0$, $a_n \neq 0$, $a_0 \neq 0$, and $\bar{a}_n \neq 0$, which is impossible.) The divisibility relationship yields the following equations over the field F

$$a_n{}' + n\,a_n\,u' = g\,a_n,$$

$$a_m{}' + m\,a_m\,u' = g\,a_m$$

where we have used the fact that the coefficients in $a(\theta)'$ take the form $\bar{a}_i = a_i' + i\,a_i\,u'$. Eliminating g from these two equations yields

$$\frac{a_n{}'}{a_n} + n\,u' = \frac{a_m{}'}{a_m} + m\,u'$$

or

$$\frac{a_n{}'}{a_n} - \frac{a_m{}'}{a_m} + (n-m)u' = 0.$$

Then

$$\left(\frac{a_n}{a_m}\theta^{n-m}\right)' = \left(\frac{a_n{}'}{a_m} - \frac{a_n \cdot a_m{}'}{a_m^2}\right)\theta^{n-m} + \frac{a_n}{a_m}(n-m)u'\theta^{n-m}$$

$$= \frac{a_n}{a_m}\theta^{n-m}\left(\frac{a_n{}'}{a_n} - \frac{a_m{}'}{a_m} + (n-m)u'\right) = 0$$

which implies that $\frac{a_n}{a_m}\theta^{n-m}$ is a constant in $F(\theta)$. In particular it belongs to F, which contradicts the assumption that θ is transcendental over F.

●

Theorem 12.4 (Differentiation of algebraic functions). Let F be a differential field and let $F(\theta)$ be a differential extension field of F. Suppose that θ is algebraic over F with minimal polynomial

$$p(z) = \sum_{i=0}^{N+1} p_i\, z^i \in F[z]$$

where $p_{N+1} = 1$ (z is a new transcendental symbol). Then the derivative of θ can be expressed in the form

$$\theta' = -\frac{d(\theta)}{e(\theta)} \in F(\theta)$$

where $d, e \in F[z]$ are the following polynomials specified in terms of coefficients appearing in the minimal polynomial

$$d(z) = \sum_{i=0}^{N} p_i{}'\, z^i\,, \quad e(z) = \sum_{i=0}^{N} (i+1)\, p_{i+1} z^i \in F[z].$$

Proof: Setting $p(\theta) = 0$ and applying the differential operator to this identity yields

$$\sum_{i=0}^{N+1} p_i{}'\,\theta^i + \sum_{i=1}^{N+1} i \cdot p_i \theta^{i-1}\theta' = 0.$$

Noting that $p_{N+1}{}' = 0$ and solving for θ' yields

$$\theta' = -\frac{\sum_{i=0}^{N} p_i{}'\theta^i}{\sum_{i=0}^{N} (i+1)\cdot p_{i+1}\theta^i}$$

which is the desired result. Note that the denominator $e(\theta) \neq 0$ because its degree is $N < \deg(p(z))$ with leading coefficient $N + 1 \neq 0$.

●

12.4. LIOUVILLE'S PRINCIPLE

We have seen from Chapter 11 that a rational function $f \in$ K(x) always has an integral which can be expressed as a transcendental elementary function. Specifically, $\int f$ can always be expressed as the sum of a rational function plus constant multiples of a finite number of logarithmic extensions. The fundamental result on elementary function integration was first presented by Liouville in 1833 and is a generalization of the above statement. It is the basis of the algorithmic approach to elementary function integration.

Theorem 12.5 (Liouville's Principle). Let F be a differential field with constant field K. For $f \in$ F suppose that the equation $g' = f$ (i.e. $g = \int f$) has a solution $g \in$ G where G is an elementary extension of F having the same constant field K. Then there exist $v_0, v_1, \ldots, v_m \in$ F and constants $c_1, \ldots, c_m \in$ K such that

$$f = v_0' + \sum_{i=1}^{m} c_i \frac{v_i'}{v_i} .$$

In other words, such that

$$\int f = v_0 + \sum_{i=1}^{m} c_i \log(v_i) .$$

We will prove the theorem in a number of stages. However, the basic idea of Liouville's Principle is quite simple to explain. If a transcendental logarithmic extension is postulated to appear in the expression for the integral, either in a denominator or in polynomial form, then we use Theorem 12.2 to show that it will fail to disappear under differentiation except in the special case of a polynomial which is linear (in the logarithmic extension) with a constant leading coefficient. Similarly, if a transcendental exponential extension is postulated to appear in the expression for the integral, then we use Theorem 12.3 to show that differentiation will fail to eliminate it. Finally, if an algebraic extension is postulated to appear in the expression for the integral, then we will use Theorem 12.4 to show that it is possible to express the integral free of the algebraic extension.

Special Case: Simple Transcendental Logarithmic Extensions

The proof of Liouville's Principle proceeds by induction on the number of new elementary extensions required to express the integral. Therefore, we examine more closely the case where only one extension θ is required, since this will reveal the crux of the induction proof. In this subsection we will consider only the case where θ is a transcendental logarithmic extension. We will assume here (as in Theorem 11.7) that the constant field K is large enough so that no new algebraic numbers are required to express the integral.

Thus, let $F = K(x, \theta_1, \ldots, \theta_n)$ be an elementary function field with constant field K and let $f \in F$. Suppose that $\int f \in G$ where G is an elementary extension of F of the form

$$G = F(\theta)$$

with $\theta = \log(u)$ for some $u \in F$ and suppose further that G has the same constant field K.

Since θ is transcendental over F, $\int f$ can be expressed in the form

$$\int f = \frac{a(\theta)}{b(\theta)}$$

where $a, b \in F[\theta]$, $GCD(a, b) = 1$, and b is monic. (Note that when θ is transcendental over the field F, the domain $F[\theta]$ can be viewed as a polynomial domain in the variable θ, with well-defined GCD and factorization operations. If θ is algebraic over the field F then such a view of $F[\theta]$ is invalid.) We can factor $b(\theta)$ into the form

$$b(\theta) = \prod_{i=1}^{\mu} b_i(\theta)^{r_i}$$

where $b_i(\theta)$ $(i \le i \le \mu)$ are distinct monic irreducible polynomials in $F[\theta]$. A partial fraction decomposition then gives

$$\frac{a(\theta)}{b(\theta)} = a_0(\theta) + \sum_{i=1}^{\mu} \sum_{j=1}^{r_i} \frac{a_{ij}(\theta)}{b_i(\theta)^j}$$

where a_0, a_{ij}, $b_i \in F[\theta]$ and $\deg(a_{ij}) < \deg(b_i)$. Differentiating both sides with respect to the integration variable x gives

$$f = a_0(\theta)' + \sum_{i=1}^{\mu} \sum_{j=1}^{r_i} \left(\frac{a_{ij}(\theta)'}{b_i(\theta)^j} - \frac{j\, a_{ij}(\theta) \cdot b_i(\theta)'}{b_i(\theta)^{j+1}} \right). \quad (12.9)$$

An important property of this equation is that the left hand side is independent of θ.

Let $b_i(\theta) \in F[\theta]$ be any particular monic irreducible polynomial with $\deg(b_i(\theta)) > 0$ appearing in the denominator on the right hand side of equation (12.9). Then, from Theorem 12.2 we know that $b_i(\theta)' \in F[\theta]$ with $\deg(b_i(\theta)') < \deg(b_i(\theta))$. Therefore $b_i(\theta)$ does not divide $b_i(\theta)'$. Also, no factors divide $b_i(\theta)$ because it is irreducible. It follows that there is precisely one term on the right hand side of equation (12.9) whose denominator is $b_i(\theta)^{r_i+1}$. Since there is no other term with which it could cancel, this term must appear on the left hand side. This is a contradiction. We therefore conclude that there can be no terms with denominators involving θ. Equation (12.9) then takes the form

$$f = a_0(\theta)'$$

where $a_0 \in F[\theta]$. Hence $a_0(\theta)'$ is independent of θ, which, by Theorem 12.2, can only happen if

$$a_0(\theta) = c\theta + d \in \mathrm{F}[\theta]$$

for some constant $c \in \mathrm{K}$ and some function $d \in \mathrm{F}$. We have proved that

$$\int f = d + c\,\log(u)$$

where $c \in \mathrm{K}$, $d, u \in \mathrm{F}$, which is the desired form for Liouville's Principle in this special case.

Special Case: Simple Transcendental Exponential Extensions

Consider now the special case where

$$\mathrm{G} = \mathrm{F}(\theta)$$

with $\theta = \exp(u)$ for some $u \in \mathrm{F}$ and with the same constant field K, that is, a transcendental exponential extension.

Since θ is transcendental over F, the same argument used in the last subsection gives f in the form of equation (12.9). As before, the $b_i(\theta)$ are distinct, monic and irreducible in $\mathrm{F}[\theta]$, and as before an important property of equation (12.9) is that the left hand side is independent of θ.

Let $b_i(\theta) \in \mathrm{F}[\theta]$ be any particular monic irreducible polynomial with $\deg(b_i(\theta)) > 0$ appearing in a denominator on the right hand side of equation (12.9). If $b_i(\theta)$ is not a monomial (i.e. a single term of the form $h\theta^k$) then Theorem 12.3 implies that $b_i(\theta)$ does not divide $b_i(\theta)'$. The same arguments used for the case of simple transcendental logarithmic extensions lead us to conclude that there can be no terms with denominators involving $b_i(\theta)$. Now if $b_i(\theta)$ is a monomial then it is simply $b_i(\theta) = \theta$ (since each $b_i(\theta)$ is monic and irreducible). It follows that equation (12.9) takes the form

$$f = \left(\sum_{j=-k}^{l} h_j \theta^j \right)'$$

where $h_j \in \mathrm{F}$ $(-k \le j \le l)$. Applying the differentiation operation on the right hand side to each term $h_j\theta^j$ yields a term $\bar{h}_j\theta^j$ with the same power of θ, where $\bar{h}_j \in \mathrm{F}$ and for $j \ne 0$, $\bar{h}_j \ne 0$ if $h_j \ne 0$ (see Theorem 12.3). Since f is independent of θ, we conclude that only the term $j = 0$ can appear

$$f = h_0',$$

in other words,

$$\int f = h_0$$

where $h_0 \in \mathrm{F}$. This is the desired form, proving that a new exponential extension cannot appear in the integral.

Special Case: Simple Algebraic Extensions

Consider now the special case where

$$G = F(\theta)$$

with θ algebraic over G.

If θ satisfies an algebraic relationship over F then we must use an argument which is very different from that used for transcendental extensions, since now $F[\theta]$ is not an ordinary polynomial domain over F in a transcendental symbol. Let the minimal polynomial defining θ be $p(z) \in F[z]$ with $\deg(p(z)) = N + 1$ (where z is a new transcendental symbol).

Suppose now that for $f \in F$,

$$\int f = a(\theta) \in F(\theta)$$

or in differentiated form

$$f = a(\theta)' .$$

Let the conjugates of θ (i.e. all the roots of $p(z) = 0$) be denoted by $\theta = \theta_0, \theta_1, \ldots, \theta_N$. Since f is independent of θ and since the differentiation operation is uniquely determined by $p(z)$ (cf. Exercise 12.4), we can conclude that

$$f = a(\theta_j)', \quad \text{for each } j = 0, 1, \ldots, N.$$

Summing over j yields

$$(N+1) f = \sum_{j=0}^{N} a(\theta_j)' .$$

Writing this in the form

$$f = h'$$

where $h = \frac{1}{N+1} \sum_{j=0}^{N} a(\theta_j)$ is a symmetric function in $F(\theta_0, \theta_1, \ldots, \theta_N)$, we conclude that $h \in F$ by Theorem 8.15. We have thus proved that

$$\int f = h$$

where $h \in F$. This is the desired form, proving that a new algebraic extension need not appear in the integral.

Example 12.5. Consider the following integral

$$\int \frac{2x^3 - 2x^2 - 1}{(x-1)^2} \exp(x^2) = \frac{\exp(x^2 + \log(x)/2)}{2(\sqrt{x} - 1)} + \frac{\exp(x^2 + \log(x)/2)}{2(\sqrt{x} + 1)}$$

which can be verified by differentiation and simplification. The integrand lies in the field $\mathbf{Q}(x, \exp(x^2))$. At first glance, it would seem that the expression for the integral violates Liouville's Principle since it involves a new exponential extension and a new algebraic extension. Upon further examination, we note that

$$\exp(x^2 + \log(x)/2) = \exp(x^2)\cdot\exp(\log(x)/2) = \sqrt{x}\exp(x^2)$$

yielding the alternative expression for the integral

$$\frac{\sqrt{x}\cdot\exp(x^2)}{2(\sqrt{x}-1)} + \frac{\sqrt{x}\cdot\exp(x^2)}{2(\sqrt{x}+1)} .$$

Still, this expression for the integral involves the algebraic extension $\sqrt{x}$. As seen in the argument for the algebraic case above, it must be possible to eliminate $\sqrt{x}$ from this expression. In this case, by simply forming a common denominator we easily see that the result can be expressed in the form

$$\int \frac{2x^3 - 2x^2 - 1}{(x-1)^2} \exp(x^2) = \frac{x}{x-1} \exp(x^2)$$

which is the form that would be obtained by our integration algorithm.

•

General Case: A Proof of Liouville's Theorem

We are now in a position to prove Liouville's theorem. The main idea was first stated by Laplace in 1820, then proved in some cases by Liouville, and subsequently generalized by Ostrowski [13] in 1946.

Proof of Liouville's Principle: The supposition is that there exist $\theta_1, \ldots, \theta_N$ such that

$$G = F(\theta_1, \ldots, \theta_N)$$

where each θ_i $(1 \le i \le N)$ is either logarithmic, exponential, or algebraic over $F_{i-1} = F(\theta_1, \ldots, \theta_{i-1})$, each extension field $F(\theta_1, \ldots, \theta_i)$ has the same constant field K, and there exists $g \in G$ satisfying the equation $g' = f$. The proof is by induction on the number N of elementary extensions appearing in G. The case N=0 is trivial since we then have $g \in F$ satisfying $g' = f$, hence $m \doteq 0$ and $f = v_0'$ with $v_0 = g$.

The induction hypothesis is that the theorem holds for any number of extensions less than N. For the case of N extensions, we may view the field $F(\theta_1, \ldots, \theta_N)$ in the form $F(\theta_1)(\theta_2, \ldots, \theta_N)$. Since $f \in F(\theta_1)$ and $g \in F(\theta_1)(\theta_2, \ldots, \theta_N)$ satisfies the equation $g' = f$, we may apply the induction hypothesis to conclude that there exist $v_i(\theta_1) \in F(\theta_1)$ $(0 \le i \le m)$ and constants $c_i \in K$ $(1 \le i \le m)$ such that

$$f = v_0(\theta_1)' + \sum_{i=1}^{m} c_i \frac{v_i(\theta_1)'}{v_i(\theta_1)} . \tag{12.10}$$

Let us denote θ_1 by the symbol θ.

Consider first the case where θ is transcendental and logarithmic over F. The proof will follow closely the argument used in the case of a simple transcendental logarithmic extension given previously. By applying the rule of logarithms $\log(v_i \cdot v_j) = \log(v_i) + \log(v_j)$, if necessary, we may assume that each $v_i(\theta)$ $(1 \le i \le m)$ is either an element of F or else is monic and irreducible in F[θ] with $\deg(v_i(\theta)) > 0$, that the $v_i(\theta)$ $(1 \le i \le m)$ are all different, and that the c_i $(1 \le i \le m)$ are nonzero. Let $v_0(\theta) \in F(\theta)$ be expressed in the form

$$v_0(\theta) = a(\theta) \,/\, b(\theta)$$

where $a, b \in F[\theta]$, $\mathrm{GCD}(a,b) = 1$, and b is monic. Factor $b(\theta)$ into the form

$$b(\theta) = \prod_{i=1}^{\mu} b_i(\theta)^{r_i}$$

where $b_i(\theta)$ $(1 \le i \le \mu)$ are distinct monic irreducible polynomials in F[θ] and $r_i \in \mathbf{Z}$, $r_i > 0$. Express $v_0(\theta)$ in a partial fraction expansion of the form

$$v_0(\theta) \;=\; a_0(\theta) \;+\; \sum_{i=1}^{\mu} \sum_{j=1}^{r_i} \frac{a_{ij}(\theta)}{b_i(\theta)^j}$$

where a_0, a_{ij}, $b_i \in F[\theta]$ and $\deg(a_{ij}) < \deg(b_i)$. Equation (12.10) then becomes

$$f = a_0(\theta)' + \sum_{i=1}^{\mu} \sum_{j=1}^{r_i} \left(\frac{a_{ij}(\theta)'}{b_i(\theta)^j} - \frac{j \cdot a_{ij}(\theta) \cdot b_i(\theta)'}{b_i(\theta)^{j+1}} \right) + \sum_{i=1}^{m} c_i \frac{v_i(\theta)'}{v_i(\theta)} \,. \tag{12.11}$$

As before, an important property of this equation is that the left hand side is independent of θ.

Since θ is logarithmic over F, there exists a $u \in F$ such that $\theta' = u' / u$. Let $p(\theta)$ be any monic irreducible polynomial in F[θ] with $\deg(p(\theta)) > 0$. Then by Theorem 12.2, $p(\theta)' \in F[\theta]$ with $\deg(p(\theta)') < \deg(p(\theta))$ and therefore $p(\theta)$ does not divide $p(\theta)'$. If $p(\theta)$ is one of the $b_i(\theta)$ appearing in a denominator in the partial fraction expansion of $v_0(\theta)$ with maximal power r_i, then the right hand side of equation (12.11) contains precisely one term whose denominator is $p(\theta)^{r_i+1}$ (note that $r_i + 1 > 1$). Since there is no other term with which it could cancel, this term must appear on the left hand side, which contradicts the fact that f is independent of θ. We therefore conclude that the middle terms of equation (12.11) (the double summation) cannot appear. Now if $p(\theta)$ is one of the $v_i(\theta)$ appearing on the right hand side of equation (12.11) then there is precisely one term whose denominator is $p(\theta)$, again a contradiction.

The previous paragraph implies that equation (12.11) takes the form

$$f = a_0(\theta)' + \sum_{i=1}^{m} c_i \, \frac{v_i'}{v_i}$$

where $a_0 \in F[\theta]$, $v_i \in F$ $(1 \le i \le m)$, and $c_i \in K$ $(1 \le i \le m)$. Since f and v_i $(1 \le i \le m)$ are independent of θ, $a_0(\theta)'$ must be independent of θ which, by Theorem 12.2, can only happen if

$$a_0(\theta) = c\,\theta + d \in \mathrm{F}[\theta]$$

for some constant $c \in$ K and some function $d \in$ F. Therefore

$$f = d' + c\,u'/u + \sum_{i=1}^{m} c_i\,v'_i/v_i$$

where $d, u, v_i \in$ F and $c, c_i \in$ K, which is the desired form.

In the same manner as above, it is easy to verify that the arguments used in the case of a simple transcendental exponential extension or a simple algebraic extension also carry over to the general case. This completes the proof of Liouville's theorem.

●

For generalizations of Liouville's theorem to include extensions such as error functions and logarithmic integrals, we refer the reader to the paper by Singer, Saunders and Caviness [22]. See also the progress report of Baddoura [1] regarding a generalization of Liouville's theorem to include dilogarithmic extensions.

12.5. THE RISCH ALGORITHM FOR TRANSCENDENTAL ELEMENTARY FUNCTIONS

In this and subsequent sections, we develop an effective decision procedure for the elementary integration of any function which belongs to a field of transcendental elementary functions (see Definition 12.4). In other words,

$$f \in \mathrm{K}(x, \theta_1, \ldots, \theta_n)$$

where the constant field K is a subfield of the field of complex numbers and where each θ_i $(1 \le i \le n)$ is transcendental and either logarithmic or exponential over the field $\mathrm{K}(x, \theta_1, \ldots, \theta_{i-1})$. The decision procedure will determine $\int f$ if it exists as an elementary function. Otherwise, it constructs a proof of the nonexistence of an elementary integral. The problem of developing an effective decision procedure for the more general field of elementary functions, involving algebraic function extensions, is more difficult. The case of algebraic function extensions is discussed in a later section.

Given an integrand f, the first step is to determine a description $\mathrm{K}(x, \theta_1, \ldots, \theta_n)$ of a field of transcendental elementary functions in which f lies (if f lies in such a field). As Examples 12.1-12.3 demonstrated, this is not necessarily a trivial step. Examples 12.1 and 12.2 showed cases where a "quick view" of the integrand led to a description involving non-transcendental extensions but where a purely transcendental description could be found. Example 12.3 showed a case where an integrand was expressed solely in terms of a logarithm and an exponential ($\exp(\log(x)/2)$) but the integrand was not transcendental over the rational functions. To handle this step in our integration algorithm, we will first convert all trigonometric (and related) functions into their exponential and logarithmic forms. Then the algebraic relationships which exist among the various exponential and logarithmic functions are determined. For the remainder of this section, we will assume that a purely transcendental description $\mathrm{K}(x, \theta_1, \ldots, \theta_n)$ has been given for the integrand f. A technical point which appeared in the theorems of the preceding section was the condition that when an extension was made to a differential field, the subfield of constants was assumed to remain the same. This condition will be handled dynamically by the integration algorithm, enlarging the constant field as necessary when determining the transcendental description $K(x, \theta_1, \ldots, \theta_n)$.

Since each extension θ_i is a transcendental symbol, the integrand may be manipulated as a rational function in these symbols. The integration algorithm for transcendental functions will follow steps that are very reminiscent of the development in the previous chapter of the rational function integration algorithm, in particular Hermite's method and the Rothstein/Trager method. Given an integrand $f \in K(x, \theta_1, \ldots, \theta_n)$, it may be viewed as a rational function in the last extension $\theta = \theta_n$

$$f(\theta) = \frac{p(\theta)}{q(\theta)} \in F_{n-1}(\theta)$$

where $F_{n-1} = K(x, \theta_1, \ldots, \theta_{n-1})$. We may assume that $f(\theta)$ is normalized such that $p(\theta), q(\theta) \in F_{n-1}[\theta]$ satisfy $\mathrm{GCD}(p(\theta), q(\theta)) = 1$ and that $q(\theta)$ is monic. (Throughout the development, we must keep in mind that $\int f(\theta)$ is integration with respect to x, not θ, and we will continue to reserve the symbol $'$ for differentiation with respect to x only, using $\frac{d}{d\theta}$ for differentiation with respect to θ. The algorithm is recursive, so that when treating $\int f(\theta)$ there will be recursive invocations to integrate functions in the field F_{n-1}. The base of the recursion is integration in the field $F_0 = K(x)$ which is handled by Algorithms 11.1 - 11.4.

12.6. THE RISCH ALGORITHM FOR LOGARITHMIC EXTENSIONS

Consider first the case where θ is logarithmic, with say $\theta' = u'/u$ and $u \in F_{n-1}$. Proceeding as in Hermite's method, apply Euclidean division to $p(\theta), q(\theta) \in F_{n-1}[\theta]$ yielding polynomials $s(\theta), r(\theta) \in F_{n-1}[\theta]$ such that

$$p(\theta) = q(\theta) \cdot s(\theta) + r(\theta) \ \text{ with } \ r(\theta) = 0 \ \text{ or } \ \deg(r(\theta)) < \deg(q(\theta)).$$

We then have

$$\int f(\theta) = \int s(\theta) + \int \frac{r(\theta)}{q(\theta)} .$$

We refer to the first integral on the right hand side of this equation as the integral of the *polynomial part* of $f(\theta)$, and to the second integral as the integral of the *rational part* of $f(\theta)$. Unlike the case of pure rational function integration, the integration of the polynomial part is not trivial (indeed, it is the harder of the two parts).

Logarithmic Extension: Integration of the Rational Part

For the rational part, we continue with Hermite's method. Compute the square-free factorization of the denominator $q(\theta) \in F_{n-1}[\theta]$

$$q(\theta) = \prod_{i=1}^{k} q_i(\theta)^i$$

where each $q_i(\theta)$ $(1 \le i \le k)$ is monic and square-free, $\mathrm{GCD}(q_i(\theta), q_j(\theta)) = 1$ for $i \ne j$, and $\deg(q_k(\theta)) > 0$. It must be remarked that all operations here are in the polynomial domain $\mathrm{F}_{n-1}[\theta]$, in particular, the definition that $q_i(\theta)$ is square-free is: $\mathrm{GCD}(q_i(\theta), \frac{d}{d\theta} q_i(\theta)) = 1$. We will require the stronger condition that $\mathrm{GCD}(q_i(\theta), q_i(\theta)') = 1$ (where $'$ denotes differentiation with respect to x), and fortunately this condition holds as we now prove.

Theorem 12.6. Let F be a differential field with differential operator $'$ satisfying $x' = 1$, where $x \in \mathrm{F}$. Let $\mathrm{F}(\theta)$ be a differential extension field of F having the same subfield of constants, with θ transcendental and logarithmic over F, specifically, $\theta' = \frac{u'}{u}$ with $u \in \mathrm{F}$. Let $a(\theta) \in \mathrm{F}[\theta]$ be a polynomial in the symbol θ with $\deg(a(\theta)) > 0$ and with $a(\theta)$ monic, such that

$$\mathrm{GCD}(a(\theta), \frac{d}{d\theta} a(\theta)) = 1$$

(i.e. $a(\theta)$ is square-free as an element of $\mathrm{F}[\theta]$). Then

$$\mathrm{GCD}(a(\theta), a(\theta)') = 1$$

where the latter GCD operation is also in the domain $\mathrm{F}[\theta]$.

Proof: From Theorem 12.2 we know that $a(\theta)' \in \mathrm{F}[\theta]$. Let the monic polynomial $a(\theta) \in \mathrm{F}[\theta]$ be factored over its splitting field F_a into the form

$$a(\theta) = \prod_{i=1}^{N} (\theta - a_i)$$

where $a_i \in \mathrm{F}_a (1 \le i \le N)$ are all distinct (because $a(\theta)$ is square-free in $\mathrm{F}[\theta]$). Then

$$a(\theta)' = \sum_{i=1}^{N} \left(\frac{u'}{u} - a_i' \right) \prod_{j \ne i} (\theta - a_j).$$

If for any i, $\frac{u'}{u} - a_i' = 0$ then a_i is a logarithm of $u \in \mathrm{F}$, in particular, the expression $\theta - a_i$ in the differential extension field $\mathrm{F}_a(\theta)$ satisfies

$$(\theta - a_i)' = 0$$

which implies that $\theta - a_i$ is a constant in $\mathrm{F}_a(\theta)$, whence

$$\theta - a_i = c \in \mathrm{F}_a,$$

contradicting the assumption that θ is transcendental over F (since $F_a(\theta)$ and F_a have the same subfield of constants – cf. Exercise 12.5). Hence $\frac{u'}{u} - a_i' \ne 0$ for $1 \le i \le N$. Now for any particular factor $\theta - a_i$ of $a(\theta)$, the expression for $a(\theta)'$ has $N - 1$ terms which are divisible by $\theta - a_i$ and one term which is not divisible by $\theta - a_i$. It follows that $a(\theta)$ and $a(\theta)'$ have no common factors.

●

Continuing with Hermite's method in this logarithmic case, we compute the partial fraction expansion of $\frac{r(\theta)}{q(\theta)} \in F_{n-1}(\theta)$ in the form

$$\frac{r(\theta)}{q(\theta)} = \sum_{i=1}^{k} \sum_{j=1}^{i} \frac{r_{ij}(\theta)}{q_i(\theta)^j}$$

where for $1 \le i \le k$ and $1 \le j \le i$, $r_{ij}(\theta) \in F_{n-1}[\theta]$ and

$$\deg(r_{ij}(\theta)) < \deg(q_i(\theta)), \text{ if } \deg(q_i(\theta)) > 0, \tag{12.12}$$

$$r_{ij}(\theta) = 0, \text{ if } q_i(\theta) = 1.$$

We then have

$$\int \frac{r(\theta)}{q(\theta)} = \sum_{i=1}^{k} \sum_{j=1}^{i} \int \frac{r_{ij}(\theta)}{q_i(\theta)^j}. \tag{12.13}$$

Hermite's reduction proceeds as follows for a particular nonzero integrand $\frac{r_{ij}(\theta)}{q_i(\theta)^j}$ with $j > 1$. By Theorem 12.6, $\mathrm{GCD}(q_i(\theta), q_i(\theta)') = 1$ so we may apply the method of Theorem 2.6 to compute polynomials $s(\theta), t(\theta) \in F_{n-1}[\theta]$ such that

$$s(\theta)\, q_i(\theta) + t(\theta)\, q_i(\theta)' = r_{ij}(\theta) \tag{12.14}$$

where $\deg(s(\theta)) < \deg(q_i(\theta)')$ and $\deg(t(\theta)) < \deg(q_i(\theta))$. (The latter inequality holds because of the inequality in (12.12).) Dividing by $q_i(\theta)^j$ in equation (12.14) yields

$$\int \frac{r_{ij}(\theta)}{q_i(\theta)^j} = \int \frac{s(\theta)}{q_i(\theta)^{j-1}} + \int \frac{t(\theta) \cdot q_i(\theta)'}{q_i(\theta)^j} \, .$$

Applying integration by parts to the second integral on the right, exactly as in Section 11.3, leads to the reduction formula

$$\int \frac{r_{ij}(\theta)}{q_i(\theta)^j} = \frac{-t(\theta)/(j-1)}{q_i(\theta)^{j-1}} + \int \frac{s(\theta) + t(\theta)'/(j-1)}{q_i(\theta)^{j-1}} \, .$$

If the numerator of the integral on the right hand side is nonzero and if $j - 1 > 1$ then the same reduction process may be repeated. Note that the numerator of the new integrand satisfies the degree constraint

$$\deg(s(\theta) + t(\theta)'/(j-1)\,) \le \max(\deg(s(\theta)), \deg(t(\theta)')) < \deg(q_i(\theta))$$

(since by Theorem 12.2, differentiation of a logarithmic polynomial either leaves the degree unchanged or else reduces it by one) which is consistent with the original numerator degree constraint expressed in (12.12). Therefore the degree constraints associated with equation (12.14) will still hold. By repeated application of this reduction process until the denominators of all remaining integrands are square-free, equation (12.13) reduces to the following form for the integral of the rational part of $f(\theta)$

$$\int \frac{r(\theta)}{q(\theta)} = \frac{c(\theta)}{d(\theta)} + \int \frac{a(\theta)}{b(\theta)}$$

where $a(\theta), b(\theta), c(\theta), d(\theta) \in F_{n-1}[\theta]$, $\deg(a(\theta)) < \deg(b(\theta))$, and $b(\theta)$ is monic and square-free. Just as in the case of rational function integration, the above result from Hermite's reduction process has the following properties (see Theorem 11.6):

$$d(\theta) = \mathrm{GCD}(q(\theta), \frac{d}{d\theta} q(\theta)) \in F_{n-1}[\theta];$$

$$b(\theta) = q(\theta)/d(\theta) \in F_{n-1}[\theta];$$

$$\deg(a(\theta)) < \deg(b(\theta)); \quad \deg(c(\theta)) < \deg(d(\theta)).$$

Therefore, one might think that Horowitz' method could apply in this case as well. However, because the underlying field is not constant with respect to x, when we specify the numerators with undetermined coefficients in Horowitz' method and then apply differentiation to remove the integral signs, what results is a system of linear *differential* equations (instead of linear algebraic equations which was the case for pure rational function integration).

It remains to compute the integral of the proper rational function $\frac{a(\theta)}{b(\theta)} \in F_{n-1}(\theta)$. As in the case of rational function integration, the Rothstein/Trager method applies here. Specifically, we compute the resultant

$$R(z) = \mathrm{res}_\theta(a(\theta) - z \cdot b(\theta)', b(\theta)) \in F_{n-1}[z].$$

Unlike the rational function case, the roots of $R(z)$ are not necessarily constants. However (see Theorem 12.7 below), $\int \frac{a(\theta)}{b(\theta)}$ is elementary if and only if

$$R(z) = \overline{R}(z)\, S \in F_{n-1}[z]$$

where $\overline{R}(z) \in K[z]$ and $S \in F_{n-1}$. Therefore we compute $\overline{R}(z) = \mathrm{pp}(R(z))$, the primitive part of $R(z)$ as a polynomial in $F_{n-1}[z]$. If any of the coefficients in $\overline{R}(z)$ is nonconstant then there does not exist an elementary integral. Otherwise, let c_i $(1 \le i \le m)$ be the distinct roots of $\overline{R}(z)$ in its splitting field $K_{\overline{R}}$ and define $v_i(\theta)$ $(1 \le i \le m)$ by

$$v_i(\theta) = \mathrm{GCD}(a(\theta) - c_i \cdot b(\theta)', b(\theta)) \in F_{n-1}(c_1, \ldots, c_m)[\theta]. \tag{12.15}$$

Then

$$\int \frac{a(\theta)}{b(\theta)} = \sum_{i=1}^{m} c_i \log(v_i(\theta)). \tag{12.16}$$

This expresses the result using the minimal algebraic extension of the constant field K.

Example 12.6. The integral

$$\int \frac{1}{\log(x)}$$

has integrand

$$f(\theta) = \frac{1}{\theta} \in \mathbf{Q}(x, \theta)$$

where $\theta = \log(x)$. Applying the Rothstein/Trager method, we compute

$$R(z) = \mathrm{res}_\theta(1 - \frac{z}{x}, \theta) = 1 - \frac{z}{x} \in \mathbf{Q}(x)[z].$$

Since $R(z)$ has a nonconstant root, we conclude that the integral is not elementary. •

Example 12.7. The integral

$$\int \frac{1}{x\ \log(x)}$$

has integrand

$$f(\theta) = \frac{1/x}{\theta} \in \mathbf{Q}(x, \theta)$$

where $\theta = \log(x)$. Applying the Rothstein/Trager method, we compute

$$R(z) = \mathrm{res}_\theta\, (\frac{1}{x} - \frac{z}{x}, \theta) = \frac{1}{x} - \frac{z}{x} \in \mathbf{Q}(x)[z].$$

Since

$$\bar{R}(z) = \mathrm{pp}(R(z)) = 1 - z$$

has constant coefficients, the integral is elementary. Specifically,

$$c_1 = 1,$$

$$v_1(\theta) = \mathrm{GCD}(\frac{1}{x} - \frac{1}{x}, \theta) = \theta,$$

and

$$\int \frac{1}{x\ \log(x)} = c_1 \log(v_1(\theta)) = \log(\log(x)).$$

•

Example 12.8. Consider the integral which appeared in the introduction to this chapter:

$$\int \frac{x(x+1)\left[\left(x^2\exp(2x^2)-\log^2(x+1)\right)^2+2x\exp(3x^2)\left(x-(2x^3+2x^2+x+1)\log(x+1)\right)\right]}{\left((x+1)\log^2(x+1)-(x^3+x^2)\exp(2x^2)\right)^2}$$

Letting $\theta_1 = \exp(x^2)$ and $\theta_2 = \log(x + 1)$, the integrand can be considered in the form

$$f(\theta_2) \in \mathbf{Q}(x, \theta_1, \theta_2).$$

The numerator and denominator of $f(\theta_2)$ are each of degree 4 in θ_2, and after normalization and Euclidean division it takes the form

$$f(\theta_2) = \frac{x}{x+1} + \frac{\frac{2x^2}{x+1}\,\theta_1^3(x-(2x^3+2x^2+x+1)\theta_2)}{(\theta_2^2 - x^2\theta_1^2)^2} .$$

The "polynomial part" with respect to θ_2 is the first term here, which in this case is simply a rational function in $Q(x)$ and it can be integrated by Algorithm 11.3 or 11.4. For the "rational part" we proceed to apply Hermite reduction. Note that the rational part is already expressed in the form

$$\frac{r(\theta_2)}{b(\theta_2)^2}$$

where $b(\theta_2) = \theta_2^2 - x^2\theta_1^2$ is monic and square-free, and $\deg(r(\theta_2)) < \deg(b(\theta_2))$, so there is no need to apply partial fraction expansion. Equation (12.14) takes the form

$$s(\theta_2)(\theta_2^2 - x^2\theta_1^2) + t(\theta_2)\left(\frac{2}{x+1}\theta_2 - 2x(2x^2+1)\theta_1^2\right) = r(\theta_2)$$

where $r(\theta_2)$ is the numerator of the rational part of $f(\theta_2)$, as expressed above. This equation has the solution

$$s(\theta_2) = \frac{-2x}{x+1}\,\theta_1,\ \ t(\theta_2) = x\theta_1\theta_2.$$

The Hermite reduction therefore yields

$$\int \frac{r(\theta_2)}{b(\theta_2)^2} = \frac{-x\theta_1\theta_2}{\theta_2^2 - x^2\theta_1^2} + \int \frac{(2x^2+1)\theta_1\theta_2 - \frac{x}{x+1}\theta_1}{\theta_2^2 - x^2\theta_1^2} .$$

We now apply the Rothstein/Trager method to the integrand on the right hand side. Denoting the numerator by $a(\theta_2)$, the resultant computation is

$$R(z) = \mathrm{res}_{\theta_2}(a(\theta_2) - z\cdot b(\theta_2)',\ b(\theta_2))$$

and after dividing out the content, we get

$$\bar{R}(z) = \mathrm{pp}(R(z)) = 4z^2 - 1.$$

Since $\bar{R}(z)$ has constant coefficients, the integral is elementary. Specifically, we compute

$$c_1 = \frac{1}{2},\ c_2 = -\frac{1}{2},$$

$$v_1(\theta_2) = \mathrm{GCD}(a(\theta_2) - c_1\, b(\theta_2)',\ b(\theta_2)) = \theta_2 + x\theta_1,$$

$$v_2(\theta_2) = \mathrm{GCD}(a(\theta_2) - c_2 b(\theta_2)',\ b(\theta_2)) = \theta_2 - x\theta_1,$$

and hence

$$\int \frac{a(\theta_2)}{b(\theta_2)} = \frac{1}{2}\log(\theta_2 + x\theta_1) - \frac{1}{2}\log(\theta_2 - x\theta_1).$$

Putting it all together, the original integral is elementary and it takes the form

$$\int f = x - \log(x+1) - \frac{x\exp(x^2)\log(x+1)}{\log^2(x+1) - x^2\exp^2(x^2)}$$
$$+ \frac{1}{2}\log(\log(x+1) + x\exp(x^2)) - \frac{1}{2}\log(\log(x+1) - x\exp(x^2)) .$$

●

Before proving the results of Rothstein and Trager, let us place their method into context as follows. Suppose that the square-free denominator $b(\theta) \in F_{n-1}[\theta]$ has factorization

$$b(\theta) = \prod_{j=1}^{m} v_j(\theta) \tag{12.17}$$

in some algebraic number extension of $F_{n-1}[\theta]$. As we saw in the case of pure rational function integration, the integral could be expressed in different forms involving more or fewer algebraic number extensions in the log terms, depending on the factorization used for the denominator. We do not want to completely factor $b(\theta)$ over its splitting field if that can be avoided, but some algebraic number extensions may be required in order to express the integral. Assuming some factorization as expressed in (12.17), we would then have the partial fraction expansion

$$\frac{a(\theta)}{b(\theta)} = \sum_{i=1}^{m} \frac{u_i(\theta)}{v_i(\theta)} \tag{12.18}$$

where $\deg(u_i(\theta)) < \deg(v_i(\theta))$ $(1 \le i \le m)$. Now suppose that for each i,

$$u_i(\theta) = c_i\, v_i(\theta)'$$

for some constant c_i. In this circumstance, the integral is readily expressed as follows

$$\int \frac{a(\theta)}{b(\theta)} = \sum_{i=1}^{m} \int \frac{c_i \cdot v_i(\theta)'}{v_i(\theta)} = \sum_{i=1}^{m} c_i \log(v_i(\theta)).$$

The Rothstein/Trager method extracts the factors $v_i(\theta)$ from $b(\theta)$ via the GCD computations (12.15). Furthermore, equation (12.15) guarantees that each factor $v_i(\theta)$ divides $a(\theta) - c_i \cdot b(\theta)'$. The latter expression takes the following form, by substituting for $b(\theta)$ according to (12.17) and for $a(\theta)$ according to (12.18)

$$a(\theta) - c_i \cdot b(\theta)' = \left(\prod_{j=1}^{m} v_j(\theta) \right) \left(\sum_{k=1}^{m} \frac{u_k(\theta)}{v_k(\theta)} \right) - c_i \left(\prod_{j=1}^{m} v_j(\theta) \right)'$$

$$= \sum_{k=1}^{m} \left(u_k(\theta) \prod_{j \neq k} v_j(\theta) \right) - c_i \sum_{k=1}^{m} \left(v_k(\theta)' \prod_{j \neq k} v_j(\theta) \right)$$

$$= \sum_{k=1}^{m} \left((u_k(\theta) - c_i\, v_k(\theta)') \prod_{j \neq k} v_j(\theta) \right).$$

Now for each term in this sum except the term $k = i$, $v_i(\theta)$ is an explicit factor. Since $v_i(\theta)$ divides the whole sum, we can conclude that

$$v_i(\theta) \mid (u_i(\theta) - c_i \cdot v_i(\theta)').$$

But $\mathrm{GCD}(v_i(\theta), v_i(\theta)') = 1$ (because $v_i(\theta)$ is square-free) so we must have

$$u_i(\theta) = c_i \cdot v_i(\theta)'.$$

This is precisely the condition noted above which allows the integral to be expressed in the desired form. Theorem 12.7 guarantees that if $\int a(\theta)/b(\theta)$ is elementary then it can be expressed in the form (12.16), gives an efficient method to determine when this form exists (by looking at the primitive part of the resultant $R(z)$), gives an efficient method to compute the factors $v_i(\theta)$ (by equation (12.15)), and guarantees that the result is expressed using the minimal algebraic extension field.

Theorem 12.7 (Rothstein/Trager Method – Logarithmic Case). Let F be a field of elementary functions with constant field K. Let θ be transcendental and logarithmic over F (i.e. $\theta' = \frac{u'}{u}$ for some $u \in$ F) and suppose that the transcendental elementary extension F(θ) has the same constant field K. Let $a(\theta)/b(\theta) \in$ F(θ) where $a(\theta), b(\theta) \in$ F[θ], $\mathrm{GCD}(a(\theta), b(\theta)) = 1$, $\deg(a(\theta)) < \deg(b(\theta))$, and $b(\theta)$ is monic and square-free.

(i) $\int \frac{a(\theta)}{b(\theta)}$ is elementary if and only if all the roots of the polynomial

$$R(z) = \mathrm{res}_\theta(a(\theta) - z \cdot b(\theta)', b(\theta)) \in \mathrm{F}[z]$$

are constants. (Equivalently, $R(z) = S \cdot \bar{R}(z)$ where $\bar{R}(z) \in \mathrm{K}[z]$ and $S \in$ F.)

(ii) If $\int \frac{a(\theta)}{b(\theta)}$ is elementary then

$$\frac{a(\theta)}{b(\theta)} = \sum_{i=1}^{m} c_i \frac{v_i(\theta)'}{v_i(\theta)} \tag{12.19}$$

where c_i $(1 \le i \le m)$ are the distinct roots of $R(z)$ and $v_i(\theta)$ $(1 \le i \le m)$ are defined by

$$v_i(\theta) = \mathrm{GCD}(a(\theta) - c_i \cdot b(\theta)', b(\theta)) \in \mathrm{F}(c_1, \ldots, c_m)[\theta].$$

(iii) Let F^* be the minimal algebraic extension field of F such that $a(\theta)/b(\theta)$ can be expressed in the form (12.19) with constants $c_i \in \mathrm{F}^*$ and with $v_i(\theta) \in \mathrm{F}^*[\theta]$. Then $\mathrm{F}^* = \mathrm{F}(c_1, \ldots, c_m)$ where c_i $(1 \le i \le m)$ are the distinct roots of $R(z)$.

Proof: Suppose that $\int \frac{a(\theta)}{b(\theta)}$ is elementary. Then by Liouville's Principle,

$$\frac{a(\theta)}{b(\theta)} = v_0(\theta)' + \sum_{i=1}^{m} c_i \frac{v_i(\theta)'}{v_i(\theta)} \tag{12.20}$$

where $c_i \in \mathrm{K}^*$ and $v_i(\theta) \in \mathrm{F}^*(\theta)$ $(0 \le i \le m)$, where K^* denotes the minimal algebraic extension of K necessary to express the integral and F^* denotes F with its constant field extended to K^*. Without loss of generality, we may assume that $v_i(\theta)$ $(1 \le i \le m)$ are polynomials in $\mathrm{F}^*[\theta]$ (by applying the rule of logarithms). Moreover, using the argument presented in the proof of Theorem 11.8, we may assume that $c_i \in \mathrm{K}^*$ $(1 \le i \le m)$ are distinct nonzero constants and that $v_i(\theta) \in \mathrm{F}^*[\theta]$ $(1 \le i \le m)$ are square-free and pairwise relatively prime.

If $v_0(\theta) \in \mathrm{F}^*(\theta)$ can be expressed in the form

$$v_0(\theta) = \frac{p(\theta)}{q(\theta)} \text{ with } p(\theta), q(\theta) \in \mathrm{F}^*[\theta], \mathrm{GCD}(p(\theta), q(\theta)) = 1,$$

and $\deg(q(\theta)) > 0$ then $v_0(\theta)'$ contains a factor in its denominator which is not square-free. (For a detailed argument about the form of the derivative, see the proof of Liouville's Principle.) Since $b(\theta)$ is square-free, we conclude that $v_0(\theta) \in \mathrm{F}^*[\theta]$. Now by Theorem 12.2, $v_0(\theta)' \in \mathrm{F}^*[\theta]$. But if $v_0(\theta)'$ is any nonzero polynomial then the right hand side of equation (12.20), when formed over a common denominator, will have a numerator of degree greater than or equal to the degree of its denominator. Since the left hand side of equation (12.20) satisfies $\deg(a(\theta)) < \deg(b(\theta))$, we conclude that $v_0(\theta)' = 0$.

We have shown that if $\int \frac{a(\theta)}{b(\theta)}$ is elementary then equation (12.19) holds where $c_i \in \mathrm{K}^*$ $(1 \le i \le m)$ are distinct nonzero constants and $v_i(\theta) \in \mathrm{F}^*[\theta]$ $(1 \le i \le m)$ are square-free and pairwise relatively prime. Applying the argument presented in the first part of the proof of Theorem 11.7, we conclude that

$$b(\theta) \mid \prod_{j=1}^{m} v_j(\theta) \text{ and } \prod_{j=1}^{m} v_j(\theta) \mid b(\theta).$$

Since $b(\theta)$ is monic, we may assume without loss of generality that $v_i(\theta)$ $(1 \le i \le m)$ are all monic and

$$b(\theta) = \prod_{j=1}^{m} v_j(\theta).$$

The rest of the argument presented in the proof of Theorem 11.7 carries through in the present case (for the polynomial domain $\mathrm{F}^*[\theta]$), yielding the conclusion that c_i $(1 \le i \le m)$ are the distinct roots of the polynomial $R(z)$ defined in part (i) of the statement of the theorem and $v_i(\theta)$ $(1 \le i \le m)$ are as defined in part (ii). We have thus proved part (ii) of the theorem and we have proved the "only if" case of part (i). (The parenthesized remark that $R(z) = S \cdot \bar{R}(z)$ where $\bar{R}(z) \in \mathrm{K}[z]$ and $S \in \mathrm{F}$ can be proved using standard algebra results; the proof is omitted here.) We have also proved part (iii) since we assumed F^* to be a minimal algebraic extension of F and then proved that the roots of $R(z)$ must appear.

To prove the "if" case of part (i), suppose that all the roots of $R(z)$ are constants. Let c_i $(1 \le i \le m)$ be the distinct roots of $R(z)$ and define $v_i(\theta)$ $(1 \le i \le m)$ by

$$v_i(\theta) = \mathrm{GCD}(a(\theta) - c_i \cdot b(\theta)', b(\theta)) \in \mathrm{F}(c_1, \ldots, c_m)[\theta].$$

Now if for $i \ne j$, $\mathrm{GCD}(v_i(\theta), v_j(\theta)) = w$ then

$$w \mid (a(\theta) - c_i \cdot b(\theta)'),$$

$$w \mid (a(\theta) - c_j \cdot b(\theta)'), \quad \text{and}$$

$$w \mid b(\theta).$$

The first two conditions imply that $w \mid (c_i - c_j) \cdot b(\theta)'$, and combining this with the third condition shows that if a nontrivial common divisor w exists, then it is a common divisor of $b(\theta)$ and $b(\theta)'$. But $b(\theta)$ is square-free, so we can conclude that $\mathrm{GCD}(v_i(\theta), v_j(\theta)) = 1$. Since each $v_i(\theta)$ divides $b(\theta)$, it follows that

$$r(\theta) = \prod_{i=1}^{m} v_i(\theta)$$

divides $b(\theta)$. In other words

$$b(\theta) = r(\theta)\, s(\theta)$$

for some $s(\theta) \in \mathrm{F}(c_1, \ldots, c_m)[\theta]$. Suppose that $\deg(s(\theta)) > 0$. Then

$$\mathrm{res}_\theta(a(\theta) - z \cdot b(\theta)',\ s(\theta))$$

is a polynomial of positive degree so let z_0 be a root. We have

$$\mathrm{GCD}(a(\theta) - z_0 \cdot b(\theta)', s(\theta)) \mid \mathrm{GCD}(a(\theta) - z_0 \cdot b(\theta)', b(\theta)) \qquad (12.21)$$

and the left side (the divisor) is nontrivial, hence

$$\mathrm{res}_\theta(a(\theta) - z_0 \cdot b(\theta)', b(\theta)) = 0.$$

It follows that z_0 is one of the c_i, say $c_1 = z_0$. But then the right side of (12.21) is $v_1(\theta)$ and we have a nontrivial common divisor of $s(\theta)$ and $v_1(\theta)$. Thus $\mathrm{GCD}(s(\theta), r(\theta)) \ne 1$ which contradicts the fact that $b(\theta)$ is square-free. This contradiction proves that $\deg(s(\theta)) = 0$. Since $b(\theta)$ is monic and $v_i(\theta)$ $(1 \le i \le m)$ are monic (by their GCD definition), we conclude that $s(\theta) = 1$ and

$$b(\theta) = \prod_{i=1}^{m} v_i(\theta).$$

Now define

$$\bar{a}(\theta) = \sum_{i=1}^{m} c_i \cdot v_i(\theta)' \prod_{j \neq i} v_j(\theta).$$

Since

$$b(\theta)' = \sum_{i=1}^{m} v_i(\theta)' \prod_{j \neq i} v_j(\theta)$$

we have, for $1 \le k \le m$,

$$\bar{a}(\theta) - c_k \cdot b(\theta)' = \sum_{i=1}^{m} (c_i - c_k) \cdot v_i(\theta)' \prod_{j \neq i} v_j(\theta)$$

from which it follows that

$$v_k(\theta) \mid (\bar{a}(\theta) - c_k \cdot b(\theta)').$$

By definition,

$$v_k(\theta) \mid (a(\theta) - c_k \cdot b(\theta)')$$

so we can conclude that $v_k(\theta) \mid (\bar{a}(\theta) - a(\theta))$. This holds for each k. Furthermore, since $\mathrm{GCD}(v_i(\theta), v_j(\theta)) = 1$ for each $i \neq j$ we have

$$b(\theta) \mid (\bar{a}(\theta) - a(\theta)).$$

Since $v_i(\theta)\ (1 \le i \le m)$ are all monic, we know from Theorem 12.2 that $\deg(v_i(\theta)') < \deg(v_i(\theta))$ and therefore $\deg(\bar{a}(\theta)) < \deg(b(\theta))$. Since also $\deg(a(\theta)) < \deg(b(\theta))$ we have $\deg(\bar{a}(\theta) - a(\theta)) < \deg(b(\theta))$ and thus $\bar{a}(\theta) - a(\theta) = 0$, i.e. $a(\theta) = \bar{a}(\theta)$. We have proved that

$$\frac{a(\theta)}{b(\theta)} = \sum_{i=1}^{m} c_i \frac{v_i(\theta)'}{v_i(\theta)}$$

and clearly $\int \frac{a(\theta)}{b(\theta)}$ is elementary in this case.

•

We remark that the Lazard/Rioboo/Trager method given in the last chapter can also be used in this case to compute the $v_i(\theta)$ in part (ii) of Theorem 12.7.

Logarithmic Extension: Integration of the Polynomial Part

Consider now the case where we are integrating the polynomial part in a logarithmic extension. For an integrand $f \in \mathrm{K}(x, \theta_1, \ldots, \theta_n)$ where the last extension $\theta = \theta_n$ is logarithmic over $\mathrm{F}_{n-1} = \mathrm{K}(x, \theta_1, \ldots, \theta_{n-1})$ (specifically, $\theta = \log(u)$ where $u \in \mathrm{F}_{n-1}$), the polynomial part is a polynomial $p(\theta) \in \mathrm{F}_{n-1}[\theta]$. Let $l = \deg(p(\theta))$ and let

$$p(\theta) = p_l \theta^l + p_{l-1}\theta^{l-1} + \cdots + p_0$$

where $p_i \in \mathrm{F}_{n-1}$ $(0 \le i \le l)$. By Liouville's Principle, if $\int p(\theta)$ is elementary then

$$p(\theta) = v_0(\theta)' + \sum_{i=1}^{m} c_i \frac{v_i(\theta)'}{v_i(\theta)}$$

where $c_i \in \overline{\mathrm{K}}$ (the algebraic closure of K) for $1 \le i \le m$ and $v_i(\theta) \in \overline{\mathrm{F}}_{n-1}(\theta)$ (the field $\mathrm{F}_{n-1}(\theta)$ with its constant field extended to $\overline{\mathrm{K}}$) for $0 \le i \le m$. Arguing as in the proof of Liouville's Principle (Theorem 12.5), we can conclude that $v_0(\theta) \in \overline{\mathrm{F}}_{n-1}[\theta]$ because a denominator dependent on θ would fail to disappear upon differentiation. Similarly, $v_i(\theta)$ $(1 \le i \le m)$ must be independent of θ. Therefore,

$$p(\theta) = v_0(\theta)' + \sum_{i=1}^{m} c_i \frac{v_i'}{v_i} \tag{12.22}$$

where $c_i \in \overline{\mathrm{K}}$ $(1 \le i \le m)$, $v_0(\theta) \in \overline{\mathrm{F}}_{n-1}[\theta]$, and $v_i \in \overline{\mathrm{F}}_{n-1}$ $(1 \le i \le m)$. Let $k = \deg(v_0(\theta))$ and let

$$v_0(\theta) = q_k\theta^k + q_{k-1}\theta^{k-1} + \cdots + q_0$$

where $q_i \in \overline{\mathrm{F}}_{n-1}$ $(0 \le i \le k)$. From Theorem 12.2 we know that $v_0(\theta)' \in \overline{\mathrm{F}}_{n-1}[\theta]$ and $\deg(v_0(\theta)') = k-1$ if $q_k \in \overline{\mathrm{K}}$ (otherwise $\deg(v_0(\theta)') = k$). It follows that the highest degree possible for $v_0(\theta)$ is $k = l+1$. Equation (12.22) takes the form

$$p_l\theta^l + p_{l-1}\theta^{l-1} + \cdots + p_0 = (q_{l+1}\theta^{l+1} + q_l\theta^l + \cdots + q_0)' + \sum_{i=1}^{m} c_i \frac{v_i'}{v_i}$$

where $p_i \in \mathrm{F}_{n-1}$ $(0 \le i \le l)$, $q_{l+1} \in \overline{\mathrm{K}}$, $q_i \in \overline{\mathrm{F}}_{n-1}$ $(0 \le i \le l)$. Applying the differentiation and equating coefficients of like powers of θ (which is a transcendental symbol over F_{n-1}) yields the following system of equations

$$0 = q_{l+1}',$$

$$p_l = (l+1)\cdot q_{l+1}\theta' + q_l',$$

$$p_{l-1} = l\cdot q_l\theta' + q_{l-1}',$$

$$\cdots$$

$$p_1 = 2q_2\theta' + q_1',$$

$$p_0 = q_1\theta' + \bar{q}_0',$$

where in the last equation we have introduced the new indeterminate $\bar{q}_0 = q_0 + \sum_{i=1}^{m} c_i \log(v_i)$.

The given coefficients are $p_i \in \mathrm{F}_{n-1}$ $(0 \le i \le l)$ and we must determine solutions for the $q_{l+1} \in \overline{\mathrm{K}}$, $q_i \in \overline{\mathrm{F}}_{n-1}$ $(1 \le i \le l)$, and $\bar{q}_0 \in \overline{\mathrm{F}}_{n-1}(\log(v_1), \ldots, \log(v_m))$. Note that in equation (12.22), m, c_i, and v_i $(1 \le i \le m)$ are unknowns, so the restriction on $\bar{q}_0$ simply states that new logarithmic extensions of $\overline{\mathrm{F}}_{n-1}$ are allowed. In contrast, q_i $(1 \le i \le l)$ must lie strictly in the field $\overline{F}_{n-1}$. These restrictions must be observed as we apply integration to solve the equations.

We can proceed to solve the equations as follows. Applying integration to both sides of the first equation yields

$$q_{l+1} = b_{l+1}$$

where $b_{l+1} \in \overline{K}$ is an arbitrary constant of integration. Substituting for q_{l+1} in the second equation and applying integration to both sides yields

$$\int p_l = (l+1) b_{l+1} \cdot \theta + q_l .$$

The integration procedure is now invoked recursively to integrate $p_l \in F_{n-1}$. In order to solve this equation for $b_{l+1} \in \overline{K}$ and $q_l \in \overline{F}_{n-1}$, the following conditions must hold for $\int p_l$:

(i) the integral is elementary;

(ii) there is at most one log extension of $\overline{F}_{n-1}$ appearing in the integral;

(iii) if a log extension of $\overline{F}_{n-1}$ appears in the integral then it must be the particular one $\theta = \log(u)$.

If one of these conditions fails to hold then the equation has no solution and we can conclude that $\int p(\theta)$ is not elementary. If conditions (i)-(iii) hold then

$$\int p_l = c_l \theta + d_l$$

for some $c_l \in \overline{K}$ and $d_l \in \overline{F}_{n-1}$. It follows that the desired solution is

$$b_{l+1} = \frac{c_l}{l+1}, \quad q_l = d_l + b_l$$

where $b_l \in \overline{K}$ is an arbitrary constant of integration. Next, substituting for q_l in the third equation and rearranging yields

$$p_{l-1} - l \cdot d_l \theta' = l \cdot b_l \theta' + q_{l-1}'$$

or, by integrating both sides,

$$\int (p_{l-1} - l \cdot d_l \frac{u'}{u}) = l \cdot b_l \theta + q_{l-1} .$$

The integrand on the left consists of known functions lying in the field $\overline{F}_{n-1}$ so we invoke the integration procedure recursively. Comparing to the right hand side, we see that the above conditions (i)-(iii) must hold for this latest integral. Otherwise, we can conclude that $\int p(\theta)$ is not elementary. If conditions (i)-(iii) hold then

$$\int (p_{l-1} - l \cdot d_l \frac{u'}{u}) = c_{l-1} \theta + d_{l-1}$$

for some $c_{l-1} \in \overline{K}$ and $d_{l-1} \in \overline{F}_{n-1}$. It follows that the desired solution is

$$b_l = \frac{c_{l-1}}{l}, \quad q_{l-1} = d_{l-1} + b_{l-1}$$

where $b_{l-1} \in \overline{\mathrm{K}}$ is an arbitrary constant of integration.

The above solution process can be continued for each equation up to the penultimate equation, when a solution has been determined of the form

$$b_2 = \frac{c_1}{2}, \quad q_1 = d_1 + b_1$$

where $b_1 \in \overline{\mathrm{K}}$ is an arbitrary constant of integration. Then substituting for q_1 in the last equation, rearranging, and applying integration yields

$$\int (p_0 - d_1 \frac{u'}{u}) = b_1 \theta + \bar{q}_0.$$

This time, the only condition on the integral is that it must be elementary. If not, we can conclude that $\int p(\theta)$ is not elementary. If it is elementary, say

$$\int (p_0 - d_1 \frac{u'}{u}) = d_0,$$

then b_1 (possibly zero) is the coefficient in d_0 of $\theta = \log(u)$ and

$$\bar{q}_0 = d_0 - b_1 \log(u).$$

In this case, the arbitrary constant of integration is the constant of integration for $\int p(\theta)$ so we leave it unspecified. This completes the integral of the polynomial part which takes the form

$$\int p(\theta) = b_{l+1}\theta^{l+1} + q_l \theta^l + \cdots + q_1 \theta + \bar{q}_0.$$

Example 12.9. The integral

$$\int \log(x)$$

has integrand

$$f(\theta) = \theta \in \mathbf{Q}(x, \theta)$$

where $\theta = \log(x)$. If the integral is elementary then

$$\int \theta = b_2 \theta^2 + q_1 \theta + \bar{q}_0$$

where the equations to be satisfied are

$$0 = b_2{}',$$

$$1 = 2b_2\theta' + q_1{}',$$

$$0 = q_1\theta' + \bar{q}_0{}'.$$

With b_2 a nondetermined constant, we consider the integrated form of the second equation

$$\int 1 = 2b_2\theta + q_1.$$

Since $\int 1 = x + b_1$ (where b_1 is an arbitrary constant), we must have

$$b_2 = 0,\ \ q_1 = x + b_1.$$

Then the third equation becomes

$$0 = (x + b_1)\theta' + \bar{q}_0{}'$$

or

$$-x\theta' = b_1\theta' + \bar{q}_0{}' .$$

Substituting $\theta' = \dfrac{1}{x}$ on the left hand side, and integrating, yields

$$\int(-1) = b_1\theta + \bar{q}_0 .$$

Since $\int(-1) = -x$ (we ignore the constant of integration in this final step), we must have

$$b_1 = 0,\ \ \bar{q}_0 = -x.$$

Hence,

$$\int \log(x) = x\log(x) - x .$$

●

Example 12.10. The integral

$$\int \log(\log(x))$$

has integrand

$$f(\theta_2) = \theta_2 \in \mathbf{Q}(x_1, \theta_1, \theta_2)$$

where $\theta_1 = \log(x)$ and $\theta_2 = \log(\theta_1)$. If the integral is elementary then

$$\int \theta_2 = b_2\theta_2^2 + q_1\theta_2 + \bar{q}_0$$

where the equations to be satisfied are

$$0 = b_2' ,$$

$$1 = 2b_2\theta_2' + q_1' ,$$

$$0 = q_1\theta_2' + \bar{q}_0{}' .$$

With b_2 an undetermined constant, we consider the integrated form of the second equation

$$\int 1 = 2b_2\theta_2 + q_1.$$

Since $\int 1 = x + b_1$ (where b_1 is an arbitrary constant), we must have

$$b_2 = 0,\ q_1 = x + b_1.$$

Then the third equation becomes

$$0 = (x + b_1)\theta_2' + \bar{q}_0'$$

or

$$-x\theta_2' = b_1\theta_2' + \bar{q}_0'\,.$$

Since $\theta_2' = \dfrac{\theta_1'}{\theta_1} = \dfrac{1}{x\log(x)}$, substituting for θ_2' on the left hand side and integrating yields

$$\int \frac{-1}{\log(x)} = b_1\theta_2 + \bar{q}_0\,.$$

From Example 12.6, we know that the integral appearing here is not elementary. Hence, we can conclude that $\int \log(\log(x))$ is not elementary. •

Example 12.11. Consider the problem

$$\int \left[x\left[\frac{\frac{1}{2}}{(x+\frac{1}{2})\log(x+\frac{1}{2})} + \frac{1+2\log(x)}{x} \right]^2 + \frac{(\frac{1}{2}\log(x+\frac{1}{2})-x)\log^2(x) - \frac{1}{4}x}{(x+\frac{1}{2})^2\log^2(x+\frac{1}{2})} \right.$$

$$\left. + \frac{((x^2+x+1)\log(x+\frac{1}{2}) + x^2 - 1)\log(x)}{(x+\frac{1}{2})^2} + \left[x - \frac{1}{x}\right]\frac{\log(x+\frac{1}{2})}{x+\frac{1}{2}} \right].$$

Letting $\theta_1 = \log(x+\frac{1}{2})$ and $\theta_2 = \log(x)$, the integrand can be considered in the form

$$f(\theta_2) \in \mathbf{Q}(x, \theta_1, \theta_2).$$

Specifically, the integrand can be expressed as the following polynomial (with respect to θ_2)

$$f(\theta_2) = \left[\frac{\frac{1}{2}\theta_1 - x}{(x+\frac{1}{2})^2\theta_1^2} + \frac{4}{x} \right]\theta_2^2 + \left[\frac{(x^2+x+1)\theta_1 + x^2 - 1}{(x+\frac{1}{2})^2} + \frac{2}{(x+\frac{1}{2})\theta_1} + \frac{4}{x} \right]\theta_2$$

$$+ \left[\frac{(x-\frac{1}{x})\theta_1}{x+\frac{1}{2}} + \frac{1}{(x+\frac{1}{2})\theta_1} + \frac{\frac{1}{2}}{x(x+\frac{1}{2})} + \frac{1}{x+\frac{1}{2}} \right].$$

Denoting the coefficient of θ_2^k by p_k for $k = 0, 1, 2$, we know that if the integral is elementary then

$$\int f(\theta_2) = b_3\theta_2^3 + q_2\theta_2^2 + q_1\theta_2 + \bar{q}_0$$

where the equations to be satisfied are

$$0 = b_3' ,$$

$$p_2 = 3b_3\theta_2' + q_2' ,$$

$$p_1 = 2q_2\theta_2' + q_1' ,$$

$$p_0 = q_1\theta_2' + \bar{q}_0' .$$

With b_3 an undetermined constant, we consider the integrated form of the second equation

$$\int p_2 = 3b_3\theta_2 + q_2.$$

Recursively applying the integration algorithm to integrate $p_2 \in \mathbf{Q}(x, \theta_1)$ yields

$$\int p_2 = 4\log(x) + \frac{x}{(x+\frac{1}{2})\log(x+\frac{1}{2})} .$$

Thus $\int p_2$ is elementary and moreover the only new log extension arising is the particular one $\theta_2 = \log(x)$. Hence we conclude that

$$b_3 = \frac{4}{3},\ q_2 = \frac{x}{(x+\frac{1}{2})\theta_1} + b_2$$

(where b_2 is an arbitrary constant). Then the third equation becomes

$$\int (p_1 - 2\frac{x}{(x+\frac{1}{2})\theta_1} \cdot \frac{1}{x}) = 2b_2\theta_2 + q_1.$$

Recursively applying the integration algorithm to the left hand side integrand, which lies in the field $\mathbf{Q}(x, \theta_1)$, yields the result

$$4 \cdot \log(x) + \frac{x^2-1}{x+\frac{1}{2}} \log(x+\frac{1}{2}) .$$

This integral is elementary, with the only new log extension being $\theta_2 = \log(x)$, so we conclude that

$$b_2 = 2,\ q_1 = \frac{(x^2-1)\theta_1}{x+\frac{1}{2}} + b_1$$

(where b_1 is an arbitrary constant). Finally, the last equation becomes

$$\int (p_0 - \frac{(x^2-1)\theta_1}{x+\frac{1}{2}} \cdot \frac{1}{x}) = b_1\theta_2 + \bar{q}_0 \,.$$

Recursively applying the integration algorithm to the left hand side integrand, which lies in the field $\mathbf{Q}(x_1, \theta_1)$, yields the result

$$\log(x) + \log\left(\log(x+\frac{1}{2})\right).$$

Hence we conclude that

$$b_1 = 1, \;\; \bar{q}_0 = \log(\theta_1)\,.$$

Putting it all together, the original integral is elementary and it takes the form

$$\int f = \frac{4}{3}\log^3(x) + \left(\frac{x}{(x+\frac{1}{2})\log(x+\frac{1}{2})} + 2\right)\log^2(x)$$

$$+ \left(\frac{(x^2-1)\log(x+\frac{1}{2})}{x+\frac{1}{2}} + 1\right)\log(x) + \log(\log(x+\frac{1}{2}))\,.$$

•

12.7. THE RISCH ALGORITHM FOR EXPONENTIAL EXTENSIONS

Suppose that the last extension θ is exponential, specifically that $\theta' / \theta = u'$ where $u \in \mathrm{F}_{n-1}$. Our problem is to compute $\int f(\theta)$ and we are given

$$f(\theta) = \frac{p(\theta)}{q(\theta)} \in \mathrm{F}_{n-1}(\theta)$$

where $p(\theta), q(\theta) \in \mathrm{F}_{n-1}[\theta]$, $\mathrm{GCD}(p(\theta), q(\theta)) = 1$, and $q(\theta)$ is monic. It is possible to proceed as in the logarithmic case by applying Euclidean division, yielding polynomials $s(\theta), r(\theta) \in \mathrm{F}_{n-1}[\theta]$ such that

$$p(\theta) = q(\theta)s(\theta) + r(\theta) \;\text{ with }\; r(\theta) = 0 \;\text{ or } \deg(r(\theta)) < \deg(q(\theta)).$$

This gives

$$f(\theta) = s(\theta) + \frac{r(\theta)}{q(\theta)}. \tag{12.23}$$

However, Hermite's method applied to integrate the rational part $r(\theta)/q(\theta)$ encounters the following difficulty. The square-free factorization of the denominator is

$$q(\theta) = \prod_{i=1}^{k} q_i(\theta)^i$$

where each $q_i(\theta)$ $(1 \le i \le k)$ is monic and square-free as an element of $\mathrm{F}_{n-1}[\theta]$. Now although for each i, $\mathrm{GCD}(q_i(\theta), \frac{d}{d\theta} q_i(\theta)) = 1$, it does not necessarily follow that

$\mathrm{GCD}(q_i(\theta), q_i(\theta)') = 1$ (where, as usual, $'$ denotes differentiation with respect to x). The latter condition is crucial in order to proceed with Hermite's method. To see that this condition may fail, consider the case $q_i(\theta) = \theta$ in which case (since $\theta' = u'\theta$ for some $u \in \mathrm{F}_{n-1}$)

$$\mathrm{GCD}(q_i(\theta), q_i(\theta)') = \mathrm{GCD}(\theta, u'\theta) = \theta.$$

Fortunately, by Theorem 12.3 we know that $q_i(\theta)$ divides $q_i(\theta)'$ only if $q_i(\theta)$ is a monomial, so the problem can be surmounted if we can remove monomial factors from the denominator.

We proceed to modify the decomposition (12.23) as follows. Let the denominator be of the form

$$q(\theta) = \theta^l\, \bar{q}(\theta) \;\text{ where }\; \theta \nmid \bar{q}(\theta)$$

(i.e. define l to be the lowest degree of all the terms in the polynomial $q(\theta)$). If $l = 0$ then (12.23) is already acceptable. Otherwise, apply the method of Theorem 2.6 to compute polynomials $\bar{r}(\theta), w(\theta) \in \mathrm{F}_{n-1}[\theta]$ such that

$$\bar{r}(\theta)\theta^l + w(\theta)\, \bar{q}(\theta) = r(\theta)$$

where $\deg(\bar{r}(\theta)) < \deg(\bar{q}(\theta))$ and $\deg(w(\theta)) < l$. Dividing both sides of this equation by $q(\theta)$ and replacing the term $\dfrac{r(\theta)}{q(\theta)}$ in equation (12.23) by the new expression arising here yields

$$f(\theta) = s(\theta) + \frac{w(\theta)}{\theta^l} + \frac{\bar{r}(\theta)}{\bar{q}(\theta)}\,.$$

We write the latter equation as the new decomposition

$$f(\theta) = \bar{s}(\theta) + \frac{\bar{r}(\theta)}{\bar{q}(\theta)}$$

where $\theta \nmid \bar{q}(\theta)$, $\deg(\bar{r}(\theta)) < \deg(\bar{q}(\theta))$, and where

$$\bar{s}(\theta) = s(\theta) + \theta^{-l}\cdot w(\theta).$$

Letting $w(\theta) = \sum_{i=0}^{l-1} w_i\theta^i$ and $s(\theta) = \sum_{i=0}^{m} s_i\theta^i$, we see that $\bar{s}(\theta)$ is an "extended polynomial" of the form

$$\bar{s}(\theta) = \sum_{i=-l}^{-1} w_{l+i}\, \theta^i + \sum_{i=0}^{m} s_i\theta^i.$$

The integration problem is now

$$\int f(\theta) = \int \bar{s}(\theta) + \int \frac{\bar{r}(\theta)}{\bar{q}(\theta)}\,.$$

Hermite's method is applicable for the integration of the rational part appearing here. The integration of the "polynomial part" is nontrivial, but the appearance of negative powers of θ does not increase the complexity. This is not surprising because if $\theta = \exp(u)$ then $\theta^{-1} = \exp(-u)$ which is simply an exponential function again.

Exponential Extension: Integration of the Rational Part

Continuing with Hermite's method for the integrand $\dfrac{\bar{r}(\theta)}{\bar{q}(\theta)}$, we compute the square-free factorization of the denominator $\bar{q}(\theta) \in F_{n-1}[\theta]$

$$\bar{q}(\theta) = \prod_{i=1}^{k} q_i(\theta)^{\,i}$$

where each $q_i(\theta)\ (1 \le i \le k)$ is monic and square-free, $\mathrm{GCD}(q_i(\theta), q_j(\theta)) = 1$ for $i \ne j$, $\deg(q_k(\theta)) > 0$, and furthermore, $\theta \nmid q_i(\theta)$ for $1 \le i \le k$. The following theorem proves that $\mathrm{GCD}(q_i(\theta), q_i(\theta)') = 1$ for each nontrivial factor $q_i(\theta)$.

Theorem 12.8. Let F be a differential field with differential operator $'$ satisfying $x' = 1$, where $x \in$ F. Let F(θ) be a differential extension field of F having the same subfield of constants such that θ is transcendental and exponential over F, specifically, $\theta'/\,\theta = u'$ where $u \in$ F. Let $a(\theta) \in$ F$[\theta]$ be a polynomial in the symbol θ such that $\deg(a(\theta)) > 0$, $a(\theta)$ is monic, $\theta \nmid a(\theta)$, and

$$\mathrm{GCD}(a(\theta), \frac{d}{d\theta}\, a(\theta)) = 1$$

(i.e. $a(\theta)$ is square-free as an element of F$[\theta]$). Then

$$\mathrm{GCD}(a(\theta), a(\theta)') = 1$$

where the latter GCD operation is also in the domain F$[\theta]$.

Proof: From Theorem 12.3 we know that $a(\theta)' \in$ F$[\theta]$. Let the monic polynomial $a(\theta) \in$ F$[\theta]$ be factored over its splitting field F_a into the form

$$a(\theta) = \prod_{i=1}^{N} (\theta - a_i)$$

where $a_i \in F_a\ (1 \le i \le N)$ are all distinct (because $a(\theta)$ is square-free in F$[\theta]$) and $a_i \ne 0\ (1 \le i \le N)$. Then

$$a(\theta)' = \sum_{i=1}^{N} (u'\theta - a_i') \prod_{j \ne i} (\theta - a_j)\,.$$

Now for any particular factor $\theta - a_i$ of $a(\theta)$, the expression for $a(\theta)'$ has $N-1$ terms which are divisible by $\theta - a_i$ and one term which is not divisible by $\theta - a_i$ unless it happens that

$$(\theta - a_i) \mid (u'\theta - a_i')\,. \tag{12.24}$$

Suppose that the divisibility relationship (12.24) holds for some i. Then since both dividend and divisor are linear polynomials in θ and considering the leading coefficients, we must have

$$u'\theta - a_i' = u'(\theta - a_i)$$

whence

$$a_i' = u' a_i .$$

Since $a_i \neq 0$, this implies that a_i is an exponential of $u \in \mathrm{F}$. In particular, the expression $\dfrac{\theta}{a_i}$ in the differential extension field $\mathrm{F}_a(\theta)$ satisfies

$$\left(\frac{\theta}{a_i}\right)' = \frac{a_i\theta' - a_i'\theta}{a_i^2} = \frac{a_i u'\theta - u' a_i \theta}{a_i^2} = 0$$

which implies that $\dfrac{\theta}{a_i}$ is a constant in $\mathrm{F}_a(\theta)$, whence

$$\frac{\theta}{a_i} = c \in \mathrm{F}_a \quad \text{(cf. Exercise 12.5)} .$$

But then $\theta = c a_i$ where $c, a_i \in \mathrm{F}_a$ which contradicts the assumption that θ is transcendental over F. This proves that the divisibility relationship (12.24) does not hold for any i, completing the proof that $a(\theta)$ and $a(\theta)'$ have no common factors.

●

Hermite's method applied to the integrand $\dfrac{\bar{r}(\theta)}{\bar{q}(\theta)}$ now proceeds exactly as in the logarithmic case, yielding the reduction

$$\int \frac{\bar{r}(\theta)}{\bar{q}(\theta)} = \frac{c(\theta)}{d(\theta)} + \int \frac{a(\theta)}{b(\theta)}$$

where $a(\theta), b(\theta), c(\theta), d(\theta) \in \mathrm{F}_{n-1}[\theta]$, $\deg(a(\theta)) < \deg(b(\theta))$, $\theta \nmid b(\theta)$, and with $b(\theta)$ monic and square-free. To complete the integration of the rational part, the Rothstein/Trager method for the integral of $\dfrac{a(\theta)}{b(\theta)}$ is almost the same as in the logarithmic case. Specifically, we compute the resultant

$$R(z) = \mathrm{res}_\theta(a(\theta) - z \cdot b(\theta)', b(\theta)) \in \mathrm{F}_{n-1}[z].$$

Then (see Theorem 12.9 below), $\displaystyle\int \frac{a(\theta)}{b(\theta)}$ is elementary if and only if

$$R(z) = S \cdot \bar{R}(z) \in \mathrm{F}_{n-1}[z]$$

where $\bar{R}(z) \in \mathrm{K}[z]$ and $S \in \mathrm{F}_{n-1}$. Therefore we compute $\bar{R}(z) = \mathrm{pp}(R(z))$, the primitive part of $R(z)$ as a polynomial in $\mathrm{F}_{n-1}[z]$. If any of the coefficients in $\bar{R}(z)$ is nonconstant then there does not exist an elementary integral. Otherwise, let c_i $(1 \le i \le m)$ be the distinct roots of $\bar{R}(z)$ in its splitting field $\mathrm{K}_{\bar{R}}$ and define $v_i(\theta)$ $(1 \le i \le m)$ by

$$v_i(\theta) = \mathrm{GCD}(a(\theta) - c_i \cdot b(\theta)', b(\theta)) \in \mathrm{F}_{n-1}(c_1, \ldots, c_m)[\theta].$$

Then

$$\int \frac{a(\theta)}{b(\theta)} = -\left[\sum_{i=1}^{m} c_i \deg(v_i(\theta)) \right] u + \sum_{i=1}^{m} c_i \log(v_i(\theta))$$

where $\frac{\theta'}{\theta} = u'$ (i.e. $\theta = \exp(u)$). Note that unlike the case where θ was logarithmic, in this case the expression for $\int \frac{a(\theta)}{b(\theta)}$ contains not only log terms but also an additional term in u. For an explanation of this, see the discussion preceding Theorem 12.9 below.

Example 12.12. Consider one of the integrals mentioned in the introduction to this chapter,

$$\int \frac{1}{\exp(x)+1} \, .$$

This has integrand

$$f(\theta) = \frac{1}{\theta + 1} \in \mathbf{Q}(x, \theta)$$

where $\theta = \exp(x)$. Applying the Rothstein/Trager method, we compute

$$R(z) = \mathrm{res}_\theta(1 - z\theta, \theta + 1) = -1 - z \in \mathbf{Q}(x)[z] \, .$$

Since

$$\bar{R}(z) = \mathrm{pp}(R(z)) = 1 + z$$

has constant coefficients, the integral is elementary. Specifically,

$$c_1 = -1, \;\; v_1(\theta) = \mathrm{GCD}(1 + \theta, \theta + 1) = \theta + 1,$$

and

$$\int \frac{1}{\exp(x)+1} = -c_1 \deg(v_1(\theta))\, x + c_1 \log(v_1(\theta)) = x - \log(\exp(x) + 1) \, .$$

•

Example 12.13. The integral

$$\int \frac{x}{\exp(x)+1}$$

is also one that was mentioned in the introduction. This time, we have the integrand

$$f(\theta) = \frac{x}{\theta + 1} \in \mathbf{Q}(x, \theta)$$

where $\theta = \exp(x)$. Applying the Rothstein/Trager method, we compute

$$R(z) = \mathrm{res}_\theta(x - z\theta, \theta + 1) = -x - z \in \mathbf{Q}(x)[z] \, .$$

Since

$$\bar{R}(z) = \mathrm{pp}(R(z)) = x + z$$

has a nonconstant coefficient, we conclude that the integral is not elementary. ●

Example 12.14. Consider the problem of Example 12.8 but this time let $\theta_1 = \log(x+1)$ and $\theta_2 = \exp(x^2)$. The integrand can be considered in the form

$$f(\theta_2) \in \mathbf{Q}(x, \theta_1, \theta_2).$$

The numerator and denominator of $f(\theta_2)$ are each of degree 4 in θ_2, and after normalization and Euclidean division it takes the form

$$f(\theta_2) = \frac{x}{x+1} + \frac{\dfrac{2}{x^2(x+1)}\,(x - (2x^3 + 2x^2 + x + 1)\theta_1)\theta_2^3}{(\theta_2^2 - \dfrac{1}{x^2}\,\theta_1^2)^2}\,.$$

The "polynomial part" with respect to θ_2 is simply a rational function in $\mathbf{Q}(x)$, as in Example 12.8, and we note that the monomial θ_2 is not a factor of the denominator of the "rational part". Proceeding with Hermite's method, we compute the square-free partial fraction expansion of the rational part, yielding

$$f(\theta_2) = \frac{x}{x+1} + \frac{r_1(\theta_2)}{b(\theta_2)} + \frac{r_2(\theta_2)}{(b(\theta_2))^2}$$

where

$$b(\theta_2) = \theta_2^2 - \frac{1}{x^2}\,\theta_1^2\,,$$

$$r_1(\theta_2) = \frac{2}{x^2(x+1)}\,(x - (2x^3 + 2x^2 + x + 1)\theta_1)\theta_2\,,$$

$$r_2(\theta_2) = \frac{2}{x^4(x+1)}\,(x\theta_1^2 - (2x^3 + 2x^2 + x + 1)\theta_1^3)\theta_2\,.$$

To apply Hermite reduction to the integral $\int \frac{r_2(\theta_2)}{(b(\theta_2))^2}$, equation (12.14) takes the form

$$s(\theta_2)(\theta_2^2 - \frac{1}{x^2}\theta_1^2) \; + \; t(\theta_2)(4x\theta_2^2 + \frac{2}{x^3}\theta_1^2 - \frac{2}{x^2(x+1)}\theta_1) = r_2(\theta_2)\,.$$

This equation has the solution

$$s(\theta_2) = 4\theta_1\theta_2, \quad t(\theta_2) = -\frac{1}{x}\theta_1\theta_2\,.$$

The Hermite reduction therefore yields

$$\int \frac{r_2(\theta_2)}{b(\theta_2)^2} = \frac{\frac{1}{x}\theta_1\theta_2}{\theta_2^2 - \frac{1}{x^2}\theta_1^2} + \int \frac{((2+\frac{1}{x^2})\theta_1 - \frac{1}{x(x+1)})\theta_2}{\theta_2^2 - \frac{1}{x^2}\theta_1^2} .$$

Combining this with the other terms of $f(\theta_2)$, we have reduced the problem to the following form

$$\int f(\theta_2) = \int \frac{x}{x+1} + \frac{\frac{1}{x}\theta_1\theta_2}{\theta_2^2 - \frac{1}{x^2}\theta_1^2} + \int \frac{\frac{1}{x^2(x+1)}(x - (2x^3+2x^2+x+1)\theta_1)\theta_2}{\theta_2^2 - \frac{1}{x^2}\theta_1^2} .$$

We now apply the Rothstein/Trager method to the third term appearing here. Denoting the numerator by $a(\theta_2)$, the resultant computation is

$$R(z) = \mathrm{res}_{\theta_2}(a(\theta_2) - z \cdot b(\theta_2)', b(\theta_2))$$

and after dividing out the content, we get

$$\bar{R}(z) = \mathrm{pp}(R(z)) = 4z^2 - 1 .$$

Since $\bar{R}(z)$ has constant coefficients, the integral is elementary. Specifically, we compute

$$c_1 = \frac{1}{2},\; c_2 = -\frac{1}{2},$$

$$v_1(\theta_2) = \mathrm{GCD}(a(\theta_2) - c_1(b(\theta_2))', b(\theta_2)) = \theta_2 + \frac{1}{x}\theta_1,$$

$$v_2(\theta_2) = \mathrm{GCD}(a(\theta_2) - c_2(b(\theta_2))', b(\theta_2)) = \theta_2 - \frac{1}{x}\theta_1,$$

and hence the integral appearing as the third term above is

$$-\left(c_1 \deg(v_1(\theta_2)) + c_2 \deg(v_2(\theta_2))\right) x^2 + c_1 \log(v_1(\theta_2)) + c_2 \log(v_2(\theta_2))$$

$$= -\left(\frac{1}{2} - \frac{1}{2}\right) x^2 + \frac{1}{2}\log(\theta_2 + \frac{1}{x}\theta_1) - \frac{1}{2}\log(\theta_2 - \frac{1}{x}\theta_1) .$$

Putting it all together, the original integral is elementary and it takes the form

$$\int f = x - \log(x+1) + \frac{\frac{1}{x}\log(x+1)\exp(x^2)}{\exp^2(x^2) - \frac{1}{x^2}\log^2(x+1)}$$

$$+ \frac{1}{2}\log(\exp(x^2) + \frac{1}{x}\log(x+1)) - \frac{1}{2}\log(\exp(x^2) - \frac{1}{x}\log(x+1)) .$$

Note that this expression for the integral is in a form different from the result obtained in Example 12.8, but the two results are equivalent modulo an arbitrary constant of integration. ●

Before proving that the Rothstein/Trager method works for the exponential case, let us examine why the expression for $\int \frac{a(\theta)}{b(\theta)}$ contains a term in u in addition to the log terms. The integrand $\frac{a(\theta)}{b(\theta)}$ is a proper rational expression but note that the derivative of a log term

$$(c_i \log(v_i(\theta)))' = c_i \frac{v_i(\theta)'}{v_i(\theta)}$$

has the property that $\deg(v_i(\theta)') = \deg(v_i(\theta))$ when $\theta = \exp(u)$. More specifically, $v_i(\theta)$ has the form

$$v_i(\theta) = \theta^{n_i} + \alpha_{n_i-1}\,\theta^{n_i-1} + \cdots + \alpha_0$$

where $n_i = \deg(v_i(\theta))$, and $v_i(\theta)'$ has the form

$$v_i(\theta)' = n_i u' \theta^{n_i} + \beta_{n_i-1}\,\theta^{n_i-1} + \cdots + \beta_0 .$$

If rather than the derivative of a pure log term we consider a modified term

$$(c_i \log(v_i(\theta)) - c_i n_i u)' = c_i \, \frac{v_i(\theta)' - n_i u' v_i(\theta)}{v_i(\theta)}$$

then the term of degree n_i in the numerator vanishes and the result is a proper rational expression. This modification to the log terms is precisely what is specified in the expression for $\int \frac{a(\theta)}{b(\theta)}$.

Theorem 12.9 (Rothstein/Trager Method – Exponential Case). Let F be a field of elementary functions with constant field K. Let θ be transcendental and exponential over F (i.e. $\theta'/\theta = u'$ for some $u \in$ F) and suppose that the transcendental elementary extension F(θ) has the same constant field K. Let $a(\theta)/b(\theta) \in$ F(θ) where $a(\theta), b(\theta) \in$ F[θ], GCD($a(\theta), b(\theta)$) = 1, $\deg(a(\theta)) < \deg(b(\theta))$, $\theta \nmid b(\theta)$, and with $b(\theta)$ monic and square-free. Then

(i) $\int \frac{a(\theta)}{b(\theta)}$ is elementary if and only if all the roots of the polynomial

$$R(z) = \mathrm{res}_\theta(a(\theta) - z \cdot b(\theta)', b(\theta)) \in \mathrm{F}[z]$$

are constants. (Equivalently, $R(z) = S \cdot \bar{R}(z)$ where $\bar{R}(z) \in$ K[z] and $S \in$ F.)

(ii) If $\int \frac{a(\theta)}{b(\theta)}$ is elementary then

$$\frac{a(\theta)}{b(\theta)} = g' + \sum_{i=1}^{m} c_i \, \frac{v_i(\theta)'}{v_i(\theta)} \tag{12.25}$$

where c_i $(1 \le i \le m)$ are the distinct roots of $R(z)$, $v_i(\theta)$ $(1 \le i \le m)$ are defined by

$$v_i(\theta) = \mathrm{GCD}(a(\theta) - c_i\, b(\theta)', b(\theta)) \in \mathrm{F}(c_1, \ldots, c_m)[\theta],$$

and where $g \in \mathrm{F}(c_1, \ldots, c_m)$ is defined by

$$g' = -\left(\sum_{i=1}^{m} c_i \deg(v_i(\theta)) \right) u'.$$

(iii) Let F^* be the minimal algebraic extension field of F such that $a(\theta)/b(\theta)$ can be expressed in the form (12.25) with constants $c_i \in \mathrm{F}^*$ and with $v_i(\theta) \in \mathrm{F}^*[\theta]$. Then $\mathrm{F}^* = \mathrm{F}(c_1, \ldots, c_m)$ where c_i $(1 \le i \le m)$ are the distinct roots of $R(z)$.

Proof: Suppose that $\int \frac{a(\theta)}{b(\theta)}$ is elementary. Then by Liouville's Principle,

$$\frac{a(\theta)}{b(\theta)} = v_0(\theta)' + \sum_{i=1}^{m} c_i \frac{v_i(\theta)'}{v_i(\theta)} \tag{12.26}$$

where $c_i \in \mathrm{K}^*$ and $v_i(\theta) \in \mathrm{F}^*(\theta)$ $(0 \le i \le m)$, where K^* denotes the minimal algebraic extension of K necessary to express the integral and F^* denotes F with its constant field extended to K^*. As in the proof of Theorem 12.7, we may assume that $c_i \in \mathrm{K}^*$ $(1 \le i \le m)$ are distinct nonzero constants and that $v_i(\theta) \in \mathrm{F}^*[\theta]$ $(1 \le i \le m)$ are polynomials which are square-free and pairwise relatively prime.

If $v_0(\theta) \in \mathrm{F}^*(\theta)$ is a rational function $v_0(\theta) = p(\theta)/q(\theta)$ with $p(\theta), q(\theta) \in \mathrm{F}^*[\theta]$, $\mathrm{GCD}(p(\theta), q(\theta)) = 1$, and if $q(\theta)$ contains a factor which is not a monomial then $v_0(\theta)'$ contains a factor in its denominator which is not square-free. (For a detailed argument about the form of the derivative, see the proof of Liouville's Principle.) Since $b(\theta)$ is square-free, we conclude that the denominator of $v_0(\theta)$ must be of the form $q(\theta) = h\theta^k$ for some $k \ge 0$ and $h \in \mathrm{F}^*$. In other words, $v_0(\theta)$ is of the form

$$v_0(\theta) = \sum_{j=-k}^{l} h_j \theta^j$$

for some $h_j \in \mathrm{F}^*$ $(-k \le j \le l)$. Then

$$v_0(\theta)' = \sum_{j=-k}^{l} \bar{h}_j \theta^j$$

where $\bar{h}_j \in \mathrm{F}^*$ and for $j \ne 0$, $\bar{h}_j \ne 0$ if $h_j \ne 0$ (see Theorem 12.3). Substituting this form into equation (12.26) and noting that $\deg(v_i(\theta)') = \deg(v_i(\theta))$, we find that if $l > 0$ then the right hand side of equation (12.26), when formed over a common denominator, will have a numerator of degree greater than the degrees of its denominator. Since the left hand side of equation (12.26) satisfies $\deg(a(\theta)) < \deg(b(\theta))$, we conclude that $l = 0$. Now if $k > 0$ then the right hand side of equation (12.26), when formed over a common denominator, has a denominator which is divisible by θ. (Note that even if $\theta \mid v_i(\theta)$ for some i, the reduced form of $\frac{v_i(\theta)'}{v_i(\theta)}$ has a denominator which is not divisible by θ.) Since the left hand side of equation

(12.26) has a denominator which is not divisible by θ, we conclude that $k = 0$.

We have shown that equation (12.26) takes the form

$$\frac{a(\theta)}{b(\theta)} = h_0' + \sum_{i=1}^{m} c_i \frac{v_i(\theta)'}{v_i(\theta)} \tag{12.27}$$

where $h_0 \in F^*$. Applying an argument similar to that presented in the proof of Theorem 11.7, we conclude that

$$b(\theta) \mid \prod_{j=1}^{m} v_j(\theta) \text{ and } \prod_{j=1}^{m} v_j(\theta) \mid b(\theta) .$$

Since $b(\theta)$ is monic, we may assume without loss of generality that $v_i(\theta)$ $(1 \le i \le m)$ are all monic and

$$b(\theta) = \prod_{j=1}^{m} v_j(\theta) .$$

Now the left hand side of equation (12.27) is a proper rational expression in $F^*(\theta)$ so the right hand side must be a proper rational expression. However, for each term in the sum we have $\deg(v_i(\theta)') = \deg(v_i(\theta))$. Let us write $h_0 \in F^*$ in the form

$$h_0 = g + h$$

for some $h \in F^*$, where g is as defined (up to an arbitrary constant) in the statement of the theorem. Then the right hand side of equation (12.27) becomes

$$h' + \sum_{i=1}^{m} c_i \left(\frac{v_i(\theta)'}{v_i(\theta)} - \deg(v_i(\theta))\, u' \right) .$$

Now each term in the latter sum is a proper rational expression (see the discussion preceding the statement of the theorem) and therefore the sum of these m terms is a proper rational expression. It follows that if h' is any nonzero element of F^* then the entire expression, when formed over a common denominator, will have a numerator of degree equal to the degree of its denominator. Hence we conclude that $h' = 0$.

We have now shown that if $\int \frac{a(\theta)}{b(\theta)}$ is elementary then equation (12.25) holds where $c_i \in K^*$ $(1 \le i \le m)$ are distinct nonzero constants, $v_i(\theta) \in F^*[\theta]$ $(1 \le i \le m)$ are monic, square-free, pairwise relatively prime polynomials, $b(\theta) = \prod_{j=1}^{m} v_j(\theta)$, and g' is as defined in the statement of the theorem. Applying an argument similar to that presented in the proof of Theorem 11.7 (it is found that the presence of the additional term g' does not complicate the argument), it can be shown that c_i $(1 \le i \le m)$ are the distinct roots of the polynomial $R(z)$ defined in part (i) of the statement of the theorem and $v_i(\theta)$ $(1 \le i \le m)$ are as defined in part (ii). We have thus proved part (ii) of the theorem and we have proved the "only if" case of part (i). We have also proved part (iii) since we assumed F^* to be the minimal algebraic extension of F and then proved that the roots of $R(z)$ must appear.

To prove the "if" case of part (i), suppose that all the roots of $R(z)$ are constants. Let c_i $(1 \le i \le m)$ be the distinct roots of $R(z)$ and define $v_i(\theta)$ $(1 \le i \le m)$ by

$$v_i(\theta) = \mathrm{GCD}(a(\theta) - c_i\, b(\theta)', b(\theta)) \in \mathrm{F}(c_1, \ldots, c_m)[\theta] .$$

The same argument as in the proof of Theorem 12.7 leads to the conclusion that

$$b(\theta) = \prod_{i=1}^{m} v_i(\theta)$$

where $v_i(\theta)$ are monic and $\mathrm{GCD}(v_i(\theta), v_j(\theta)) = 1$ for $i \ne j$. Now define

$$\bar{a}(\theta) = g' \cdot b(\theta) + \sum_{i=1}^{m} c_i\, v_i(\theta)' \prod_{j \ne i}^{m} v_j(\theta)$$

where g' is as defined in the statement of the theorem. Since

$$b(\theta)' = \sum_{i=1}^{m} v_i(\theta)' \prod_{j \ne i}^{m} v_j(\theta)$$

we have, for $1 \le k \le m$,

$$\bar{a}(\theta) - c_k \cdot b(\theta)' = g' \cdot b(\theta) + \sum_{i=1}^{m} (c_i - c_k) \cdot v_i(\theta)' \prod_{j \ne i}^{m} v_j(\theta)$$

from which it follows that

$$v_k(\theta) \mid (\bar{a}(\theta) - c_k\, b(\theta)').$$

By definition,

$$v_k(\theta) \mid (a(\theta) - c_k \cdot b(\theta)')$$

so we can conclude that $v_k(\theta) \mid (\bar{a}(\theta) - a(\theta))$. This holds for each k. Furthermore, since $\mathrm{GCD}(v_i(\theta), v_j(\theta)) = 1$ for $i \ne j$ we have

$$b(\theta) \mid (\bar{a}(\theta) - a(\theta)).$$

Using the definition of g', the formula defining $\bar{a}(\theta)$ takes the form

$$\bar{a}(\theta) = \sum_{i=1}^{m} c_i \prod_{j \ne i}^{m} v_j(\theta)\, (v_i(\theta)' - \deg(v_i(\theta))\, u'\, v_i(\theta))$$

from which it follows that $\deg(\bar{a}(\theta)) < \deg(b(\theta))$ (see the discussion preceding the statement of the theorem regarding cancellation of the leading term of $v_i(\theta)'$). Since also $\deg(a(\theta)) < \deg(b(\theta))$ we have $\deg(\bar{a}(\theta) - a(\theta)) < \deg(b(\theta))$ and thus $\bar{a}(\theta) - a(\theta) = 0$, i.e. $a(\theta) = \bar{a}(\theta)$. We have proved that

$$\frac{a(\theta)}{b(\theta)} = g' + \sum_{i=1}^{m} c_i \frac{v_i(\theta)'}{v_i(\theta)}$$

and clearly $\int \frac{a(\theta)}{b(\theta)}$ is elementary in this case.

●

As mentioned in the case of logarithmic extensions, the $v_i(\theta)$ in part (ii) can be computed using the Lazard/Rioboo/Trager method of the previous chapter.

Exponential Extension: Integration of the Polynomial Part

The "polynomial part" of an integrand $f \in K(x, \theta_1, \ldots, \theta_n)$, in the case where the last extension $\theta = \theta_n$ is exponential over $F_{n-1} = K(x, \theta_1, \ldots, \theta_{n-1})$, is an "extended polynomial"

$$\bar{p}(\theta) = \sum_{j=-k}^{l} p_j \theta^j, \quad p_j \in F_{n-1} \tag{12.28}$$

which may contain both positive and negative powers of θ. We now consider the problem of computing $\int \bar{p}(\theta)$ where $\theta = \exp(u)$ for some $u \in F_{n-1}$. By Liouville's Principle, if $\bar{p}(\theta)$ has an elementary integral then

$$\bar{p}(\theta) = v_0(\theta)' + \sum_{i=1}^{m} c_i \frac{v_i(\theta)'}{v_i(\theta)}$$

where $c_i \in \bar{K}$ (the algebraic closure of K) for $1 \le i \le m$ and $v_i(\theta) \in \bar{F}_{n-1}(\theta)$ (the field $F_{n-1}(\theta)$ with its constant field extended to $\bar{K}$) for $0 \le i \le m$. Arguing as in the proof of Liouville's Principle (Theorem 12.5), we may assume without loss of generality that each $v_i(\theta)$ $(1 \le i \le m)$ is either an element of $\bar{F}_{n-1}$ or else is monic and irreducible in $\bar{F}_{n-1}[\theta]$. Furthermore, we conclude that $v_0(\theta)$ cannot have a non-monomial factor in its denominator because such a factor would remain (to a higher power) in the denominator of the derivative. Similarly, $v_i(\theta)$ $(1 \le i \le m)$ cannot have non-monomial factors. It follows that each $v_i(\theta)$ $(1 \le i \le m)$ is either an element of $\bar{F}_{n-1}$ or else $v_i(\theta) = \theta$ (the only monic, irreducible monomial in $\bar{F}_{n-1}[\theta]$). However, if $v_i(\theta) = \theta$ then the corresponding term in the summation becomes

$$c_i \frac{v_i(\theta)'}{v_i(\theta)} = c_i\, u'$$

which can be absorbed into the term $v_0(\theta)'$. The conclusion is that if $\bar{p}(\theta)$ has an elementary integral then

$$\bar{p}(\theta) = \left(\sum_{j=-k}^{l} q_j \theta^j \right)' + \sum_{i=1}^{m} c_i \frac{v_i'}{v_i} \tag{12.29}$$

where $q_j \in \bar{F}_{n-1}$ $(-k \le j \le l)$, $c_i \in \bar{K}$ $(1 \le i \le m)$, and $v_i \in \bar{F}_{n-1}$ $(1 \le i \le m)$. The fact that the index of summation in the $v_0(\theta)$ term on the right hand side of equation (12.29) must have the same range $-k \le j \le l$ as the summation defining $\bar{p}(\theta)$ in equation (12.28) follows from Theorem 12.3.

Noting that

$$(q_j\theta^j)' = (q_j' + j\,u'q_j)\theta^j, \quad -k \le j \le l,$$

by equating coefficients of like powers of θ equations (12.28)-(12.29) yield the following system of equations:

$$p_j = q_j' + j\,u'q_j \ \text{ for } -k \le j \le -1 \text{ and } 1 \le j \le l,$$

$$p_0 = \bar{q}_0'$$

where we have introduced the new indeterminate $\bar{q}_0 = q_0 + \sum_{i=1}^{m} c_i \log(v_i)$. The given coefficients are $p_j \in \mathrm{F}_{n-1}$ $(-k \le j \le l)$ and we must determine solutions $q_j \in \bar{\mathrm{F}}_{n-1}$ $(j \ne 0)$ and $\bar{q}_0 \in \bar{\mathrm{F}}_{n-1}(\log(v_1), \ldots, \log(v_m))$. The solution for the case $j = 0$ is simply

$$\bar{q}_0 = \int p_0 .$$

If $\int p_0$ is not elementary then we can conclude that $\int \bar{p}(\theta)$ is not elementary. Otherwise, the desired form for $\bar{q}_0$ has been determined. For each $j \ne 0$, we must solve a differential equation for $q_j \in \bar{\mathrm{F}}_{n-1}$. This particular form of differential equation, namely

$$y' + f\,y = g$$

where the given functions are $f, g \in \mathrm{F}_{n-1}$ and we must determine a solution $y \in \bar{\mathrm{F}}_{n-1}$, is known as a *Risch differential equation.* At first glance, it might seem that we have replaced our original integration problem by a harder problem, that of solving a differential equation. However, since the solution to the differential equation is restricted to lie in the same field as the functions f and g (possibly with an extension of the constant field), it is possible to solve the Risch differential equation or else to prove that there is no solution of the desired form. If any of the Risch differential equations in the above system fails to have a solution then we can conclude that $\int \bar{p}(\theta)$ is not elementary. Otherwise

$$\int \bar{p}(\theta) = \sum_{j \ne 0} q_j\theta^j + \bar{q}_0 .$$

The Risch differential equation was initially studied by Risch [15] in 1969. The first algorithm for solving this equation can be found in this paper. Subsequent algorithms were given by Rothstein [21] in 1976, Davenport [10] in 1986 and Bronstein [4] in 1990. The latter article presents a simple, readable description of this problem and its solution. It also presents a step-by-step algorithm which determines whether the equation does indeed have an elementary solution, and if so, it computes the solution.

Example 12.15. The integral

$$\frac{\sqrt{\pi}}{2}\mathrm{erf}(x) = \int \exp(-x^2)$$

has integrand $f(\theta) = \theta \in \mathbf{Q}(x, \theta)$ where $\theta = \exp(-x^2)$. If the integral is elementary then

$$\int \theta = q_1 \theta$$

where $q_1 \in \mathbf{Q}(x)$ is a solution of the Risch differential equation

$$q_1' - 2x\,q_1 = 1\,.$$

Now suppose that this differential equation has a solution $q_1(x) = a(x)/b(x)$ with $\deg(b(x)) > 0$. Plugging this rational function into the differential equation, we find that the left hand side will be a rational function with a nontrivial denominator (because differentiation of $a(x)/b(x)$ yields a new rational function which, in reduced form, has a denominator of degree greater than $\deg(b(x))$). Since the right hand side has no denominator, we conclude that the only possible rational function solution must be a polynomial. Finally, if we postulate a polynomial solution $q_1(x)$ with $\deg(q_1(x)) = n$ then the left hand side will be a polynomial of degree $n+1$ (because differentiation decreases the degree while the product $2x\,q_1(x)$ increases the degree by one). Since the right hand side is a polynomial of degree zero, we conclude that the differential equation has no solution in $\mathbf{Q}(x)$.

Hence we conclude that $\int \exp(-x^2)$ is not elementary.

●

Example 12.16. The integral $\int x^x$ can be written with integrand

$$\exp(x \cdot \log(x)) = \theta_2 \in \mathbf{Q}(x, \theta_1, \theta_2)$$

where $\theta_1 = \log(x)$ and $\theta_2 = \exp(x\theta_1)$. If the integral is elementary then

$$\int \theta_2 = q_1 \cdot \theta_2$$

where $q_1 \in \mathbf{Q}(x, \theta_1)$. Differentiating both sides gives

$$\theta_2 = q_1' \theta_2 + q_1(\theta_1 + 1)\,\theta_2$$

which after equating coefficients simplifies to

$$1 = q_1' + (\theta_1 + 1)\,q_1\,.$$

Since θ_1 is transcendental over $\mathbf{Q}(x)$, equating coefficients of this equation as polynomials in θ_1 gives

$$1 = q_1' + q_1, \text{ and } 0 = q_1,$$

which has no solution. Thus $\int x^x$ is not elementary.

●

Example 12.17. The integral

$$\int \frac{(4x^2+4x-1)(\exp(x^2)+1)(\exp(x^2)-1)}{(x+1)^2}$$

has integrand

$$f(\theta) = \frac{4x^2+4x-1}{(x+1)^2}(\theta^2-1) \in \mathbf{Q}(x,\theta)$$

where $\theta = \exp(x^2)$. If the integral is elementary then

$$\int f(\theta) = q_2\theta^2 + \bar{q}_0$$

where the equations to be satisfied are

$$q_2' + 4x\,q_2 = \frac{4x^2+4x-1}{(x+1)^2},$$

$$\bar{q}_0' = -\frac{4x^2+4x-1}{(x+1)^2}.$$

The latter equation can be integrated to yield $\bar{q}_0$ which is elementary. The differential equation for $q_2 \in \mathbf{Q}(x)$ can be solved using Bronstein's algorithm yielding

$$q_2 = \frac{1}{x+1}.$$

Hence the original integral is elementary and it takes the form

$$\int f = \frac{1}{x+1}\exp^2(x^2) - \frac{(2x+1)^2}{x+1} + 4\log(x+1).$$

•

For a Risch algorithm which allows for extensions other than just exponential, logarithmic, or algebraic, we refer the reader to the work of Cherry [7, 8] which allows for extensions such as error functions and logarithmic integrals.

12.8. INTEGRATION OF ALGEBRAIC FUNCTIONS

As the reader has undoubtedly noticed in the previous sections, the algorithms for integration of exponential and logarithmic transcendental extensions differ only slightly from those required for integration of rational functions. However, as one can clearly see from the proof of Liouville's Principle (Theorem 12.5), the mathematics available in algebraic extensions is markedly different from that available in a transcendental extension. While the results of Liouville's Principle were identical for both types of extensions, the arguments had little in common in the accompanying proofs.

In this section we describe an algorithm, initially presented in the Ph.D. thesis of Trager [23] in 1984, for the integration of a function in an algebraic extension. This is the problem of integrating

$$\int \frac{p(x,y)}{q(x,y)} dx$$

with y algebraic over the function field K(x) with K the field of constants. While the algorithm follows more or less the same outline as that of the Rothstein/Trager approach for transcendental extensions, the mathematical tools differ greatly. Our description of this algorithm in this section will be little more than a summary. Many details are left out; more specifically, the computational algebraic geometry background which could easily comprise two long chapters, is not included.

Let $f \in K(x,y)$ with y algebraic over K(x). Let $F(x,y) \in K(x)[y]$ be an irreducible polynomial satisfying

$$F(x,y) = 0.$$

By a simple change of coordinates, if necessary, we may assume that F is monic in y and has coefficients from the domain K[x]. For example, if the algebraic equation is

$$F(x,y) = x \cdot y^4 - \frac{1}{x} \cdot y + 1 = 0$$

then this represents the same equation as

$$\hat{F}(x,\hat{y}) = \hat{y}^4 - x \cdot \hat{y} + x^3 = 0$$

where $\hat{y} = x \cdot y$.

Rational polynomials in a transcendental unknown z have a simple unique representation as $a(z)/b(z)$ with GCD$(a,b) = 1$. Normalizing the denominator to be monic creates a unique representative in each case. In the case of algebraic extensions, the representation issue is not so straightforward. There are, however, some natural simplifications. For $f \in K(x,y)$ with

$$f = \frac{p(x,y)}{q(x,y)}$$

we can combine denominators and numerators so that we can assume $p(x,y), q(x,y) \in K[x][y]$ (rather than K(x,y)). The irreducibility of $F(x,y)$, combined with the extended Euclidean algorithm, implies the existence of polynomials $s(x,y), t(x,y) \in K(x)[y]$ such that

$$s(x,y) \cdot q(x,y) + t(x,y) \cdot F(x,y) = 1.$$

Clearing denominators gives

$$\hat{s}(x,y) \cdot q(x,y) + \hat{t}(x,y) \cdot F(x,y) = r(x) \in K[x]$$

so that we can represent f by

$$f = \frac{p(x,y)}{q(x,y)} = \frac{\hat{s}(x,y) \cdot p(x,y)}{r(x)} = \frac{c(x,y)}{d(x)}.$$

If $\deg_y(F) = n$, we can further simplify our representation by assuming that $\deg_y(c) < n$, since otherwise we can simply divide F into c (both considered as polynomials in y) and achieve this representation.

For all intents and purposes, the above representation is natural and seems to provide a suitable framework. However, for the purposes of symbolic integration there is one major failing. Because of the logarithmic terms which we know exist in the integral (by Liouville's Principle), we wish to have knowledge of the poles of the integrand. In the transcendental case, such information was straightforward: the poles of a reduced rational function are the zeros of the denominator and vice versa. We would like to have the same property of our representation in the case of a rational function over an algebraic extension. Our present representation does not satisfy this criterion. For example, if

$$F(x,y) = y^4 - x^3 \tag{12.30}$$

then we represent $f = \frac{x^2}{y^2}$ as

$$f = \frac{x^2}{y^2} = \frac{y^2}{x}.$$

As such, f appears to have one pole of order 1 at the point $x = 0$. However, we have

$$f^2 = \frac{y^4}{x^2} = \frac{x^3}{x^2} = x.$$

Therefore f^2 (and hence also f) has no poles. What is desired is to represent any $f \in K(x,y)$ by $c(x,y)/d(x)$ where $c(x,y) \in K(x,y)$, $d(x) \in K[x]$, and where the poles of f are the same as the zeros of $d(x)$. Consequently $c(x,y)$ cannot have any pole.

The set

$$\{\ c(x,y) \in K(x,y) :\ c(x,y) \text{ has no poles }\}$$

is called the *integral closure* of $K[x,y]$ in $K(x,y)$. A basis for such a set is called an *integral basis*. Thus, we will obtain a suitable representation for our functions if we construct such an integral basis. This is our initial trip into computational algebraic geometry. Algorithms for the construction of such bases are highly nontrivial in most cases. For a complete description of one such algorithm, we refer the reader to the original thesis of Trager; additional algorithms can also be found in the thesis of Bradford [2]. There is, however, one example where the computation of an integral basis is relatively easy computationally. This is the case when we have a simple radical extension, that is, an extension $K(x,y)$ with

$$F(x,y) = y^n - h(x) = 0.$$

In this case, an integral basis can be determined from the square-free factorization of $h(x)$. For example, for F given by (12.30) the following is an integral basis:

$$\{\ 1\ , y, \frac{y^2}{x}, \frac{y^3}{x^2}\ \}.$$

(The reader should verify that these four functions indeed have no poles.) Thus, in the above case our unique representation for an $f \in K(x,y)$ is given by

$$f = \frac{a_0(x) + a_1(x)\cdot y + a_2(x)\cdot \frac{y^2}{x} + a_3(x)\cdot \frac{y^3}{x^2}}{b(x)} .$$

In the special case where $h(x)$ is square-free, an integral basis is given by

$$\{ 1, y, \ldots, y^{n-1} \}.$$

Example 12.18. Let

$$f = \frac{2x^4+1}{(x^5+x)\cdot\sqrt{x^4+1}} = \frac{2x^4+1}{(x^5+x)\cdot y}$$

where

$$F(x,y) = y^2 - x^4 - 1 = 0 .$$

Multiplying numerator and denominator by y gives

$$f = \frac{(2x^4+1)\cdot y}{x^9+2x^5+x} .$$

Since F is of the form $y^2 - (x^4+1)$ and x^4+1 is square-free, an integral basis is given by $\{1, y\}$ and so we have the desired representation for f.

●

We remark that there are a number of integral bases to choose from at any one time. Trager further requires that the integral basis chosen be *normal at infinity*. This is an integral basis which, in some well-defined sense, is the best behaved at infinity.

Once we have an integral basis $\{w_1, \ldots, w_n\}$ suitable for our purposes, we have a representation which can be "Hermite reduced" just as in the transcendental case. Thus, if

$$f = \frac{a_1(x)\cdot w_1 + \cdots + a_n(x)\cdot w_n}{d(x)}$$

is our representation, then we obtain the square-free factorization of d:

$$d = d_1\cdot d_2^2 \cdots d_k^k = g\cdot h^k .$$

In the transcendental case, the Hermite reduction proceeds by solving via Theorem 2.6 and integration by parts, an equation of the form

$$\int \frac{a}{d} = \int \frac{a}{g\cdot h^k} = \frac{b}{h^{k-1}} + \int \frac{c}{g\cdot h^{k-1}} .$$

The reductions continue until we have a square-free denominator. In the algebraic case, this process simply proceeds component-wise:

$$\int \frac{\sum_{i=1}^{n} a_i w_i}{d} = \int \frac{\sum_{i=1}^{n} a_i w_i}{g\cdot h^k} = \frac{\sum_{i=1}^{n} b_i w_i}{h^{k-1}} + \int \frac{\sum_{i=1}^{n} c_i w_i}{g\cdot h^{k-1}} .$$

We refer the reader to the thesis of Trager for the details, including a proof that the system always has a solution.

Example 12.19. Consider the integrand f from the previous example. In this case, the square-free factorization of the denominator is given by

$$d(x) = x\cdot(x^4+1)^2$$

and the Hermite reduction results in

$$\int \frac{(2x^4+1)\,y}{x^9+2x^5+x}dx = -\frac{y}{2(x^4+1)} + \int \frac{y}{x^5+x}dx .$$

●

Thus, just as in the case of transcendental extensions, we have a method to reduce our problem to the case where there are only log terms in the integral. As was the case previously, we begin determining these log terms by first determining the constants which will appear. By Liouville's Principle, we know that we may write

$$\int f = \sum_{i=1}^{m} a_i \log(v_i) \qquad (12.31)$$

for some constants $a_i \in \mathrm{K}'$ and functions $v_i \in \mathrm{K}'(x,y)$, where K′ is some extension field of K. Of course the representation (12.31) for the integral is not unique. Let $V = \{ \sum_{i=1}^{m} r_i a_i : r_i \in \mathbf{Q} \}$. Then V can be considered as a vector space over **Q**, and as such we can determine a basis for V over **Q**. If $\beta_1, \ldots, \beta_q$ is one such basis, then for each i we have

$$a_i = n_{i1}\cdot\beta_1 + \cdots + n_{iq}\cdot\beta_q$$

where, by rescaling the β_l's if necessary, we may assume $n_{ij} \in \mathbf{Z}$. Thus we have

$$\begin{aligned} \int f &= \sum_{i=1}^{m}\sum_{j=1}^{q} n_{ij}\beta_j \log(v_i) \\ &= \sum_{j=1}^{q} \beta_j \sum_{i=1}^{m} n_{ij} \log(v_i) \\ &= \sum_{j=1}^{q} \beta_j \log(u_j) \end{aligned} \qquad (12.32)$$

where $u_j = \prod_{i=1}^{m} v_i^{n_{ij}}$.

Of course, the above is useful only if we know at least one set of a_i and v_i. Trager, however, points out that if $\beta_1, \ldots, \beta_q$ is a basis over **Q** of the *residues* of f, then we also obtain equation (12.32) with the $u_j \in \mathrm{K}'(x,y)$. Here K′ is K extended by all the residues of f. Thus, as in the transcendental case, we desire a method to determine the residues of f.

Let

$$f = \frac{a(x,y)}{d(x)}$$

where $d(x)$ is square-free, and where the zeros of $d(x)$ are the same as the poles of f. By clearing denominators of $a(x,y)$, we may write

$$f = \frac{g(x,y)/h(x)}{d(x)} = \frac{g(x,y)}{d(x)\cdot h(x)} .$$

The problem now is to determine the residues of f at its poles. Let $\hat{x}$ be a pole of f and let $\hat{y}$ be a value such that

$$F(\hat{x}, \hat{y}) = 0 . \tag{12.33}$$

Then c is a residue of f at the point $(\hat{x}, \hat{y})$ if and only if

$$c = \frac{a(\hat{x}, \hat{y})}{d'(\hat{x})} = \frac{g(\hat{x}, \hat{y})}{d'(\hat{x})\cdot h(\hat{x})}$$

that is, if and only if c solves

$$0 = g(\hat{x}, \hat{y}) - z\cdot d'(\hat{x})\cdot h(\hat{x}) . \tag{12.34}$$

For a given $\hat{x}$ we wish to determine all solutions to (12.34) as $\hat{y}$ runs through the solutions of (12.33). Thus, for a given $\hat{x}$ we are interested in all the solutions of

$$0 = \mathrm{res}_y(g(\hat{x}, y) - z\cdot d'(\hat{x})\cdot h(\hat{x}), F(\hat{x}, y)) . \tag{12.35}$$

We wish to find all solutions of (12.35) as $\hat{x}$ varies over all the poles of f, that is, over all the zeros of d. Removing the content from (12.35) to avoid any false zeros, we find that we want the roots of

$$R(z) = \mathrm{res}_x(\mathrm{pp}_z(\mathrm{res}_y(g(x, y) - z\cdot d'(x)\cdot h(x), F(x, y))), d(x)) \tag{12.36}$$

where pp_z denotes taking the primitive part of the polynomial with respect to z. If any of the roots of (12.36) are non-constant, then the integral is not elementary.

Example 12.20. Consider the integral from the previous example. In this case our integrand is $y/(x^5 + x)$, with $g(x,y) = y$, $h(x) = 1$, $d(x) = x^5 + x$, and $F(x,y) = y^2 - x^4 - 1$, so

$$\begin{aligned} R(z) &= \mathrm{res}_x(\mathrm{pp}_z(\mathrm{res}_y(y - z(5x^4 + 1), F(x, y))), x^5 + x) \\ &= \mathrm{res}_x(25z^2x^8 + (10z^2 - 1)x^4 + (z^2 - 1), x^5 + x) \\ &= 65536\,(z^2 - 1)\,z^8 \end{aligned}$$

which has $z = 1$ and $z = -1$ as its nonzero roots.

●

Example 12.21. The integral

$$\int \frac{1}{\sqrt{(1-x^2)(1-k^2x^2)}} = \int \frac{1}{y} = \int \frac{y}{(1-x^2)(1-k^2x^2)}$$

where $y^2 = (1-x^2)(1-k^2x^2)$ is known as an elliptic integral of the first kind. For $k \neq 0, 1, -1$, the denominator is square-free so that the integral is already Hermite reduced. In this case $R(z)$ is seen to be

$$R(z) = 256z^8(k-1)^8(k+1)^8k^8 .$$

In particular, $R(z)$ has no nonzero roots and hence the integral is not elementary.

●

If we were dealing with the transcendental case, we would now obtain our logarithm sum

$$\sum_{i=1}^{m} c_i \log(v_i)$$

by calculating

$$v_i = \mathrm{GCD}(a - c_i b', b) .$$

The function v_i can also be thought of as the product of $x - r$ as r varies over all the poles of a/b for which c_i is the residue. This cannot work in its present form for the algebraic case. We immediately run into problems simply because our domain is not a UFD. As such, the notion of a GCD is not even well-defined. However, the theory of ideals was originated by Kummer and Dedekind precisely to provide a framework to deal with this problem (cf. Hecke [11] – in particular, Chapter 5). In the case of integration, the theory revolves around the formalism of *divisors* of an algebraic function.

Let $\beta_1, \ldots, \beta_q$ be a basis for the vector space over $\mathbf{Q}$ spanned by the roots of $R(z)$. For each root c_i we have

$$c_i = n_{i1}\cdot\beta_1 + \cdots + n_{iq}\cdot\beta_q$$

where, as before, we may assume $n_{ij} \in \mathbf{Z}$. Associated to each such root is a formal sum

$$D_i = n_{i1}\cdot P_1 + \cdots + n_{iq}\cdot P_q \tag{12.37}$$

where P_j is the point(s) where f has residue c_j. Each point(s) $P_j = (x_j, y_j)$ may be identified with the ideal

$$P_j = \langle x - x_j, y - y_j \rangle$$

while each formal sum (12.37) may be identified with the ideal

$$D_i = P_1^{n_{i1}} \cdots P_q^{n_{iq}} .$$

The problem of determining a u_j for equation (12.32) is then the problem of finding a function that has poles of order n_{ij} at the point P_i. This is the same as finding a principal generator for the ideal D_i (or a principal divisor if we are using the formalism of divisors). Of course, being unable to find a principal divisor does not imply that our integral is not elementary. We may instead change our basis for the residues to $\beta_1/2, \ldots, \beta_q/2$, in which case we are looking for a principal divisor for $2 \cdot D_j$. If this fails, we search for a principal divisor for $3 \cdot D_j$, etc.

Example 12.22. Continuing with our previous example, let P_1 be the point $(x=0, y=1)$ and P_2 be the point $(x=0, y=-1)$. The vector space spanned by the residues 1 and -1 is generated by $\beta_1 = 1$. The function f has residue 1 at P_1 and residue -1 at P_2, so we are looking for a principal divisor for $D = P_1 - P_2$. This is the same as finding a principal generator for the ideal

$$P_1 \cdot P_2^{-1} \tag{12.38}$$

where we have identified the points P_1 and P_2 with the ideals

$$P_1 = \langle x, y-1 \rangle, \quad P_2 = \langle x, y+1 \rangle$$

that they generate. Finding a generator of (12.38) is the same as finding a function $u \in \mathbf{Q}(x,y)$ having a zero of order 1 at P_1 and a pole of order 1 at P_2 and no other poles or zeros (the reader should verify this). No such principal generator exists in this case, so we consider the ideal

$$P_1^2 \cdot P_2^{-2} \tag{12.39}$$

(i.e. the divisor $2D$) and search for a generator. If this fails, we search $P_1^3 \cdot P_2^{-3}$, etc. In the case of our example, there is a generator of (12.39) given by

$$u = \frac{x^2}{1+y} .$$

Thus, our initial integral has the solution

$$\int \frac{2x^4+1}{(x^5+x)\sqrt{x^4+1}} dx = -\frac{1}{2\sqrt{x^4+1}} + \frac{1}{2}\log\left(\frac{x^2}{1+\sqrt{x^4+1}}\right) .$$

●

Determining whether a divisor or some multiple of it is principal, is a problem of finding a principal generator of the corresponding ideal or one of its powers. The problem reduces to constructing an integral basis for the ideal and checking to see if all of the basis elements have a pole at infinity. If they all do then there is no principal generator; otherwise a generator may be constructed. Finally, it is not possible simply to search for possible generators amongst the infinite set D_j, $2D_j$, $3D_j$, etc. Somewhere there must be a multiple such that, if no principal divisor has been found by that point, then no such principal divisor exists and the integral is not elementary. Such a multiple is obtained by reducing our computations modulo two different primes and determining principal divisors for multiples there (this process does have a terminal point). The solution to the problem of deciding whether there is some multiple of a divisor which is principal was a significant achievement, and indeed was thought to be unsolvable at the turn of the century.

The above description lacks many details and glosses over many computational difficulties. Nonetheless, it gives some idea of the similarity of the approach used to solve this problem compared with the methods used earlier in the chapter. Also, we described only the purely algebraic case and not the case of mixed transcendental and algebraic extensions. The case where the tower of field extensions contains some transcendental extensions followed by an algebraic extension, was not completely solved until 1987 when Bronstein [5] gave an algorithm in his Ph.D thesis. The case where an algebraic extension is followed by a transcendental extension is essentially a matter of applying the algorithm for transcendental extensions as described in this chapter. However, in such a case one must be able to solve Risch differential equations over algebraic extension fields. A description of Risch's methods for this problem can be found also in Bronstein's thesis.

Exercises

1. Using the algebraic definition of exponentials (Definition 12.1), prove that
   (a) $\exp(f+g) = \exp(f)\exp(g)$;
   (b) $\exp(nf) = \exp(f)^n$ where $n \in \mathbf{Z}$.
2. Using the structure theorem (Theorem 12.1), decide if the following extensions are transcendental or algebraic:
   (a) $\mathbf{Q}(x, \theta_1, \theta_2, \theta_3)$ from Example 12.1;
   (b) $\mathbf{Q}(x, \theta_1, \theta_2, \theta_3, \theta_4)$ from Example 12.2.
3. Let $\mathrm{F} = \mathbf{Q}(x, \exp(x), \exp(x^2), \log(p(x)), \log(q(x)))$ where $p(x)$ and $q(x)$ are polynomials in $\mathbf{Q}[x]$. Show that $\exp(x^3)$ is a monomial over this field.
4. Prove the following. Let F be a differential field. If θ is transcendental over F then prove that:
   (a) for specified $u \in \mathrm{F}$, there is one and only one way to extend the differential operator from F to the extension field $\mathrm{F}(\theta)$ such that $\theta' = u'/u$;
   (b) for specified $u \in \mathrm{F}$, there is one and only one way to extend the differential operator from F to the extension field $\mathrm{F}(\theta)$ such that $\theta'/\theta = u'$.

   Furthermore, if θ is algebraic over F then:

   (c) there is one and only one way to extend the differential operator from F to the extension field $\mathrm{F}(\theta)$; in particular, the differential operator on $\mathrm{F}(\theta)$ is completely determined by the minimal polynomial defining θ and by the differential operator defined on F. *Hint:* Show that there is a unique representation for derivatives of polynomials from the domain $\mathrm{F}[\theta]$. Why is this enough to show uniqueness over $\mathrm{F}(\theta)$?

5. In the proof of Theorem 12.6 we used the fact that $F(\theta)$ and F have the same subfield of constants implies that $F_a(\theta)$ and F_a also have the same subfield of constants. Prove this.

6. Prove that the Risch differential equation

$$y' - 2xy = 1$$

can have no rational solution (cf. Example 12.15). *Hint*: First prove that any potential solution must be a polynomial.

7. Let $K(\theta)$ be a transcendental logarithmic extension of the field K. Then any $f(\theta) \in K(\theta)$ may be written as

$$f(\theta) = p(\theta) + \frac{r(\theta)}{q(\theta)}$$

where $p, q, r \in K[\theta]$ with $\mathrm{GCD}(r,q) = 1$ and $\deg(r) < \deg(q)$. Prove the following *decomposition lemma*: if f has an elementary integral, then each of p and r/q have elementary integrals.

8. Decide if

$$\int \frac{\log^3(x) - \log^2(x) - x^2\log(x) + \frac{1}{x}\log(x) + x^2 + x}{\log^4 x - 2x^2\log^2(x) + x^4}$$

is elementary and, if so, determine the integral.

9. Decide if

$$\int \log^3(x)$$

is elementary and, if so, determine the integral. Do this also for

$$\int \frac{\log^4(x)}{x}.$$

10. Decide if

$$\int (1+\log(x)) \cdot x^x$$

is elementary and, if so, determine the integral.

11. Decide if

$$\int \log(x)\left\{\log(x+1) + \frac{1}{(x+1)}\right\}$$

is elementary and, if so, determine the integral.

12. State and prove a decomposition lemma similar to the one in Exercise 7 for the case of transcendental exponential extensions.

13. Decide if

$$f = \frac{2\exp(x)^2 + (3 - \log^2(x) + 2\,\dfrac{\log(x)}{x})\exp(x) + 2\,\dfrac{\log(x)}{x} + 1}{\exp(x)^2 + 2\exp(x) + 1}$$

considered as an element of $\mathbf{Q}(x, \log(x), \exp(x))$ has an elementary integral. If the answer is yes, then determine the integral.

14. Repeat Exercise 13, but this time consider f as an element of $\mathbf{Q}(x, \exp(x), \log(x))$.

15. Decide if

$$\int \frac{x^2}{\sqrt{(1-x^2)(1-k^2x^2)}}$$

(where $k^2 \neq 1$) is elementary.

Just as we have defined exponentials and logarithms, we can also define tangents and arctangents. Thus if F is a differential field and G is a differential extension field then $\theta \in$ G is *tangent* over F if there exists $u \in$ F satisfying

$$\theta' = (1 + \theta^2)\cdot u'.$$

We write $\theta = \tan(u)$ in this case. If there exists $u \in$ F such that

$$\theta' = \frac{u'}{1 + u^2}$$

then θ is *arctangent* over G and we write $\theta = \arctan(u)$. Tangent and arctangent extensions are useful in cases where it is desired to avoid converting to complex exponentials. There is a corresponding structure theorem (cf. Bronstein [3]) along with algorithms for integration. Similar definitions also exist for hyperbolic tangents and arctangents.

16. (cf. Bronstein [6]) (Differentiation of tangent polynomials.) Using the above definitions prove:
Let F be a differential field and let F(θ) be a differential extension field of F having the same subfield of constants. If θ is transcendental and tangent over F then
   (a) for $a \neq 0 \in$ F, $n \neq 0 \in \mathbf{Z}$, $a\,(1 + \theta^2)^{n\prime} = \bar{a}\,\theta\,(1 + \theta^2)^n$ with $\bar{a} \neq 0 \in$ F;
   (b) for $a(\theta) \in$ F(θ), $a(\theta)' \in$ F(θ) and $\deg(a(\theta)') = \deg(a(\theta)) + 1$.

17. Suppose t is transcendental and tangent over F with $t = \tan(u)$, $u \in$ F. Suppose a is *algebraic* over F and satisfies the same differential equation as t, that is

$$a' = (1 + a^2)\cdot u'.$$

Show that if F(a) and F(a)(t) have the same subfield of constants, then $a^2 = -1$. *Hint*: Look at $c = \dfrac{(a - t)}{(1 + t\,a)}$ (cf. Bronstein[6]).

18. (cf. Bronstein [6]) Let F be a differential field and let F(θ) be a differential extension field of F having the same subfield of constants. Suppose θ is transcendental and tangent over F. Let $a(\theta) \in F[\theta]$ with

$$\mathrm{GCD}(a(\theta), \frac{d}{d\theta}a(\theta)) = 1, \quad \text{and } \mathrm{GCD}(a(\theta), \theta^2 + 1) = 1 .$$

Then

$$\mathrm{GCD}(a(\theta), a(\theta)') = 1 .$$

19. State and prove the equivalent theorems of Exercises 16 - 18 for F(θ) a hyperbolic tangent extension of F.

References

1. J. Baddoura, "Integration in Finite Terms and Simplification with Dilogarithms: a Progress Report," pp. 166-171 in *Proc. Computers and Math.*, ed. E. Kaltofen, S.M. Watt, Springer-Verlag (1989).

2. R.J. Bradford, "On the Computation of Integral Bases and Defects of Integrity," Ph.D. Thesis, Univ. of Bath, England (1988).

3. M. Bronstein, "Simplification of Real Elementary Functions," pp. 207-211 in *Proc. ISSAC '89*, ed. G.H. Gonnet, ACM Press (1989).

4. M. Bronstein, "The Transcendental Risch Differential Equation," *J. Symbolic Comp.*, **9**(1) pp. 49-60 (1990).

5. M. Bronstein, "Integration of Elementary Functions," *J. Symbolic Comp.*, **9**(2) pp. 117-173 (1990).

6. M. Bronstein, "A Unification of Liouvillian Extensions," *Appl. Alg. in E.C.C.*, **1**(1) pp. 5-24 (1990).

7. G.W. Cherry, "Integration in Finite Terms with Special Functions: the Error Function," *J. Symbolic Comp.*, **1** pp. 283-302 (1985).

8. G.W. Cherry, "Integration in Finite Terms with Special Functions: the Logarithmic Integral," *SIAM J. Computing*, **15** pp. 1-21 (1986).

9. J.H. Davenport, "Integration Formelle," IMAG Res. Rep. 375, Univ. de Grenoble (1983).

10. J.H. Davenport, "The Risch Differential Equation Problem," *SIAM J. Computing*, **15** pp. 903-918 (1986).

11. E. Hecke, *Lectures on the Theory of Algebraic Numbers*, Springer-Verlag (1980).

12. E. Kaltofen, "A Note on the Risch Differential Equation," pp. 359-366 in *Proc. EUROSAM '84, Lecture Notes in Computer Science* **174**, ed. J. Fitch, Springer-Verlag (1984).

13. A. Ostrowski, ''Sur l'integrabilite elementaire de quelques classes d'expressions,'' *Commentarii Math. Helv.*, **18** pp. 283-308 (1946).

14. R. Risch, ''On the Integration of Elementary Functions which are built up using Algebraic Operations,'' Report SP-2801/002/00. Sys. Dev. Corp., Santa Monica, CA (1968).

15. R. Risch, ''The Problem of Integration in Finite Terms,'' *Trans. AMS*, **139** pp. 167-189 (1969).

16. R. Risch, ''The Solution of the Problem of Integration in Finite Terms,'' *Bull. AMS*, **76** pp. 605-608 (1970).

17. R. Risch, ''Algebraic Properties of the Elementary Functions of Analysis,'' *Amer. Jour. of Math.*, **101** pp. 743-759 (1979).

18. J.F. Ritt, *Integration in Finite Terms*, Columbia University Press, New York (1948).

19. M. Rosenlicht, ''Integration in Finite Terms,'' *Amer. Math. Monthly*, **79** pp. 963-972 (1972).

20. M. Rosenlicht, ''On Liouville's Theory of Elementary Functions,'' *Pacific J. Math*, **65** pp. 485-492 (1976).

21. M. Rothstein, ''Aspects of Symbolic Integration and Simplification of Exponential and Primitive Functions,'' Ph.D. Thesis, Univ. of Wisconsin, Madison (1976).

22. M. Singer, B. Saunders, and B.F. Caviness, ''An Extension of Liouville's Theorem on Integration in Finite Terms,'' *SIAM J. Computing*, pp. 966-990 (1985).

23. B. Trager, ''Integration of Algebraic Functions,'' Ph.D. Thesis, Dept. of EECS, M.I.T. (1984).

NOTATION

Operations

$a\,/\,b$	a divided by b
$\mathrm{sign}(a)$	sign of a
$\mathrm{n}(a)$	normal part of a
$\mathrm{u}(a)$	unit part of a
$\lvert a \rvert$	absolute value of a
$\lVert \mathbf{a} \rVert_\infty$	sup norm of **a**
$\mathrm{quo}(a, b)$	quotient of a with respect to b
$\mathrm{rem}(a, b)$	remainder of a with respect to b
$\mathrm{GCD}(a, b, \ldots, c)$	greatest common divisor of $a, b, \ldots, c$
$\mathrm{LCM}(a, b, \ldots, c)$	least common multiple of $a, b, \ldots, c$
$\deg(p)$	[total] degree of polynomial p
$\deg_x(p)$	degree of polynomial p in variable x
$\deg_i(p)$	degree of polynomial p in variable x_i
$\partial(p(\mathbf{x}))$	degree vector of multivariate polynomial p
$\mathrm{lcoeff}(p)$	leading coefficient of [univariate] polynomial p
$\mathrm{lcoeff}_x(p)$	leading coefficient of polynomial p in variable x
$\mathrm{lcoeff}_i(p)$	leading coefficient of polynomial p in variable x_i
$\mathrm{res}(p)$	resultant of [univariate] polynomial p
$\mathrm{res}_x(p)$	resultant of polynomial p in variable x
$\mathrm{res}_i(p)$	resultant of polynomial p in variable x_i
$\mathrm{cont}(p)$	content of p
$\mathrm{pp}(p)$	primitive part of p
$\mathrm{tcoeff}(p)$	trailing coefficient of polynomial p
$\mathrm{ord}(a)$	order of power series a
$\det(A)$	determinant of matrix A
A^{adj}	adjoint of matrix A
A^T	transpose of matrix A
$[x]$	congruence (residue) class of x
$x + \mathrm{I}$	residue class of x modulo ideal I
$\binom{k}{r}$	k, r binomial coefficient

Sets and Objects

$\mathbf{N}$	set of natural numbers (0, 1, 2, ...)
$\mathbf{Z}$	set of integers
$\mathbf{Q}$	set of rational numbers
$\mathbf{C}$	set of complex numbers
$\mathbf{Z}_p$	set of integers modulo p
$\mathrm{R}[x]$	ring of polynomials in variable x over R
$\mathrm{R}[\mathbf{x}]$	ring of polynomials in variables $\mathbf{x}$ over R
$\{ a : b \}$	set of a such that b
$\emptyset$, $\{ \}$	empty set
$( )$	empty ordered set
$(x_1, \ldots, x_n)$	ordered set of n elements
$[x_1, \ldots, x_n]$	n-dimensional row vector
$[x_1, \ldots, x_n]^T$	n-dimensional column vector
$\mathbf{a}$	ordered set or vector with entries a_i
$[a_{ij}]$	matrix with entries a_{ij}
$<p>$	principal ideal generated by p
$<P>$	ideal generated by set P
$A \times B$	Cartesian product of sets A and B
A / B	quotient set of A modulo B
$A / \sim$	quotient set of A modulo relation $\sim$
$\mathrm{T}_{\mathbf{x}}$	set of terms in $\mathbf{x}$

Relations

$a \mid b$	a divides b
$a \nmid b$	a does not divide b
$\subset$	subset
$\subseteq$	improper subset
$\sim$	is equivalent to
$\equiv$	is congruent to
$\leftrightarrow$	corresponds to
$<_T$	less than with respect to T
$\leq_T$	less than or equal with respect to T
$\mapsto$	reduces to
$\overset{+}{\mapsto}$	reduces transitively to
$\overset{*}{\mapsto}$	reduces fully to

INDEX